WATER TREATMENT PLANT DESIGN
For the Practicing Engineer

edited by

Robert L. Sanks

Professor of Civil Engineering and
    Engineering Mechanics
    Montana State University, and
Senior Engineer
    Christian, Spring, Sielbach & Associates

BUTTERWORTH PUBLISHERS

Boston • London
Sydney • Wellington • Durban • Toronto

An Ann Arbor Science Book

Ann Arbor Science is an imprint of Butterworth Publishers.

Copyright © 1978 by Ann Arbor Science Publishers.
All rights reserved.

Library of Congress Catalog Card. No. 77-76914
ISBN 0-250-40183-5

Butterworth Publishers
80 Montvale Avenue
Stoneham, MA   02180

10   9   8   7   6

Manufactured in the United States of America

PREFACE

This book is intended to be as practical and current as possible for consulting engineers (including those who have limited experience in this specialty) who design water treatment plants. Each author was chosen because he is an expert in his subject, and each chapter includes only the most authoritative theory and practice, although there may sometimes be conflicting viewpoints.

It is assumed that the reader is a graduate engineer who has some familiarity with the subject and has access to several "standard" works (some of which are mentioned herein) and to manufacturers' literature. Hence, this book is not intended to be encyclopedic, but rather to stress (a) those subjects not completely covered in other works, (b) subjects in which more up-to-date theory and practice are now available, (c) advice that is valuable to the practitioner, and (d) discussion that enhances or improves decisions and helps the engineer to put lines on drawings and to write good specifications for the design of water treatment plants.

Robert L. Sanks
Bozeman, Montana

Dedicated to the Memory of
Warren J. Kaufman
1922-1973
Professor of Sanitary and Radiological Engineering
and
Director of the Sanitary Engineering Research Laboratory
University of California at Berkeley

ACKNOWLEDGMENTS

About half of the material in this book was presented in a preliminary form at the Fifth Environmental Engineers' Conference held at Big Sky of Montana in June 1976 and sponsored jointly by Montana State University and the Bureau of Water Quality, Department of Health and Environmental Sciences. All of the original papers were completely revised for the book and, of course, half of the chapters are new. All contributors were selected by the editor.

The authors and the editor are especially indebted to the following: Claiborne W. Brinck (Chief, retired, Montana Water Quality Bureau), Donald G. Willems (Chief, Montana Water Quality Bureau), Gerry Shell (Gerry Shell Environmental Engineers, Nashville, Tennessee), and from Montana State University, A. Amirtharajah (Associate Professor, Civil Engineering and Engineering Mechanics), Theodore T. Williams (Head of the Department of Civil Engineering and Engineering Mechanics), Byron D. Bennett (Dean, College of Engineering), and Jean Julian, Jackie McGuire and Etta Grinwis (secretarial staff).

Robert L. Sanks received his BS from the University of California at Berkeley, his MS from Iowa State College and, after two decades of experience in consulting, research and teaching, obtained his PhD from the University of California at Berkeley. He is the originator and coordinator of the Environmental Engineering and Environmental Health Engineering programs at Montana State University.

Dr. Sanks has published a textbook and numerous articles and papers in national and international journals and also edited the Ann Arbor Science book entitled **Land Treatment and Disposal of Municipal and Industrial Wastewater.** He is a Fellow of the American Society of Civil Engineers, and a member of the American Water Works Association, Water Pollution Control Federation, and Association of Environmental Engineering Professors. He now divides his time equally between teaching and research at Montana State University and the consulting firm, Christian, Spring, Sielbach & Associates where he is Senior Engineer.

Cover: Montage of two water treatment plants and drawing of a typical pipe gallery, all designed by James M. Montgomery, Consulting Engineers, Inc., Pasadena, California.

CONTENTS

CHAPTER 1

EFFECTS OF FEDERAL REGULATIONS
ON WATER TREATMENT PLANT DESIGN

Jack W. Hoffbuhr, P. E.
Chief, Water Supply Section

Dean R. Chaussee
Water Supply Engineer
U.S. Environmental Protection Agency
Region VIII
1860 Lincoln Street
Denver, Colorado 80295

The design of a water treatment plant for a small water system has always been a challenge. Too often this challenge has not been met, as evidenced by the number of treatment plants unable to produce consistently safe drinking water because they were designed with little consideration given to critical factors such as raw water quality variations, small town resources and the capabilities of the operator. The time has passed when a standard design for a large plant can be pulled from a file drawer and simply scaled down. The discovery of new health concerns in water supplies, such as organics and asbestos, plus the implementation of federal drinking water regulations make it essential that the design engineer consider much more than the hardness, pH, alkalinity and iron content of the water. Small communities deserve more. The general impacts of the federal drinking water regulations on water system design are discussed in this chapter.

BACKGROUND

Federal guidelines governing the quality of drinking water have existed since 1914. These have been periodically updated, with the most recent version being the 1962 Public Health Service Drinking Water Standards.[1] Although these guidelines were adopted by many states as regulations, others have had no regulations at all on the quality of drinking water. This, in addition to an apathetic public, has led to a lack of concern about drinking water problems. The apparent increase in severe waterborne disease outbreaks since 1955 certainly indicates that drinking water problems do exist. Other more complex problems involving chemical contaminants are also being uncovered in water supplies. For example, studies have indicated that the process of chlorinating water supplies is causing the formation of potentially carcinogenic compounds, such as chloroform and bromoform. In fact, over 700 separate organic compounds have been isolated in water supplies across the country. The discovery of high concentrations of asbestos fibers in a number of major water systems is creating additional concerns regarding their possible health effects. These concerns led to the passage of the Safe Drinking Water Act of 1974.[2]

SAFE DRINKING WATER ACT

The Safe Drinking Water Act (SDWA) charged the U.S. Environmental Protection Agency (EPA) with the responsibility of developing drinking water regulations to protect the public health. After a lengthy public review process, the Interim Primary Drinking Water Regulations[3] were finalized and published in the *Federal Register* on December 24, 1975. They became effective on June 24, 1977.

Administration of the Safe Drinking Water Act

Congress, when it passed the SDWA, envisioned that drinking water programs would be administered at the state level. However, there is no requirement that the states assume primary enforcement responsibility (primacy). In those states that do not accept this responsibility, EPA must implement the SDWA. Therefore, when considering the design of a water treatment plant, one of the first steps is to ascertain the responsible water supply agency. On most federal and all Indian lands, the Act will be administered by EPA.

The minimum federal regulations specify only the final quality of drinking water to be achieved, and do not spell out the methods by which the water must be treated. However, most states have specific design regulations, and it is very important that the design engineer become familiar with these. Federal regulations do specify that the responsible water supply agency must review all design plans for new or modified water treatment systems. Therefore, consultants should work closely with the responsible water supply agency and plan to have their designs reviewed before finalizing them.

National Interim Primary Drinking Water Regulations (IPR's)

These regulations are referred to as "interim" because they will be revised based upon the results of research and special studies being conducted by the National Academy of Sciences. The Revised Primary Regulations will be proposed for comment in 1978 and probably become effective sometime in 1979. They are referred to as the "primary" regulations because they concern only those contaminants that are believed to have a direct effect on the public health.

Definitions

The following are key definitions of the types of water systems covered by the IPR's.

Public Water Supply: A system that has at least 15 service connections or that regularly serves an average of 25 people daily for at least 60 days each year. Contained within the group of public water supplies are two types—community and noncommunity.

Community Water Supply: A public water system that serves a residential (year-round) population. Examples would be cities, towns, subdivisions and mobile home communities.

Noncommunity Water Supply: A public water system that serves not residents but only intermittent users such as tourists. This category includes campgrounds, motels, restaurants and roadside rest stops having their own water supply.

Maximum Contaminant Levels (MCL's)

The maximum contaminant levels specified in the IPR's are summarized in Tables I through V.

Grants

Unlike the existing water pollution control legislation (PL 92-500), the Safe Drinking Water Act did not establish a grant program for construction of water treatment facilities. Therefore, EPA does not have

Table I. Maximum Contaminant Levels (MCL) for Inorganic Chemicals

Chemical	MCL (mg/l)
Arsenic	0.05
Barium	1.
Cadmium	0.010
Chromium	0.05
Lead	0.05
Mercury	0.002
Nitrate (as N)	10.
Selenium	0.01
Silver	0.05
Fluoride[a]	
Air Temperature ($^\circ$F)	
53.7 and below	2.4
53.8 to 58.3	2.2
58.4 to 63.8	2.0
63.9 to 70.6	1.8
70.7 to 79.2	1.6
79.3 to 90.5	1.4

[a] Limits apply only to naturally occurring fluoride. When adjusting fluoride level with addition of fluoride-containing compounds, state health department recommendations should be followed. Limits are set according to annual average of the maximum daily air temperatures.

Table II. Maximum Contaminant Levels (MCL) for Organic Chemicals

Chemical	MCL (mg/l)
Chlorinated Hydrocarbons	
Endrin	0.0002
Lindane	0.004
Methoxychlor	0.1
Toxaphene	0.005
Chlorophenoxys	
2,4-D	0.1
2,4,5-TP (Silvex)	0.01

funds available for this purpose. The only grants authorized by the Act are program grants to be given to the states to improve their water supply supervision programs and grants to be given to other agencies for research, training or demonstration projects.

Table III. Maximum Contaminant Levels (MCL) for Radionuclides

Chemical	MCL
Natural	
Gross Alpha Activity	15 picocuries/l
Radium-226 + Radium-228	5 picocuries/l
Man-Made	
	Annual dose cannot exceed 4 millirems/yr to body or any organ.

Table IV. Maximum Contaminant Levels (MCL) for Turbidity

Monthly Average:	1 turbidity unit (TU) or
	5 TU's with state approval provided it does not interfere with:
	1. Disinfection
	2. Maintenance of Chlorine Residual
	3. Bacteriological Testing
2-Day Average:	5 TU's

Table V. Maximum Contaminant Level for Microbiological Contaminants[a]

		Individual Sample Basis	
Coliform Method	Monthly Basis	Fewer Than 20 Samples/Month	More Than 20 Samples/Month
	Number of Coliform Bacteria Shall Not Exceed		
Membrane Filter (100-ml portions)	1/100-ml avg density	4/100 ml in more than one sample	4/100 ml in 5% of samples
	Coliform Bacteria Shall Not Be Present in More Than		
Multiple-Tube Fermentation (10-ml portions)	10% of portions	3 portions in more than one sample	3 portions in 5% of samples

[a]Failure of either the monthly basis *or* the individual sample basis constitutes a violation of the MCL.

Variances and Exceptions

If a water system cannot meet one of the specified MCL's, it can apply to the responsible water supply agency for relief from the regulations for a certain period of time. If there is no practical treatment for removal of the contaminant in the water, an application should be made for a variance. If a treatment process for removal of the contaminant does exist but the community cannot afford it, an application for an exemption should be made. An exemption cannot be issued to a water system that came into existence after June 24, 1977, and it must expire June 24, 1981. Both variances and exemptions must be accompanied by a compliance schedule agreed upon by both the water system manager and the responsible water supply agency. Neither variances nor exemptions will be issued for monitoring requirements or bacteriological problems, and neither must result in a threat to public health.

Laboratories

The IPR's specify that analyses, except for turbidity and chlorine residual, must be conducted in a laboratory certified by the state or EPA. Since the approved test procedures for the inorganic, organic and radiological chemicals require complex, expensive equipment, they probably would not be conducted at the water treatment plant except for very large water systems. Therefore, for small systems, the design consultant should specify the normal operational test equipment, plus a chlorine residual test kit and, for a system using surface water as a source, a nephelometric turbidimeter.

It is important to note that certain methods are also specified for analysis of chlorine residual and turbidity. For chlorine residual the DPD method is specified. However, in a laboratory, amperometric titration is also acceptable. The nephelometric method is specified to measure turbidity. A nephelometer measures the light scattered from a sample rather than the light transmitted through the sample. Nephelometers are produced by several manufacturers in the United States and Canada and include not only laboratory models but also on-line units that can be installed following the filters for continuous monitoring. Portable battery-operated units are also available for field uses such as watershed surveys. Turbidity is the only test that is measured at the point at which the water leaves the treatment plant rather than at the consumer's tap.

Secondary Regulations

In addition to regulations covering contaminants that affect the public health, EPA was also given the responsibility for developing guidelines on those contaminants that affect the esthetic qualities of drinking water. The Secondary Regulations[4] are intended to serve as guidelines and are not federally enforceable. Table VI shows the limits proposed on March 31, 1977. Although design engineers are not bound by these recommendations, they should be aware that keeping the quality of drinking water within these guidelines makes it much more acceptable to the consumer, thereby decreasing complaints and, possibly, later expensive modifications.

Table VI. Secondary Regulations

Chemical	Recommended Limit
Chloride	250 mg/l
Color	15 color units (cu)
Copper	1 mg/l
Corrosivity	Noncorrosive
Foaming Agents	0.5 mg/l
Hydrogen Sulfide	0.05 mg/l
Iron	0.3 mg/l
Manganese	0.05 mg/l
Odor	3 Threshold Odor Number (TON)
pH	6.5-8.5
Sulfate	250 mg/l
Total Dissolved Solids (TDS)	500 mg/l
Zinc	5 mg/l

Total dissolved solids, including sulfates and chlorides, generally become a problem only in deep well sources. These constituents impart a mineral taste to the water and can cause gastrointestinal discomfort to those not accustomed to them. Since they are not effectively removed by conventional treatment, selection of a source with low levels of these constituents is quite important.

Iron and manganese, which can cause staining of laundry and plumbing fixtures, are also a problem with some ground water sources. Generally they are removed by oxidation through aeration or the addition of chemicals such as chlorine or potassium permanganate followed by filtration (see Chapter 21).

High copper and zinc concentrations, which cause unpleasant tastes, are generally related to corrosion of household plumbing by aggressive water. The only effective approach to this problem is to stabilize the water through the addition of chemicals (see Chapter 23).

Surface water supplies frequently experience color and odor problems that may result from a variety of causes including industrial discharges, decaying vegetation or sediment materials stirred up in the process of reservoir turnover in the spring and fall. More importantly, these problems may indicate the presence of organic materials that, when combined with chlorine, could form undesirable compounds such as trihalomethanes. Additional examination of the water should be conducted to determine the extent of the organic content. Although some of the organic material may be removed by conventional treatment processes, additional treatment may be required (see Chapters 21 and 22).

The presence of foaming agents is also an indicator of a more important problem: recent industrial or domestic pollution. A sanitary survey of the watershed should be conducted to locate and eliminate, if possible, the source of pollution.

Siting Requirements

The Interim Regulations specify certain siting requirements in order to protect the water supply from extensive damage due to natural or man-made disasters. These requirements state "Before a person may enter into a financial commitment for or initiate construction of a new public water system or increase capacity of an existing public water system, he shall notify the state and, to the extent practicable, avoid locating part or all of the new or expanded facility at a site which (a) is subject to a significant risk—earthquake, floods, fires or other disasters which could cause a breakdown in the public water system or a portion thereof or (b) except for intake structures, is within the 100-year flood plain or is lower than any recorded high tide where appropriate records exist." In addition to these general siting requirements, the responsible water supply agency reviewing the plans will obviously be interested in potential pollution hazards and their effect upon the location of water supply intakes and reservoirs. EPA will eventually be issuing more complete guidelines on the siting of water facilities.

Operation and Maintenance (O & M)

The Safe Drinking Water Act requires that the primary regulations "contain criteria and procedures to ensure proper operation and maintenance of the system." Although O&M regulations have not been included in the IPR's, they will be included in the Revised Primary Regulations, which are expected to be published sometime in 1978. Preliminary indications are that the requirements will contain provisions concerning disinfection of newly installed water mains, minimum pressures in the distribution system, cross-connection control, use of acceptable chemical and protective coatings within the system and emergency operation plans.

Regardless of the requirements of the responsible water supply agencies, the design engineer should take into account the operation and maintenance of the water system before the design is completed. Too often, a treatment plant that looks great on paper turns out to be an operation and maintenance burden for the community. The engineer must remember that small water system operators have much less knowledge of water chemistry and the overall water treatment process than he does.

The design firm should certainly provide a good O&M manual written in "operator" language. This should include the suggested maintenance schedule for each piece of equipment in the plant together with the manufacturers' names and addresses for all equipment and material supplied. The manual should contain some suggested record forms to assist in the day-to-day operation of the system. As a companion to the manual, the design consultant should provide detailed plans and specifications that show the "as built" construction of the plant. The design consultant should also plan to spend time with the operator during and after start-up in order to assist him in learning the processes and tests needed for effective operation.

Public Notification

One of the key sections of the SDWA requires that water suppliers notify their customers when their water systems are in violation of the IPR's. The purpose is to increase public awareness of the problems that water systems face and the true cost of providing safe drinking water. Hopefully, the consumers will be more aware of their responsibility to pay these costs. Table VII indicates the public notification requirements for various situations.

Table VII. Public Notification Requirements

Condition	Mail	Newspaper	Broadcast
Violation of an MCL	X	X	X
Failure to Monitor	X		
Failure to Follow Compliance Schedule	X		
Failure to use Approved Testing Procedure	X		
Having a Variance or Exemption	X		

Consulting engineers should become familiar with these requirements since they may be called upon to write a public notification, especially when a variance or an exemption has been granted. There are no hard and fast rules on how to write a public notice. However, it should not be overly technical or negative. An easily read, positive statement will go far to enlist the aid of the community.

DESIGN CONSIDERATIONS

The Interim and Revised Regulations will have an impact on all phases of water system development from source selection to treatment design. The consulting engineer must design a system that can consistently produce water meeting the regulations or the community could face tremendous add-on costs in the future.

Source Selection

The most effective approach for assuring that a water system meets the regulations is to select a good quality source. This approach also keeps treatment costs to a minimum, which is always important, especially for small communities.

The development of ground water as a source should be a prime consideration for small water systems. Compared to surface water, a good quality ground water system costs less in terms of treatment, monitoring, and operation. However, the ground water must be developed properly to prevent contamination and to enhance the efficiency of the well or little is gained. Universal Oil Products Company's *Ground Water and Wells,*[5] and EPA's *Manual of Water Well Construction Practices,*[6] are excellent references.

Regardless of the source type, extreme care should be taken to locate all potential sources of contamination. For ground water in particular, it is important to determine if industrial or municipal waste lagoons and landfills (abandoned as well as existing) exist near the proposed well site. With surface waters, the possible influence of point and nonpoint sources of contamination should be explored. Irrigation return flows, for example, can carry high concentrations of pesticides and nitrates.

Regardless of source quality, however, two types of treatment should always be considered:

1. disinfection
2. filtration for all surface sources.

Treatment Selection

Since the IPR's cover chemical contaminants in addition to bacteria and turbidity, proper treatment selection now includes more complex factors than have been considered traditionally. Moreover, the unique problems of small communities must be kept in mind or they will have plants they do not know how (or cannot afford) to operate.

Inorganic Chemicals

This is the most difficult group to deal with since conventional treatment techniques generally are not effective. Therefore, first consideration should be given to alternative sources and blending of sources. Fortunately, the inorganics usually are not present in concentrations anywhere near the MCL's. They are generally found naturally only in certain ground waters or only when industrial contamination is present. However, lead and cadmium contamination can occur as a result of an aggressive water attacking plumbing systems where lead or galvanized piping is still in use. This problem can best be handled by stabilizing the water through chemical addition.

When treatment must be considered, choices are somewhat limited, as shown in Table VIII. In fact, no full-scale experience exists for removal of selenium, cadmium, chromium, lead, mercury and silver. In choosing a treatment method, the chemistry of the water to be treated cannot be taken lightly. The valence state of some of the contaminants and the pH are critical to the treatment process chosen. Since the common chemical analyses only measure the concentration of the total element, detailed lab work is sometimes needed. Bench- and pilot-scale work using the water to be treated may also be required in order to select the most effective treatment method.

Organic Chemicals

Because of the widespread use of the six pesticides included in the IPR's, there is definitely a potential for contamination of drinking water. Generally, this is only a concern with surface water due to the influence of runoff. For this reason ground water systems do not have to monitor for pesticides unless requested to do so by the responsible water supply agency.

The most effective methods for removing these compounds are to use powdered (PAC) or granular activated carbon (GAC). The engineer must choose the most cost-effective approach by weighing the relative advantages and disadvantages of both types of carbon. For example, one important consideration is that most small systems will not be able to monitor

Table VIII. Treatment Methods for Inorganic Contaminant Removal[7]

Contaminant	Methods	% Removal
Arsenic		
As^{+3}	Oxidation to As^{+5} required	> 90
As^{+5}	Ferric sulfate coagulation, pH 6-8	> 90
	Alum coagulation, pH 6-7	> 90
	Lime softening, pH 11	> 90
Barium	Lime softening, pH 10-11	> 80
	Ion exchange	> 90
Cadmium[a]	Ferric sulfate coagulation, > pH 8	> 90
	Lime softening >pH 8.5	> 95
Chromium[a]		
Cr^{+3}	Ferric sulfate coagulation, pH 6-9	> 95
	Alum coagulation, pH 7-9	> 90
	Lime softening, pH > 10.5	> 95
Cr^{+6}	Ferrous sulfate coagulation, pH 6.5-9 (pH may have to be adjusted after coagulation to allow reduction to Cr^{+3})	> 95
Fluoride	Ion exchange with activated alumina or bone char media (limited full-scale experience), pH slightly above 7	> 90
Lead[a]	Ferric sulfate coagulation, pH 6-9	> 95
	Alum coagulation, pH 6-9	> 95
	Lime softening, pH 7-8.5	> 95
Mercury[a]		
Inorganic	Ferric sulfate coagulation, pH 7-8	> 60
Organic	Granular activated carbon	> 90
Nitrate	Ion exchange (limited full-scale experience)	> 90
Selenium[a]		
Se^{+4}	Ferric sulfate coagulation, pH 6-7	70-80
	Ion exchange	> 90
	Reverse osmosis	> 90
Se^{+6}	Ion exchange	> 90
	Reverse osmosis	> 90
Silver[a]	Ferric sulfate coagulation, pH 7-9	70-80
	Alum coagulation, pH 6-8	70-80
	Lime softening, pH 7-9	70-90

[a]No full-scale experience.

frequently enough to determine when the PAC should be added. Therefore, using GAC either in beds or as caps on gravity filters would provide the best protection.

Pesticides account for only a few of the more than 700 organic compounds identified in drinking water supplies throughout the United States. These compounds result from such sources as industrial and municipal discharges, urban and rural runoff and natural decomposition of vegetative and animal matter. Compositions and concentrations vary from virtually nil in protected ground water supplies to substantial levels in many surface waters. Due to lack of data on health effects and the difficulties of detecting these compounds, MCL's were not established. However, before long, EPA will be introducing a regulation establishing an MCL for a group of organics known as the trihalomethanes, which are of concern since they are formed when chlorine reacts with certain natural organic compounds (humic acids in particular) during the disinfection process. In addition, chloroform, the most common trihalomethane found in drinking water, has been implicated as causing cancer in test animals.

The MCL will probably be 0.10 mg/l for the total concentration of chloroform, bromoform, bromodichloromethane and dibromochloromethane. Therefore, the engineer will have to examine the raw water source, particularly if it is surface water, and the treatment processes to determine the most cost-effective procedure to meet the MCL. Perhaps moving the prechlorination point to follow sedimentation will be sufficient since many of the trihalomethanes may settle with the floc. On the other hand, treatment changes such as the addition of GAC beds or using disinfectants other than chlorine (such as ozone or chlorine dioxide) may be necessary More complete information can be found in *Interim Treatment Guide for the Control of Chloroform and Other Trihalomethanes.*[8]

Turbidity

Although the treatment techniques to remove turbidity have been known and used for years, it will probably be the MCL violated most often. This is due to two major problems:

1. Many communities using surface water do not use filtration. This is particularly true of systems serving fewer than 10,000 people.
2. Many systems (especially small ones) have poorly designed and/or operated filtration plants.

Filtration of all surface sources is a must with the increasing degradation of formerly high quality waters and the higher incidence of chlorine-resistant organisms, such as *Giardia lamblia*. However, poorly designed facilities offer little protection. The major problems are related to misapplication of pressure and direct filtration. Before beginning the design, a thorough evaluation of the turbidity loading over at least a full year should be undertaken. This will indicate if direct filtration is applicable or if more complete treatment is needed to meet the MCL of 1 TU consistently.

Microbiological Contaminants

The treatment methods to control microbiological contaminants have also been well known for years. However, waterborne outbreaks still occur frequently due to lack of or inadequate disinfection. All drinking water, regardless of quality, should be disinfected. Carrying a residual throughout the distribution system provides the additional benefit of keeping the general bacterial population low, which reduces the incidence of taste, odor and appearance complaints.

The choice of disinfectant depends on a number of factors including raw water quality and amount of water to be treated. For example, if the raw water contains enough organic compounds to result in formation of trihalomethanes, a disinfectant other than chlorine may be chosen. However, if this is done, an evaluation should be made to define the proper operating parameters to ensure proper disinfection. A key point, regardless of the process chosen, is that disinfection is far more effective if applied to the highest quality water possible.

Radiological Chemicals

The IPR's include two different groups of radiological chemicals: natural and manmade.

The natural radionuclides (alpha emitters) are present primarily in ground waters as a naturally occurring element. Lime, lime-soda ash and ion exchange softening processes are effective in removing radium. Reverse osmosis can also be used successfully. Factors influencing the final selection include the raw water quality, current treatment in use and the costs involved. Another important factor is the ultimate disposal of the treatment wastes. In fact, this may be the deciding factor because all the methods have either a troublesome radium-containing sludge or brine.

The manmade radionuclides (beta and gamma emitters) result from such human-controlled activities as nuclear fallout and power plants. Therefore, it is not expected that these MCL's will be exceeded very often. The purpose of these MCL's is to prevent degradation of public water supplies by stressing source control. Treatment studies have shown that the methods discussed above would also be effective for beta and gamma emitters. However, pilot studies should be conducted to determine the relative efficiencies of the methods.

SUMMARY

The Interim and Revised Primary Drinking Water Regulations will certainly have an impact on the design of new or modified treatment facilities. Since communities serving fewer than 10,000 people will experience the most difficulty in meeting the regulations, the consulting engineer faces a definite challenge. Effective treatment methods must be designed which small communities can afford to operate and maintain because some of the contaminants require sophisticated and untried treatment methods. Unfortunately, cost-effective water treatment processes for small water systems have never really been addressed. However, it is time for the researchers, consultants and funding agencies to work together to develop, design and construct effective and practical treatment systems for small communities. In this manner, the goal of safe drinking water for all Americans can be achieved.

REFERENCES

1. *Public Health Service Drinking Water Standards* (Washington, D.C.: U.S. Department of Health, Education and Welfare, 1962).
2. The Safe Drinking Water Act of 1974. Public Law No. 523, 93rd Congress (December 16, 1974).
3. U.S. Environmental Protection Agency. "National Interim Primary Drinking Water Regulations," *Federal Register* 40:59566-59588 (December 24, 1975).
4. U.S. Environmental Protection Agency. "National Secondary Drinking Water Regulations," *Federal Register* 42:17144-17147 (March 31, 1977).
5. *Ground Water and Wells,* Johnson Division, Universal Oil Products Co., St. Paul (1972).
6. *Manual of Water Well Construction Practices,* Office of Water Supply (Washington, D.C.: U.S. Environmental Protection Agency, 1975).
7. *Manual of Treatment Techniques for Meeting the Interim Primary Drinking Water Regulations,* Office of Research and Development (Cincinnati: U.S. Environmental Protection Agency, 1975).
8. *Interim Treatment Guide for the Control of Chloroform and Other Trihalomethanes,* Office of Research and Development (Cincinnati: U.S. Environmental Protection Agency, June 1976).

CHAPTER 2

PUBLIC HEALTH ASPECTS OF WATER SUPPLIES

Alfred W. Hoadley, Ph.D.
Professor of Civil Engineering
School of Civil Engineering
Georgia Institute of Technology
Atlanta, Georgia 30332

It is the responsibility of the engineer in water supply to provide adequate and safe water to consumers through proper protection of water sources and design, maintenance and operation of treatment facilities and distribution systems. In order to do so, he should understand potential threats to the health of consumers, standards of quality by which we judge safety, and techniques by which we evaluate quality.

It is the intent in the present discussion to consider some aspects of public health related to water supplies that seem to be emerging as matters of concern. Recent information on waterborne infectious diseases, including emerging waterborne pathogens, special pathogens which may be of importance in particular circumstances, some questions relating to indicator systems, and bacteriological methods are reviewed. Finally, questions relating to trace metals, asbestos, and organic chemicals in drinking water and their relation to health are examined.

WATERBORNE INFECTIOUS DISEASES IN THE UNITED STATES

Extent of Waterborne Infectious Disease

Summaries of reported waterborne disease for the periods 1946-1970 and 1961-1970 have been published by Craun and McCabe,[1] for 1971-1972 by Merson *et al.,*[2] and for 1971-1974 by Craun, McCabe and Hughes.[3] Yearly reports are now included in the *Annual Summary of Foodborne and Waterborne Disease Outbreaks* published by the Center for Disease Control.[4-6] Between 1971 and 1974 the number of reported outbreaks has varied from 18 to 29, and the number of cases from 1638 to 8413.

Ordinarily, the waterborne disease most commonly reported has been acute gastrointestinal illness. This represents illness of unknown etiology, and probably is often of viral origin. In general, shigellosis ranks second in frequency of cases. The ranking of other waterborne infectious diseases varies, primarily because they occur occasionally in large outbreaks involving municipal systems. Furthermore, because such outbreaks are investigated extensively, reporting of cases is more reliable than generally is true of smaller outbreaks. The outbreak of *Salmonella typhimurium* in Riverside, California, in 1965, affecting some 16,000 persons, and the outbreak of giardiasis in Rome, New York, in 1974-1975, involving perhaps 4,800 persons, were large outbreaks that contributed more cases than were reported for other diseases during the years in which they occurred. In addition, typhoid fever, which normally is rare, achieved prominence in 1973 as a result of the outbreak at a migrant labor camp in Dade County, Florida, involving 210 persons.

Several years ago we made very crude estimates of the incidence of waterborne salmonellosis, shigellosis, and hepatitis A. The estimates of salmonellosis and shigellosis were based on studies of the reporting of disease outbreaks and of cases. It was estimated that only about 25% of *Salmonella* outbreaks and about 1% of cases were reported to the Center for Disease Control, and that only about 8% of shigella outbreaks and 1% of cases were reported. On this basis, it was concluded that the actual yearly incidence of waterborne salmonellosis and shigellosis was nearly 35,000 and over 200,000 cases, respectively. An even more precarious estimate of waterborne hepatitis A indicated that there might occur over 75,000 cases per year. Total costs of the three diseases, including lost work time, hospitalization, mortality, and production losses, were approximately $6,500,000, $35,000,000, and $115,000,000, respectively.

A cost/benefit analysis was undertaken on the basis of costs of the three diseases calculated independently and costs of providing adequate chlorination, salary adjustments and surveillance estimated from the Community Water Supply Study.[7] Results of the analysis justified expenditures for adding and upgrading disinfection, for increased surveillance and for limited annual salary adjustments.

Waterborne Bacterial Pathogens

The classical waterborne pathogens, *Salmonella* and *Shigella*, need be considered only briefly here. It was pointed out previously that reported and estimated outbreaks and cases of shigellosis are more frequent than those of salmonellosis. A large single outbreak of acute gastrointestinal illness caused by *Shigella sonnei* and involving about 1200 persons served by a municipal supply in Florida occurred in 1974. A second outbreak in 1974 involved 600 persons. The prevalence of waterborne shigellosis over salmonellosis may be attributable to the relative infectious doses reported for shigellae, which lie between 10 and 100 cells, in contrast to those reported for salmonellae (including the typhoid bacillus) which are about 10^5.[8] However, the potential for large outbreaks of salmonellosis is demonstrated by the Riverside outbreak.

Four outbreaks caused by enteropathogenic *Escherichia coli* occurred during 1961-1979. Certain *E. coli* strains produce cholera-like enterotoxins that either cause a diarrheal syndrome or are capable of penetrating epithelial cells causing a shigella-like illness. While enteropathogenic strains have been known for some time to cause infantile gastroenteritis, water- and foodborne outbreaks among adults have been reported in Europe and America in recent years. One interesting outbreak occurred in 1967 at a conference center near Washington, D.C. This center was served by well supplies, one of which was located close to a drainage field and another of which drew its water from a stream fed by a contaminated lake. Identical enterpathogenic *E. coli* strains were isolated from seven samples of water from these wells. The two contaminated wells were disconnected from the system and chlorination was instituted. Following these measures, no additional cases occurred.[9]

The etiology of the third largest outbreak of waterborne disease in 1974 was not established with certainty, but *Yersinia enterocolitica* was strongly implicated. This outbreak, involving over 600 persons (41% of guests) occurred at a ski resort in Montana between December, 1974, and January, 1975. The genus *Yersinia*, which includes *Yersinia pestis*, the etiologic agent of plague, is generally considered to be a member of the family *Enterobacteriaceae*. In the outbreak in Montana, contamination of one of two wells that served as the source of supply was implicated as the source of the organisms causing illness, and termination of the outbreak coincided with chlorination of the water supply. Bacteriological investigations indicated that *Y. enterocolitica* was prevalent in well waters and in the distribution system. The colonies appeared brick red on m-Enterococcus agar, but the best medium for their isolation was M-Endo medium used for the enumeration of coliforms in the water.[10] *Yersinia enterocolitica* constituted a large part of the nonsheen background population on M-Endo medium, and the numbers of *Yersinia* colonies were not correlated with numbers of coliforms in the water. Other large outbreaks of possible waterborne *Y. enterocolitica* gastroenteritis have been reported in Japan[11] and a waterborne case in New York state was related to the consumption of spring water.[12]

Investigations of *Y. enterocolitica* in rivers, reservoirs, and private wells in Colorado[13] confirmed the presence of the bacteria in about 17% of river and reservoir samples and 1.8% of wells. Among isolates, only 6% consisted of serotypes known to cause disease in man, however, Similar findings have been reported in Wisconsin[14] by Saari and in Scandinavian countries.[15] Harvey *et al.*[16] have demonstrated the species in 10 of 34 lakes and streams examined. Only one strain was serotypable. Investigators should be aware of the possible presence of *Y. enterocolitica* as well as coliforms on Endo medium when routinely examining samples of potable water.[10]

Pseudomonas aeruginosa is a pathogen of man and animals, and it can cause spoilage and deterioration of a variety of materials. This organism differs from the *Enterobacteriaceae* by exhibiting oxidative, rather than fermentative, carbohydrate metabolism, being oxidase positive and possessing polar flagella. It also produces distinctive fluorescent and blue pigments.

Of special interest is the role of *P. aeruginosa* in hospital infections. It has been demonstrated that water supplies may be the source of the organisms causing outbreaks of diarrhea in infants, and foods may carry the organisms to burns units and cancer wards where they represent a serious threat to life. In swimming pools, the species may cause outer ear infections among swimmers, and in whirlpool baths it may cause pruritic, pustular skin rashes. The organism is able to grow in distilled water, and naturally occurring cells appear to be more resistant to chlorine than cells grown on laboratory media. Some investigators also have reported growth of *P. aeruginosa* on nonmetallic materials with which waters come into contact.

Pseudomonas aeruginosa usually is reported only rarely from distribution systems. Some reports of its frequent isolation from drinking waters have appeared, however, and where it has been demonstrated, most workers recommend that routine analyses for its presence be employed. Hungary probably is the only country presently with a *P. aeruginosa* standard, but in view of the generally weak correlation between this species and *E. coli* or coliforms, Shubert[16a] in Germany has recommended testing for it in sanitary evaluations of drinking water samples. Other workers in England and other European countries have developed a strong interest in this and related fluorescent pseudomonads in drinking waters.

Finally, before departing from the subject of waterborne bacterial pathogens, attention should be directed to antibiotic resistance transfer factors. In recent years, considerable interest has been demonstrated in extrachromosomal determinants of antibiotic resistance, which have increased in frequency and importance with the widespread use of antibiotics in medicine and agriculture. Extrachromosomal resistance determinants may be transferred from cell to cell, not only within a single bacterial species, but from species to species, and they may occur between species of more remotely related genera, such as *P. aeruginosa* and *E. coli*. The importance of transfer becomes clear when it is recognized that transfer may occur between harmless *E. coli* cells and pathogens.

Environmental sources of bacteria carrying resistance transfer factors are of some interest in view of speculation that it is the resident flora of man which is the source of resistance in infections failing to respond to antibiotic therapy. Since drinking water may be a source of organisms becoming established in the resident flora, more interest may be shown in the future in antibiotic resistance among bacteria in drinking waters.

Waterborne Protozoan Pathogens

While outbreaks of waterborne *Entamoeba histolytica* were reported in the years 1961-1970, none have been reported in the United States since 1970. On the other hand, outbreaks of waterborne disease caused by the flagellated protozoan *Giardia lamblia* have become more frequent. Three outbreaks of giardiasis were reported between 1961 and 1970, three in 1970-1971, two in 1973, and seven in 1974. The organism received much attention in the fall of 1974 when it became public knowledge that such infections had been acquired by tourists in Leningrad.

Many of the outbreaks have occurred in the Rocky Mountain states, sometimes as a result of consuming water from mountain streams. The highest attack rates observed in studies conducted in Colorado occurred in counties bordering the mountains, rates being somewhat lower in the mountains and lowest in the eastern plains. However, the largest reported outbreak of waterborne giardiasis occurred in Rome, New York, in 1974. The results of interviews and stool examinations indicated that the overall attack rate was 10.6% and that over 4800 persons may have been ill between November, 1974 and June, 1975. The attack rate was highest among persons utilizing city water, and the distribution of cases suggested that the water supply was contaminated at its source.

Sediments from raw water samples were fed to beagle puppies and caused disease in two of ten animals employed. A *Giardia lamblia* cyst was demonstrated in one sample. Human and animal carriers in the watershed were thought to constitute the source of the parasites. The cysts of *Giardia lamblia* are known to survive in tap water at 8°C for as long as 16 days and infect human volunteers. Furthermore, disease may be caused in man by 10 to 25 cysts. The attack rate in the Rome outbreak was significantly higher among persons consuming more than one glass of water per day than among persons consuming less than one glass per day.

Over the previous two years, coliform counts in the finished water were routinely low, with occasional counts ranging from 8,000 to 30,000/100 ml. After chlorination, a total residual of 0.8 mg chlorine per liter was maintained, but no free chlorine residual was present. No further treatment was provided.

Waterborne Viruses

In the early part of the discussion, estimates of waterborne salmonellosis, shigellosis, and hepatitis A were compared. While the crudeness of the estimates should be stressed, the figures do suggest the possible importance of viruses in water supplies relative to the bacterial pathogens. The costs assigned to the waterborne viral disease were nearly three times those assigned to the two waterborne bacterial diseases. Furthermore, if a substantial portion of the waterborne gastroenteritis of unknown etiology is caused by viruses, and if the enteric viruses transmitted by water cause disease in man, we should devote some effort to understanding the behavior of these agents in water supplies and to their elimination.

Even very small numbers of virus particles appear to be capable of causing infections in man, although perhaps not disease. Infected individuals may excrete large numbers of viruses and may cause infections and symptoms in their contacts. Berg[17] has suggested that the failure to demonstrate virus transmission by the water route may be a consequence of the use of clinical illness and not infection rates as criteria. In view of these observations, Berg[18] has suggested that any virus in drinking or recreational waters that can be detected in cell culture constitutes a hazard to those drinking the water.

At the International Conference on Viruses in Water held in Mexico City in 1974, Shuval[19] arrived at a recommended tentative virus standard for drinking water of no virus particles in 100 gallons. He arrived at this standard by first demonstrating that it was possible to achieve in a properly operating water treatment plant. This would result in a concentration of 1 pfu/1,000 liters. Employing further assumptions relating to ingestion and disease, Shuval estimated that this would lead to an attack of 1 per 100,000 persons at risk per day, which could not be detected epidemiologically. Addressing the question of detection, Shuval pointed out that it is possible to sample 100 gallons of finished water, and therefore a standard of no virus particles in 100 gallons is attainable and can be monitored. This recommendation and the earlier recommendation of 1 pfu/100 gallons made by Melnick[20] must be faced by the conscientious plant operator and governmental agencies.

Recent Questions Relating to Bacterial Indicator Systems

Several questions relating to currently employed indicator systems and their application have received some attention recently. The first relates to the definition of a coliform. If fermentation tubes are employed to detect and enumerate lactose fermenters, high counts may result if bacteria of the genus *Aeromonas* are present. These bacteria differ from members of the *Enterobacteriaceae,* being polarly flagellated and oxidase-positive. Furthermore, aeromonads frequently occur in surface waters and are of little sanitary significance. Thus, in some surface waters and raw drinking waters, total coliform counts obtained by the most probable number technique may be grossly overestimated, especially during the summer months. The exclusion of oxidase-positive organisms from the total coliform group should eliminate this group of organisms. There may be some pressure exerted from water bacteriologists to include the oxidase test during the confirmation of total coliforms in the future.

A second point that deserves attention is the inability of injured bacteria to recover on selective bacteriological media. It is now well established that viable bacteria injured during suspension in water or during chlorination may fail to form colonies on membrane filters, on the usual selective media, or even in lactose broth.[21-26] Injured cells may recover and form colonies on rich agar media, but fail to do so in the selective system. Preincubation of membrane filters on rich media at optimum temperatures has been employed by Lin[23,26] to resuscitate injured bacteria. He compared various enrichment media and conditions on the recovery of fecal coliforms from chlorinated secondary effluents, concluding that preenrichment on phenol red lactose broth for 4 hr at 35°C prior to transfer to m-FC medium yielded excellent correlation of counts with most probable numbers. Rose *et al.*[27] employed a double layer method for the enumeration of fecal coliforms in chlorinated wastewaters, reservoirs, and river waters. A basal layer of m-FC agar was overlaid with lactose broth containing agar. Membrane filters were placed on the layer of rich medium and incubated for 2 hr at 35°C to permit repair of injury. Following incubation at 35°C, plates were transferred to 44.5°C for 24 hr. This technique resulted in improved recoveries of fecal coliforms. Hartman *et al.*[28] employed a double layer direct plating technique for coliforms. Preenrichment has been employed by Rose and Litsky[29] and by Lin[30] to improve the recovery of fecal streptococci from water and chlorinated effluents on membrane filters. Lin, however, also improved recoveries from chlorinated effluents by increasing incubation time from two to three days. It should be noted, however, that while such techniques may double to triple recoveries, this probably is not the result of improved recoveries of stressed cells.[31]

Finally, Geldreich *et al.*[32] developed a rationale for the establishment of a standard plate count limit of 500 bacteria per ml in distribution systems. The frequency of total and fecal coliforms has been shown to increase until plate counts of 500/ml are reached. Above plate counts of 1000/ml, the frequency of isolation of coliforms decreases as a result of the antagonistic activity of other bacterial species. Geldreich *et al.* proposed that such plate counts could be attained if chlorine residuals of 0.3 mg/l and if turbidities of less than 1 unit were maintained. The rationale for the limit of 1 TU has been explained more fully recently by Symons and Hoff,[33] citing studies that demonstrated the reduced efficacy of chlorination at higher turbidities, the breakthrough of viruses as turbidities exceeded 0.5 TU, reduced taste and odor problems, and reduced nutrients for bacterial regrowth in the distribution system incorporated into the Interim Drinking Water Standards. If chlorine residuals and turbidities could be maintained, the need for bacteriological measurements in the distribution system would be less critical. Plate counts could be monitored every three months or so, although plant personnel would have to be alert to unusual circumstances which would indicate more frequent sampling. These limitations were given consideration in the "Statement of Basis and Purpose for the Proposed National Interim Primary Drinking Water Standards."

The use of a standard plate count would serve also as a measure of deterioration in water quality within the distribution system. Finished waters containing floc particles or unfiltered turbid waters receiving only chlorination may carry many bacteria into the distribution system where growth may occur. Organisms also may enter the distribution system for short periods following backwashing of filters.[34] The effect of growth in a distribution system on coliform counts and total plate counts was described recently by Becker,[35] who showed that monthly mean counts approached 500/ml during the summer months, often exceeding that limit in individual samples where adequate chlorine residuals could not be maintained.

The limit on turbidity has been incorporated into the National Interim Primary Drinking Water Regulations,[36] but the standard plate count has not. The coliform test remains. Coliform bacteria have been included in the drinking water standards since the first Public Health Service Drinking Water Standards in 1914. The value of this indicator system in the control of drinking water quality and waterborne infectious bacterial disease is well established. Fecal inhabitants among the group are of obvious significance. Other coliforms that occur in soil but which are not common in fecal material may better survive the stress of suspension in water or of chlorination, but should not persist after treatment of drinking waters. In view of the value of coliforms as indicators of the adequacy of treatment as well as of fecal contamination, coliform standards remain in the drinking water standards. The value of the coliform standard in providing assurance that a finished water is free of viruses is open to serious question.

The National Interim Primary Drinking Water Regulations include maximum contaminant levels for coliforms determined both by the most probable number and the membrane filter techniques. Included also are frequencies for sampling, requirements regarding notification of state authorities when the presence of coliform bacteria has been confirmed in check samples following initial failure to meet maximum contamination levels, and notification of state authorities and the public when maximum contamination levels are exceeded. The volume of samples to be filtered through membrane filters has been increased to 100 ml from 50 ml, and may be increased further in the future in view of increased protection against virus contamination. It was felt that the increase in volume was realistic in view of recommendations in *Standard Methods* and its current use in many state health laboratories.

The National Interim Drinking Water Regulations include the provision that a supplier of water may, with the approval of the state and based upon a sanitary survey, substitute the use of free chlorine residuals for not more than 75% of the required bacteriological samples. Such samples must be representative of conditions within the distribution system and at least four determinations of chlorine residual must be substituted for each bacteriological sample. Furthermore, it is required that no less than 0.2 mg/l of free chlorine be maintained throughout the distribution system.

The inclusion of a residual chlorine requirement has been made in an attempt to improve on the records of community water supplies, 85% of which failed to collect the required number of samples and 5% of which failed to meet the bacteriological quality standard during the Community Water Supply Study,[7] and in hopes that the residual chlorine determination might be a more practical means by which to assure safety. In an effort to establish the validity of residual chlorine requirements, Buelow and Walton[37] reviewed available data from Cincinnati and the Community Water Supply Study. On the basis of their investigations, they concluded that a change from combined to free chlorine residual may reduce monthly mean coliform counts substantially, that the probability of finding coliform bacteria in a sample from a distribution system decreases as the residual chlorine concentration of the water increases, that a chlorine residual must be maintained throughout the distribution system, and that a majority of bacteriological samples should be taken in known problem areas including deadends and the peripheries of systems. In order to improve bacteriological quality in Cincinnati over that achieved by maintaining 0.2 mg/l free chlorine residual, a very high residual was required. A free residual of 0.2 mg/l thus was probably optimum.

White[38] concluded that because the value of coliforms as indicators of viruses and therefore of water quality was questionable, and because contemporary disinfection practices may not eliminate unnecessary risks associated with viruses, it is doubtful that coliform bacteria constitute a suitable indicator of the efficacy of disinfection. Rather, disinfection should be based upon a combination of contact time and concentration of disinfectant providing a 99.6-100% inactivation of a resistant virus. On the other hand, such proposals should be viewed with caution in view of recent suggestions that chlorinated organics may be formed that may be carcinogenic. A search should be made for improved or alternative disinfection techniques that would be effective but would not lead to the formation of chlorinated organics.

In Chicago the use of free chlorine residuals maintained close to 0.75 mg/l when leaving the treatment plant, with combined chlorine being about 0.12 mg/l, have been employed.[39] Automated flushing of deadends to maintain free residual chlorine concentrations above 0.2 mg/l has proven successful. In other cities, deteriorating quality in distribution systems as a result of microbial activity has been reported. In Sioux Falls, S.D., free chlorine residuals of about 1 mg/l have been maintained in the water leaving the water treatment plant, but have been observed to drop rapidly at many points in the distribution system.[40] It has been suggested that increased chlorination, rechlorination and lowered pH might prove useful in controlling bacterial growth in distribution systems. In addition, the control of bacterial growth might be achieved through control of organic matter and nitrogen in the finished water. The prospects of such an effort should, however, be viewed with caution because of the well-established ability of some bacteria to grow even in distilled water.[41] Such organisms may be highly resistant to stresses such as chemical disinfection.[42]

CHRONIC DISEASES RELATED TO WATER

Trace Elements and Health

The most extensive literature on the connection between trace elements in water and human health relates to the relationship between water hardness and cardiovascular disease. Cardiovascular diseases are responsible for a larger proportion of deaths in this country than any other group of chronic diseases. Furthermore, life expectancy of white males over 45 years of age has increased little since 1900. This may be attributed to the slow progress in the fight against cardiovascular-renal diseases, and it has been concluded that no progress has been made in the control of cardiovascular heart disease since 1950.

In 1959 Sauer and Enterline[43] examined the geographic distribution of cardiovascular heart disease and observed considerable differences in rates in some instances. Notable geographical differences in the incidence of death from cerebral hemorrhage in Japan led Kobyashi[44] to consider the role of environmental factors. He found that river waters in areas that exhibited high mortality rates had excessively high sulfate to carbonate ratios, and concluded that the correlation between mortality and chemical composition of waters was a result of the abnormally acidic nature of many Japanese rivers. Schroeder[45] investigated mortality from all heart disease in the same areas of Japan and found an even higher correlation between this parameter of water quality and mortality. Subsequently, Schroeder examined the geographic variation in cardiovascular mortality and several parameters used to describe the chemical and physical characteristics of potable water in the U.S.[46] He found highly significant negative correlations between mortality from cardiovascular heart disease and magnesium, calcium, bicarbonate, sulfate, fluoride, dissolved solids, specific conductance, and pH. The most significant correlations were between hardness and cardiovascular heart disease. He concluded that some factor present in hard water, or missing from or entering soft water, appeared to affect death rates from degenerative cardiovascular diseases.

These findings stimulated similar investigations in other countries during the next five years. Morris *et al.*[47] reported a highly significant correlation between water hardness and mortality from cardiovascular heart disease in 83 county boroughs in England and Wales. A notable difference between the British and the U.S. studies was the absence of a correlation between magnesium and mortality in the latter study. Muss[48] investigated mortality from cardiovascular heart disease during the years 1936 through 1940 in health districts in New York City served by two different water sources, one of which was soft and the other relatively hard. He reported that the mortality rate was ten percent lower in the hard water districts. In Sweden, Biorck *et al.*[49] investigated death rates in 34 cities with respect to geology, water source, and 23 chemical parameters. Highly significant negative correlations between male (25-44 years) mortality resulting from degenerative diseases of the heart and water hardness, between male (25-64) mortality and calcium, and female (65-74 years) mortality resulting from cerebrovascular disease were correlated with total water hardness at the 1% level of significance. In general, correlations between magnesium or other sclerotic heart disease for any age-sex group were not significant. A later study of mortality in Sweden's three largest cities by Bostrom and Wester[50] indicated that decreased cardiovascular mortality among women, but not men, was associated with hard waters.

Anderson *et al.*[51,52] found that death rates from ischemic heart disease were higher in soft water areas of Ontario than in hard water areas. An investigation by Peterson *et al.*[53] in the U.S. supports these findings. Masironi[54] reviewed death rates in two hard water and two soft water river basins. Mortalities from hypertensive heart disease and ischemic heart disease were lower by 41% and 25%, respectively, in the hard water river basins than in the soft water river basins. Studies of populations served by water supplies subjected to substantial changes in the hardness[55,56] confirm the relationship. Crawford *et al.*[56] found that mortality from cardiovascular heart disease in county boroughs that changed from hard water to soft water was twice that in county boroughs that did not change their source supply.

Schroeder has suggested that trace metals in water may be an important factor in cardiovascular mortality. The corrosiveness of water is a potentially important factor. Schroeder proposed that cadmium, a known factor in high blood pressure, may be mobilized from galvanized pipes in the home. Schroeder and Kraemer[57] found highly significant inverse correlations between 15 constituents of water (12 were metals) and cardiovascular mortality. They also examined the corrosiveness of waters as measured by the Langlier Index and demonstrated a direct correlation with atherosclerotic heart disease at high levels of significance. They pointed out, furthermore, that the chemical composition of tap water may be a more important etiologic factor than that of finished water at the treatment facility. Neri *et al.*[58] also examined the relationship between water hardness and sudden death and suggested the possible significance of trace metals.

While most investigations tend to confirm the existence of an inverse correlation between mortality from cardiovascular heart disease and a "water factor," several epidemiological studies fail to demonstrate such a relationship. Mulcahy[59] examined the 15 largest urban centers in Ireland but was unable to establish any significant correlation between hardness and heart disease. In another study of the 20 largest urban areas in the Republic of Ireland and Northern Ireland, he found a consistent inverse relationship; however, it was not statistically significant.[60] An investigation by Lindemand and Assenzo[61] in Oklahoma also failed to establish a significant inverse relationship. As part of the International Atherosclerosis Project, Strong et al.[62] investigated the possible association between the mineral content of drinking waters and atherosclerotic lesions in the aorta and coronary arteries. Clinical examination of samples from 14 countries revealed an insignificant negative association between water hardness and the prevalence and extent of atherosclerosis.

An analysis has been undertaken at the University of Florida to determine the benefits that might accrue from improvements in drinking water quality, *i.e.,* decreases in cardiovascular heart disease attributable to increases in total dissolved solids. In their crude analysis of Standard Metropolitan Statistical Areas, an increase in total solids by the addition of 20 mg/l of lime would reduce the mortality rate from cardiovascular heart disease by 0.5%, and assuming a similar reduction in morbidity, benefit-cost ratios of from 3.04 in 1962 to 5.61 in 1971 were derived. As a result of this study, Hudson and Gilcrease[63] have pointed out that annually about 82 deaths/100,000 population are associated with variations in hardness of drinking waters, a rate four times that of typhoid deaths in 1910. The costs of lost income as a result of premature death amount to about $3 billion annually.

Further studies of trace elements include investigations of the relationship of lithium in drinking water to schizophrenia, psychosis, neurosis, personality problems, and homicidal tendencies explored by Dawson et al.[64,65] in Texas, who demonstrated a quantitative relationship between admissions to state mental hospitals and lithium ingestion via tap water.

An additional study of fluorosis in Uttar Pradesh[66] in India confirmed a protective effect of calcium against the uptake of fluoride, but suggested also that magnesium was antagonistic to the protective calcium effect. There appeared also to be a relationship between fluorosis and goiter, retarded growth and delayed puberty, and renal stones.

The National Interim Primary Drinking Water Regulations include maximum contaminant levels for eight metals and nitrate as shown in Table I. The toxicity of arsenic is well established and the ingestion of small amounts may cause severe effects. Arsenic is easily absorbed and slowly excreted, and chronic exposure may result in harmful effects. The limit of 0.05 mg/l has been maintained from the 1962 Drinking Water Standards. The limit for barium is derived from the threshold limit for air,

Table I. Maximum Contaminant Levels for Inorganic Chemicals in the National Interim Primary Drinking Water Regulations

Contaminant	Limit (mg/l)
Arsenic	0.05
Barium	1.0
Cadmium	0.010
Chromium	0.05
Lead	0.05
Mercury	0.002
Nitrate-N	10.0
Selenium	0.01
Silver	0.05

making allowances for absorption with a safety factor of two. Barium exerts serious effects on the nervous and the cardiovascular systems, but does not accumulate appreciably. Cadmium has been implicated in severe episodes of poisoning and in renal hypertension. The limit of 0.01 mg/l in drinking water is based on the contribution from water and food with a safety factor of four. It does not take into account cigarette smoking, however. Less is known of the effects of chromium in the diet. However, it was considered that the limit of 0.05 mg/l should provide protection from adverse effects on the health of consumers and bathers with a reasonable factor of safety.

The limit of 0.05 mg/l for lead was established to minimize the intake, especially in children, in view of the fact that the amount absorbed from food and the air is very close to the limit recommended by the World Health Organization. The total daily intake of mercury in some situations is approaching the maximum dietary intake, allowing a safety factor of 10. The limit of 0.002 mg/l for mercury minimizes the contribution to total intake, and is seldom exceeded in drinking waters. Nitrate is converted by bacteria to nitrite, which in turn converts hemoglobin to methemoglobin, which fails to carry oxygen in the blood. The disease that follows the formation of hemoglobin in the blood, methemoglobinemia, occurs primarily in infants who are predisposed to the disease. Few cases of the disease occur where nitrate-nitrogen concentrations fall below 10 mg/l. Data on the toxicity of selenium are not as readily available as for other elements. Based on intakes that apparently cause toxic effects, the limit of 0.01 mg/l provides a safety factor of three. The standard of 0.05 mg/l for silver was based upon indefinite storage of silver once absorbed.

Maximum contaminant levels for fluoride are linked, in the National Interim Drinking Water Regulations, to air temperature. These levels are twice the optimum level for the control of dental caries since concentrations substantially exceeding the optimum levels may cause mottling of teeth. Levels decrease in relation to air temperature to compensate for increased water intake at higher temperatures. It should be noted that decreases below optimum concentrations result in rapidly decreasing effectiveness in the prevention of dental caries.

Cancer

In view of the widely accepted relationship between environment and cancer, it is not surprising that the demonstration of suspected carcinogens in drinking water should receive considerable attention. Recently Page et al.[67] demonstrated statistically significant relationships between drinking water from the Mississippi River and all cancer and cancers of the urinary organs and the gastrointestinal tract. Asbestos fiber contamination of drinking waters, as well as contamination by polynuclear aromatic hydrocarbons, polychloro-biphenyls, chloroform, carbon tetrachloride and many other suspected carcinogens has been demonstrated.[68,69] It has been estimated that the intake of polynuclear aromatic hydrocarbons from water typically is only about 0.1% of the amount consumed in food. Andelman and Suess[70] recommended a maximum permissible concentration of 0.017 μg/l, but the WHO standards recommend a limit of 0.2 μg/l and Borneff,[71] who has published extensively on these compounds in water, recommended flocculation, ozonation, filtration, and activated carbon treatment of raw waters containing from 0.1-1.0 μg/l polynuclear aromatic hydrocarbons, and rejection of raw waters containing over 1.0 μg/l.

In response to the many unanswered questions and speculation concerning carcinogenic compounds in drinking waters, the Environmental Protection Agency has initiated studies on the character and extent of the contamination of water supplies, health effects, sources, control and costs of control. A study of asbestos fibers in water supplies in relation to gastrointestinal tumors will be initiated in Connecticut, where a tumor registry provides extensive data on the incidence of tumors by township.

A recent report of Sonstegard[72] is of special interest, and is worth mentioning here. Sonstegard studied tumors in fish from the Great Lakes, demonstrating increases in incidence from 0% in 1950 to 90% in 1975 in the Detroit River. From the studies of tumors and their geographical distribution, questions arise as to causative factors in the environment, especially carcinogens in the water, and Sonstegard recommended intensive studies to determine the incidence, etiology and distribution of tumors in fishes in the Great Lakes. The monitoring of tumors in fish offers great promise as an indicator system for a broad spectrum of carcinogens and antagonistic and synergistic interactions in source waters.

Maximum contaminant levels for certain chlorinated hydrocarbons are included in the National Interim Primary Drinking Water Regulations. Levels were based upon concentrations exhibiting minimal or no toxic effects and safety factors of 100 or 500 applied to animal data depending upon the adequacy of human data. Maximum contaminant levels for the chlorophenoxy compounds 2,4-D and 2,4,5-TP were based upon concentrations having minimal or no effects on animals and a safety factor of 500.

REFERENCES

1. Craun, G. F. and L. J. McCabe. "Review of the Causes of Waterborne Disease Outbreaks," *J. Am. Water Works Assoc.* 65:74 (1973).
2. Merson, M. H., W. H. Barker, Jr., G. F. Craun and L. J. McCabe. "Outbreaks of Waterborne Disease in the United States, 1971-1972," *J. Infect. Dis.* 129:614 (1974).
3. Craun, G. F., L. J. McCabe and J. M. Hughes. "Waterborne Disease Outbreaks in the U.S.–1971-1974," *J. Am. Water Works Assoc.* 68:420 (1976).
4. Center for Disease Control. "Foodborne and Waterborne Disease Outbreaks," Annual Summary, U.S. Public Health Service, Atlanta (1974).
5. Center for Disease Control. "Foodborne and Waterborne Disease Outbreaks," Annual Summary, U.S. Public Health Service, Atlanta (1975).
6. Center for Disease Control. "Foodborne and Waterborne Disease Outbreaks," Annual Summary, U.S. Public Health Service, Atlanta (1976).
7. Bureau of Water Hygiene. "Community Water Supply Study; Analysis of National Survey Findings," Bureau of Water Hygiene, U.S. Public Health Service (1970).
8. DuPont, H. I. and R. B. Hornick. "Clinical Approach to Infectious Diarrheas," *Medicine* 52:265 (1973).
9. Schroeder, S. A., J. R. Caldwell, T. M. Vernon, P. C. White, S. I. Granger and J. V. Bennett. "A Waterborne Outbreak of Gastroenteritis in Adults Associated with *Escherichia coli,*" *Lancet* i:737 (1968).
10. Highsmith, A. K., J. C. Feeley, B. T. Wood, P. Skaliy, J. G. Wells and M. L. Rosenberg. "Isolation of *Yersinia enterocolitica* from Well Water," Abstracts, Annual Meeting of the American Society for Microbiology, Atlantic City (1976).
11. Asakawa, Y., S. Akahane, N. Kagata, M. Noguchi, R. Sakazaki and K. Tamura. "Two Community Outbreaks of Human Infection with *Yersinia enterocolitica,*" *J. Hyg.* 71 : 715 (1973).

12. Keet, E. E. *"Yersinia enterocolitica* Septicemia," *N.Y. State J. Med.* 74:2226 (1974).
13. Saari, T. N. and T. J. Quan. "Waterborne *Yersinia enterocolitica* in Colorado," presented at the Annual Meeting of the American Society for Microbiology, Atlantic City (1975).
14. Saari, T. N. and T. J. Quan. "Waterborne *Yersinia enterocolitica* in Colorado," Abstracts, Annual Meeting of the American Society for Microbiology, Atlantic City (1975).
15. Lassen, J. *"Yersinia enterocolitica* in Drinking Water," *Scand. J. Infect. Dis.* 4:125 (1972).
16. Harvey, S., J. R. Greenwood, M. J. Pickett and R. A. Mah. "Recovery of *Yersinia enterocolitica* from Streams and Lakes of California," *Appl. Environ. Microbiol.* 32:352 (1976).
16a. Schubert, R. H. W. and U. Blum. "Zur Frage der Enweiterung der hygienischer Wassercontrolle auf den Nachweis von Pseudomonas aeruginosa," *GWF-Wasser/Abwasser* 115:224 (1974).
17. Berg, G. "An Integrated Approach to the Problem of Viruses in Water," in *National Specialty Conference on Disinfection,* American Society of Civil Engineers (1970), pp. 339-364.
18. Berg, G., R. M. Clark, D. Berman and S. L. Chang. "Aberrations in Survival Curves," in *Transmission of Viruses by the Water Route,* G. Berg, Ed. (New York: Interscience Publishers, 1966), pp. 235-240.
19. Shuval, H. I. "Water Needs and Usage: The Increasing Burden of Enteroviruses on Water Quality," in *Viruses in Water,* G. Berg, H. L. Bodily, E. H. Lennette, J. L. Melnick and T. G. Metcalf, Eds. (Washington, D.C.: American Public Health Association, 1976), pp. 12-25.
20. Melnick, J. L. "Detection of Virus Spread by the Water Route," in *Viruses and Water Quality,* V. L. Snoeyink, Ed. (University of Illinois, 1971).
21. Hoadley, A. W. and C. M. Cheng. "The Recovery of Indicator Bacteria on Selective Media," *J. Appl. Bacteriol.* 37:45 (1974).
22. Bissonnette, G. C., J. J. Jezeski, G. A. McFeters and D. G. Stuart. "Influence of Environmental Stress on Enumeration of Indicator Bacteria from Natural Waters," *Appl. Microbiol.* 29:186 (1975).
23. Lin, S. D. "Evaluation of Coliform Tests for Chlorinated Secondary Effluents," *J. Water Poll. Control Fed.* 45:498 (1973).
24. Braswell, J. R. and A. W. Hoadley. "Recovery of *Escherichia coli* from Chlorinated Secondary Sewage," *Appl. Microbiol.* 28:328 (1974).
25. Mowat, A. "Most Probable Number versus Membrane Filter on Chlorinated Effluents," *J. Water Poll. Control Fed.* 48:724 (1976).
26. Lin, S. D. "Membrane Filter Method for Recovery of Fecal Coliforms in Chlorinated Sewage Effluents," *Appl. Environ. Microbiol.* 32:547 (1976).
27. Rose, R. E., E. E. Geldreich and W. Litsky. "Improved Membrane Filter Method for Fecal Coliform Analysis," *Appl. Microbiol.* 29:532 (1975).
28. Hartman, P. A., P. S. Hartman and W. W. Lanz. "Violet Red Bile 2 Agar for Stressed Coliforms," *Appl. Microbiol.* 29:537 (1975).

29. Rose, R. E. and W. Litsky. "Enrichment Procedure for Use with the Membrane Filter for the Isolation and Enumeration of Fecal Streptococci in Water," *Appl. Microbiol.* 13:106 (1965).

30. Lin, S. "Evaluation of Fecal Streptococci Tests for Chlorinated Secondary Sewage Effluents," *J. Environ. Eng. Div.* ASCE 100:253 (1974).

31. Hoadley, A. W. and C. M. Cheng. "The Recovery of Indicator Bacteria on Selective Media," *J. Appl. Bacteriol.* 37:45 (1974).

32. Geldreich, E. E., H. D. Nash, D. J. Reasoner and R. H. Taylor. "The Necessity of Controlling Bacterial Populations in Potable Waters: Community Water Supply," *J. Am. Water Works Assoc.* 64:596 (1972).

33. Symons, J. M. and J. C. Hoff. "Rationale for Turbidity Maximum Contaminant Level," in *Proceedings AWWA 1975 Water Quality Technology Conference*, paper 2A-Za.

34. Boorsma, H. J., C. H. Elzanga and J. C. van der Vlugt. "Deterioration of Water Quality in the Distribution System. International Standing Committee on Water Quality and Treatment," Subject 2, (1974), pp. L7-L12.

35. Becker, R. J. "Bacterial Regrowth within the Distribution System," in *Proceedings AWWA 1975 Water Quality Technology Conference*, paper 2B4 (1976).

36. U.S. Environmental Protection Agency. "National Interim Primary Drinking Water Regulations," *Federal Register* 40:59566 (1975).

37. Buelow, R. W. and G. Walton. "Bacteriological Quality vs Residual Chlorine," *J. Am. Water Works Assoc.* 63:28 (1971).

38. White, G. C. "Disinfection: The Last Line of Defense for Potable Water," *J. Am. Water Works Assoc.* 67:410 (1975).

39. Willey, B. F., C. M. Duke and J. Rasho. "Chicago's Switch to Free Chlorine Residuals," *J. Am. Water Works Assoc.* 67:438 (1975).

40. O'Connor, J. T., L. Hash and A. B. Edwards. "Deterioration of Water Quality in Distribution Systems," *J. Am. Water Works Assoc.* 67:113 (1975).

41. Favero, M. S., L. A. Carson, W. W. Bond and N. J. Petersen. "*Pseudomonas aeruginosa*: Growth in Distilled Water from Hospitals," *Science* 173:836 (1971).

42. Carson, L. A., M. S. Favero, W. W. Bond and N. J. Petersen. "Factors Affecting Comparative Resistance of Naturally Occurring and Subcultured *Pseudomonas aeruginosa* to Disinfectants," *Appl. Microbiol.* 23:863 (1972).

43. Sauer, H. I. and P. E. Enterline. "Are Geographic Variations in Death Rates for Cardiovascular Diseases Real?" *J. Chron. Dis.* 10:513 (1959).

44. Kobyashi, J. "On Geographical Relationship Between the Chemical Nature of River Water and Death Rate of Apoplexy," *Ber. Ohara Inst. Lanur. Forech.* 11:12 (1957).

45. Schroeder, H. A. "Degenerative Cardiovascular Disease in the Orient," *J. Chron. Dis.* 8:312 (1958).

46. Schroeder, H. A. "Relation Between Mortality from Cardiovascular Disease and Treated Water Supplies," *J. Am. Med. Assoc.* 172:1902 (1960).

47. Morris, J. N., M. D. Crawford and J. A. Heady. "Hardness of Local Water Supplies and Mortality from Cardiovascular Disease," *Lancet* i:860 (1961).
48. Muss, D. L. "Relationship Between Water Quality and Deaths from Cardiovascular Disease," *J. Am. Water Works Assoc.* 54:1371 (1962).
49. Biorck, G., H. Bostrom and A. Widstrom. "On the Relationship Between Water Hardness and Death Rate of Cardiovascular Diseases," *Acta Med. Scand.* 178:239 (1965).
50. Bostrom, H. and P. O. Wester. "Trace Elements in Drinking Water and Death Rate in Cardiovascular Disease," *Acta Med Scand.* 181:465 (1967).
51. Anderson, T. W., W. H. LeRiche and H. S. MacKay. "Sudden Death and Ischemic Heart Disease Correlation with Hardness of Local Water Supply," *New England J. Med.* 280:805 (1969).
52. Anderson, T. W. and W. H. LeRiche. "Sudden Death from Ischemic Heart Disease in Ontario and its Correlation with Water Hardness and Other Factors," *Can. Med. Assoc. J.* 105:155 (1971).
53. Peterson, D. R., D. J. Thompson and J.-M. Nam. "Water Hardness, Arteriosclerotic Heart Disease and Sudden Death," *Am. J. Epidemiol.* 92:90 (1970).
54. Masironi, R. "Cardiovascular Mortality in Relation to Radioactivity and Hardness of Local Water Supplies in the U.S.A.," *Bull. WHO* 43:687 (1970).
55. Robertson, J. S. "The Water Story," *Lancet* i:1160 (1969).
56. Crawford, M. D., M. J. Gardner and J. N. Morris. "Changes in Water Hardness and Local Death Rates," *Lancet* ii:327 (1971).
57. Schroeder, H. A. and L. A. Kraemer. "Cardiovascular Mortality, Municipal Water and Corrosion," *Arch. Environ. Health* 28:303 (1974).
58. Neri, L. C., D. Hewitt and J. S. Mandel. "Risk of Sudden Death in Soft Water Areas," *J. Epidemiol.* 94:101 (1971).
59. Mulcahy, R. "The Influence of Water Hardness and Rainfall on the Incidence of Cardiovascular and Cerebrovascular Mortality in Ireland," *J. Irish Med. Assoc.* 55:17 (1964).
60. Mulcahy, R. "The Influence of Water Hardness and Rainfall on Cardiovascular and Cerebrovascular Mortality in Ireland," *J. Irish Med. Assoc.* 59:14 (1966).
61. Lindeman, R. D. and J. R. Assenzo. "Correlations Between Water Hardness and Cardiovascular Deaths in Oklahoma Counties," *Am. J. Pub. Health* 54:1071 (1964).
62. Strong, J. P., P. Correa and L. A. Solberg. "Water Hardness and Atherosclerosis," *Lab. Invest.* 18:620 (1968).
63. Hudson, H. E., Jr. and F. W. Gilcreas. "Health and Economic Aspects of Water Hardness and Corrosiveness," *J. Am. Water Works Assoc.* 68:201 (1976).
64. Dawson, E. B., T. D. Moore and W. J. McGanity. "The Mathematical Relationship of Drinking Water Lithium and Rainfall to Mental Hospital Admission," *Dis. Nerv. Syst.* 31:811 (1970).
65. Dawson, E. B., T. D. Moore and W. J. McGanity. "Relationship of Lithium Metabolism to Mental Hospital Admission and Homicide," *Dis. Nerv. Syst.* 33:546 (1972).

66. Teotia, S. P. S. and M. Teotia. "Dental Fluorosis in Areas with a High Natural Content of Calcium and Magnesium in Drinking Water— An Epidemiological Study," *Fluoride* 8:34 (1975).

67. Page, T., R. H. Harris and S. S. Epstein. "Drinking Water and Cancer Mortality in Louisiana," *Science* 193:55 (1975).

68. Andelman, J. B. and J. E. Snodgrass. "Incidence and Significance of Polynuclear Aromatic Hydrocarbons in the Water Environment," CRC Critical Reviews in Environmental Control, January (1974), pp. 69-83.

69. Environmental Protection Agency. "Preliminary Assessment of Suspected Carcinogens in Drinking Water," Interim report to Congress, U. S. Environmental Protection Agency (1975).

70. Andelman, J. B. and M. J. Suess. "Polynuclear Aromatic Hydrocarbons in the Water Environment," *Bull. WHO* 43:479 (1970).

71. Borneff, J. "Elimination of Carcinogenic Polycyclic Aromatic Compounds during Water Purification," *Gas Wasserfach* 110:29 (1969).

72. Sonstagard, R. A. "Virological and Epizootiological Studies of Fish Neoplasms in Polluted and Nonpolluted Waters of the Great Lakes," unpublished report, Environment Canada (1975).

CHAPTER 3

PREDESIGN STUDIES

R. Rhodes Trussell, Ph.D., P.E.
Vice President
James M. Montgomery, Consulting Engineers, Inc.
555 East Walnut Street
Pasadena, California 91101

INTRODUCTION

This chapter is written to inform the potential water treatment plant design engineer of methods for establishing treatment requirements for a particular water supply. Processes normally encompassed by the term "water treatment" include, for example, turbidity removal, softening, removal of iron and manganese, H_2S removal, color removal, treatment for tastes and odors and disinfection. The most widely employed process, however, is turbidity removal, and it is that process with which this discussion will deal. First, water quality standards are discussed, including a few comments on the current status of water quality regulations where organic materials are concerned. Then the situations most often faced by the design engineer when he is retained to do a study preliminary to a water treatment plant design are evaluated. Finally, some of the bench-, pilot-, and field-scale tests that have been found useful in evaluating treatment alternatives are described.

WATER QUALITY STANDARDS

Tables I, II and III summarize the levels of water quality specified by four different agencies: The World Health Organization,[1] the U.S. Environmental Protection Agency,[2] the American Water Works Association,[3] and the U.S. Public Health Service.[4] The WHO standards are discussed because they represent an internationally recognized standard of fairly recent vintage. The

Table I. Drinking Water Criteria–Inorganic Chemicals (all in mg/l)

Constituent	World Health Organization (1971)			EPA Interim Primary Regulation (1975)	American Water Works Association (1968)	USPHS (1962)	
	Recommended	Acceptable	Tolerance	Maximum Contaminant Limit	Goal	Recommended	Mandatory
Aluminum	-	-	-	-	0.05	-	-
Arsenic	-	-	0.05	0.05	-	0.01	0.05
Barium	-	-	-	1.0	-	-	1.0
Cadmium	-	-	0.01	0.01	-	-	0.01
Chromium (total)	-	-	-	0.05	-	-	-
Chromium (VI)	-	-	-	-	-	-	0.05
Chloride	200	600	-	-	-	250	-
Copper	0.05	1.5	-	-	0.2	1.0	-
Cyanide	-	-	0.05	-	-	0.01	0.2
Fluoride	a	1.0	-	a	-	a	a
Iron	0.1	1.0	-	-	0.05	0.3	-
Lead	-	-	0.1	0.05	-	-	0.05
Manganese	0.05	0.5	-	0.05	-	0.05	-
Mercury	-	-	0.001	0.002	0.01	-	-
Nitrate (as N)	-	10	-	10	-	10	-
Selenium	-	-	0.01	0.01	-	-	0.01
Silver	-	-	-	0.05	-	-	0.05
Sulfate	200	400	-	-	-	250	-
Zinc	5.0	15	-	-	1.0	5	-
Hardness	-	-	-	-	80-100	-	-
TDS	-	-	-	-	200	500	-

aThe acceptable fluoride concentration is described as a function of ambient temperature. Values range from 0.6 to 2.4 mg/l.

Table II. Drinking Water Criteria—Organic Chemicals

Constituent	World Health Organization (1971)		EPA Interim Primary Regulation (1975)	American Water Works Association (1968)	USPHS (1962)	
	Recommended	Acceptable Tolerance	Maximum Contaminant Limit	Goal	Recommended	Mandatory
Carbon-Chloroform Extract	-	-	-	0.04	0.2	-
Foaming Agents (MBAS)	0.2	1.0	-	0.2	0.5	-
Aldrin	-	-	a	-	-	-
Chlordane	-	-	a	-	-	-
DDT	-	-	a	-	-	-
Dieldrin	-	-	a	-	-	-
Endrin	-	-	0.0002	-	-	-
Heptachlor	-	-	a	-	-	-
Heptachlor Epoxide	-	-	a	-	-	-
Lindane	-	-	0.004	-	-	-
Methoxychlor	-	-	0.1	-	-	-
Toxaphene	-	-	0.005	-	-	-
2,4-D	-	-	0.1	-	-	-
2,4,5-TP (Silvex)	-	-	0.01	-	-	-
Chloroform	-	-	b	-	-	-
Phenols	0.001	0.002	-	-	0.001	-

a Interim primary regulations are under preparation.
b Regulation being seriously considered.

Table III. Drinking Water Criteria—Physical, Radiological and Microbiological Parameters

Constituent	World Health Organization (1971)			EPA Interim Primary Regulation (1975)	American Water Works Association (1968)	USPHS (1962)	
	Recommended	Acceptable	Tolerance	Maximum Contaminant Limit	Goal	Recommended	Mandatory
Physical							
Color (CU)	5	50	-	-	3	15	-
Suspended Solids (mg/l)	-	-	-	-	0.1	-	-
Taste	a	-	-	-	a	a	-
Turbidity (TU)	5	25	-	1[b]	0.1	5	-
Radiological[c] (pc/l)							
Gross Alpha	3	10	-	15	-	-	-
Gross Beta	30	100	1000	-	100	-	1000
Radium 226 & 228	3	-	-	5	-	-	3
Strontium 90	30	-	-	8	-	-	10
Bacteriological[c]							
Coliform (org/100 ml)	1	-	-	1	None	-	1
Plankton Count (org/ml)	-	-	-	-	None	-	-
Virus (pfu/l)	-	1	-	-	-	-	-

[a] Not objectionable.
[b] Maximum contaminant level is average of 1 TU, but may be increased to 5 TU under special circumstances.
[c] Here the standards have been somewhat simplified to allow a straightforward presentation. The original documents should be referred to.

USEPA interim primary regulations are included because they are the most recently developed and because they will probably be of direct concern to most people reading this chapter. The AWWA water quality goals and the 1962 USPHS standards are included for purposes of comparison.

Generally, the inorganic chemicals of concern consist of a number of metals of established health significance and few additional anions, some of which are also associated with health effects whereas others have a general aesthetic effect. Most of these compounds are rarely encountered in potential water supplies. Although no systematic information has been gathered recently regarding raw water supplies, the inorganic chemicals most often exceeding standards are probably nitrate, fluoride and arsenic. Excessive levels of these constituents generally appear in ground water supplies. Where fluoride and arsenic are concerned, high levels are generally the result of natural phenomena, although incidents of manmade pollution also occur. High nitrate levels, on the other hand, result principally from agricultural and municipal pollution of ground water basins. Recent studies in Seattle and Boston have demonstrated that undesirably high levels of cadmium and lead may also appear at the consumer's tap when a water is corrosive. Organic compounds singled out by these regulations consist of the more common pesticides and herbicides and some crude indices of the total content of organic and detergent materials.

A great deal of research is presently taking place on organics in water supplies, and indications are that new requirements may soon be set by the EPA. Among the most likely is a limitation on the trihalomethanes, *e.g.,* chloroform. Others likely to be considered are aldrin, chlordane, DDT, dieldrin, heptachlor, heptachlor epoxide and PCBs because studies on these materials have just been completed. A limitation on the overall organic content, such as the total organic carbon, has also been suggested. However, it seems unlikely that such a limit will be set because in most water supplies the bulk of the TOC is composed of natural aquatic humus materials having no known direct health effects.

Many have felt that the considerable attention paid to the organohalides in recent months is unjustified, and the EPA has generally taken the position that no standards should be set until their significance can be evaluated. Several events have occurred during the past three years, however, that may impact this situation. Principal among these is a study conducted by the Environmental Defense Fund in New Orleans,[5] which concluded: "Presumptive epidemiological evidence . . . suggests a significant relationship between cancer mortality among white males and drinking water from the Mississippi River." Another epidemiological study conducted in Holland[6] has shown also that death rates from cancer were higher among those persons whose municipal system derived water from a river than among those who derived their water from wells.

About this same time, two studies, one conducted in Rotterdam by Rook[7] and the other conducted in New Orleans by Bellar, Lichtenberg and Kroner,[8] demonstrated that certain chlorinated organics (mostly chloroform) appeared regularly in water treatment plant effluents at much higher levels than could be found in the water influent to the same plants. Subsequent research[9,10] showed that the formation of these trihalomethanes resulted primarily from the reactions occurring between natural aquatic humus and the chlorine used for disinfection. Subsequently, (1) EPA's National Organics Reconnaissance Survey (NORS) showed that chloroform appears in virtually every chlorinated water supply,[11] (2) the National Cancer Institute released the results of a study showing that chloroform can cause tumors in mice and rats,[12] (3) the EPA announced a blanket ban on the use of chloroform in food and drugs[13] and (4) the EPA published an Advanced Notice of Proposed Rule Making (ANPRM) proposing alternative regulations for monitoring organics.[14] The EPA also published an interim treatment guide reviewing techniques for controlling trihalomethanes.[15] Present indications are that some sort of trihalomethane regulation is imminent.

In the past, all of the physical parameters in Table III have been given recommended rather than mandatory limits, and their effects have been described as being largely aesthetic in nature. Recent events would indicate that this situation is rapidly changing. In 1972, California[16] established a mandatory requirement for filtration to a turbidity of 0.5 TU for all waters exposed to a significant sewage hazard and, as of June, 1977, the EPA interim standards will require that all surface water supplies meet a turbidity of 1 TU. These mandatory limits have been established because research has shown that particulate matter may protect pathogens from the disinfecting agent.[17]

It is likely that color is another physical parameter that may, in effect, be limited soon. Just as turbidity is a crude measure of particulate material, color appears to be a crude measure of the humic substances that seem to be the primary precursors to the organohalides.[7-11]

Radioactivity of significance in water supplies occurs in one of three forms, each of them much different where health significance is concerned. Alpha radiation results from the emission of large, positively charged particles that are highly damaging. Each alpha particle takes the form of a helium atom stripped of its electrons and traveling at speeds as high as 10^6 m/s. Because of their mass, alpha particles cannot penetrate the epidermis; however, when ingested, an alpha source is very dangerous. Beta radiation results from the emission of high energy electrons that travel near the speed of light. Because of its smaller size, a beta particle is more penetrating but less damaging than an alpha particle. Gamma radiation is electromagnetic radiation with tremendous penetrating power but limited effect, similar to X-rays.

The standards described here set limits on the gross quantity of alpha and beta radiation and on radium 226, radium 228, and strontium 90. All of the above isotopes are beta emitters and the two radium isotopes emit alpha and gamma rays as well.

Historically, the microbiological standards are the most important standards presently being used. Present standards use the coliform organism as an indicator of sewage contamination. This method is chosen because it is not practical to assay for specific pathogens as it is too costly, the assay methods are too slow, there are too many pathogens to be evaluated, and the results are only available after the fact of disease exposure. The coliform organism is chosen as a good indicator of contamination because it is found in the human gastro-intestinal tract at very high levels, because coliforms have been shown to have resistance similar to most pathogens, and because the coliform test is a sensitive, economical test. The WHO standards only suggest that a water which does not show one plaque-forming unit (pfu) in a liter is safe from the standpoint of viral contamination. Regular assays for virus are not recommended.

WATER SOURCE EVALUATION

Evaluating a potential water source is a matter of determining the amount of water available both on a long-term basis (safe yield) and on a short-term basis and determining the quality of water available. The latter should usually include a sanitary survey. More often than not, a number of studies are conducted to fulfill these needs. The last phase is an evaluation of the water treatment needs. It is this last phase of the investigation that is discussed in detail here.

In evaluating the chemical and bacteriological quality of a water source where turbidity removal is the matter of primary concern, it is desirable to include as much long-term data on particulate matter and microbiological parameters as possible, specifically turbidity, suspended solids, plankton and coliform. If any special water quality problems exist, a history should be developed for these parameters as well.

General mineral quality is also an important aspect of understanding the characteristics of a water supply. This is especially true where water treatment is concerned because certain aspects of the mineral quality, especially the alkalinity and pH, have an important influence on the performance of treatment processes. Table IV summarizes the eight minerals most often encountered in water supplies in a hypothetical mineral analysis. Also shown in the table are illustrative values for the concentration of each mineral ion and an illustrative calculation of the "Deviation from Electroneutrality." The latter deviation is a balance between the positive and negative charges in the solution and should be checked in every mineral analysis used in evaluating a water supply.

Table IV. Mineral Analysis

Cations	mg/l	meq/l	Anions	mg/l	meq/l
Calcium	20	1.00	Alkalinity	50	1.00
Magnesium	10	0.83	Sulfate	50	1.04
Sodium	25	1.09	Chloride	35	0.99
Potassium	5	0.13	Nitrate	5	0.08
		Tcat = 3.05			Tan = 3.11

$$\text{Deviation from electroneutrality} = \frac{\text{Tcat - Tan}}{\text{Tcat + Tan}} = 0.010$$

According to the principal of electroneutrality, these charges should be equal and the deviation should be zero. As a practical matter, water analyses should be rejected whenever the deviation is greater than 0.03.

Data presentation is another important aspect of evaluating the quality of a water supply. Although arithmetic averages are useful in many instances, a statistical presentation is usually more useful. Figures 1 and 2 show such a presentation for the turbidity and coliform count of a raw water supply. The former is best plotted on an arithmetic-probability scale and the latter on a logarithmic-probability scale. The method for constructing these diagrams can be found in a number of texts.[18]

Generally, in describing data of this sort, there are two aspects of data analysis that are most important. These are the means of describing the central tendency of the data (average, geometric mean, median) and the degree of variability in the data (standard deviation, 90% value). Plots of the sort shown in Figures 1 and 2 give a complete pictorial description of both of these phenomena.

BENCH-SCALE TESTS

Because the phenomena of water treatment are mostly of a physiochemical nature, bench-scale testing often gives meaningful insight to full-scale results. Among the most famous of these tests is the "jar test," a technique first devised by Langelier five and a half decades ago[19] and developed little since. There are a number of variations in equipment set up which may be used to conduct this test and these are described in the literature[20-22]; however, tests are most frequently conducted on a commercially available, variable-speed jar tester such as that manufactured by Phipps-Bird.* This equipment can simultaneously vary the speed of all six of its 1"x 3" paddles between approximately 8 and 170 rpm. Although some standardization of the test would probably

---

*Phipps-Bird jar-tester Model No. 7790-300, Phipps-Bird, Inc., Richmond, Va.

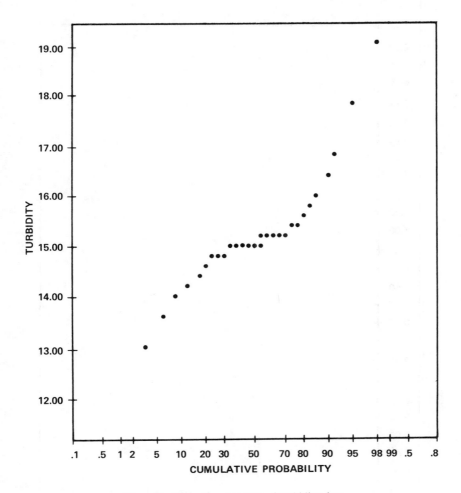

Figure 1. Normal probability of turbidity data.

be useful, little standardization has developed so far. With this need in mind, a summary is hereby given of a technique that has suited the author well in a number of water treatment evaluations:

1. Six 1.5-liter pyrex beakers are filled with 1 liter of the water to be tested and placed under the stirrer.

2. The stirrer is turned to its maximum speed (170 rpm).

3. The appropriate coagulant dosages are made. (A convenient reagent concentration for alum is 1 g/l.)

4. Stirring is continued for 2 min and then the speed is lowered to 80 rpm.

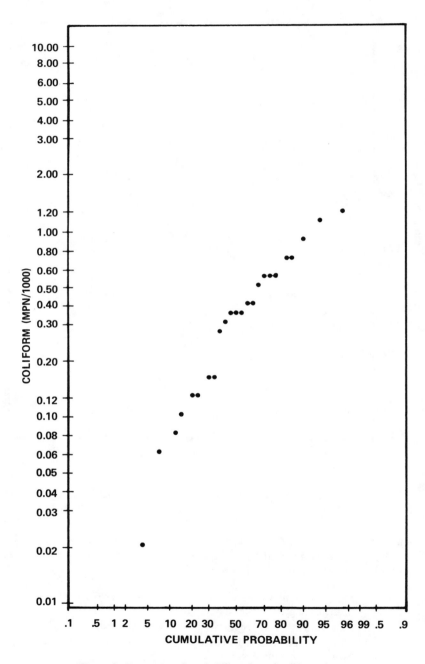

Figure 2. Log–normal probability plot of coliform data.

5. Stirring is continued for 8 min and the stirrer is turned off, the paddles are carefully pulled from the solution, and the samples are allowed to settle quietly.

6. After 15 min of sedimentation, a sample is carefully withdrawn via a pipette with a ground-off tip, and appropriate measurements are made (turbidity, pH).

In a typical jar test, either an Al(III) or an Fe(III) salt is used in varying amounts and no effort is made to control the pH. A much more complete description of the coagulation characteristics of a raw water can be developed, however, if a turbidity topogram is prepared. A turbidity topogram is a plot of iso-turbidity lines on an axis of log (Al or Fe) vs pH. To construct such a diagram a series of four or more jar tests are usually used. Because of the large quantities of water involved, the tests are best conducted onsite. A typical procedure is as follows:

1. A large volume of raw water (60 liters or more) is collected in a plastic container (a plastic garbage can is useful) and constantly agitated. A small submersible pump serves the latter purpose well.

2. A 10-liter portion of the raw water is withdrawn and adjusted to a particular pH. This is best accomplished by titrating 1 liter with $0.02\ N\ H_2SO_4$ and determining the amount of acid necessary to adjust the remainder of the 10-liter volume.

3. The pH-adjusted sample is now used as the raw water for the jar test described earlier. Care should be taken to record the alum dose used, the final pH and the residual turbidity.

4. The residual turbidity is noted at the intersection of the appropriate final pH and coagulant dose on a log-log plot.

5. Steps 2, 3 and 4 are repeated for each of the pH ranges of interest. For purposes of a clear representation, pH and coagulant doses, which are well beyond those deemed practical or economical, should be used in the investigation.

Figure 3 represents such a topogram developed from data on a mixture of Colorado River water and a local southern California surface supply. The coagulant dose distinctly drops as the pH decreases throughout the pH range of 7 to 9. As alum is not necessarily the least expensive chemical available for the purpose of pH adjustment, the possibility of pH adjustment with a less expensive chemical, such as sulfuric acid, should be considered. Figure 4 is such a plot, showing the cost of acid and alum vs the pH. Such a plot is constructed by selecting a particular turbidity isopleth from Figure 3, in this example the 0.5 TU isopleth. Using the same data necessary to construct Figure 3, the amount of acid and the amount of alum required to reach the chosen turbidity level can be determined for a range of pH values. As shown in Figure 4, cost curves can then be constructed.

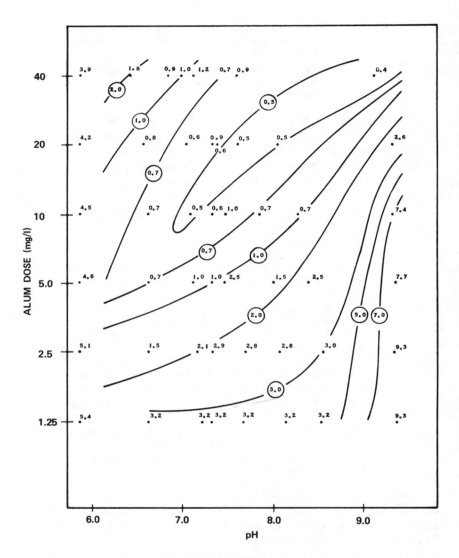

Figure 3. Turbidity topogram.

Meaningful tests of the energy input necessary to accomplish the flocculation task can also be conducted with this apparatus. However, extensive extrapolations are probably not justified. In this regard, the jar test's batch design is its most significant shortcoming. For the specific arrangement discussed

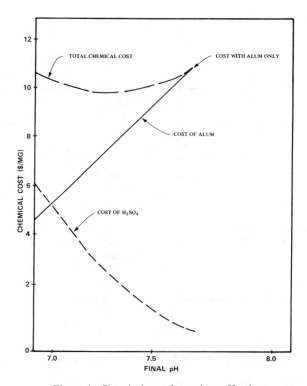

Figure 4. Chemical cost for various pH values.

herein (1 liter of sample in a 1.5-liter beaker with a 1″ x 3″ paddle), the following equation was developed by TeKippe[23]:

$$G = 0.084 \ N^{3/2} \qquad (1)$$

where G = root mean square velocity gradient, sec^{-1}
       N = speed of rotation (rpm)

Figure 5 shows that, within a reasonable range, this equation may be extrapolated to different flocculator volumes (V, liters) using the following relation:

$$G = 0.084 \ N^3/V^{1/2} \qquad (2)$$

Although the jar test affords a very simple means of ascertaining the suitability of various coagulation/flocculation/sedimentation treatments for a given raw water, an equivalent means of evaluating various coagulation/filtration or coagulation/flocculation/filtration treatments has not been accepted. In this regard, the author will review a test method that he and his colleagues developed

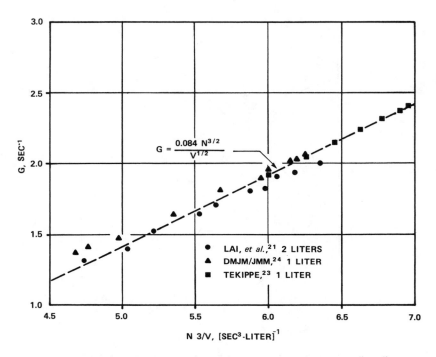

Figure 5. Velocity gradient calibration for jar test unit (based on use of 3 $''$ x 1 $''$ paddle).

and are currently evaluating. Essentially, this device consists of a series of six stirring compartments, each with its own variable speed stirrer. Each compartment is relatively large in volume (6 liters) and is arranged so that it can be decanted, at a controlled rate, through a 10-in.-deep layer of filter sand (0.2 mm) in a 1-in.-diameter column. Performance is measured via evaluation of the last liter of four passed through the filter. Figure 6 shows a photo of the apparatus and Figure 7 shows a plot of turbidity vs chemical dose as determined by this short-term bench test along with a similar plot determined with several long-term pilot runs. Although these results show the test may be helpful for establishing optimum chemical dose, they do not show as much precision as might be desired. Optimizing chemical doses for direct or in-line filtration remains a fruitful area of research.

PILOT STUDIES

Although it may appear that the processes to be used in a proposed water treatment plant are fairly straightforward and well established after decades of experience, pilot studies are still of great value and can often be used to

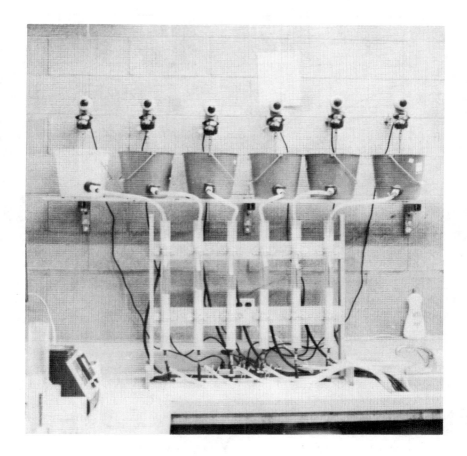

Figure 6. Bench-scale test unit.

produce designs that have improved performance and/or reduced cost. This is especially true when direct filtration or very high filtration rates are being considered. Even in a full-scale operating plant, pilot filters can be extremely useful in evaluating various treatment alternatives.

In general, the filtration and flocculation processes can be very easily modeled on a pilot scale and the sedimentation process cannot. The significance of good initial mixing of coagulant can also be studied effectively on a pilot scale. However, the method of initial mixing is often difficult to scale up. The design of pilot units for each of these unit processes will be briefly discussed.

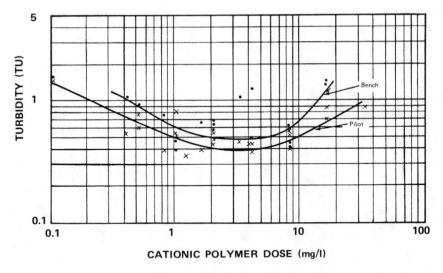

Figure 7. Comparison of results from bench- and pilot-scale filters. (After reference 24.)

INITIAL MIXING

The term "initial mixing" was coined in the classroom by Kaufman to describe the process whereby the coagulant is initially mixed with the water to be treated. The term is used to distinguish between the concept of initial mixing and various methods employed to accomplish it. Wilson,[25] Vrale and Jordan,[26] Stenquist and Kaufman,[27] and Letterman, Quon and Gemmell[28] have all recently completed studies bringing into view the importance of initial mixing in flocculation.

The studies at Berkeley began when Wilson, conducting flocculation studies for Kaufman using the apparatus of Argaman,[29] replaced the traditional flash mixer with a tubular mixer (a venturi-type arrangement) and found that orthokinetic flocculation was considerably improved. Wilson observed that the alkalinity and turbidity of the raw water as well as the amount of alum added all influenced the amount of benefit obtained from his alternate initial-mixing configuration. Figure 8 from the data of Wilson[25] shows the sort of benefit obtained when comparing a traditional back-mixing flash mixer with a specifically designed in-line mixing device. It does not appear that efficient initial mixing is equally important in every instance. As a result, parallel runs with a rapid mixer of traditional design and a more efficient in-line mixer are often justified. Figure 9 shows cross sections of three alternate in-line designs found to be effective in pilot studies. The efficiency of each is affected by the rate of flow.

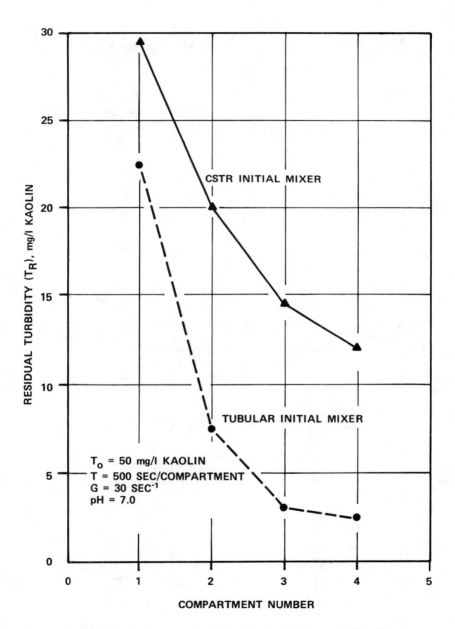

Figure 8. Importance of initial mixing step. (After reference 25.)

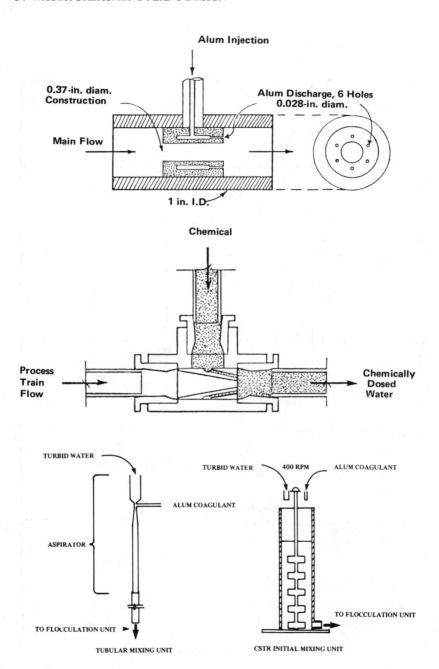

Figure 9. Pilot-scale initial mixers. (After references 24, 25 and 26.)

PILOT FLOCCULATION

Many of the early studies on flocculation were conducted using the jar test as the principal test device. However, a number of recent flocculation studies have been conducted using pilot-scale facilities,[25,27,29,30] and it appears that most of the parameters developed should scale up very easily. Important factors that should be studied in pilot-scale flocculation facilities are the appropriate chemical dose, the effect of mixing energy and the effect of mixing time. The effect of chemical dose is best studied first with the jar test and then confirmed on a pilot scale.

At one time, conventional wisdom suggested that the use of energy in conjunction with chemical coagulation for water clarification was undesirable.[19] However, Langelier demonstrated that the opposite was clearly true and even obtained a U.S. patent on a variation of the process.[31] Later Camp and Stein[32] demonstrated that the mean velocity gradient, G, was a good parameter for evaluating the energy input necessary for flocculation, and Camp[33] demonstrated that the dimensionless parameter GT also proved useful within certain bounds. Subsequent to this work, Argaman and Kaufman[29] developed a relationship between flocculation performance, G, and T, which demonstrated that performance improves with increasing G up to a certain point where floc break-up occurs. Their relationship can be written as follows:

$$\frac{N}{N_0} = \frac{1 + K_b G^2 T}{1 + K_a GT} \tag{3}$$

where
N_o = concentration of primary particles in influent (particles/ml)
N = concentration of primary particles in effluent (particles/ml)
K_a = aggregation constant (dimensionless)
K_b = break-up constant (sec)
G = mean velocity gradient ($\sec^{-1}$)
T = hydraulic detention time (sec)

When alum is used as the sole coagulant for typical inorganic turbidity, K_a is usually about 10^{-4} and K_b is usually about 10^{-7}. Constants K_a and K_b can be determined via pilot tests if a sufficiently broad range of G is used. It should be noted that the value of K_a is also proportional to the floc volume concentration. As a result, changes in coagulant dose or raw water turbidity will result in changes in K_a. For low values of G, the relationship simplifies to:

$$\frac{N}{N_o} = \frac{1}{1 + K_a GT} \tag{4}$$

Thus, K_a can be determined from a plot of N_o/N vs G. Usually this relationship alone is adequate for values of G below 60 sec^{-1}. It is felt that the Argaman-Kaufman relationship represents the best description of G, T, and flocculator performance currently available.

It is very important that a pilot flocculator be well compartmentalized (at least four compartments) with the ability to bypass the flow from any of the compartments. There are two basic reasons for this. First, compartmentalization is important in the performance of the flocculation reaction itself. Kaufman[34] described its importance as early as 1953 and it has been subsequently demonstrated a number of times.[29,30] An example replotted from the data of Argaman and Kaufman is shown in Figure 10. It will be observed in the

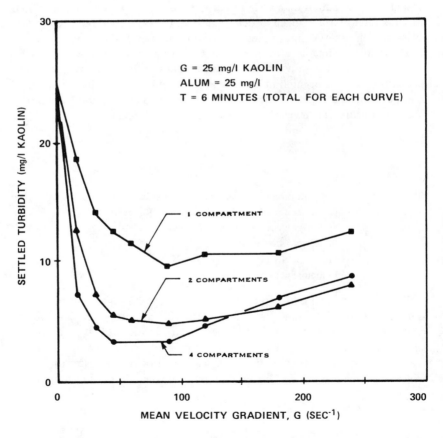

Figure 10. Influence of compartmentalization of flocculation efficiency. (After reference 29.)

figure that increased compartmentalization also reduced energy requirements. Second, additional compartments greatly increase the flexibility of pilot flocculators, allowing for various hydraulic detention times.

Previous experience with pilot-scale flocculation units has shown that they should be designed with three or more compartments, each having a detention time of about 5 min and the capability of varying the G value independently in each compartment over a wide range. Plexiglass flocculation chambers are desirable as they allow the investigator to view floc formation.

Variation of G is best provided by placing individual laboratory-type stirrers on each compartment, each individually adjustable by its own solid-state controller. Figure 11 shows a typical pilot unit of this type. The unit shown

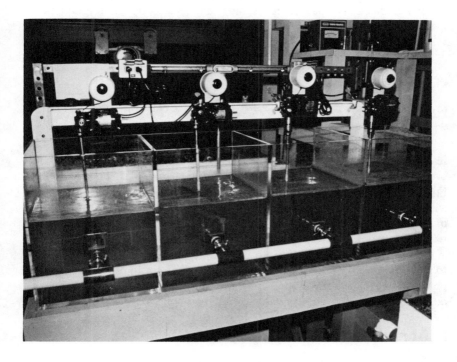

Figure 11. Typical pilot flocculators.

has three parallel flocculation trains, and it is plumbed so that one, two, or all three compartments can be used. When a new setup is being built, it should be calibrated by the use of a torquemeter, a stroboscope and a tachometer. For this sort of calibration, the following formula by Camp[33] proves useful:

$$G = \frac{\pi N T^{\frac{1}{2}}}{30 \, V \, \mu} \qquad (5)$$

where N = paddle speed (rpm)
       T = input torque (dyne - cm)
       V = mixing chamber volume (cm^3)
       μ = absolute viscosity of fluid (g/cm - sec)

The G that can be developed in a particular volume depends on the paddle configuration and on the paddle area. Paddle configuration is important because it places basic limitations on the level of G that can be accomplished before undesirable vortexing occurs and because it influences the distribution of the velocity gradient. The paddle area is important because at a given rotational speed, the energy input is proportional to its square root:

$$G \propto \sqrt{a} \qquad (6)$$

where a = paddle area.

PILOT SEDIMENTATION

A number of different types of configurations have been used for pilot-scale tests of sedimentation. If traditional design parameters are used (*e.g.*, overflow rate and detention time), the device that evolves may be very unwieldy. For example, a circular clarifier designed to treat a flow of 5 gpm (0.32 1/s) with an overflow rate of 1000 gpd/ft^2 (41 m^3/m^2 day) and a detention time of 2 hr, would be 3 ft (0.92 m) in diameter and 11.2 ft (3.4 m) deep! As a result of some of the scale-up problems that occur, great care should be used in designing these facilities on a pilot scale. Although such units may be used to study the relative effect of certain design proposals or to provide adequate pretreatment for subsequent processes, they rarely produce the same effluent quality or sludge characteristics as their full-scale counterparts.

PILOT FILTRATION

Filtration is the single most important step in any effective water treatment process train. It is the filtration step that provides insurance that particulate matter which might interfere with disinfection[17] is consistently removed. In a recent survey conducted on 20 different water treatment plants located in California, Nevada and Utah,[35] 15 of the plants reduced the number of particles greater than 2.5 μ in the influent by one order of magnitude or more and 9 reduced the particle count by two orders or more. Significantly, few plants showed any particles of 10μ or more coming through the filters. The median effluent had only 50 particles/ml greater than 2.5μ. This study also demonstrated the significance of chemical conditioning, because the only plant not engaging in this practice had a particle count more than four times greater than its nearest competitor.

The object of studies with pilot filters should be to ensure that all the necessary facilities are included in the full-scale design to meet designated goals for effluent turbidity while maximizing the net water production per unit filter area. Turbidity goals must be met because they are the fundamental objective of the whole process. Maximum net water production is desirable because it minimizes capital costs. The principal factors affecting the net water production of a filter are the filtration rate, the length of filter run and the amount of water required for backwash. These considerations are also discussed by Cleasby and Baumann.[36]

The filter area required for a given plant capacity is determined by the net or effective filtration rate (Re), that is, the net amount of product water generated per unit of filter area. The effective filtration rate is to be contrasted with the design filtration rate (Rd) which is the maximum rate at which the filter is designed to pass water. The difference between the two is related to the volume of water that passes through each unit of filter area during the course of the run (gal/ft²-run), or the Unit Filter Run Volume (UFRV) and the volume of backwash water required per unit of filter area, or the Unit Backwash Volume (UBWV). The following relationship can be developed to relate these parameters:

$$Re = Rd \ \frac{UFRV - UBWV}{UFRV} \qquad (7)$$

Figure 12 shows the relationship between the production efficiency (Re/Rd) and the Unit Filter Run Volume for various Unit Backwash Volumes. Inspection of this figure reveals that rapid reductions in production efficiency result when the UFRV drops below 5000 gal/ft² (203 m³/m²). In order to develop a design having adequate production efficiencies, it is recommended that a UFRV goal of between 5000 and 10,000 gal/ft² be set for any pilot studies. This corresponds to between 16.7 and 33.3 hr at a rate of 5 gpm/ft² (0.2 m³/m²). With water backwash and water surface wash, typical UBWVs range from 100 to 200 gal/ft². If air backwash is to be used and reduced UBWVs are observed, then UFRV goals can be set somewhat lower.

Ordinarily the number of filters used in a given pilot study is significantly affected by the budget. However, it is always desirable to use as large a number of filters as possible. This will greatly reduce the time consumed in running a particular test. More importantly, the use of a number of filters enables the

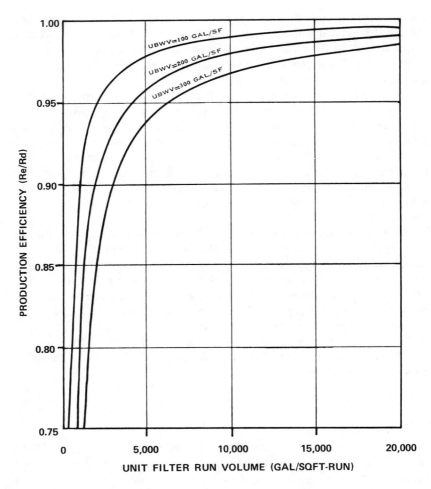

Figure 12. Influence of Unit Filter Run Volume and Unit Backwash Volume on production efficiency.

simultaneous evaluation of various treatment options even with a variable influent quality. When two different treatment modes are tested over different time periods, variations in influent quality often prevent reasonable comparisons. A minimum of two filters should be used if any options of pretreatment, filtration rate or media type are to be evaluated.

There are, of course, a number of designs that may be used for pilot filters. However, during a number of recent pilot studies, the author and his colleagues have developed what they believe to be a very simple straightforward design.

Figure 13 is a representation of such a design. Basically the filter's hydraulics are controlled by pumping at a constant rate with a tubing-type pump, and the minimum water surface is kept above the media by an airbreak (control weir) in the effluent line. As the head loss builds up, it is continuously recorded with a pressure recorder. Filter media is placed in an interchangeable "filter bottom" and new medias to be tested are first prepared off-line using a spare filter bottom that is then switched with a filter bottom presently in use. Not shown is a single surface wash nozzle that is designed according to traditional criteria for the Baylis system and is very effective in eliminating some of the backwashing problems peculiar to small filter columns. Experience has shown that as long as effluent quality comparable to full-scale plants is the primary concern, filters as small as 2 in. are effective, but much larger filters should be used if the effectiveness of backwashing is to be tested. Air backwashing itself may be tested for effectiveness on large pilot columns. However, alternate air backwash systems are best studied on plant scale.

A number of mistakes in the design of pilot filters occur often enough to deserve special mention. First, engineers often do not bother to calculate the head loss in tubing and associated fittings and as a result, operation is severely limited. In related fashion, pilot plants are often designed with insufficient piping for disposing of backwash water. Possible tubing airlocking is often not considered. Finally, filters are often designed so that either the influent falls several feet from the top of the filter to the media, resulting in airbinding, or the tubing is extended down to the media with no airbreak so that the head differential on the feed to the filter varies continuously throughout the run.

SUMMARY

This chapter outlines some of the most important considerations in establishing treatment requirements and design criteria for a particular water supply. These are summarized as follows:

1. Applicable water quality standards and their meaning have been discussed.
2. Methods for evaluating water sources are reviewed, with special emphasis on meaningful methods of data analysis.
3. Appropriate bench-scale tests have been discussed, including a recommended procedure for the jar test, a method for comprehensive evaluation of jar test data (turbidity topogram), and a bench-scale method for conducting preliminary direct filtration studies.
4. The conduct of pilot studies is discussed in some detail, including comments on pilot flocculation, pilot sedimentation and pilot filtration.

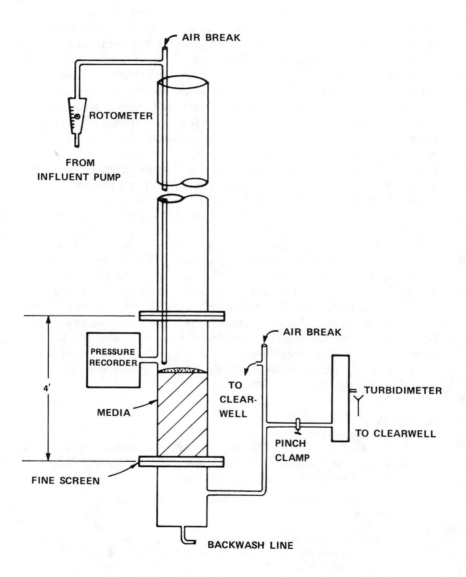

Figure 13. General sketch of JMM pilot filter design.

REFERENCES

1. World Health Organization. "International Standards for Drinking Water," WHO Geneva (1971).
2. U.S. Environmental Protection Agency. "National Interim Primary Drinking Water Regulations," *Federal Register* 40:59566 (1975).
3. American Water Works Association. "Quality Goals for Potable Water," *J. Am. Water Works Assoc.* 60:1317 (1968).
4. U.S. Public Health Service. "Public Health Service Drinking Water Standards—1962," PHS Pub. No. 956 (Washington, D.C.: U.S. Government Printing Office, 1962).
5. Harris, R. *The Implications of Cancer-Causing Substances in Mississippi River Water*, the Environmental Defense Fund (November 1974).
6. Tromp, S. W. "Possible Effects of Geophysical and Geochemical and Geographical Distribution of Cancer," *Schweiz, J. Pathol.* 18:929 (1955).
7. Rook, J. J. "Formation of Haloforms During Chlorination of Natural Waters," *J. Water Treatment Exam.* 23:234 (1974).
8. Bellar, T. A. and J. J. Licktenberg. "Determining Volatile Organics at Microgram-per-Liter Levels in Water by Gas Chromatography," *J. Am. Water Works Assoc.* 66:1739 (1974).
9. Rook, J. "Formation and Occurrence of Chlorinated Organics in Drinking Water," *Proceedings 95th American Water Works Association*, Paper No. 32-4 (1975).
10. Stevens, A., C. Slocum, D. Seeger and G. Robeck. "Chlorination of Organics in Drinking Water," *Proceedings Conference on the Environmental Impact of Water Chlorination*, Oak Ridge, Tennessee (1975).
11. Symons, J., T. Bellar, J. Carswell, J. Demarco, K. Knapp, G. Robeck, D. Seeger, C. Slocum, B. Smith and A. Stevens. "National Organics Reconnaissance Survey for Halogenated Organics in Drinking Water," *J. Am. Water Works Assoc.* 67:634 (1975).
12. National Cancer Institute. "Report on Carcinogenesis Bioassay of Chloroform," Carcinogenesis Program, Division of Cancer Cause and Prevention, Bethesda (March 1, 1976).
13. Food and Drug Administration. "Chloroform as an Ingredient of Human Food and Cosmetic Products," *Federal Register* (June 29, 1976).
14. Environmental Protection Agency. "Advance Notice of Proposed Rule-Making," *Federal Register* (July 14, 1976).
15. Symons, J. "Interim Treatment Guide for Control of Chloroform and Other Trihalomethanes," EPA Water Supply Research Division M.E.R.L. Cincinnati, Ohio (April 1976).
16. California Department of Health. "Laws and Regulations Relating to Domestic Water Supplies—Quality and Monitoring," Title 17, California Administrative Code, Part 1, Chapter 5, Subchapter 1 (1972).
17. Symons, J. M. and J. C. Hoff. "Rationale for Turbidity Maximum Contaminant Level," *Proceedings of Third Water Quality Technology Conference*, American Water Works Association, Atlanta (December 8-10, 1975).
18. Fair, G. M., J. C. Geyer and D. A. Okun. *Water and Wastewater Engineering*, Vol. I (New York: Wiley Interscience, 1966), pp. 4-34-4-39.

19. Langelier, W. F.. "Coagulation of Water with Alum by Prolonged Agitation," *Eng. News-Record* 86:924 (1921).
20. Black, A. P. *et al.* "Review of the Jar Test," *J. Am. Water Works Assoc.* 49:1414 (1957).
21. Tai, R., H. Hudson and J. Singley. "Velocity Gradient Calibration of Jar Test Equipment," *J. Am. Water Works Assoc.* 67:553 (1975).
22. ASTM D2035-74. "Standard Recommended Practice for Coagulation-Flocculation Jar Test of Water," *ASTM 1976 Annual Book of Standards,* Part 31, WATSR, p. 865.
23. TeKippe, R. J. "The Control of Coagulation for Turbidity Removal," PhD Thesis, University of Wisconsin (1969).
24. DMJM/JMM. *Los Angeles Aqueduct Water Quality Improvement Program Draft Report Task 3 Alternative Process Systems, Volume II: Pilot Plant Studies,* Figure 3-2 (March 1977).
25. Wilson, G. E. "Initial Mixing and Turbulent Flocculation," PhD Thesis, University of California, Berkeley (March 1972).
26. Vrale, L. and R. Jordan. "Rapid Mixing in Water Treatment," *J. Am. Water Works Assoc.* 63:52 (1971).
27. Stenquist, R. and W. Kaufman. *Initial Mixing in Coagulation Processes,* University of California, Berkeley, S.E.R.L. Report No. 72-2.
28. Letterman, R., J. Quon and R. Gemmell. "Influence of Rapid-Mix Parameters on Flocculation," *J. Am. Water Works Assoc.* 65:716 (1973).
29. Argaman, Y. and W. Kaufman. "Turbulence and Flocculation," *J. San. Eng. Div., Am. Soc. Civil Eng.* 96:223 (1970).
30. Harris, H., W. Kaufman and R. Krone. "Orthokinetic Flocculation in Water Purification," *J. San. Eng. Div., Am. Soc. Civil Eng.* 92:95 (1966).
31. Langelier, W. F. U.S. Patent No. 1,456,137 (1917).
32. Camp, T. R. and P. C. Stein. "Velocity Gradients and Internal Work in Fluid Motion," *J. Boston Soc. Civil Eng.* October 1943, p. 219.
33. Camp, T. R. "Flocculation and Flocculation Basins," *Am. Soc. Civil Eng. Trans.* 120:1 (1955).
34. Kaufman, W. J. "The Effect of Compartmentalization on the Efficiency of Flocculation Basins," October 1953, unpublished.
35. Tate, C. H. and R. R. Trussell. "Survey of Particle Removals by Water Treatment Plants," March 1976, unpublished.
36. Cleasby, J. and R. Baumann. *Wastewater Filtration,* EPA Technol. Transfer Bulletin (1975).

CHAPTER 4

COAGULATION

Charles R. O'Melia, Ph.D., PE
Professor, Department of Environmental
Sciences and Engineering
University of North Carolina
Chapel Hill, North Carolina 27514

INTRODUCTION

Coagulation is a process for combining small particles into larger aggregates. It is an essential component of accepted water treatment practice in which coagulation, sedimentation and filtration processes are combined in series to remove particulates from water. It is very important to recognize that this conventional treatment system does much more. Concentrations of pollutants (inorganic, nonliving organic and biological) are in most instances much higher in suspended solids than in the waters with which these solids are associated. Hence, water treatment for the removal of particulates also accomplishes the removal of many harmful substances from water. Some examples follow.

Clays are a major portion of natural "turbidity" in raw water supplies. With the possible exception of asbestos, they are not directly responsible for harmful effects on human health. There is some possibility that they can exert important health effects by adsorption, transport and release of inorganic and organic toxic substances, viruses and bacteria. Direct evidence for the removal of clays from water supplies using coagulation is lacking, since clay particles are not detected directly in routine water analysis. However, laboratory studies are numerous. For example, Black and Hannah[1] and Packham[2] have demonstrated effective coagulation of kaolinite, montmorillonite and other clay suspensions using alum under conditions similar to those found in potable water treatment.

65

Epidemiological studies have shown that occupational exposure to asbestos dust can lead to cancers in the gastrointestinal tract. This has led to the suggestion that asbestos particles in water supplies can cause similar problems. There is as yet no definitive and conclusive evidence that asbestos in water supplies is a significant health hazard, but there is sufficient evidence to provide a basis for serious concern. Asbestos concentrations in water supplies exposed to asbestos sources are in the range of 10 million fibers per liter, while unpolluted or background concentrations are typically less than 1 million fibers per liter. Results of pilot-scale coagulation and filtration experiments for the removal of asbestos fibers from the Duluth water supply have been reported by Black and Veatch[3] and Logsdon and Symons.[4] Two types of asbestos fibers were found, amphibole and chrysotile. The amphibole fibers were larger and probably negatively charged; the chrysotile fibers were smaller and may have been positively charged. The majority of the particles of both minerals were less than 2 μ in length. The amphibole fibers were easily removed by coagulation and filtration, while considerable difficulties were encountered in removing the chrysotile fibers. Since conventional filters are able to remove very small particles,[5] and since conventional pretreatment chemistry in coagulation is designed for negative particles, it is plausible that changes in coagulant addition would provide good removals of positively-charged chrysotile particles.

Coagulation is effective in the removal of color and many other organic macromolecules and particulates from water. Here again, laboratory studies provide the bulk of the available evidence. Hall and Packham[6] have studied the coagulation of humic and fulvic acids by iron(III) and aluminum(III) salts. These organic materials have natural origins in the decay of plant materials and are ubiquitous in natural waters. These investigators found that humic acids were readily removed whereas a significant fraction of the fulvic acids were not removed by either coagulant. Similar results have been reported by others, including Black and Willems[7] and Rook.[8] The ability of conventional coagulation processes to provide for removal of humic substances is important in a new perspective. These substances have been shown to be reactants in the formation of chloroform and other halomethanes by chlorine during disinfection. Since humic substances can be removed directly by coagulation and may also be adsorbed on clays that are also effectively treated by coagulation, it is reasonable to conclude that coagulation as conventionally practiced or with minor modifications can provide significant decreases in haloform production from chlorination.

Large microorganisms including algae and amoebic cysts are readily removed by coagulation and filtration. Bacterial removals of 99% are also achievable.[9] More than 98% of poliovirus type 1 was removed by conventional coagulation and filtration.[10] Several recent studies have shown that bacteria and viral agents are attached to organic and inorganic particulates.[11,12] Hence, removal of these particulates by conventional coagulation and filtration is a major component of effective treatment for the removal of pathogens.

Finally, many toxic organic substances such as PCB and DDT and also many inorganic toxic materials are adsorbed on naturally occurring inorganic and organic particulates. Removal of particulates will also provide removal of these hazardous substances.

PROCESS DESCRIPTION

A coagulation process has two separate and distinct components or steps. First, particles in the water must be treated chemically to make them "sticky" or unstable. This involves the addition of one or more chemicals in rapid mix tanks. Second, these destabilized particles must be brought into contact with each other so that aggregation can occur. This is done by gentle stirring of the water in flocculation tanks. If either one of these steps is not properly designed, the coagulation process will not function, *i.e.*, aggregation will not occur.

A schematic flow diagram of a coagulation process is shown in Figure 1. The water to be treated is brought to the rapid mixing tank where destabilizing chemicals are added; vigorous mixing occurs for a short time, normally 1 min or less. Usually one mixing tank is used, although occasionally two tanks may be used in series when two coagulants requiring separate addition are employed. Destabilization is fast and essentially complete after this rapid mixing. The water and its destabilized particles are then introduced into the flocculation tank, where gentle fluid motion brings the particles into contact so that aggregates can form. This gentle mixing is usually done mechanically, although hydraulic mixing is sometimes employed using baffled tanks. Detention times of 1 hr are typical.

The rapid mixing and flocculation tanks together bring about aggregation, and comprise the coagulation process. No materials are removed from the water in these tanks. In fact, materials are added in the form of coagulant chemicals. Solids are removed in subsequent settling and filtration facilities. These solid-liquid separation processes must remove the particles present in the original raw water and the chemicals added to bring about coagulation. These solids leave the treatment system in sludge from the settling tanks and in backwash water from the filters. Disposal of these water treatment plant wastes is a problem in itself. Here it is important to note that the characteristics of these wastes (solids concentration, quantity, dewatering capability) are functions not only of the raw water supply but also of the materials used as coagulants.

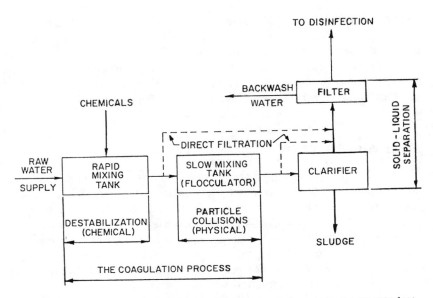

Figure 1. Schematic diagram of a coagulation process in a water treatment plant.

COAGULANTS, FLOCCULANTS AND COAGULANT AIDS

Chemicals added to bring about or to assist in destabilizing suspended particles include salts of hydrolyzing metal ions such as aluminum sulfate and ferric chloride, synthetic organic polyelectrolytes, activated silica and various clays. A brief description of each of these chemical types follows.

Hydrolyzing Metal Ions

Let us consider aluminum sulfate, $Al_2(SO_4)_3$. When this salt is added to water it dissolves readily. The sulfate ions disperse throughout the liquid simply as SO_4^{-2}. The aluminum ions, however, react with the water, or hydrolyze. In its simplest form the aluminum ion is completely surrounded by six water molecules and has three positive charges. This ion is represented as $Al(H_2O)_6^{3+}$ or simply Al^{3+}. However, under the pH conditions used in all water treatment processes, the trivalent aluminum ions react (hydrolyze) immediately to form many different species such as the following:

$$Al^{3+} + H_2O = AlOH^{2+} + H^+ \tag{1}$$

$$Al^{3+} + 2 H_2O = Al(OH)_2^+ + 2 H^+ \tag{2}$$

$$7 Al^{3+} + 17 H_2O = Al_7(OH)_{17}^{4+} + 17 H^+ \tag{3}$$

$$Al^{3+} + 3 H_2O = Al(OH)_3(s) + 3 H^+ \tag{4}$$

The ions $AlOH^{2+}$ and $Al(OH)_2^+$ shown in Reactions 1 and 2 are mono-mers, *i.e.*, they contain one aluminum atom. Although their charges are less than +3, they are more effective coagulants for negative colloids because they are readily adsorbed on the surface of many solids. Ions such as $Al_7(OH)_{17}^{4+}$ may resemble polymers, since they contain several aluminum ions. They are probably adsorbed very strongly on most nega-tive colloids and are good coagulants. $Al(OH)_3(s)$ denotes the solid, amorphous, gelatinous precipitate of aluminum hydroxide that is formed in most coagulation plants. Note that alum is an acid; protons are liber-ated into solution in Reactions 1 to 4. Finally, the aluminum species that are formed when alum is added to water depend primarily on the pH of the water and on the amount of alum added.

Alum can act as a coagulant in two ways. In most waters, enough alum is added to precipitate $Al(OH)_3(s)$. This coats the colloids with a gelatinous and "sticky" sheath. It also provides additional targets for the original solids to "hit" in the flocculation tank, thereby accelerating the flocculation of these particles into large aggregates. These targets may be necessary in coagulating waters having a low turbidity, since excessive flocculation times are needed to aggregate the primary solids alone. This mode of coagulation, in which considerable quantities of aluminum hydrox-ide are formed, has been termed "sweep coagulation" by some.

A second mechanism of coagulation by alum is adsorbing positively charged aluminum monomers and polymers on negative colloids, thereby neutralizing the original charge on these particles and rendering them sticky or unstable so that aggregates are formed when contacts occur. This type of coagulation can only be used for high turbidity waters, since few additional solids are added to the water. In many cases less coagulant may be needed than for low turbidity waters, since a precipi-tate is not needed. Sludges formed in sweep coagulation are typically gelatinous and difficult to dewater; sludges formed when coagulation is achieved with adsorption of small aluminum species are more compact and presumably should dewater more easily.

Alum can be purchased in dry or liquid form, and may be prepared in a variety of strengths as stock solutions for use in treatment plants. The strength of a stock solution can affect the results obtained when alum is mixed with the water to be coagulated, because the actual chemi-cal species present in a stock solution depend on its strength. This is illustrated in Figure 2. These results are calculations based on the addi-tion of pure alum to pure water. Species composition and solution pH are plotted as functions of pAl_T or $-log[Al_T]$, where $[Al_T]$ is the total aluminum concentration in mol/l. When $[Al_T]$ is 1 mol/l, pAl_T is 0 and the corresponding alum concentration would be 333 g/l as $Al_2(SO_4)_3 \cdot 18H_2O$.

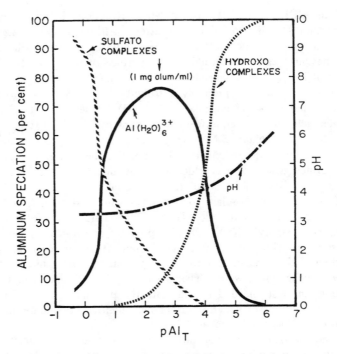

Figure 2. Species composition of alum stock solutions.

For very dilute alum solutions (*e.g.*, pAl_T of 4 or larger), the principal soluble species are hydroxo complexes, such as $AlOH^{2+}$. These solutions, corresponding to 33 mg/l or less of alum, are too dilute for use as stock solutions. In concentrated systems (*e.g.*, pAl_T of 0.5 or smaller), the principal soluble aluminum species are sulfato complexes such as $AlSO_4^+$. In the intermediate range ($0.5 < pAl_T < 4$), the simple aquo complex, Al^{3+} or $Al(H_2O)_6^{3+}$, predominates. At $pAl_T = 2.5$, the fraction of the aluminum species that exists as Al^{3+} is a maximum. This also corresponds to a stock solution of 1 mg alum/ml, a concentration frequently used for stock solutions in jar tests.

Some liquid alum solutions may contain excess sulfuric acid, have a lower pH, and contain more sulfato complexes than the solutions described in Figure 2. The addition of dry alum to raw water containing alkalinity can produce a higher pH and more hydroxo complexes. The resulting compositions can be calculated if the acidity or alkalinity is known. It is not yet possible to determine, however, the actual effects on coagulation. It is only possible to say that the reactions of alum in water can depend on the aluminum species added to it, and hence on the strength of the stock solution. The "best" strength is still a matter for additional research.

Synthetic Polymers

A polymer is a chain of small subunits or monomers. Polymers may be large or small; *i.e.*, they may have molecular weights ranging from a few hundred to more than 10 million. Although some researchers call small polymers "coagulants" and large ones "flocculants," here all polymers will be considered coagulants, with the term "flocculation" used for those tanks (Figure 1) in which fluid motion produces contacts among suspended particles.

Polymers can have positive (cationic) or negative (anionic) charges, or they may not be charged at all (nonionic). Polymer charge can arise from many different functional groups (*e.g.*, carboxyl, sulfonate) and can vary with the pH of the water being treated.

For a polymer to be useful as a destabilizing chemical, it must adsorb on the suspended particles to be removed. Hence there is usually a direct relationship (sometimes termed a stoichiometry) between the concentration of particles to be destabilized and the concentration of a polymer needed to bring this destabilization about. Higher turbidities require higher polymer doses if the solids are similar in type. At very low turbidities, polymers are ineffective coagulants. This is not because they do not adsorb and produce destabilization, for they can. They are ineffective because contact opportunities in flocculation tanks with reasonable detention times are insufficient to yield large aggregates. Some other solids (targets) can be added, including alum [to produce $Al(OH)_3(s)$] and clay, so that the turbidity of the water is made high enough to permit effective flocculation in a reasonable time.

Polymers destabilize particles in two ways. Both require adsorption of the polymer on the suspended particles. First, when the polymer and the particle are opposite in charge (*e.g.*, a cationic polymer and a negatively charged coliform organism), adsorption of the polymer can neutralize the charge on the particles. In these situations, molecular weight is not an important factor, and small polymers can be used. Second, when simple charge effects are not significant (*e.g.*, destabilization of biological sludges with anionic or nonionic polymers), then the polymers must "bridge" a small region between the particles within which the particles repel each other. In this case the polymer must be large enough to form the bridge, so high-molecular-weight polymers are needed.

Typical results of jar tests using polymers are represented by the data in Figure 3. Here a negatively charged suspension (kaolinite) has been

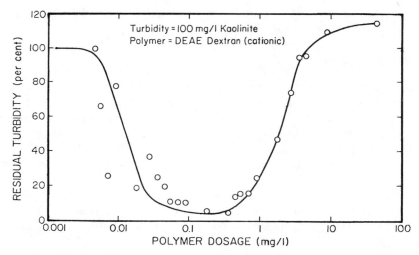

Figure 3. Jar test results of coagulation using a polymer.

destabilized by a cationic polymer and then flocculated and settled in the jars. Three regions are of interest. First, at low polymer doses (less than 0.01 mg/l), insufficient polymer has been added to produce destabilization, probably because insufficient positive charges have been added and adsorbed to produce charge neutralization. Second, a region of optimum dosage is seen, centering at a dose of 0.1 mg/l. Here, effective charge neutralization has occurred; it is possible that some bridging also was involved. Finally, it is possible to add too much polymer; *i.e.,* a region of overdosing occurs at polymer additions greater than about 1 mg/l. This is because destabilization involves adsorption. When too many positive charges (or, in other cases, too many bridges) are adsorbed, restabilization occurs. In this case the restabilized suspension is made up of positively charged particles. Although not shown in Figure 3, it is useful to remember that stoichiometry is involved; higher clay concentrations would require higher optimum polymer dosages.

Coagulant Aids

Coagulant aids assist coagulation; this assistance is provided in several ways. Activated silica, for example, is prepared by acidifying concentrated solutions of sodium silicate and then diluting them in such a way that a stable "activated silica" is formed.[13] Depending on the method of preparation, this material might be considered as a solution of an anionic polyelectrolyte or as a negatively charged colloidal suspension. As an anionic polymer, activated silica can act as a primary coagulant for the destabilization of some suspensions,[13] although it is not generally used in this way in water treatment. Added before alum in treating low turbidity waters, activated silica can provide additional targets for flocculation, which sometimes permits a reduction in alum dose, usually produces a denser floc, and "aids" coagulation. Added after alum in coagulation, activated silica can help to bind small alum flocs together into larger and denser aggregates. Alum flocs often have a slight positive charge at pH values less than 7, so that the destabilizing properties of activated silica as an anionic polymer are useful here. Activated silica is not likely to be effective if it is added concurrently with alum since they react with each other.

A good description of the preparation of activated silica and its use in coagulation is given by Packham and Saunders.[14] Preparation involves (1) acidification of a concentrated sodium silicate solution, (2) aging for a definite time, and (3) dilution. Suitable acids include hydrochloric acid, sulfuric acid, alum, sodium bicarbonate and hypochlorous acid (aqueous chlorine). Conditions for acidification, aging and dilution are frequently specified by the supplier of the sodium silicate solution. In one procedure for laboratory jar tests, Packham and Saunders[14] diluted 30.9 ml of a sodium silicate solution (29% SiO_2) to 1 liter to yield a 1.25% SiO_2 solution. Eighty ml of this solution were added to a 100-ml volumetric cylinder, acidified with a predetermined quantity of HCl to lower the pH to 5, and then diluted to 100 ml. This colloid (1.0% SiO_2) was aged or activated for 30 min and then used directly.

Other materials are also used as coagulant aids for similar reasons. Clays may be added to provide targets and also produce denser flocs. Synthetic organic polymers may be used to destabilize partially stable alum flocs. The order of addition is important. These substances should not be added together with alum in a single mixing tank; two or more mixing tanks should be used in series. Clay "targets" should be added before alum, while polymers to destabilize alum flocs should be added after alum addition.

COAGULANT SELECTION

Some general considerations about coagulant selection are presented here, based on characteristics of the water to be treated. Actual selection of the type and dosage of chemicals to be used *always* requires experiments; simple jar tests are sufficient in most cases. This overview is given to assist in deciding what to investigate in jar tests and how to interpret some of the results obtained. Coagulation for the removal of turbidity, color and the simultaneous removal of both are considered.

Coagulation of Turbidity

Turbidity in water is caused by the presence of suspended matter such as clay, silt, nonliving organic particulates, plankton and other microscopic organisms. Turbid waters are classified into four types:

1. *High turbidity, high alkalinity:* This is the easiest system to treat because many coagulants are effective. Cationic polymers provide good destabilization, and the large concentration of particles permits easy flocculation into aggregates. It is likely that some anionic and nonionic polymers may be effective, possibly at lower costs. High-molecular-weight materials should be considered. However, trial-and-error with experiments such as jar tests is needed. Alum and ferric salts also generally prove effective. Frequently, alum is effective in the pH region 6 to 7, and ferric salts are useful in the pH region 5 to 7. When these metal salts are used here, they frequently do not need other coagulant aids or addition of base for pH control.

2. *High turbidity, low alkalinity:* Polymers function here as in Type 1. Alum and ferric salts are also effective, but care must be taken to measure pH in testing. The typical useful pH ranges are the same as in Type 1 waters. Addition of base may be needed to prevent pH from falling below levels at which aluminum or ferric polymers are formed.

3. *Low turbidity, high alkalinity:* Polymers cannot work alone for these waters. Additional particles must be added, usually before the polymer. Clays are suitable targets. Alum and ferric salts are effective in relatively large doses, so that $Al(OH)_3(s)$ or $Fe(OH)_3(s)$ is precipitated. The precipitates are gelatinous and may be slightly stable. Clay or activated silica added before the alum may reduce the alum dose and should produce a more settleable and dewaterable floc. Polymers (often anionic) or activated silica added after the alum may produce a more settleable floc.

4. *Low turbidity, low alkalinity:* These are the most difficult waters to coagulate. Polymers do not work alone, again because of the low turbidity. Clays or other targets may be added as in Type 3. Alum and iron(III) salts used alone are

also usually ineffective, since the pH can be lowered below the neutral range where $Al(OH)_3(s)$ and $Fe(OH)_3(s)$ are produced and sweep coagulation is achieved. Similarly, the flocculation rate is too low to permit aggregation if metal polymers are formed to achieve charge neutralization. Options include: (a) adding lime or other base to bring the water to Type 3, or (b) adding clay or its equivalent to transform the water to Type 2.

The effects of raw water turbidity on coagulation with alum are illustrated schematically in Figure 4. This example also shows the two mechanisms by which alum can act in coagulation. Residual turbidity in jar tests is plotted as a function of alum dose for a high turbidity water (Figure 4-A) and for a low turbidity water (Figure 4-B). The results may be characterized by four zones.

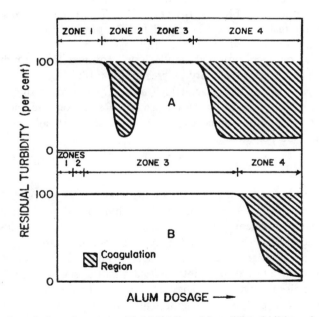

Figure 4. Coagulation of waters with high (A) and low (B) turbidities using alum.

Consider the coagulation of a high turbidity water (Figure 4-A). At low alum dosages (Zone 1), insufficient alum has been added to destabilize the turbidity by adsorption of positively charged aluminum polymers. At somewhat higher dosages (Zone 2), effective destabilization is achieved because sufficient adsorbable positive charges have been added to neutralize the negative charge on the colloids. The optimum dosage often (but not always) corresponds to a zeta potential that is at or near zero. At still higher dosages (Zone 3), a region of overdosing may occur. (But this is not always observed.) This arises because adsorption of aluminum polymers is sufficient to produce a positively charged and stable suspension. The region of highest alum dosage (Zone 4) corresponds to the precipitation of large amounts of solid aluminum hydroxide and concurrent removal of the original turbidity by adsorption on or enmeshment in these solids.

For a low turbidity water (Figure 4-B), only one region of turbidity removal is observed (Zone 4). Removal after destabilization by adsorption of aluminum polymers (Zone 2) is not possible because insufficient contact opportunities are available to permit aggregation after destabilization occurs. In some cases the alum dosage for sweep coagulation (Zone 4) may be larger for a low turbidity water than for a high turbidity water (compare Zones 4 in Figures 4-A and 4-B). This occurs because particles in the water may also serve as nuclei for the precipitation of $Al(OH)_3(s)$ and increase the rate of precipitation.

To place numbers on "high" and "low," consider low turbidity as less than 10 TU and high turbidity as greater than 100 TU. Low alkalinity is less than 50 mg/l as $CaCO_3$ and high alkalinity is greater than 250 mg/l as $CaCO_3$. These suggestions for coagulation selection and these classifications are generalities; some exceptions will occur. In all situations, experiments must be performed.

Color

Color in water is due primarily to the presence of natural organic matter. This, in turn, is composed largely of humic and fulvic acids. Humic acid is that component which is soluble in a strong base but insoluble in a strong acid; fulvic acid is soluble in both acid and base. Humic acids have molecular weights ranging from several hundred to a few thousand. They can aggregate naturally into colloidal particles with much higher apparent molecular weights. They are composed primarily of aromatic compounds with carboxyl, phenolic and quinoid groups. Fulvic acids are similar in structure to humic acids but have lower molecular weights. Because of their carboxylic and phenolic groups, humic substances have high cation exchange capacities and can concentrate or bind metal ions. Because of their organic character, they can also accumulate hydrophobic organic compounds, including several pesticides. Their role as precursers in the formation of halomethanes has been noted previously.

Several authors[6-8] have shown good removals of color by coagulation with aluminum and ferric salts. A fraction of the humic substances is not readily removed. Hall and Packham[6] have shown that some fulvic acids can be difficult to coagulate, and they suggest that this material may be deficient in carboxyl groups needed to complex with the coagulant ions. Rook[8] has shown good removal of color and fair removal of TOC from the Rotterdam water supply by coagulation with Fe(III). Bowie[15] has investigated the removal of color from several ground waters in Mississippi and shown effective treatment with high doses of alum. Black and co-workers[7,16] demonstrated effective color removal using alum and ferric sulfate. Alum was most effective at pH values about 5; ferric sulfate was most effective at about pH 4. A definite stoichiometry between color concentration and coagulant dose was demonstrated.[16] The jar test was confirmed as a useful tool for selecting the types and dosages of chemicals to be added.

These results suggest the following interpretation. Alum is considered as illustrative of hydrolyzing metal salts. Humic substances react with positively charged monomeric and polymeric aluminum hydrolysis products (see Reactions 1 to 3) to form colloidal precipitates. These are then aggregated further into large flocs by additional aluminum species that act as destabilizing agents. Because of the stoichiometry involved in precipitation and in coagulation by adsorption, the alum dose required is normally dependent on the concentration of color to be removed. Underdosing and overdosing can occur. Because these reactions are all pH-dependent, special care should be given to measuring and controlling this parameter.

Synthetic cationic polymers have been used successfully for color removal. Black and Willems[7] obtained good removal by treating a water containing 160 color units with 60 mg/l of cationic polymer at a pH of about 5. Higher and lower pH values yielded less efficient removals. Similar results are reported in more recent experiments by Narkis and Rebhun[17] and by Bowie.[15] In most cases, cationic polymers appear to act in the same way as alum. A polymer is needed to precipitate the negatively charged humic substances. Stoichiometry is observed[15]; underdosing and overdosing can occur. The pH is also very significant.

Simultaneous Coagulation of Color and Turbidity

Color and turbidity react differently in coagulation, so that treatment of a water containing objectionable concentrations of both can be difficult. A recent paper by Narkis and Rebhun[17] describing the coagulation of such mixtures by a cationic polymer illustrates some problems and provides a basis for coagulant selection. Consider clay as turbidity and humic substances as color. When sufficient clay is present to adsorb all of the humic substances from the water (and this is probably true in the majority of raw water supplies), the presence of the adsorbed organic materials increases the polymer dosage required to remove the turbidity. Since all of the organics are adsorbed on the clay, turbidity removal also accomplishes the removal of the organics. When an excess of humic substances exists, the turbidity is coated with organics, and free humic and fulvic acids remain in solution (this occurs in many water supplies in the southeastern United States). The polymer then reacts first with the soluble organics to form colloidal precipitates. After this is completed, additional polymer reacts with the turbidity and the colloidal organics. Polymer dosages required are higher than for turbidity removal alone.

Similar extensive investigations using hydrolyzing metal ions are not known to this author. However, because hydrolyzing metal ions can act as cationic polymers under appropriate conditions, similar results can be anticipated. In other words, coagulant dosages are higher than the removal of either color or turbidity alone. The effect of pH, although not extensively investigated, is important.

RAPID MIX UNITS

The primary purpose of rapid mix units is to accomplish destabilization of the particles in the raw water. When alum and ferric salts are used, it is important that vigorous mixing be provided because many hydrolysis reactions are almost instantaneous and the products that are formed (e.g., Reactions 1 to 4) are very dependent on pH and coagulant concentration. Poor mixing permits local regions of high dose and uncertain pH, so that coagulant is wasted. Values of the mean velocity gradient (G) from 700 to 1000 sec^{-1} have been suggested.[9]

A recent investigation by Letterman et al.[18] has demonstrated the existence of an optimum detention time for rapid mix units. This is to be expected because some finite time is needed for destabilization, and prolonged mixing at high speeds will destroy flocs that are formed. These authors show that optimum times of a few minutes were required, considerably longer than the few seconds needed for adsorption and destabilization. This suggests that some time should be provided in the rapid mix unit to permit the precipitation of $Al(OH)_3(s)$ when sweep coagulation is used.

When synthetic polymers are used to achieve destabilization, such vigorous mixing is not required. Here the destabilizing polymers are added directly to the system, not formed within it. Local regions of high concentration are still undesirable, since some suspended solids will be overdosed while others are underdosed. Detention times of a few tenths of seconds should be effective.

A good summary of mixing practice in water treatment is provided in "Water Treatment Plant Design," by ASCE, AWWA and CSSE.[9]

FLOCCULATION TANKS

Flocculation tanks are designed to produce collisions between particles. Generally these contacts occur by fluid motion. Differences in velocity occur from point to point in all flowing or stirred water. Because of these velocity gradients, particles suspended in the water also have different velocities, and hence can come into contact. A mean velocity gradient (G) can be determined that depends primarily on the power input into the water. Dimensions are ft/sec/ft, or simply sec^{-1}.

Velocity gradients are induced by mechanical or hydraulic mixing. Hydraulic mixing is accomplished using baffled flocculation tanks that can be horizontal (around the end) or vertical (over and under). Mechanical mixing can be provided in several ways. Typically, installations use horizontal rotating paddles mounted either perpendicular to or along the axis of flow. Short circuiting is a problem, and a minimum of three tanks in series should be used.

The basic principles for the design of flocculation basins were published by Camp,[19] who showed that the total number of particle collisions is proportional to Gt. Here G is the mean velocity gradient (sec^{-1}) and t is the hydraulic detention time (sec). Values for the dimensionless product in water treatment using alum vary by about one order of magnitude, from 10^4 to 10^5. Typically G may be from 10 sec^{-1} to 100 sec^{-1}; the upper limit represents the fact that flocs can shear apart at high velocity gradients.

A more complete approach to flocculation includes two other factors that have major effects on flocculation. These are the floc volume fraction (ϕ) and the collision efficiency or sticking factor (α), so that effective flocculation can be represented as a dimensionless product $\alpha\phi Gt$. If all parameters could be determined, this product would have a value ranging from 0.1 to 1.

The floc volume fraction is the volume of solid material in the water per unit volume of suspension. For concentrated suspensions, ϕ is high and the rate at which contacts occur is also high. Hence, the time required to produce a given degree of aggregation can be reduced. This is why flocculation tanks are not used in the chemical conditioning (coagulation) of sludges in wastewater treatment plants. The high floc volume fraction allows almost instantaneous aggregation if proper chemical addition has been provided. This is also why upflow units can produce coagulation and sedimentation with short detention times; the solids in the sludge blanket are good targets for the particles to be removed. In a very real sense this is how filters function. The sand beds provide a concentrated system of targets (a high floc volume fraction) for the removal of suspended materials.

If proper chemical pretreatment has not been provided, collisions can occur in the flocculation tank but aggregation may not occur. The collision efficiency or sticking factor (α) represents the fraction of the total number of collisions that are successful in producing aggregation. Complete success (complete destabilization) corresponds to a collision efficiency factor of 1.0. Complete failure in pretreatment (no destabilization) corresponds to α equal to zero.

Consider the product $\alpha \phi Gt$. From right to left, the terms are given in decreasing order of measurability. Hydraulic detention time can be calculated easily from tank volume and flow rate. The mean velocity gradient is more difficult to determine. Some assumptions must be made. However, the equations given by Camp[19] provide a sound basis for engineering estimates. The floc volume fraction depends on the particulates in the water supply and, when metal hydroxides are precipitated, on the coagulant dose. It is rarely measured and is usually not even estimated. Finally, the collision efficiency factor has only been measured in the laboratory, and even there only with great difficulty. However, successful flocculation requires some serious attention by the design engineer to all of these parameters.

The following recommendations are made:

1. Use Camp's guidelines[19] to design the mixing and detention characteristics of the flocculation tanks, remembering to subdivide the total detention time into at least three basins in series. Hence, use $10^4 \leqslant Gt \leqslant 10^5$ and $10 \text{ sec}^{-1} \leqslant G \leqslant 100 \text{ sec}^{-1}$

2. Use jar tests or other experimentation to provide for ϕ and α. In other words, the chemicals added must provide sufficient destabilization (α) and targets (ϕ) to permit good flocculation in tanks designed to produce the conditions described in step 1.

REFERENCES

1. Black, A. P. and S. A. Hannah. "Electrophoretic Studies of Turbidity Removal by Coagulation with Aluminum Sulfate," *J. Am. Water Works Assoc.* 53:438-452 (1961).
2. Packham, R. F. "Some Studies of the Coagulation of Dispersed Clays with Hydrolyzing Salts," *J. Coll. Sci.* 20:81-92 (1965).

3. Black and Veatch, Consulting Engineers. "Direct Filtration of Lake Superior Water for Asbestiform Fiber Removal," Report No. EPA-670/2-75-0500, Environmental Protection Agency, National Environmental Research Center, Cincinnati, Ohio (1975), 107 pp.

4. Logsdon, G. S. and J. M. Symons. "Removal of Asbestiform Fibers by Water Filtration," Water Supply Research Laboratory, Environmental Protection Agency, Cincinnati, Ohio (1975), 29 pp.

5. Yao, K-M, M. T. Habibian and C. R. O'Melia. "Water and Waste Water Filtration: Concepts and Applications," *Environ. Sci. Technol.* 5:1105-1112 (1971).

6. Hall, E. S. and R. F. Packham. "Coagulation of Organic Color with Hydrolyzing Coagulants," *J. Am. Water Works Assoc.* 57:1149-1166 (1965).

7. Black, A. P. and D. G. Willems. "Electrophoretic Studies of Coagulation for Removal of Organic Color," *J. Am. Water Works Assoc.* 53:589-604 (1961).

8. Rook, J. J. "Haloforms in Drinking Water," *J. Am. Water Works Assoc.* 68:168-172 (1976).

9. ASCE, AWWA, CSSE. *Water Treatment Plant Design* (New York: American Water Works Association, 1969).

10. Robeck, G. G., N. A. Clarke and K. A. Dostal. "Effectiveness of Water Treatment Processes in Virus Removal," *J. Am. Water Works Assoc.* 54:1275-1292 (1972).

11. Britton, G. and R. Mitchell. "Protection of *E. coli* by Montmorillonite in Seawater," *Proc. Am. Soc. Civil Eng., J. Environ. Div.* 100 (EE6):1310-1313 (1974).

12. Britton, G. and R. Mitchell. "Effect of Colloids on the Survival of Bacteriophages in Sea Water," *Water Res.* 8:227-229 (1974).

13. Stumm, W., H. Huper and R. L. Champlin. "Formation of Polysilicates as Determined by Coagulation Effects," *Environ. Sci. Technol.* 1:221-227 (1967).

14. Packham, R. F. and M. Saunders. "Laboratory Studies of Activated Silica," *Proc. Soc. Water Treatment Exam.* 15:245-270 (1977).

15. Bowie, J. E., Jr. "Removing Organic Color from Groundwaters," paper presented at 95th Annual Conference of the American Water Works Association, Minneapolis, Minnesota, June 8-13, 1975.

16. Black, A. P., J. E. Singley, G. P. Whittle and J. S. Maulding. "Stoichiometry of the Coagulation of Color-Causing Organic Compounds with Ferric Sulfate," *J. Am Water Works Assoc.* 55:1347-1366 (1963).

17. Narkis, N. and M. Rebhun. "The Mechanism of Flocculation Processes in the Presence of Humic Substances," *J. Am. Water Works Assoc.* 67:101-108 (1975).

18. Letterman, R. D., J. E. Quon and R. S. Gemmell. "Influence of Rapid-Mix Parameters on Flocculation," *J. Am. Water Works Assoc.* 65:716-722 (1973).

19. Camp, T. R. "Flocculation and Flocculation Basins," *Trans. Am. Soc. Civil Eng.* 120:1-16 (1955).

CHAPTER 5

SLUDGE DISPOSAL AND
THE A. P. BLACK PROCESS

A. T. DuBose, Ph.D.

Superintendent, Wastewater Treatment Plants
Gainesville Alachua County Regional Electric
    Water & Sewer Utilities Board
P.O. Box 490
Gainesville, Florida 32602

INTRODUCTION

In March 1969 the American Water Works Association Research Foundation sponsored a conference on the Disposal of Wastes from Water Treatment Plants. The conference committee on Plant Operation Needs[1] summarized its conclusions as follows: "The principal needs are to find effective and economical means to dispose of water treatment plant wastes by direct treatment of sludge, or by eliminating undesirable chemicals, such as alum, through changes in water treatment methods."

Today, a process is available that meets these criteria fully. Basically, one can now view a softening plant as a chemical plant producing three valuable by-products: carbon dioxide, lime and a new coagulant, magnesium carbonate. The latter two compounds can now be reclaimed, recycled to the front of the treatment plant and reused for production of a superior quality water.

The developmental work for this system of water treatment began in the 1950s in Dayton, Ohio. Black,[2] working with the City of Dayton to reduce the waste sludge from the water treatment plant, developed a technique for the separation of magnesium hydroxide from the calcium carbonate component of the sludge. Dayton's water supply is obtained from clear well water very high in calcium and magnesium hardness. The softening sludge produced is very high in magnesium hydroxide, which had to be separated prior to lime recalcination. Carbon dioxide, produced in lime recalcination, is used to dissolve selectively the magnesium hydroxide as the soluble bicarbonate. The clear magnesium bicarbonate solution is separated by thickening and discharged to a nearby watercourse.

Lime recalcination with sludge carbonation, begun in 1957, has been operating very successfully and resulted in the elimination of waste sludge discharge and at the same time greatly reduced the chemical cost of water treatment.[3]

In 1968 Dayton was advised by the State of Ohio that because of the high dissolved solids, this clear magnesium bicarbonate discharge represented a pollution problem that must be eliminated. This impetus led to the discovery of a relatively simple and inexpensive method of recovering the magnesium as a carbonate. After extensive laboratory and pilot-scale work, Black[4] found that extremely pure magnesium carbonate could be easily and inexpensively precipitated from the magnesium bicarbonate liquor. Another discovery was that the magnesium carbonate produced from the Dayton plant was an excellent coagulant for water, one that could be recovered and recycled. During this same time, Black discovered that froth flotation provided a highly selective method of separating relatively pure calcium carbonate from clay, silt or other common raw water contaminants. He found this to be true only if the coagulant used had been removed prior to the flotation process.

These four basic discoveries—separation of magnesium hydroxide from calcium carbonate, flotation of calcium carbonate from raw water impurities, the use of magnesium as a recycled coagulant, and the production of magnesium carbonate from the sludges of waters high in magnesium concentration—meshed to produce an entirely new system of water treatment. This coagulation system is a unique combination of water softening and conventional coagulation.

Briefly, the Black process is as follows. Thickened sludge, produced from the lime-softening of hard, turbid and/or colored waters, is pumped to a series of sludge carbonation cells where it is mixed with scrubbed kiln gas containing about 20% CO_2. The magnesium hydroxide is dissolved from the calcium carbonate and turbidity factors. This slurry then passes to a thickener from which the clear supernatant containing the magnesium, now in the form of soluble magnesium bicarbonate, overflows and passes to a heat exchange unit where it is warmed to 40°C. The solution then flows to aeration cells equipped with mechanical stirrers. Precipitation of the magnesium carbonate as the trihydrate is rapid and complete in 90 min. The snow white product is then vacuum filtered, dried and bagged for shipment. The turbidity factors are removed from the $CaCO_3$ by froth flotation, and the purified $CaCO_3$, thickened to about 70% solids, is calcined to a high quality quicklime. For every 1 ton of lime fed, 1.2-1.4 tons are recovered. The excess is derived from the calcium carbonate in the raw water.

Because it is impossible to separate the coagulant from the calcium carbonate, this process is not feasible using aluminum or iron coagulants. However, the magnesium hydroxide is dissolved by the CO_2 for easy separation from the calcium carbonate. During 1970 coagulation studies were carried out at the University of Florida's Environmental Engineering Laboratory by Black and Thompson.[5-7] These studies compared magnesium carbonate and alum as coagulants for organic color and turbidity. Waters from approximately 20 major cities, along with several synthetic mixtures, were evaluated utilizing the jar test procedure. These waters represented a wide range in physical and chemical characteristics. In summation, the following conclusions were reached:

1. Magnesium carbonate is superior to alum and iron salts for the removal of both turbidity and color.
2. The flocs formed are larger, heavier and settle faster than the alum flocs. Therefore, the capacity of plants is often increased.
3. Color is much more significant than turbidity in determining the necessary chemical dosages.
4. Release of the coagulated color during the sludge carbonation step is not a problem when the color of the water is less than 150.
5. The use of magnesium carbonate produces a treated water with superior physical and chemical characteristics compared to alum treated water.

At the present time, three EPA-funded projects related to the overall magnesium process have been completed. The first project was the laboratory study conducted at the University of Florida.[5] The second was a demonstration project,[8] the initial objectives of which were concerned with the treatment of a very soft, low magnesium water at Montgomery, Alabama, with no consideration given for magnesium production. In an extension of this project in Melbourne, Florida, the application of the process for treatment of a much harder, highly colored, low turbidity water was studied. The third project, in Gainesville, Florida, utilized a 72,000 gpd sewage treatment package plant to study the physical-chemical treatment of domestic wastewaters.[9]

A fourth EPA research project is still underway to study the application of the lime and magnesium recovery aspects of this process on a hard, high-magnesium, turbid surface water at Johnson County, Kansas.[10]

PROCESS ADVANTAGES

The use of magnesium salts has probably not been considered previously for economic reasons. Both the chloride and the sulfate cost more per pound than alum, and their use, in conjunction with lime, would increase the noncarbonate hardness of the water being treated in direct ratio to the makeup dosage required. Magnesium carbonate trihydrate ($MgCO_3 \cdot 3 H_2O$) is currently unavailable because there are no commercial uses for it. The available "basic carbonate," $4 MgCO_3 \cdot Mg(OH)_2 \cdot 5 H_2O$, has many commercial uses, costs 28-31¢/lb (13-14¢/kg), but it is not a coagulant because of its low solubility and bulk density. However, three factors now indicate, perhaps mandate, that the widespread use of the trihydrate be given careful consideration:

1. As stated previously, treatment of water plant wastes is becoming mandatory. The dewatering and at some plants complete recovery of this sludge is an integral part of the magnesium recovery process.

2. Recovery and recycling of a coagulant had not proven practical in the past. An economical, easily controlled process is now available for the recovery of magnesium carbonate from the sludge.

3. A new low-cost source of magnesium carbonate trihydrate will soon be available. It will be recovered and sold at a very low cost (5¢/lb; 2.3¢/kg) from the sludges produced by major plants softening high magnesium waters. Such plants will be able to reduce substantially their chemical treatment costs by allowing (a) the elimination of the use of alum, (b) the sale of recovered magnesium carbonate—estimated at over 150,000 ton/yr (136,000 t/yr) from only 20 cities, and (c) the recalcination, recycling, and reuse of lime and sale of excess production. Perhaps of equal importance is the fact that in so doing, these plants will eliminate their individual sludge pollution problems. When the lime is not recalcined, sludge can be thickened to 40% solids, filtered to a dense, low moisture cake, and disposed of as landfill or soil conditioner.

Magnesium carbonate hydrolyzed with lime has been shown to be more effective than alum for the removal of organic color and turbidity. This new system of water treatment has been shown to be superior to the present treatment methods in many ways. The recovery of the magnesium by carbonation allows dewatering of the sludge for lime recovery or disposal as landfill in the smaller plants, eliminating one of the greatest problems of the water industry today. The flocs formed are large and heavy and settle rapidly, which makes it possible to increase significantly the capacity of many plants.

The high pH of coagulation, usually above 11.0, should (a) provide complete disinfection and virus inactivation where adequate contact time is available, (b) eliminate the need for prechlorination in many plants, and (c) provide essentially complete removal of iron and manganese, where present. Recycling and reusing the coagulant and recovering the sludge water reduces chemical treatment costs. Sludge water, which is rarely recovered in a conventional system, is about 90% recovered as treated water with this process. For soft surface waters, the new technology produces soft waters with a calcium alkalinity sufficient to permit stabilization by pH control of recarbonation and reduces or eliminates corrosion. Waters high in carbonate hardness are softened by the process.

PROCESS DESCRIPTION

The use of this new process in large water treatment plants allows for the recovery and reuse of both lime and magnesium carbonate, with carbon dioxide supplied by recalcining the lime.

The practice of lime recalcination is most attractive for large plants treating moderately hard to hard waters. For other plants (since the CO_2 produced in recalcination greatly reduces the treatment costs) it would be advantageous to locate an inexpensive source near the plant. In many cities such a source exists in the form of diesel or natural gas engine exhaust, stack gases from incinerators or power plants or some other acceptable industrial emitter. A 50-hp natural gas engine provides sufficient CO_2 in the exhaust for a 1-mgd water plant.

This new system of water treatment is a unique combination of water softening and conventional coagulation. Lime is used to precipitate the magnesium as magnesium hydroxide. The hydroxide concentration of the water can be increased to the necessary level only after converting all of the CO_2 and bicarbonate alkalinity to carbonate alkalinity. These well-known softening reactions are:

$$CO_2 + Ca(OH)_2 \rightarrow CaCO_3 + H_2O \tag{1}$$

$$Ca(HCO_3)_2 + Ca(OH)_2 \rightarrow 2\ CaCO_3 + 2\ H_2O \tag{2}$$

Magnesium bicarbonate is converted to magnesium carbonate and then to magnesium hydroxide on further addition of lime:

$$Mg(HCO_3)_2 + Ca(OH)_2 \rightarrow MgCO_3 + CaCO_3 + 2\ H_2O \tag{3}$$

$$MgCO_3 + Ca(OH)_2 \rightarrow Mg(OH)_2 + CaCO_3 \tag{4}$$

If the magnesium in the water is present as noncarbonate hardness, there is no net change in total hardness, only an exchange of calcium for magnesium as:

$$Mg^{2+} + SO_4^{2-} + Ca(OH)_2 \rightarrow Mg(OH)_2 + Ca^{2+} + SO_4^{2-} \tag{5}$$

The lime dosages required for the removal of free CO_2 and carbonate hardness are stoichiometric, but the amount to be added for the precipitation of magnesium depends upon a number of factors. This variability may be illustrated by the following example. Water X, containing both organic color and turbidity, contains 8 mg/l magnesium as the ion (33 mg/l magnesium as $CaCO_3$). Jar tests show that the minimum effective amount of $Mg(OH)_2$ to be precipitated is provided by a dosage of 60 mg/l $MgCO_3 \cdot 3 H_2O$ (44 mg/l as $CaCO_3$). There are two alternatives: (1) supply the entire dosage with $MgCO_3 \cdot 3 H_2O$ and (2) add lime to precipitate the magnesium ion in solution and make up the remainder needed with $MgCO_3 \cdot 3 H_2O$. The more economical treatment is to add all of the magnesium required with a dosage of 60 mg/l $MgCO_3 \cdot 3 H_2O$ and to use none of the magnesium ion originally present in the water. There are four reasons for this choice.

1. The dosage of lime required to precipitate the necessary amount of $Mg(OH)_2$ from 60 mg/l $MgCO_3 \cdot 3 H_2O$ is significantly less than that required to precipitate the same amount of $Mg(OH)_2$ from the magnesium ion present naturally plus the 11 mg of magnesium to be added. Furthermore, the pH is significantly lower.

2. The amount of CO_2 required to redissolve the $Mg(OH)_2$ from the sludge is the same for both choices, but the amount of CO_2 required to recarbonate the high pH water of the second procedure is much greater than the amount required for the lower pH water of the first procedure.

3. After a few cycles, a pH value is identified at which complete recovery and recycling of coagulant is achieved, after which no new coagulant is added.

4. The unused 33 ppm of magnesium remaining in the water represents reserve coagulant, which can be used if necessary by increased lime addition.

Although this example of a water containing a significant amount of magnesium is not typical of surface waters as a class, there are many large U.S. cities, including Chicago, Dallas, Cleveland, Philadelphia, Pittsburgh, Cincinnati and Washington, D.C., and many smaller-sized cities treating such waters. All of them should be able to reduce substantially their chemcial treatment costs and at the same time eliminate or greatly reduce the extent of their sludge problems by adopting the new technology.

A flow diagram of the unit operations involved is shown in Figure 1. The sludge from the settling basin, containing $Mg(OH)_2$, $CaCO_3$, turbidity, color and other pollutants is 1-5% total solids, suspended in water containing 50-70 mg/l hydroxide alkalinity. Thickening the sludge allows recycling of the clear, highly alkaline supernatant, which has a volume equal to two-thirds of the sludge flow. Such processing reduces the

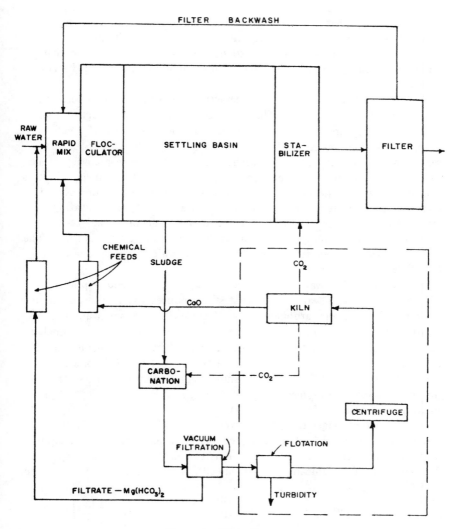

Figure 1. Flow diagram.

required lime dosage, the amount of carbon dioxide required for sludge carbonation, and the size of the carbonation basin. These advantages must be weighed against the capital costs and operational problems resulting from the additional unit required.

Carbon dioxide is introduced with rapid mixing in the carbonation cells, which have a retention time of approximately 75-90 min. The magnesium is completely solubilized as $Mg(HCO_3)_2$. Vacuum filtration separates the solid and liquid phases. The magnesium bicarbonate solution is then recycled to make up the dose required for coagulation. If lime is not recovered, this filter cake (40-50% solids) represents a plant waste that can be handled easily and disposed of in agricultural use as a pH stabilizer for soil.

The broken-line box in Figure 1 shows the additional unit operations necessary for lime recovery. The filter cake is slurried and factors such as clay turbidity are thereby removed by flotation in a line of small flotation cells. This process, which has been shown to be successful at a very minimum cost, purifies the calcium carbonate before recovery. The small volume of rejects, which can easily be dewatered and disposed of as landfill, represents the only waste produced from the entire process. The relatively pure calcium carbonate is then dewatered by centrifugation and burned in a lime kiln producing calcium oxide (quicklime) and carbon dioxide. The quicklime is slaked and reused and any excess is sold. The carbon dioxide is used for stabilization of the water and carbonation of the sludge.

Two-stage settled water stabilization is often recommended prior to filtering the finished water. The first stage reduces the pH of minimum solubility for calcium carbonate from 11.3 to 10.3. A small amount of preformed calcium carbonate must be present to act as a seed or nucleus for precipitation and to increase particle size. Preformed calcium carbonate represents recycled $CaCO_3$ sludge in plant practice. Precipitation at pH 10.3 prevents $CaCO_3$ supersaturation, high alkalinities, and scale build-up on mechanical equipment. This first stage is necessary only for plants treating hard waters (Tables I and II).

In the second stage, the pH is lowered from 10.3 to the pH_s of the water. The term pH_s is the calculated pH at which a water is in equilibrium with calcium carbonate, and it is calculated from Equations 6 or 7 or determined graphically from Caldwell-Lawrence diagrams of the water (see Chapter 23).

$$pH_s = pK_2 - pK_s + pCa + p\left[Alk + (H^+) - \frac{K_w}{(H^+)}\right] + \log\left[1 + \frac{2K_2}{(H_s^+)}\right] \quad (6)$$

where p   =   the negative logarithm
    K_2 =   the dissociation "constant" for HCO_3
    K_s =   =   the solubility product "constant" for $CaCO_3$
    (H^+) =   the molar concentration of hydrogen ion
    K_w =   the dissociation "constant" for water.

Values for these "constants" are given by Langelier.[1]

If pH_s falls between 6.5 and 9.5, Equation 6 can be simplified with little error to Equation 7:

Table I. Examples of Cities in Category 2 (Medium Hardness)

City	Source of Supply	Chemical Characteristics			Turbidity	
		Mg^{++}	CH	TH	Average	Maximum
Cleveland, Oh.	Lake Erie	7	94	127	9	140
Philadelphia, Pa.	Delaware R.	6	34	67	36	13
Philadelphia, Pa.	Schuylkill R.	15	65	153	27	85
Washington, D.C.	Potomac R.	8	70	101	49	600
Louisville, Ky.	Ohio R.	10	74	131	101	800
Dallas, Tex.	Lakes	7	110	152	49	1120
Cincinnati, Oh.	Ohio R.	9	40	137	70	1100

Table II. Examples of Cities in Category 3 (High Hardness and Mg)

City	Source of Supply	Chemical Characteristics			Turbidity	
		Mg^{++}	CH	TH	Average	Maximum
Dayton, Oh.	Wells	33	256	356	0	0
Des Moines, Ia.	Racoon R.	33	244	331	50	1330
Minneapolis, Mn.	Mississippi R.	16	158	185	7	60
Columbus, Oh.	Scioto R.	26	159	272	40	110
Columbus, Oh.	Big Walnut Creek	15	92	152	13	28
Oklahoma City, Ok.	Lake Hefner	26	143	246	6	6
Topeka, Kn.	Kansas R.	23	203	292	912	1120

$$pH_s = pK_s + pCa + pAlk \qquad (7)$$

The scale-forming tendency of the water can be crudely calculated from the Ryznar Index:

$$RI = 2\ pH_s - pH \qquad (8)$$

where pH is the actual or measured pH. If RI is 6.5 ± 0.3, the water is balanced or stable. As the RI increases from 6.8, the tendency for corrosion or scale-dissolving increases. As the RI decreases from 6.2, the scale-forming tendency increases. A more accurate assessment of water conditioning is obtained by using Caldwell-Lawrence Diagrams as shown in Chapter 23. The second stage is all that is required for plants treating soft waters (Table III). These plants precipitate magnesium in the pH range 10.4 to 10.7, reclaim and recycle the magnesium, and discard the lime cake if recalcining is not found to be economical.

Table III.   Examples of Cities in Category 1 (Soft Waters)

City	Source of Supply	Chemical Characteristics			Turbidity	
		Mg^{++}	CH	TH	Average	Maximum
Baltimore, Md.	Three R. (imp.)	2-3	35	43	1	3
Tulsa, Ok.	Imp. Creek	2	86	86	7	13
Newark, N. J.	Imp. R.	3	17	19	-	-
Richmond, Va.	James R.	2	34	40	44	274
Atlanta, Ga.	Chattahoochee R.	1	14	14	27	200
Birmingham, Al.	Lake Purdy imp.	1	7	7	-	-
Shreveport, La.	Cross Lake	4	25	42	17	27

PROCESS ECONOMICS

The economic value depends primarily on the character of the water being treated or a critical sludge disposal situation. The cost is most favorable for hard, high-magnesium waters. The feasibility of lime recovery and recalcination or, in the absence of recalcination, the availability of an acceptable CO_2 source near the plant would contribute significantly to the adoption of the Black process. In operating the new process, the potential savings include:

1. Reduction of treatment costs by
   a. using magnesium rather than alum or iron salts,
   b. using magnesium recovery and recycle,
   c. using lime recalcining to reduce costs by 50%, and
   d. using CO_2 from kiln stack gas to recover magnesium and stabilize the water.
2. Reduction or elimination of sludge disposal.
3. Recovery of sludge water.
4. Reduction of chlorine consumption.
5. Increased plant capacity.
6. Production of soft, stable water.
7. Sale of excess magnesium, lime, and calcium carbonate.

A 1973 study was conducted by a consulting engineering firm to convert the Melbourne, Florida, water plant to the Black process. The capital costs required to convert the plant to the magnesium process vs the capital costs for adding sludge filtration facilities are shown in Table IV.

Melbourne, Florida, has converted to the magnesium process with the following initial benefits:

Table IV. Cost to Convert Melbourne Plant to the Black Process

Capital Cost Comparison	Cost ($)
Addition of Sludge Filtration	1,548,000
Conversion to Magnesium	
• With Lime Recalcination	1,513,000
• Without Recalcination	613,000
The savings in operation costs for this 10-mgd plant were:	
Operation Costs Per Year	
Present Plant with Sludge Filtration	336,700
Black Process	
• With Lime Recalcination	217,000
• Without Recalcination	270,000

1. Upflow reactors had a rise rate of 0.75 gpm/ft^2 (0.03 m^3/m^2 min) with alum floc. The same reactors now operate at 1.5-2.0 gpm/ft^2 (0.06-0.8 m^3/m^2 min) rise rate with the coprecipitated $Mg(OH)_2$ + $CaCO_3$ floc.

2. Two water plants were in operation prior to the conversion. The increased capacity, as a result of the heavy floc, has permitted the closing of one plant. The remaining plant now produces water in excess of the amount formerly produced by two plants.

A procedure for determining the economics of this new technology as applied to hard, turbid waters has been developed and is shown below.

A. To estimate the approximate reduction in chemical treatment costs:
   1. Take the sum of the following:
      a. cost of alum or ferric salt used as coagulant
      b. 50% of the cost of the lime presently used
      c. cost of the fuel used to produce CO_2 used for stabilization, or cost of liquid CO_2 if purchased
      d. 90% of the production cost of all treated water now lost as sludge.
   2. Subtract from the above total the cost of the additional *recalcined lime* that must be added to reduce the amount of magnesium present in the raw water to the figure desired in the treated water.
B. The following additional savings are obtained:
   1. sale of excess lime produced
   2. sale of magnesium values produced
   3. cost of an alternate method of sludge disposal
   4. from these should be deducted the cost of disposing of the clay-calcium carbonate cake as landfill.

C. Other potential savings include:
1. the possibility of reducing the amount of chlorine used for prechlorination or perhaps eliminating it entirely
2. production of a softer treated water
3. increase in treatment plant capacity
4. production of a more stable water and elimination of corrosion.

Plant Applications

The magnesium carbonate process of water treatment has replaced alum in a portion of two water plants in full-scale studies conducted over the past 2.5 yr. This new water treatment technology was compared to the presently used alum process in parallel treatment in Montgomery, Alabama, and Melbourne, Florida.

The results of these studies indicate that this new process offers a number of significant advantages over the alum process. The primary advantage is that the existing problem of sludge disposal at Melbourne is completely eliminated and at Montgomery is greatly reduced. All water is recycled within the process along with the three basic water treatment chemicals: lime, magnesium bicarbonate and carbon dioxide. Other advantages found were: increased floc settling rates, simplicity of operation and control, reduced costs when sludge treatment and disposal costs are considered, and more complete disinfection. At Melbourne, considerable energy is conserved by on-site lime recovery.

In addition to the two full-scale studies, a number of special studies were conducted in Montgomery using a 50-gpm pilot plant. These studies showed that added cadmium was almost completely removed by the highly adsorptive $Mg(OH)_2$ flocs and that it was not released during sludge carbonation and magnesium recycling.

Both Montgomery and Melbourne still use the conventional alum process in a portion of their plants. The raw waters treated vary considerably as shown in Table V.

Table V. Raw Water Quality at Montgomery and Melbourne

Source	Alkalinity (mg/l CaCO$_3$)	Hardness (mg/l CaCO$_3$)	Magnesium (mg/l Mg)	Color (CU)	Turbidity (TU)
Tallapoosa River Montgomery	10-22	10-22	0-5	5-60	2-300
Lake Washington Melbourne	25-100	40-200	6-50	60-300	5

Montgomery typically uses from 20 to 30 mg/l of alum and coagulates in the pH range of 6.0 to 7.0. Melbourne uses as much as 130 mg/l of alum and coagulates in the 5.1 to 5.3 pH range. In addition, because of high dissolved organics (color) in the Melbourne water, large dosages of activated carbon and chlorine are required. The plant facilities at Melbourne differed from Montgomery as follows:

1. Vertical flocculators, Melbourne.
   Horizontal flocculators, Montgomery.
2. Upflow clarifiers, 0.5 gpm/ft^2 (0.02 m^3/m^2 min), Melbourne.
   Horizontal settling basins, 6-hr retention, Montgomery.
3. Single-stage finished water stabilization, Melbourne.
   Two-stage finished water stabilization, Montgomery.

Table VI summarizes the design criteria determined from both the Montgomery and Melbourne studies.

Table VI. Design Summary from Montgomery and Melbourne Studies

Unit	Design Parameters	Comments
Rapid Mix	1-3 min	Short rapid mix appears to be desirable in color removal applications.
Flocculation	15-30 min	Floc forms rapidly. However, contact time increases color adsorption.
Clarifier	1.5 gpm/ft^2 (0.06 m^3/m^2 min)	The use of polymers (0.1-0.3 mg/l) prevents excessive Mg(OH)$_2$ floc carryover.
Filtration	2.0-4.0 gpm/ft^2 (0.08-0.16 m^3/m^2 min)	Filter rates up to 4.0 gpm/ft^2 evaluated in Montgomery. Increased rates improved performance normally.
Settled Water Carbonation	10-30 min	Two stage, separated by time, shown as design parameters.
Sludge Carbonation		
Purchased 100% CO$_2$	15 min	
Kiln Gas 18% CO$_2$	90 min	
Sludge Thickening	40-50 lb/ft^2 day (200-240 kgf/m^2 day)	
Sludge Dewatering	25-35 lb/ft^2 day (120-170 kgf/m^2 day	

The use of magnesium bicarbonate as a recycled coagulant has been found to equal or exceed the results obtained by the use of alum in every aspect of water treatment including water quality produced, operational characteristics, economy and adaptability over a wide range of raw water qualities. Plant personnel demonstrated their proficiency in the operation of the process. The nature of the process is such that pH control at three critical points is the primary method of ensuring adequate treatment and producing excellent quality over a wide range of influent quality. The process was found to be easily automated and controlled. The full-scale use of this new process is compatible with most existing water treatment plants requiring a minimum of land area and capital cost. The conversion of most existing alum treatment plants to this new technology involves few internal process changes and only minor piping changes to add the necessary recovery and recycling units.

PROCEDURE FOR EVALUATING
THE BLACK PROCESS

The use of a six-station magne-drive laboratory stirrer is strongly recommended. The stirrer is manufactured by Coffman Industries, Inc., 22 North 6th Street, Kansas City, Kansas 66101. A dual drive unit is available and it should be purchased if alum and magnesium carbonate flocculants are to be compared. Several useful accessories are also available.

Either 1.0 or 1.5 liters of raw water is added to each of the six jars. Then using jar 1 as an example, remove about 180 ml of water from the jar using a 200-ml beaker; place it on the table and in front of the jar. Along side it are placed two 50-ml Pyrex beakers, calibrated 10-50 and labeled 1-M and 1-L. The 1-M beaker contains, in powder form, the desired dosage of *finely ground* $MgCO_3 \cdot 3 H_2O$, which has been accurately weighed using an analytical balance. Small beaker 1-L contains the accurately weighed dosage of *pure reagent grade* $Ca(OH)_2$. If pure material is not available, a correction is made to compensate. It should also be finely ground and free of lumps before weighing. All six jars are so prepared, which completes the first step.

Stirring is begun at 100 rpm, and about 30 ml of the water in the 200-ml beaker is added to the small beaker 1-M, the contents thoroughly slurried with a rubber tipped glass stirring rod and rapidly transferred to jar 1. The small beaker is thoroughly washed with two successive 30-ml portions of water from the 200-ml beaker and the washings added to jar 1. This series of operations is now repeated for jars 2-6. With a little practice, all six dosages can be quantitatively transferred to jars 1-6 in about 2 min, using about one-half of the water in the 200-ml beaker.

Rapid mix is used for at least 3 more minutes to make sure that all of the $MgCO_3 \cdot 3 H_2O$ is dissolved.

At the end of the 3-min period for solution of the $MgCO_3 \cdot 3 H_2O$ the same series of operations is used for the addition of the $Ca(OH)_2$ dosages. More care should be used for these since very little of the $Ca(OH)_2$ will dissolve. Any lumps, however small, should be broken up and every trace of solid material transferred to the jars, as well as all of the water in the 200-ml beaker.

Time and Speed of Rapid Mix

Rapid mix is carried out at 100 rpm. The time used may vary from as little as 30 sec to as long as 10 min, depending upon the type of water being treated and the type of flocs formed. When treating soft turbid and/or colored waters, the lime dosage will be low and the flocs will contain little $CaCO_3$. For such water, an anionic flocculant may be used. However, with hard waters, usually containing considerable magnesium, lime dosages will be much higher and the flocs will be large and heavy, loaded with $CaCO_3$. The duration of rapid mix is not critical since flocculation is carried out at speeds considerably higher than those usually used for alum. For the first jar test of water, a rapid mixing time of 3 min after the addition of the $Ca(OH)_2$ is suggested.

In calculating lime dosages, keep in mind that no magnesium precipitates below pH 10.3 but all of it precipitates at pH 11.3. The usual working range is pH 10.7 to pH 11.3.

Flocculation

Flocculation is the critical step in a jar test since it must (a) complete the reaction between the $MgCO_3$ and the $Ca(OH)_2$ and (b) build the flocs to maximum size and density. The speed may vary from as low as 10 rpm to as high as 25 rpm, the criterion being that no settling should take place during this stage. *Sludge on the bottom of the jar is performing no useful purpose.* The minimum time is 25 min.

Where recalcination is not to be employed and no flotation step is involved, a flocculant may be used. If one is used, it should be added 1 min before the speed is reduced from rapid mix to flocculation.

Tests on all types of waters have shown that for waters of medium to high alkalinity and total hardness, a dosage of about 2.0 mg/l of activated silica is the preferred flocculant when one is required. For soft turbid and/or colored waters, where little $CaCO_3$ is present to "load" the flocs, a low dosage, 0.5 mg/l of alum or 0.35 mg/l of potato starch, may give better results.

Collection of Samples for Analysis

The following procedure has been found to be simple and satisfactory for collecting samples for analysis. Six numbered 500-ml beakers are prepared to receive the samples, which are collected as follows. Have a glass blower heat the tubing above the top of the bulb of a 100-ml pipette and bend it in a U so that when the empty pipette is carefully lowered into the settled water in the jar, the top of the tube points directly upward and is approximately 2 in. below the level of the settled water. Carefully lower it into each jar, taking care not to disturb the settled sludge at the bottom of the jar, and withdraw a 100-ml portion and determine "settled turbidity." Transfer four more 100-ml portions to 500-ml beaker No. 1.

If this procedure is carefully followed for all jars, the samples so collected yield values for settled turbidity that correctly represent the dosages used for the jars. However, the remainder of each 500-ml portion should be filtered before withdrawing samples for chemical analysis. The sludge from each jar should be transferred into numbered beakers for use in subsequent jar tests as "returned sludge."

Stabilization of Treated Water

After determining pH and differential alkalinity of the filtered samples, add about 0.5 g of analytical reagent grade $CaCO_3$ to the remainder of each sample, immerse a pH electrode and blow through a pipette into each sample until the pH is reduced to 8.75-9.0, which is recorded. With $MgCO_3$ as the coagulant, the pH of the settled water will be 10.7-11.3 and caustic alkalinity will be present. Stabilization with CO_2 results in the formation of $CaCO_3$. Unless a small amount of $CaCO_3$ is present in the sample, supersaturation and high alkalinities will be found. The presence of a little $CaCO_3$, representing return sludge in plant practice, prevents this.

The samples are filtered, and differential and total alkalinity, calcium, magnesium and color (if present) are determined. It is very important that final color not be determined until the sample has been stabilized and filtered since organic color is an indicator that changes with pH. A sample of settled, colored water coagulated with alum in the pH range 5.2-5.5 will *increase* in color when stabilized with lime to final pH, whereas the same water, coagulated with $MgCO_3$ in the pH range 10.7-11.3 will *decrease* from its settled value following recarbonation.

A form for recording the results of jar tests is shown in Figure 2.

Jar No.	Dosage in ppm			pH	Color	Turbidity	Mobility	Alkalinity			Stability to pH	Color	Alkalinity		Hardness			Magnesium as $CaCO_3$		
								OH	CO_3	HCO_3			CO_3	HCO_3	C	NC	T			

Characteristics of raw water

Alkalinity as $CaCO_3$ __________

Total Hardness as $CaCO_3$ __________

pH __________

Organic Color __________

Turbidity __________

Calcium as $CaCO_3$ __________

Magnesium as $CaCO_3$ __________

Figure 2. Form for recording jar test data.

Addition of $MgCO_3 \cdot 3 H_2O$ Dosages

These dosages should always be weighed and added as a slurry, as directed. Water solutions of the material slowly decompose to form the relatively insoluble "basic" carbonate, $4 MgCO_3 \cdot Mg(OH)_2 \cdot 5 H_2O$.

Caution Concerning Magnesium Carbonate Coagulant

Magnesium carbonate trihydrate, $MgCO_3 \cdot 3 H_2O$, is the new coagulant that, when hydrolyzed with lime, is the preferred coagulant for the treatment of all types of waters, surface and ground, hard and soft. It is a snow white powder of very fine particle size with a bulk density of 40 lb/ft^3 (640 kg/m^3), the same as "light" soda ash. The air-dried product contains about 4% moisture and on a dry basis is 99.7% pure.*

"Magnesium carbonate" purchased from any source in this country is the "basic carbonate" $4 MgCO_3 \cdot Mg(OH)_2 \cdot 5 H_2O$. Its low solubility (only 90 mg/l), its slowness to dissolve, and its extremely low bulk density (only 5-8 lb/ft^3) make it unsatisfactory for practical use. *It should NOT be used for jar tests.*

Coagulation with Alum or Ferric Sulfate

The operator should use the procedures that he has found to yield the best and most reproducible results. However, a few suggestions are given to prevent errors that are often made.

Preparation of solutions: Fresh stock solutions (10 g/l) should be prepared weekly and stored in a refrigerator when possible. For adding dosages for coagulation, a 1:10 working solution should be prepared daily (1 ml = 1.0 mg of coagulant).

For adding the very small dosages of alum [0.5 mg/l to be used as a floc aid for $Mg(OH)_2 \cdot CaCO_3$ flocs referred to elsewhere in these directions], a second 1:10 dilution is used, so that 1 ml = 0.10 mg of coagulant. This is prepared daily.

Experience with hydrolyzing coagulants has shown that the use of 3-5 min rapid mix at 100 rpm and 15-20 min flocculation at 10-15 rpm usually gives excellent results.

---

*A sample for testing may be obtained upon request by addressing: Mr. Thomas Saygers, Assistant Superintendent, Division of Water Supply and Treatment, 1044 Ottawa, Dayton, Ohio 45102.

Calculation of Dosage of 100% Ca(OH)$_2$ for Jar Tests

The following equation is suggested as the simplest for calculating the theoretical dosage of pure Ca(OH)$_2$ for any type of treatment:

$$\text{mg/l } 100\% \text{ Ca(OH)}_2 = [\text{CO}_2 + \text{Alk} + \text{Mg} + \text{(OH)}_2] \times 0.74$$

The four components in the [ ] *are expressed as CaCO$_3$*. "Mg" means the *amount to be precipitated as Mg(OH)$_2$*. "(OH)" means the causticity, expressed as CaCO$_3$, *needed to precipitate the Mg*. When "Mg" is 0, pH is 10.3 and (OH) is 0. When *all Mg* is to be removed, pH is 11.3 and (OH) is 70.

SIMULTANEOUS COMPARISON OF MAGNESIUM CARBONATE AND ALUM

When evaluating the effectiveness of two different coagulants for a given water, one may often make valuable comparisons by observing their actions simultaneously. This may, of course, be done to a limited extent by a team of two workers, each utilizing three jars of a six-unit jar test unit. The procedure is begun by worker no. 1 slurrying his MgCO$_3$ dosages and adding them to jars 1-3, as described. This mixing time has no significance, since it is not a hydrolyzing coagulant and does not react with any of the constituents in the water except possibly free CO$_2$, which would not affect the results of the test.

At a stirring speed of 100 rpm, both workers now begin adding their respective dosages of lime and alum. This should be completed within 2 min. (They should work together jar by jar.) Rapid mix is continued for the chosen period of time and the stirring speed is then reduced to that chosen for flocculation so that speed is continued for the chosen period of time. If a flocculant is to be used with either or both coagulants, it should be added at least 1 min before rapid mix is reduced to flocculation speed.

However, this procedure does not provide each of the two coagulants with the conditions at which they perform best. The Mg(OH)$_2$ flocs are made dense and heavy by the coprecipitated CaCO$_3$ and require flocculation speeds substantially higher than can be tolerated by the weaker and lighter alum flocs.

SUGGESTIONS FOR PLANNING JAR TESTS

Waters to be treated fall into one of three categories.

Soft Waters

Category 1 (Table III) includes soft, turbid and/or colored waters containing little magnesium ion (usually in the range 1-5 mg/l). For such waters, make-up magnesium carbonate is needed to replace that lost in filter cake washing. The jar tests should be planned to determine the magnesium carbonate dosage needed for coagulation, taking into consideration seasonal or other variations in raw water quality. For plants of small or medium size, where recalcination is not practical, the use of a flocculant should also be studied. Magnesium is, of course, recycled and reused.

Medium Hard Waters

Category 2 (Table I) includes turbid and/or colored waters of medium hardness and magnesium ion in the range 6-11 mg/l. The size of the city and the desired hardness of the finished water are the determining factors. Assume, as an example, City X, whose raw water has TH = 140, Mg^{++} as $CaCO_3$ = 45, NCH = 20 and average turbidity 65 TU. First, jar tests are run to determine the amount of $MgCO_3 \cdot 3 H_2O$ to be added for turbidity removal. Let us assume that it is 20 mg/l. Since the raw water contains the equivalent of 45 x 1.38 = 62 mg/l $MgCO_3 \cdot 3 H_2O$, the total dosage as $MgCO_3 \cdot 3 H_2O$ is 62 + 20 = 82 mg/l. At least 70 mg/l $MgCO_3 \cdot 3 H_2O$ is to be added as recycled Mg on the second pass, leaving only 12 mg/l to be supplied from the water, so that the lime dosage must be *reduced* to precipitate only that amount, leaving about one-half (or less if desired) of the magnesium as hardness in the finished water. The final TH would be approximately (22 mg/l Mg as $CaCO_3$ + 45 mg/l alkalinity + 20 NCH) = 87 mg/l as $CaCO_3$. This treatment is excellent for waters low in NCH.

From the standpoint of plant operation, this substantial "bank" of unused coagulant serves as a safety factor because in times of high turbidity the operator simply increases his lime dosage as needed. However, the cost of lime has dramatically increased and will probably remain high. If this is a large plant, the recovery and sale of both excess lime and $MgCO_3 \cdot 3 H_2O$ would substantially reduce chemical treatment costs and the cost of sludge disposal. Jar tests should, therefore, be run removing most or all of the magnesium, after which overall mass balance calculations can be compared for both alternatives.

Hard Waters

Category 3 (Table II) includes waters high in both total hardness and Mg. The hardness varies from about 200 to 350 mg/l and magnesium ion varies from 13 to 33 mg/l. These cities benefit by recycling and recovering both lime and magnesium and selling the excess production of each.

SUMMARY

In the opinion of Dr. A. P. Black, few plants are too small to use the new magnesium process. The undesirable physical, chemical and bacteriological characteristics of the nation's raw water supplies, the sludge disposal problems, and/or the mandated improvements in finished water quality have already determined the need for the magnesium technology. The Black process holds great potential for most water treatment systems because of the flexibility of the process with respect to multiple sources of carbon dioxide and magnesium salts plus the ability to recover, recycle and reuse the chemical reactants.

REFERENCES

1. American Water Works Association Research Foundation. "Disposal of Wastes from Water Treatment Plants," Water Pollution Control Research Series (August 1969).
2. Black, A. P., and F. A. Eidsness. "Carbonation of Water Softening Plant Sludge, *J. Am. Water Works Assoc.* 49:1343 (1957).
3. Black, A. P. "Split-Treatment Water Softening at Dayton," *J. Am. Water Works Assoc.* 58(1):97 (1966).
4. Black, A. P., B. B. Shuey and P. J. Fleming. "Recovery of Calcium and Magnesium Values from Lime-Soda Softening Sludges," *J. Am. Water Works Assoc.* 63(10):616 (1971).
5. Black, A. P., and C. G. Thompson. "Magnesium Carbonate—A Recycled Coagulant for Water Treatment," Environmental Protection Agency, Washington, D.C., Project Number 12120, ESW (June 1971).
6. Thompson, C. G., J. E. Singley and A. P. Black. "Magnesium Carbonate—A Recycled Coagulant," *J. Am. Water Works Assoc.* 64(1): 11 (1972).
7. Thompson, C. G., J. E. Singley and A. P. Black. "Magnesium Carbonate—A Recycled Coagulant, Part II," *J. Am. Water Works Assoc.* 64(1):94 (1972).
8. Black, A. P., and C. G. Thompson. "Plant Scale Studies of the Magnesium Carbonate Water Treatment Process," Environmental Protection Agency, Washington, D.C., Project Number 12120 HMZ (May 1975).
9. Black, A. P., A. T. DuBose and R. P. Vogh. "Physical-Chemical Treatment of Municipal Wastes by Recycled Magnesium Carbonate," Environmental Protection Agency, Washington, D.C., Grant Number 12130 HRA (June 1974).
10. Black, A. P., and C. G. Thompson. "Recovery, Recycle, and Reuse of Carbon Dioxide, Lime, and Magnesium in Potable Water Treatment," Environmental Protection Agency, Washington, D.C., Project Number S803196-01-2 (in press).

CHAPTER 6

FLOTATION PROCESS USED FOR CALCIUM CARBONATE RECOVERY FROM WATER TREATMENT SLUDGES

Harmel A. Dawson

President
Dawson Metallurgical Laboratories, Inc.
5217 Major Street
Murray, Utah 84107

INTRODUCTION

The results of investigations initiated by Black during the development of his water treatment process, which is described in Chapter 5 by DuBose, has shown that froth flotation can be economically and effectively used to produce a relatively high grade calcium carbonate, with a good recovery from the sludges obtained, when using magnesium carbonate with lime for flocculation in water treatment. The calcium carbonate recovered can be calcined to give lime for recycle to the treatment process as well as by-product lime for marketing. In addition, removal of the calcium carbonate, the major ingredient in sludge from water softening treatment plants, reduces the disposal problem to a fraction of that previously required.

In order for the flotation to be effective, the sludge must be recycled to increase the particle size of the calcium carbonate. This recycle also gives heavier floc in the settling basins, which results in faster settling rates. For the flotation to be effective, the magnesium hydroxide in the sludge must also be completely dissolved by the carbon dioxide and washed out.

In the froth flotation process (used to separate the calcium carbonate from the clay, silt, or other water contaminants), an aqueous slurry of the sludge is first conditioned (mixed) with soda ash and sodium silicate to disperse the clay and adjust the pH. Then the slurry is conditioned with a fatty acid soap which selectively coats the calcium carbonate particles with an insoluble soap making them hydrophobic and collectable. In the flotation machine, which mixes as well as disperses fine air bubbles into the slurry, the coated particles attach to the bubbles so that they can float to the surface and be removed from the machine. The clay and silt, which are still water-wetted, remain in the slurry.

FROTH FLOTATION PROCESS

Flotation Defined

Froth flotation, an important and common process in the present mineral industry, is a physico-chemical method for separating types of minerals or compounds present as discrete particles in an aqueous slurry. The process depends on being able to coat selectively the surface of the particles so as to give them an affinity for air. Then, with the addition of a frothing agent and fine air bubbles introduced into the mixed slurry, the coated particles are collected on the air bubbles and carried off with the froth. The uncoated wetted particles remain in the slurry.

Brief History of Flotation

An invention by Haynes in 1860 is the earliest patent relating to flotation. In this patent, which is the basis of a number of "bulk-oil" flotation processes, the dry, ground ore is mixed with one-fifth to one-ninth as much fatty or oily agent. When agitated with water, the sulfide minerals segregate with the fatty or oily agent from the earthy matter and water.

The next stage of development was the skin flotation process in which the dry, ground ore was gently brought into contact with the water. The metallic sulfides tended to float more than earthy matter (gangue).

Froth flotation had its beginnings in 1901 to 1905, when patents were obtained in which gas was used as a buoyant medium. The gas was produced by reacting acid with suspended carbonates and sulfides in the slurries. The process was called "flotation by chemical generation." The process for vacuum generation of bubbles, patented in 1904, has had some industrial success and is similar to the dissolved air flotation process presently used for removing sludges obtained in sewage treatment. In this process, air is dissolved in the slurry under pressure and when the pressure is released, fine gas bubbles form throughout the slurry. The sludge particles attach to the gas bubbles and are carried to the surface for separation.

The modern method of flotation began in 1923 with Keller's discovery of xanthates for selectively coating (collecting) the sulfide minerals. Through the years, a variety of effective collectors, conditioning agents, activating agents and depressants have been found to aid in improving the froth flotation process.

Early flotation efforts were directed toward the recovery of metallic sulfides of lead, zinc and copper. Subsequently, effective separations were found for the recovery of other minerals and the field expanded tremendously. The extension of froth flotation into the nonmetallic and industrial minerals has been spectacular, and large tonnages are now being treated by froth flotation. These include phosphate, coal, fluorite, feldspar, glass sand and limestone.

According to data collected by the U.S. Bureau of Mines, there were 202 plants in 1960 processing 198 million tons per year to recover 20 million tons of product that contained about $1 billion dollars in recoverable materials. In 1975 over 422 million tons were treated.

Many uses for flotation have also developed outside of the mineral industry for separations in the chemical, paper and food processing industries. Much of it is related to the recovery or removal of materials from process waste streams such as oils, fats, greases and fibers. Examples of such flotation processes would include removal of ink from waste paper, separation of wheat hulls from kernels, and recovery of cellulose from pulp mill waste streams.[1-3]

Mechanism of Flotation

Many theories have been advanced concerning the mechanisms involved in surfacing the mineral particles so as to create a hydrophobic hydrocarbon film on the mineral surface, and many investigations have been carried out to define these mechanisms. When froth flotation is used in an aqueous medium that carries the solids to be separated (together with dispersed air bubbles and possibly an organic liquid) a three- or possibly a four-phase system must be considered. In most froth flotation processes, the solid particles are initially completely water-wetted, and the solid-liquid interface must be replaced by a solid-gas interface by using suitable reagents. Studies of the changes in thermodynamic properties such as free energy and chemical potentials have done much to advance the understanding of flotation.[4]

Whether the adsorption of the reagent at the surface of the solid is by physical adsorption, chemisorption or chemical reaction, there is a definite correlation between the flotability of most minerals and the solubility of the compound formed by the collector agent and the compound to be floated. Theory postulates the formation of an insoluble metal organic compound at the particle surface. For example, in the use of xanthates for the collection of sulfides, it has been shown that the insoluble metal xanthate formed with lead and copper results in a floatable particle. With the more soluble zinc xanthate, zinc sulfide is not floatable unless it is activated by the addition of copper sulfate.[5] Although the actual reactions may be more complex, the following simplified reactions of an organic compound with a polar-nonpolar configuration such as a xanthate are indicative:

$$2 \ RCOSSNa + PbS \rightarrow Pb(RCOSS)_2 + Na_2S \qquad (1)$$

$$ZnS + CuSO_4 \rightarrow CuS + ZnSO_4 \qquad (2)$$

$$2 \ RCOSSNa + CuS \rightarrow Cu(RCOSS)_2 + Na_2S \qquad (3)$$

The organic portion, R, in the xanthate is generally obtained from alcohols ranging from ethyl to amyl (C_2H_5OH to $C_5H_{11}OH$).

The mechanism for surfacing (collecting) the carbonate and oxide compounds by the use of fatty acid or fatty acid soaps can also be described by the formation of insoluble organic metallic compounds at the surface of the particles. The use of an oleic acid soap ($Na_2O_2H_{33}C_{18}$) for the flotation of limestone ($CaCO_3$) demonstrates the basic chemistry.

$$2 \ NaO_2H_{33}C_{18} + CaCO_3 \rightarrow Ca(O_2H_{33}C_{18})_2 + Na_2CO_3 \qquad (4)$$

Classification of Flotation Reagents

In order to obtain conditions in an aqueous slurry to create a selective surfacing of particles and to maintain a stable, but brittle, froth that carries the attached particles to the surface for separation, a variety of chemical compounds (reagents) are used depending on the type and character of the separations involved. Selective flocculation of the collected particles is indicative of good collection. The selection of an effective reagent combination is based on the art, as well as science, of flotation. No hard and fast rules can be set down as to the floatability of a mixed slurry even after an examination and chemical analyses. Much depends on the inherent knowledge of and testing by a competent person.

Reagents used in flotation are classified according to their general effects in operation. Some reagents fit into more than one classification.

Collectors

Collector reagents are used to provide a water repellent surface on the particles to be floated so as to obtain adherence to the air bubbles. The collectors are classified according to their cationic or anionic reaction and type of minerals to be floated.[6] Examples of the types are given in Table I. The anionic types of collectors react with the metal portion (cation) in the compound to be floated, whereas the cationic types react with anion portions.

Table I. Collector Reagents

Reagent	General Formula	Type
Xanthates	ROCSSNa (R-short chain alcohol)	Anionic for sulfides
Fatty Acids	RCO_2H (R-long chain fatty acid)	Anionic for oxides, carbonates
Alkyl Amines	RNH_2 (R-long chain fatty acid)	Cationic for silicates, acidic minerals

In the froth flotation process, the collector reagents required vary from less than 0.1 lb/ton of solid feed for some sulfide separations to +4 lb/ton for some oxide mineral separations. Although mineral types indicate the type of collector required, ore characteristics vary so much that the reagent requirements must be tailored for each ore and may require adjustments on ores taken from different areas on the same ore body.

Modifiers

Modifying agents may act as selective depressants, selective activators, pH regulators, or they may reduce the harmful effects of colloidal material or soluble salts. Often one compound may perform several functions.

Depending on the particle separation desired and the character of the slurry, the pH required may be from 1.0 to 12.5 or higher. Lime, soda ash, caustic or acids are used for pH adjustment. The pH adjustment may act as an activator to aid in collector surfacing or as a selective depressant by preventing collector surfacing. As an example, soda ash (Na_2CO_3) may be an activator for some sulfides, a depressant if present in excess in calcite flotation, or a pH regulator.

Depressants

Depressants act to prevent the surfacing of the collector on a particle. An example is the use of zinc sulfate in preventing zinc sulfide from floating while allowing lead sulfide to be collected. Sodium silicate not only aids in the dispersion of slimes or colloidal material, but depresses silica and silicates. The cyanide ion aids in selectively assisting in the activation of lead sulfide but depressing pyrite (iron sulfide) and zinc sulfide. Lime concentrations at a relatively high pH (10-12) depress pyrite, allowing the copper or zinc sulfide mineral to float.

Organic colloids such as starch, glue and tannins (quebracho) act as dispersants, and an excess can prevent any collection. In controlled amounts, they are used to depress carbonaceous material, clays, talc and calcium carbonate.

Activators

Activators selectively react with particles to cause the collector to surface. The classic example, as mentioned above, is the use of copper sulfate for the activation of zinc sulfide so that it can be collected by standard sulfide mineral collectors. Another example is the surfacing of lead carbonate, copper carbonate and copper oxide with the use of sodium sulfide so that collection is also possible by the sulfide collectors.[7,8]

Froths

A frothing agent is used to form a stable yet brittle froth at the surface of the flotation machine so that the froth can be removed from the slurry along with the attached particles, thus accomplishing a separation of the hydrophobic and wetted particles. The froth generated should be able to support the particles, but should also readily break down when removed from the flotation machine so as not to interfere with subsequent processing.

Frothing agents function by reducing the surface tension of the water. Compounds used are generally heteropolar. They contain an organic nonpolar radical that repels water and a polar portion that attracts water. Used in limited quantities, the heteropolar molecules are aligned at the gas-liquid interface with the polar end toward the water and the nonpolar end toward the air. This mono-molecular film tends to retain the size of the bubble formed by the flotation machine and prevents the bubbles from breaking as they burst through the top of the water layer.

A wide variety of organic compounds could function to a greater or lesser degree as frothing compounds, but the number that are low in cost, readily available, effective in low concentrations, and essentially free of collector properties is limited. Initially, most frothers contained a hydroxyl (OH) polar group and had limited water solubility. Compounds included in this group are: amyl alcohol ($C_5H_{11}OH$), cresol ($CH_3C_6H_4OH$) in cresylic acid, and terpineol ($C_{10}H_{17}OH$) in fine oil. Within the last few years, a group of water-soluble frothers have been developed that are very effective. They are made from polypropylene glycol compounds. It is theorized that the location of the oxygen atoms in the molecule creates the hydrophilic portion on the hydrocarbon.[9]

The type and quantity of frother required depends upon the application. Quantities are generally less than 0.15 lb/ton of dry feed and in some cases as little as 0.015 lb/ton (0.008 kg/ton) are required. The chemical structure of collectors often gives them frother properties to a greater or lesser degree. In most nonmetallic mineral flotation with anionic collectors, such as soap or petroleum sulfanates, no additional frother is required.

Flotation Processing

In order for the concentration of minerals by froth flotation to be effective, discrete particles must exist and the particles must be fine enough to be buoyed by the air bubbles. Usually, the particles should be less than 35 mesh (500 μ), but separations up to 10 mesh (2000 μ) are possible in a few special applications. In mineral applications, froth flotation is preceded by a series of crushing, grinding and sizing steps in order to accomplish the desired particle liberation and sizing. The equipment needed for this preparation is generally the main capital cost of a processing plant and the major operating cost item.

In the selective froth flotation, one stage of treatment (rougher flotation) is not usually sufficient to obtain the desired product grade. The concentrate from the rougher flotation contains entrained gangue (waste) as well as other mineral species floated along with the desired mineral or minerals. Further treatment (cleaning), with or without additional reagents, is normally required. Cleaning stages can number from only one to as many as six or seven. These cleaning steps are usually set up as a closed circuit operation in which the cleaner tailing flows counter to the concentrate. The diagrammatic flow scheme in Figure 1 demonstrates the closed circuit operation where the rougher concentrate is pumped to the first cleaner and the first cleaner tailing combines with the feed to the rougher flotation.

The concentrate, or final product, obtained by flotation is in a slurry form that varies from 10 to over 30% solids. Final recovery of the product is the liquid-solid separation that consists of thickeners (settling tanks), filters and driers. In plant operation, this is an important section and should not be neglected in design and operation.

Another item that should not be neglected is the handling and disposal of the tailings (waste) slurry. At most plants, they are pumped (or flow

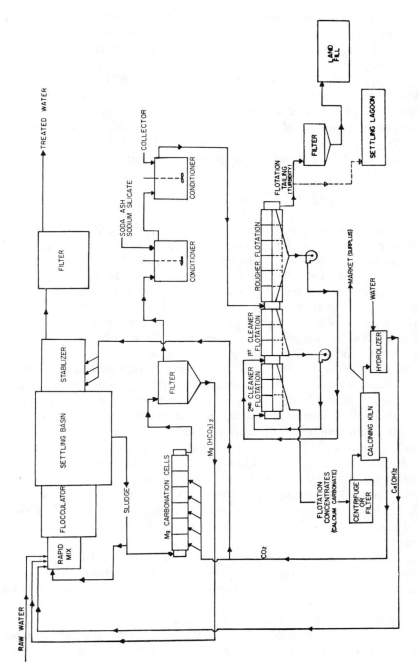

Figure 1. Flow diagram of flotation process for calcium carbonate recovery from water treatment sludge.

by gravity) to a pond where they are impounded and the water overflows for possible recycle. However, filtering and hauling to a landfill area may be required.

FLOTATION OF CALCIUM CARBONATE FROM SLUDGE

Development of Process

In the initial laboratory studies conducted in 1968, the results indicated that the calcium carbonate precipitated during water treatment could not be selectively recovered from the clays and other contaminants by flotation. However, these sludges were flocculated using alum [Al$_2$(SO$_4$)$_3$ 14 H$_2$O], which indiscriminately coated the calcium carbonate particles and prevented selective coating with the collecting agent.

At Black's request in 1971, additional testing was conducted to determine if precipitated calcium carbonate could be selectively floated from the clay-lime fraction. The first tests were conducted using a synthetic mix of pure precipitated calcium carbonate and clay. The results of these tests indicated that separation is possible. Tests were then conducted on a sludge derived from the Black process and obtained from the St. Louis County Water Department. The magnesium flocculant had been removed by treatment with carbon dioxide, filtration and washing. Although the flotation appeared to be selective, the insoluble portion was still high (13-15%) in the final product. The insoluble portion is residue silica left after conventional acid digestion. The clay or silica appeared to be physically locked with the calcium carbonate.

In 1973, sludge from the Johnson County, Kansas, pilot plant was obtained. Initial testing on this material was negative. A nonselective product was floated. Further examination showed that soluble magnesium bicarbonate, which dissolved during flotation, remained in the sludge. It consumed the reagents and also deposited an insoluble magnesium soap indiscriminately on the surface of the particles. Repulp washes prior to flotation dramatically improved the selectivity.

Results of subsequent laboratory and pilot plant testing were not consistent and a variety of pretreatments of the sludge were tried. The studies now show that the particle size of the precipitated calcium carbonate has a marked effect on the recovery and grade in the final product. Recycle of sludge from the settling basins not only aids in the operation of the treatment plant but also causes the particle size of the calcium carbonate to grow. Particle size increases of more than ten diameters have been obtained by recycle, which results in greatly improved flotation after the magnesium flocculant has been removed.

Some testing has been conducted that indicates that if the sludges from a water treatment plant are lagooned for a time prior to the flotation of the calcium carbonate, a selective float can be made even though the flocculants were not removed prior to lagooning.

FLOTATION TEST RESULTS DURING PROCESS DEVELOPMENT

Flotation testing of a variety of sludges from the Black process are shown in Tables II through VI.

Table II. Synthetic Mix: 80% Precipitated Pure Calcium Carbonate, 20% Clay; Two Stages of Retreatment (Cleaning) Required

Product	Percentage of Weight	$CaCO_3$, Percentage Assay	$CaCO_3$, Percentage Distribution
#2 Cleaner Concentrate	77.3	97	89
Combined 1st and 2nd Cleaner Tailing	14.7	57	10
Rougher Tailing	8.0	12	1
Head (Feed)	100.0	84	100

Table III. St. Louis County Water Department Sludge; Four Stages of Retreatment (Cleaning)

Product	Percentage of Weight	$CaCO_3$, Percentage Assay	$CaCO_3$, Percentage Distribution
#4 Cleaner Concentrate	59.6	79.5	77.9
Combined 1st Through 4th Cleaner Tailings	28.5	37.5	17.6
Rougher Tailing	11.9	23.0	4.5
Head (Feed)	100.0	60.8	100.0

Table IV. Sludge from Johnson County, Kansas, Water Treatment Pilot Plant; Magnesium Bicarbonate Washed out Prior to Flotation, Three Stages of Cleaning

Product	Percentage of of Weight	CaCO$_3$, Percentage	
		Assay	Distribution
#3 Cleaner Concentrate	56.2	87	63
Combined #1 Through #3 Cleaner Tailing	33.8	65	30
Rougher Tailing	10.0	52	7
Head (Feed)	100.0	77	100

Table V. Sludge Recycled in Johnson County, Kansas, Water Treatment Pilot Plant

Product	Percentage of Weight	CaCO$_3$, Percentage	
		Assay	Distribution
Concentrate	80.6	98.4	91.9
Tailings	19.4	34.4	8.1
Head (Feed)	100.0	86.0	100.0

Table VI. Lagoon Sludge from the Johnson County, Kansas, Water Treatment Plant

Product	Percentage of Weight	CaCO$_3$, Percentage	
		Assay	Distribution
#2 Cleaner Concentrate	64.7	96.0	72.0
#2 Cleaner Tailing	1.0	76.8	1.0
#1 Cleaner Tailing + Rougher Tailing	34.3	67.7	27.0
Head	100.0	86.1	100.0

SUMMARY OF THE CALCIUM CARBONATE FLOTATION PROCESS

During the investigations conducted on the development of the flotation process for the flotation of the calcium carbonate in water treatment sludge, the results have shown that the proper operation of the treatment plant and effective removal of the magnesium hydroxide and magnesium bicarbonate are vital for the separation to be effective.

The use of alum for flocculation produces flocs with indistinct particles. Therefore, a physical separation is not possible and flotation is not possible. With the use of magnesium carbonate for flocculation, the flocculant is removed for recycle, which results in distinct calcium carbonate particles that can be selectively coated for flotation. However, the complete removal of any magnesium compound either in solution or made water-soluble during flotation is necessary for effective flotation. Insoluble magnesium compounds in the clays or carbonates do not affect flotation except that the magnesium carbonate would be collected and floated along with the calcium carbonate. Soluble lime and magnesium form insoluble soaps with the collector, which tend to coat all surfaces indiscriminately, reduce selectivity and increase collector requirements.

Open circuit operation during water treatment results in very fine calcium carbonate particles. Selective dispersion of the clays and collection of the calcium carbonate have proven to be impossible. With recycle of sludge during water treatment, the particle size of the calcium carbonate can be increased from about $1\,\mu$ to larger than $10\,\mu$. The larger particles result in improved product grades and recovery, as well as reduced collector requirements. The collector required varies as a function of the particle surface or as an inverse function to the particle size.

Impurities in the raw water vary from place to place as well as with the seasons and result in modifications in the water treatment. Hence, variations in the sludge, which would require adjustments in the flotation, would also be anticipated. Therefore, laboratory investigations are required to develop a flow scheme for each plant, keeping in mind the variations that must be anticipated due to seasonal changes in the raw water. The balance of reagents required for optimum flotation conditions could vary over a wide range throughout the year, and proper adjustments depend on the experience of the operators. Chemical analyses of the sludge may serve as a guide, but they do not show the physical and other characteristics of the slurry that can profoundly affect flotation. However, the same reagents and basic flow scheme are used for the flotation of a majority of the sludges.

The flotation of calcium carbonate is conducted in a slurry at about 10% solids. First the slurry is conditioned (mixed) with modifiers to aid in dispersing the clays, depressing the gangue (silica, silicates) and adjusting the pH to 9.0-9.5 for effective soap flotation. Next, the collector is conditioned into the slurry prior to flotation. The initial separation in flotation produces a final tailing containing the gangue and a calcium carbonate rougher concentrate with some impurities. The rougher concentrate is then cleaned (retreated) two or three times to remove these entrained impurities and produce the final product. The number of cleaner stages depends on the character of the sludge. In plant operation, the flotation circuit is generally arranged for countercurrent flow of the cleaner tailing from each retreatment stage. The first cleaner tailing combines with the conditioned feed slurry to the rougher flotation, the second cleaner tailing combines with the rougher concentrate slurry to the first cleaner stage, and the third cleaner tailing (if a third stage is required) combines with the first cleaner concentrate to the second cleaner stage. The arrangement of the flotation circuit effectively and simply to obtain the countercurrent flow is shown in Figure 1. Note that the concentrates are pumped to subsequent cleaning stages and the cleaner tailings flow by gravity counter to the concentrates.

Soda ash (Na$_2$CO$_3$) and sodium silicate (Na$_2$SiO$_3$) have proven to be effective modifiers prior to flotation. Soda ash not only adjusts the pH to the desired 9.0-9.5 pH, but aids in precipitating soluble salts (such as calcium, magnesium, iron) that could interfere with flotation and aid in dispersion of the clays. Sodium silicate also aids in selective dispersion of the gangue, and prevents surfacing of the gangue with the collector (depressant). A variety of anionic nonsulfide collectors are available under various trade names for the flotation of the calcium carbonates. Sodium soaps of fatty acids, or tall oil, appear to be the most effective. Soaps made from unsaturated fatty acids such as oleic and linoleic are best. Saturated fatty acid soap (such as stearic) is poor for flotation. Tall oil is a fatty acid-rosin acid by-product obtained from the paper pulp industry. Sodium salts of petroleum sulfonates can also be used for the flotation of the calcium carbonate. They may be used in conjunction with the fatty acid soap to improve flotation.

The collectors used generally create sufficient froth so that no additional frothing agents are required to carry the concentrates from the flotation machines.

The general procedure for flotation of calcium carbonate from the sludges is outlined as follows. NOTE: Reagents additions are lb/ton (1.0 lb/ton = 0.5 kg/ton):

1. Repulp of filter cake after magnesium removal to about 10% solids.
2. Condition 2 to 5 min. Add 1.0-4.0 lb (0.50-2.0 kg) soda ash to 9.0-9.5 pH plus 0 to 2.0 lb (0 to 1.0 kg) sodium silicate.
3. Condition 2 to 5 min. Add 0.5-2.0 lb (0.25-1.0 kg) fatty acid soap equivalent.
4. Rougher flotation: 8-12 min.

5. First cleaner flotation: 6-8 min. Add: water to dilute to 8-10% solids; 0-1.0 lb (0-0.5 kg) soda ash; 0-0.5 lb (0-0.25 kg) sodium silicate; 0-0.5 lb (0-0.25 kg) fatty acid soap equivalent.

6. Second cleaner flotation: 6-8 min. Add: water to dilute to 8-10% solids. 0-0.5 lb (0-0.25 kg) soda ash.

7. Third cleaner flotation: 6-8 min. Add: water to dilute to 8-10% solids.

8. Pump rougher tailings to settling lagoon or filter and truck to landfill.

9. Filter or centrifuge concentrates.

FLOTATION MACHINES

A flotation machine is a mechanism designed to create conditions that will cause the hydrophobic particles to attach to air bubbles and rise to the surface of the slurry for removal. Inherent in this mechanism is the ability to handle wide variations in flow in order to give continuous operation.

The flotation machine is designed to:

- mix the slurry to maintain a suspension of particles
- generate and disperse air bubbles into the slurry
- remove the particle-laden froth from the slurry
- allow for continuous slurry feed and discharge.

Incorporated in the design of the flotation machine should be features that give low cost, high performance flotation.

- low power requirement per ton of ore processed
- low maintenance costs
- minimum floor space for flotation equipment
- minimum number of operators for large tonnages
- operation to give low reagent costs
- operation to give optimum mineral (particle) separation.

The AGITAIR® flotation machine* is one that includes the desirable requirements and features.

General Operating Principles and Details

In the AGITAIR flotation machine, shown in Figure 2, the slurry is mixed and the air is dispersed by an impeller-stabilizer combination. The impeller (Figure 3) is connected to a vertical hollow shaft that rotates the impeller and feeds low pressure air under the impeller plate. As the air

*Manufactured by The Galigher Company, P. O. Box 209, Salt Lake City, Utah 84110.

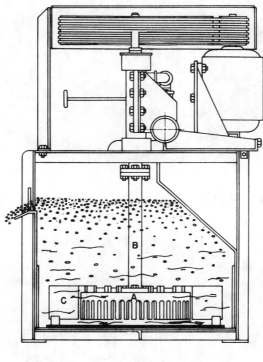

A. Impeller
B. Hollow Impeller Shaft
C. Stabilizer

Figure 2. Cross section of Galigher AGITAIR flotation machine.

flows between the rotating impeller plate and the stabilizer (Figure 4) it is broken into fine bubbles by the shearing action and is dispersed throughout the tank. As the pulp contacts the impeller, it is intensely agitated and aerated. The flow pattern directs the bubbles to the surface. The values collected in the froth column at the surface are discharged at the froth overflow lips.

Each cell is provided with an individually controlled air valve. Air pressure can be set between 0.75 psig (5.17 kN/m²) and 2 psig (13.79 kN/m²) depending on the depth and size of the machine and the pulp density. The pulp level can be adjusted to suit particular treatment requirements, but it is generally set from 2-12 in. (5-30 cm) below the top of the froth weir level.

Figure 3. Standard AGITAIR flotation machine impeller. Several other types are used.

Figure 4. A typical stabilizer.

Depending on the character of the pulp and the desired separation, changes in the impeller speed, diameter and length of the fingers may be desired. The amount of circulation and air dissemination can be varied over a wide range by adjusting the impeller speeds from 900 to 1750 ft/min (274 to 533 m/min). Impellers can be adjusted to meet individual requirements and special flotation problems.

The air can be supplied by positive displacement or centrifugal blowers, with the pressure controlled with a pressure relief valve. For economy, the air should be maintained only at adequate level, generally less than 2 psi (13.79 kN/m^2). The quantity of air required depends on the particle separation being made, which in turn influences the type of froth obtained. The volume of air measured by volume per unit of surface area of the flotation cell varies from about 1-3 cfm/ft^2 (1-3 m^3/m^2 min).

The power required depends on the rotational speed of impeller and percentage of solids of the pulp, as well as the particle size and specific gravity of the solids in the slurry. Increased solids, larger particles and high specific gravities require increased power. Only about 80% of the installed power is generally required for cell operation. The surplus power is required for starting under load, changes in feed, and process variances. The operating power is used for circulating pulp and dispersing air since the pulp flows by gravity and is not pumped within the machine.

Since capacities vary for different materials, many factors enter into calculations for estimating the number of cells required to treat a given tonnage of a particular ore in 24 hr. Before a final selection is made, all facts, based on the process development test work, pilot plant or actual plant operation, should be carefully evaluated. The number of cells used per row should give enough stages to ensure optimum recovery of values. Generally, at least six cells should be used in series.

Sizes of Machines Available

The AGITAIR flotation machine is available in many sizes to suit the varied demands of the industry from the laboratory units (Figure 5) to the largest plant units (Figure 6). The flotation machines are bolted together in order to allow for straight line flow of pulp through a properly proportioned row of cells. The flow is produced by a gravity head. The pulp enters through a feed box that distributes the flow over the entire width of the cell and leaves from the last cell through a tailing (discharge) box, which has an adjustable weir overflow for controlling pulp level and dart-type sand bleeder valves that remove oversized material. The dart valves can be manually or automatically controlled. Automatic control serves as a means for continuous pulp level control.

Some flotation circuits require a break in continuity for such purposes as (a) staging reagents, (b) separating cleaning from roughing cells and roughing from scavenger cells, and (c) reducing the effect of short circuiting. In these instances, a junction box is installed between cells in the row.

Gas Diffusion in Slurries

The design of a flotation machine must effectively disperse fine air bubbles throughout a slurry. This feature also makes the unit ideal for diffusion and adsorption of gases in a slurry.

Figure 5. The 3-in-1 Model LA 500 laboratory AGITAIR flotation machine with 500, 1000, and 2000 g interchangeable cells.

The flotation machine has the ability to mix and disperse a high volume flow of low pressure gas into an aqueous slurry with low power input. The gases, not diluted by air, can be fed through the hollow impeller shaft or via a pipe extending to the bottom of the tank. This feature has a number of possible applications in water and waste treatment, as well as cleaning and recovering material from industrial waste gases. A variety of gases can be handled, including air, oxygen, ozone, carbon dioxide, sulfur dioxide and ammonia.

SUMMARY

Although the flotation process may seem complex (and it can be for ore dressing), it is relatively simple, easily controlled, and practical for the recovery of magnesium carbonate in large water treatment plants. Although

Figure 6. No. 144X650 four-spindle AGITAIR flotation machine.

the magnesium carbonate process is applicable for coagulation in the smallest plants, the recovery of magnesium carbonate should be limited to plants that require at least a ton (0.9 ton) of coagulant per day.

REFERENCES

1. *Flotation Fundamentals and Mining Chemicals* (Midland, Michigan: The Dow Chemical Company, 1976), p. 3.
2. Gaudin, A. M. *Flotation* (New York: McGraw-Hill Book Company, 1932), pp. 1-7.
3. Fuerstenau, D. W. *Froth Flotation* (The American Institute of Mining, Metallurgical, and Petroleum Engineers, Inc., 1962), pp. 2, 39-42.
4. Fuerstenau, D. W. *Froth Flotation* (The American Institute of Mining, Metallurgical, and Petroleum Engineers, Inc., 1962), p. 91.
5. *Flotation Fundamental and Mining Chemicals* (Midland, Michigan: The Dow Chemical Company, 1976), p. 4.
6. *Flotation Fundamentals and Mining Chemicals* (Midland, Michigan: The Dow Chemical Company, 1976), p. 8.

7. *Flotation Fundamentals and Mining Chemicals* (Midland, Michigan: The Dow Chemical Company, 1976), pp. 11-13.
8. Fuerstenau, D. W. *Froth Flotation* (The American Institute of Mining, Metallurgical, and Petroleum Engineers, Inc., 1962), p. 59.
9. *Flotation Fundamentals and Mining Chemicals* (Midland, Michigan: The Dow Chemical Company, 1976), p. 9.

CHAPTER 7

CHEMICAL HANDLING AND FEEDING

Herbert E. Hudson, Jr., P.E.

President
Water and Air Research, Inc.
P.O. Box 1121
Gainesville, Florida 32602

INTRODUCTION

Previous chapters have touched on chemical handling and feeding, particularly in conjunction with the method of addition of the coagulant, discussed in Chapter 4. The topic of chemical handling and feeding has been discussed thoroughly in other publications and it is not necessary to present it here.

One important concept in analyzing chemical handling requirements is whether the chemical under consideration is one whose application may be interruptible. Those materials used for corrosion control, taste-and-odor control and fluoridation can have their application interrupted, and therefore less rigorous criteria apply to their storage space and handling equipment. On the other hand, substances such as coagulants and chlorine must not be interrupted under any circumstances. Adequate storage space, handling facilities, and adequate standby provisions for feeding are required for them.

CAPACITY AND RESERVE PROVISIONS

Depending on the number of application points required for each chemical, the number of feeding units is established. For example, in a

125

large water treatment plant, where it is necessary to have three or four application points for coagulant chemicals, it should be anticipated that a coagulant feeder would be provided for each application point, plus approximately 50% surplus units or not less than two standby units, whichever is least. Similarly with chlorinators, an adequate reserve should be provided, so that service can be maintained regardless of equipment servicing or maintenance requirements. A good general rule is that there should still be a unit in reserve, even when one of the largest units is out of service.

For interruptible materials, less rigorous standards may be applied. Some reserve equipment is desirable, but it need not meet all the criteria stated above.

Capacity of feeding equipment is based on two simultaneous requirements: capability to meet the maximum dosage required, and capability to feed that dosage under the maximum demand to be expected on the plant while still maintaining reserve units. When considered in connection with minimum feed rates for chemical feed equipment (usually about 10% of the maximum feed rate), these requirements may call for a greater range of chemical feed than a single unit can provide. In these circumstances, a larger number of units may be needed in order to cover the range requirement. Frequently, it is preferable to plan chemical feeding equipment for a shorter time of growth than the life of the entire plant. Additional units may then be added a few years after plant construction, as plant demands grow. (The same kind of range requirements apply to flow metering devices.)

STORAGE PROVISIONS

Storage provisions are outlined in Table I, which states them in terms of days of supply for both interruptible and noninterruptible materials. Table II indicates the storage characteristics that apply in design.

The arrangement of storage and chemical handling should make these operations as convenient and easy as possible. Use of hand or fork lift trucks should be maximized, with design eliminating lifting of heavy loads by operators. For example, personnel should not have to carry bagged materials up stairs.

LIQUID VS DRY CHEMICAL HANDLING

In the infancy of water purification, most chemical handling was in liquid form. Dry feeding equipment was then developed and used intensively for many decades. In the last decade or two, there has been a

Table I. Chemical Storage Space Criteria

Class of Chemicals	Noninterruptible	Interruptible
Examples of class	All chemicals used for disinfection; chemicals used for coagulation in treatment plants where raw water is polluted; softening chemicals.	Chemicals used for corrosion control; taste and odor; fluoridation.
Minimum stock to be maintained, in days	30	10
Additional allowance based on shipping time, in days[a]	2 times shipping time.	1.5 times shipping time.

[a]Additional allowance must be large enough to accommodate maximum size shipment (truckload, carload, fractional shipload).

Table II. Chemical Storage Type Criteria

Type of Storage	Dry	Wet
Handling Requirements	1. Allow for access corridors between stacks of packaged chemicals. 2. Palletize and use fork lift truck only in large installations.	1. Provide agitation for slurries such as carbon or lime: not less than 1 hp mixing per $100\ ft^3$. 2. Check manufacturers of feed equipment for pumps and pipe sizing.
Safety and Corrosion Requirements	1. Provide separated storage spaces for combustibles and for toxic chemicals, such as carbon or chlorine gas. 2. Provide ample space between stores of materials that may interact, such as ferrous sulfate and lime.	1. Double-check corrosion resistance of bulk storage linings. 2. Isolate hazardous or toxic solutions such as fluosilicic acid. 3. Prefer below-ground or outdoor storage.

return to the use of liquid chemicals rather than dry for several reasons. First, the equipment for handling liquids is commonly simpler and less costly. Second, handling chemicals in liquid form eliminates numerous serious dust problems.

Systems for handling dry chemicals are well known and adequately described in manufacturers' literature. Systems for handling wet chemicals are not as widely known or described. They include variable-head orifice-type feeders, usually regulated by means of float valves; constant-head orifice feeders in which the orifices may be adjustable; and various types of rotating dipper units. Additionally, chemicals that do not cause coatings or encrustation can be metered through tapered flow tubes, or rotameters. In a number of recent installations, the chemical has been transferred from bulk storage solution to a pressurized day tank, from which it is fed. A pneumatic drive may then be applied to the day tank, to force the liquid through a rotameter, or through some other type of flow device such as a magnetic flow meter. Systems of this type lend themselves to flow automation, enabling the chemical feed to be proportioned to the dosage required and to the flow through the plant.

With either liquid or dry feed chemical equipment, it has been fairly common to pace the rate of feed of the chemical by starting and stopping the feeding equipment. This can be a disadvantage when used for coagulant or coagulant aid feeds, since the feed may be interrupted frequently and continuously, thus overdosing one stream of water and underdosing a succeeding one. This can have serious adverse effects on chemical treatment reactions.

HANDLING CHLORINE

Until recent years, it was a widespread practice to meter chlorine gas under vacuum at the chlorinator, and to dissolve it in water passing through an eductor located at the feeder. The strong chlorine solution was then piped under pressure to the application point. In recent years, the practice has changed, and the vacuum is produced in an eductor located very close to the application point. The gas is still metered under vacuum, but is transported to the eductor as a gas, under vacuum. This eliminates the effects of the corrosiveness of the solutions, which necessitated transport of the solutions in large costly rubber-lined pipe. The gas under vacuum can be transported safely in smaller diameter plastic pipes. This cannot be done with chlorine gas under pressure because it diffuses outward through plastics. Dry gas under pressure is handled in steel pipes.

Recently there has also been a tendency to change from liquid chlorine to hypochlorites. This greatly reduces the hazards of leaks in the plant or hazards in transport from factory to the plant.

LAYOUT OF CHEMICAL HANDLING SPACE

Adequate working space should be provided around all feeding devices so that equipment can be maintained easily. Piping should be arranged to facilitate access between units and around them. In general, the feeding equipment should be housed in a room separate from all other functions. Adequate aisle space should be provided in storage areas to enable easy access and to make it possible to handle the chemicals on a first-in first-out basis.

Such facilities should be arranged so that the chemicals are easily moved into storage, out of storage, to the chemical feeding day tanks or dry feeders, and from there as quickly and expeditiously as possible, and by the shortest route, to the application point. This means that every treatment plant is quite likely to have unique features in the arrangement and layout of the chemical handling and feeding equipment, depending on the plant arrangement dictated by site consideration.

Additionally, chemical feed facilities should be placed at a location where they can be frequently and readily checked by supervisory personnel responsible for the operation of the entire plant.

CONTROL OF CHEMICAL FEEDS

For important chemicals like coagulants and chlorine, a double-check system on feed rates should be employed. The cumulative feed rate should be checked at one-hour intervals, and no less than once per shift. The momentary feed rate should also be checked at least hourly; for example, if alum solution is being fed, the change in level in the day tank should be regularly recorded, and, at even more frequent intervals, samples of the quantity of solution being fed should be collected over accurately timed intervals. Similarly, the chlorine feeder rate and the cylinder weight loss should also be checked. All such readings should be recorded and preserved on suitable forms and, ultimately, balanced against shipments received and inventory on hand. Such primary readings are much preferable to reliance on secondary standards such as pH or residual chlorine attained. The latter determinations should, however, also be made and preserved.

Where dry feed equipment is used, timed pan catches should be made regularly and similarly balanced against feed hopper weights, weights dumped into the feed hopper, and ultimately against purchases and inventory.

Most of the foregoing is less rigorously applied to most interruptible materials, with the exception of fluoride materials. With them, tight control of dosages and measurement of residuals is important to establish that there is no overdose.

Clearly, feeding equipment should be designed to make all of the above operations convenient and easy, so that they can be carried out quickly and without excessive burden on operating personnel.

REFERENCES

1. American Water Works Association. *Water Quality and Treatment,* 3rd ed. (New York: McGraw-Hill Book Company, 1971), Chapter 17.
2. American Society of Civil Engineers, American Water Works Association and Conference of State Sanitary Engineers. *Water Treatment Plant Design* (New York: American Water Works Association, Inc., 1969), Chapter 14.

CHAPTER 8

DESIGN OF RAPID MIX UNITS

A. Amirtharajah, Ph.D.
Associate Professor
Department of Civil Engineering and Engineering Mechanics
Montana State University
Bozeman, Montana 59717

INTRODUCTION

The rapid mix unit is often the first step in a water treatment process train, and essentially the unit provides the hardware necessary for the initial stages of the coagulation process. Ideally, engineers would like to design rapid mixers from a knowledge of the basic hydrodynamic parameters, geometry, and molecular properties of the water to be treated, and a knowledge of the kinetics of the chemical processes that occur. However, the complexity of the turbulent fluid field in most rapid mixers has made quantitative treatment of rapid mixing difficult, and the rapid mix unit is largely designed on empiricism, even though we have considerable insight into the coagulation process.[1-3]

The coagulation-flocculation process is considered by O'Melia in Chapter 4 wherein the basic requirements of rapid mix units are noted. An understanding of the process of coagulation-flocculation is an essential prerequisite for the design of the components of a rapid mix unit. This chapter presents some theoretical considerations for the rapid mixing operation and summarizes the available alternatives in design, including the magnitude of the parameters recommended in various references. Typical illustrative design calculations are included.

SOME THEORETICAL CONSIDERATIONS

The major consideration in rapid mixing has been for uniform dispersion of the coagulant with the raw water in order to avoid over and under treatment of the water. This rationale has been developed intuitively and qualitatively. Recently, some attempts[4,5] have been made to quantify these considerations in terms of the kinetics of the reactions and the hydrodynamics of the fluid field.

It is necessary to examine rapid mixing in terms of the coagulation reactions. Figure 1 summarizes in schematic form the predominant mechanisms in coagulation. An analysis of the kinetics of the reactions is especially important. Coagulation in water treatment occurs predominantly by two mechanisms: (a) adsorption of the soluble hydrolysis species on the colloid and destabilization or (b) sweep coagulation where the colloid is entrapped within the precipitating aluminum hydroxide. The reactions in adsorption-destabilization are extremely fast and occur within microseconds without formation of polymers, and within 1 sec if polymers are formed.[1,4,5] Sweep coagulation is slower and occurs in the range of 1-7 sec.[6] The difference between these two modes of coagulation in terms of rapid mixing is not delineated in the literature.

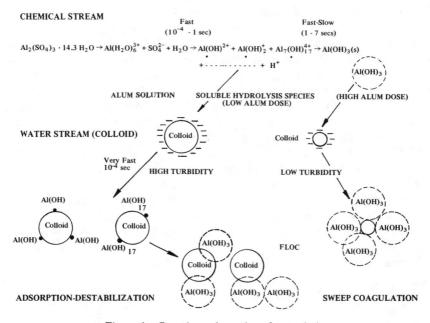

Figure 1. Reaction schematics of coagulation.

An analysis of the two modes of destabilization implies that for adsorption-destabilization it is imperative that the coagulants be dispersed in the raw water stream as rapidly as possible (less than 0.1 sec) so that the hydrolysis products that develop in 0.01 to 1 sec will cause destabilization of the colloid. Based on this theoretical analysis, Hudson and Wolfner[7] recommended instantaneous in-line blenders for rapid mixing. Vrale and Jorden[8] showed that for this type of coagulation, a plug flow tubular reactor with the chemicals injected into the fluid stream was the most efficient. They also showed that a typical mechanical mixer, which is classified as a backmix reactor, is comparatively inefficient. Recent work[9,10] at the University of California at Berkeley also confirms the superiority of the diffuser-injection system on a pilot plant scale.

If a careful analysis of sweep coagulation is made, it is evident that extremely short dispersion times are not as important as in adsorption-destabilization since coagulation is due to colloid entrapment amidst the aluminum hydroxide precipitate. A typical backmix reactor would perform quite well under these conditions. Stumm and Morgan[11] have estimated that the rapid formation of precipitates occurs when the solution is oversaturated by two orders of magnitude. For the reaction

$$Al(OH)_3(s) = Al^{3+} + 3 OH^- \quad K_s = 10^{-33}$$

the extent of oversaturation is given by the ratio

$$\frac{[Al^{3+}][OH^-]^3}{10^{-33}}$$

which must be $\sim$ 100 or greater. At a pH = 6.5, the concentration of alum to produce an extent of oversaturation of 100 is 0.001 mg/l [as $Al_2(SO_4)_3 \cdot 14.3 H_2O$]. Thus, minimum alum doses under normal water treatment practice ($\sim$ 10 mg/l) are considerably oversaturated, nearly 10^4 times beyond the oversaturation needed for rapid precipitation, and can easily produce sweep floc.

Most of the experiments of Letterman et al.[6] were under conditions similar to sweep coagulation. They found that the optimum time of mixing was in minutes in contrast to the formation of the aluminum hydroxide precipitate which occurred in 1 to 7 sec. Thus, instantaneous rapid mixing was not crucial to good flocculation. In addition, these experiments would also imply that not only is the time for the formation of the precipitate important, but also the initial stages of flocculation need to be enhanced by rapid mixing.

Reported studies do not make this distinction for the time of rapid mixing in terms of the predominant mode of coagulation. Failure to realize this important distinction and the fact that there is little fundamental understanding of the interaction of the chemical kinetics and the hydro-dynamics of turbulent mixing has led to a preponderance of conflicting recommendations in the literature. Some[7,8,12] suggest instantaneous mixing (< 1 sec) based on chemical theories of adsorption-destabilization as best, while others[6,13] recommend detention times of minutes as promoting best floc characteristics for subsequent flocculation and settling. Probably, both mechanisms occur simultaneously in typical water treat-ment practice, which causes considerable difficulty in identifying the pre-dominant mechanism and relating it to the optimum rapid mixing operation. Some states[14] require a minimum of 30 sec rapid mix, which does not meet either the need for the adsorption-destabilization reactions nor the extended time for sweep coagulation and initiation of flocculation. Based on present day theory, this seems to be an extremely poor recommendation.

EQUATIONS FOR DESIGN

It is generally recognized that the velocity gradient or G-value concept is a gross, simplistic and totally inadequate parameter for design of rapid mixers. However, until a more detailed understanding of rapid mixing is possible, this parameter will continue to be used as the only available means for designing the hardware of a rapid mix unit.

The power requirements are calculated from the equation given by Camp and Stein.[15]

$$G = \left(\frac{P}{\mu V}\right)^{1/2} \tag{1}$$

in which G = velocity gradient in sec^{-1}
P = power input in ft lb/sec
μ = viscosity in lb sec/ft^2
V = volume in ft^3

In metric units,

G = sec^{-1}
P = watts, *i.e.*, N m/s
μ = Pa.s *i.e.* N s/m^2
V = m^3

For turbulent conditions, Rushton[16] developed the following mathe-matical relationship for the power requirements:

$$P = \frac{k}{g} \rho N^3 D^5 \tag{2}$$

in which k = constant
ρ = density of fluid in lb/ft^3
N = revolutions/sec
D = diameter of impeller in ft

The values of k range from 1.0 for a three-bladed propeller to 6.30 for a turbine with six flat blades.

From Equations 1 and 2, it is seen that

$$G \propto N^{3/2} \tag{3}$$

Letterman et al.[6] recently suggested the following empirical relationship for optimizing the rapid mix operation.

$$GT_{opt}C^{1.46} = 5.9 \times 10^6 \tag{4}$$

in which T_{opt} = optimum rapid mix period, in sec, and C = concentration of alum dose in mg/l. The equation was developed for alum coagulation of an activated carbon colloid suspension and had a correlation coefficient of 0.92. While recognizing the fact that the equation has not been tested under a variety of combinations of colloid-coagulant systems, it should be noted that the equation does provide a first step in the optimum design of rapid mix units.

The above four mathematical relations enable the design of a mechanical rapid mixer to be made. They can be used to determine the design of the tank and the power requirements of the motor. Their use is illustrated in the example to follow.

It should be appreciated that the above equations have limitations: they do not describe the process or operation of rapid mixing in realistic and well understood terms; however, they are the best presently available models in environmental engineering.

ALTERNATIVES IN DESIGN

A review of typical designs and the available literature suggest that the following alternative units can be used for rapid mixing.

Mechanical Mixer

The most commonly used unit is a mechanical mixer with a propeller impeller. It is effective, has little head loss, and is unaffected by flow variations. The requirements in Table I are suggested in *Water Treatment Plant Design*.[17]

Table I. Contact Time and Velocity Gradient for Rapid Mixing

Contact Time, sec	20	30	40	>40
Velocity Gradient, G, sec^{-1}	1000	900	790	700

This text[17] also notes that typical practice in design is to provide 10 to 30 sec contact time. Gemmell,[18] in *Water Quality and Treatment*, stated that no clear-cut guidelines exist for determining the power dissipation or detention time required to disperse chemicals in a flow of water, and suggested 20 sec or less detention time with a power input of 1 to 2 hp (0.75-1.5 kW) for each cfs (0.028 m³/sec) of flow. He agreed with the guidelines in *Water Treatment Plant Design*.[17] Some typical mechanical rapid mix units are shown in Figure 2A.

Camp's studies,[13] made with Boston tap water at ferric sulfate doses of 15 mg/l at a pH of 6, clearly approximated sweep floc precipitation conditions and indicated that with velocity gradients of 700 to 1000 sec⁻¹ using a flat-bladed propeller mixer, the floc volume concentration reached a maximum value after 2 to 2.5 min of rapid mixing and remained constant for mixing periods extending to 20 min. The results are significant because precipitation of the ferric hydroxide floc was measured to occur within 8 sec.

Letterman *et al*.,[6] in an extensive recent study with alum as coagulant, also found that the optimum period of rapid mix was 2.5 min at a G value of 1000 sec⁻¹ for turbidity produced by activated carbon concentrations of 20 to 1000 mg/l. They developed Equation 4 already presented. While this equation has not been tested under many different conditions, it does provide a first step toward optimum design of rapid mix units. The dominant mechanism in this study appeared to be sweep coagulation based on estimations of colloid surface area.

Vrale and Jorden[8] tested several flow-through type rapid mixers and concluded that the mechanical type backmix reactor was very inefficient for rapid mixing in comparison with a tubular reactor for adsorptive destabilization.

In designing a rapid mix unit with mechanical mixers, the available limited literature evidence indicates:

- superiority in performance of a square vessel over that of a cylindrical vessel[6]
- stator baffles provide improvement[13,16]
- a flat-bladed impeller performs better than a fan or propeller impeller[6]
- chemicals introduced at agitator blade level enhance coagulation.[19,20]

The mechanical mixer is also most amenable to variation in operation by changing the rotation speed. Thus, it can be easily operated for lower G values, which is especially important for the addition of polyelectrolytes and variations in raw water quality.

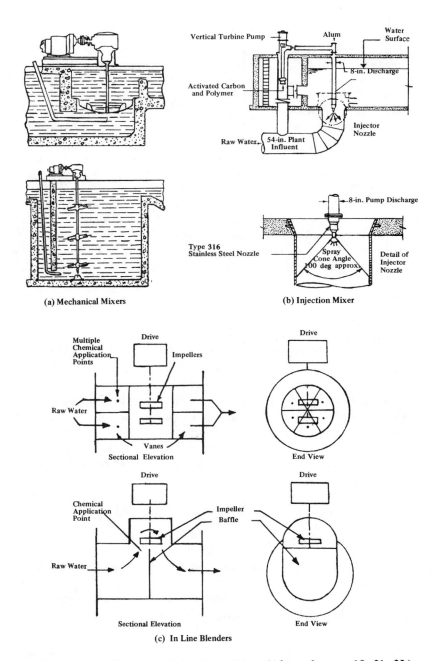

(a) Mechanical Mixers

(b) Injection Mixer

(c) In Line Blenders

Figure 2. Rapid mixers—design alternatives. (After references 18, 21, 22.)

In-Line Blenders

The rapid rate of the adsorption-destabilization reactions was the rationale for instantaneous rapid mix using in-line blenders. Hudson and Wolfner[7] suggested a rapid mix time of less than 30 sec with high-powered mixing devices. A typical example was a 24-in. (0.61 m) unit applying 10 hp (7.45 kW) to a flow of 20 mgd (876 liter/sec) in a fraction of a second. The value of G developed was in the range of 3000 to 5000 sec[-1]. In a recent review Hudson[21] suggests using in-line blenders with residence times of 0.5 sec and water hp of 0.5 (0.37 kW) per mgd (43.8 liter/sec) of flow. Head losses in the manufactured units range from 1 to 3 ft (0.3 to 0.9 m).

Kawamura[12,22] favors the use of in-line blenders for three reasons: (1) they provide good instantaneous mixing with little short-circuiting, (2) there is no need for head loss computations, and (3) cost can be reduced by omitting a conventional rapid mix facility. Figure 2C shows diagrammatically some typical manufactured units.*

Hydraulic Mixing

European practice tends to favor the use of hydraulic jumps for mixing. The plant flow is metered at a Parshall flume and a hydraulic jump is incorporated immediately downstream of the flume by an abrupt drop. The coagulants are introduced upstream of the standing wave, and typical residence times are 2 sec with G-values of 800 sec[-1].

Mathematical relations for determining the relative height of the drop required to stabilize a jump for any given combination of discharge, upstream depth, and downstream depth are given by Chow.[23] Typical head losses are 1 ft (0.3 m) or greater. The chief advantage of the hydraulic jump rapid mixer is the absence of any mechanical equipment, which should always be considered for design of treatment plants in developing nations where expertise for maintenance of mechanical devices is limited and replacement parts are difficult to obtain.

Diffusers and Injection Devices

In a recent basic study, Stenquist and Kaufman[9] used a rectangular grid of diffusers consisting of a series of tubes with orifices to develop the turbulence for mixing in downstream wakes. The coagulants were fed through the orifices. This comparative study indicated that the

---

*Walker Process Equipment, Chicago Bridge and Iron Company, Aurora, Illinois, and Philadelphia Gear Corporation, King of Prussia, Pennsylvania.

diffuser system was superior to mechanically stirred flash mixers. A minimum orifice density of 1/in.^{2} was necessary for significant improvement in mixing. Wilson[10] showed that a venturi-type tubular mixer performed much better than a mechanical mixer in promoting subsequent orthokinetic flocculation. Vrale and Jorden[8] also showed that a tubular orifice reactor proved superior in performance to the backmix reactor. The reactor functioned with low head loss at an approximate G value of 1000 sec^{-1}.

Typical units used in pilot-scale laboratory studies are shown in Figure 3. These mixing devices have significant theoretical advantages that merit their use in practice. However, their utilization and effectiveness in plant-scale use has still to be demonstrated. Possible disadvantages that may be limiting constraints for practical use are clogging of the orifices and inflexibility in operation since the turbulence cannot be varied.

Kawamura[22] describes the design of an injection nozzle-type flash mixing unit. Design criteria suggested were:

- G = 750 to 1000 sec^{-1}
- dilution ratio at maximum alum dosage = 100:1
- velocity at injection nozzles = 20 to 25 ft/sec (6-7.6 m/sec)
- mixing time = 1 sec.

A valve on the chemical injection pump discharge line enabled flexibility in adjusting the discharge rate to changing raw water conditions or coagulants. Dimensions and calculation procedures for an 82-mgd (3592 liter/sec) facility are given by Kawamura.[22] The mixer is shown in Figure 2B.

DATA NEEDED FOR DESIGN

The amount of data needed for design would vary with variability of raw water characteristics and magnitude of the project, and either greater variability or larger size would require an increase of data. The following represents minimum requirements for the design of a rapid mix unit:

- raw water analysis representative of best and worst conditions in quality
- maximum and minimum flow rates
- typical jar test results of all water samples.

The collection of detailed predesign data is described by Trussell in Chapter 3.

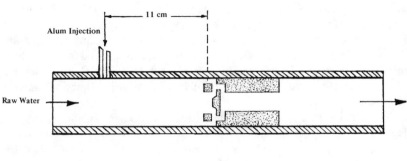

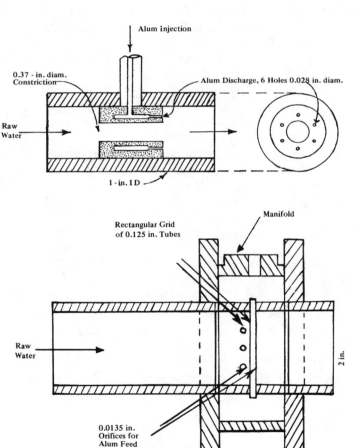

Figure 3. Schematics for diffuser-injection mixers. (After references 8,9.)

ILLUSTRATIVE DESIGN PROBLEMS

Example 1

The water supply for a town is taken from a river with considerable variations in quality. The raw water analyses for a year are shown in Table II, with typical coagulant doses obtained from jar tests. The design flows for 24-hr operation are: average day = 1.0 mgd (43.8 liter/sec), maximum day = 1.5 mgd (65.7 liter/sec). Design the rapid mix unit for the treatment plant.

Table II. Data for Design in Example 1 River Source

Occurrence, %	11	34	28	13	9	3	2
Raw Water Analyses							
Turbidity, TU	28	45	81	126	140	180	> 200
Alkalinity, mg/l as $CaCO_3$	86	132	92	84	93	77	82
Temperature, °C	13	10	8	18	18	16	
Jar Tests							
Alum,[a] mg/l as alum	18	27	36	28	41	31	36
pH	6.4	6.7	6.4	6.4	6.4	6.2	6.4

[a]Most jar tests indicated an optimum alum dose and possible charge reversal and restabilization.

Design

The raw water turbidity for 62% of the year ranged from 45 to 81 TU, and 27% of the time it was > 126 TU.

According to the following stoichiometry,

$$\underbrace{Al_2(SO_4)_3 \cdot 14.3H_2O}_{600} + \underbrace{3Ca(HCO_3)_2}_{3 \times 100\ CaCO_3} \rightarrow 3CaSO_4 + 2Al(OH)_3 + 6CO_2 + 14.3H_2O$$

the optimum alum doses in the range 27 to 36 mg/l would require alkalinity (as $CaCO_3$) of 300/600 x (27 to 36), *i.e.*, 13.5 to 18.0 mg/l. Sufficient alkalinity is available in the raw water. Alum would be the only coagulant, as confirmed by the jar tests. The pH range for optimum alum coagulation from the jar tests was approximately 6.4.

The raw water turbidity was high (> 126 TU) for 27% of the year and the jar tests indicated that there was an optimum coagulant dose and probable restabilization by charge reversal. *Therefore, adsorption-destabilization was the predominant mechanism. Hence, instantaneous mixing or an in-line blender would be an appropriate design.*

In English Units

Choose a typical commercial unit at a water hp of 0.5/mgd of flow. For a maximum day demand of 1.5 mgd, use an in-line blender of 1.5 x 0.50/1.0 = 0.75 water hp.

Assume a residence time of 0.4 sec in blender. (Check on typical times of commercial unit). The lowest temperature gives the highest viscosity and hence the lowest velocity gradient.

Power input:
  P = 0.75 hp x 550 ft lb/sec hp = 412.5 ft lb/sec
Volume treated:
  V = 1.5 mgd x 1.547 cfs/mgd = 2.32 cfs
Volume to which power is supplied at residence time of 0.4 sec:
  V = 2.32 cfs x 0.4 sec = 0.93 ft^3
Viscosity at 8°C:
  μ = 1.387 centipoises x 2.088 x 10^{-5} lb sec/ft^2/centipoise = 2.90 x 10^{-5} lb sec/ft^2

Therefore, the velocity gradient is:

$$G = \left(\frac{P}{\mu V}\right)^{1/2} = \left(\frac{412.5}{2.90 \times 10^{-5} \times 0.93}\right)^{1/2} = 3900 \text{ sec}^{-1}$$

Similarly, at a temperature of 18°C,

$$G = \left(\frac{412.5}{2.2 \times 10^{-5} \times 0.93}\right)^{1/2} = 4500 \text{ sec}^{-1}$$

The range of G values lies within the recommended values of 3000 to 5000 sec^{-1}. If the residence time in the blender is not given, an alternate method of calculation is to estimate the effective volume of blender from dimensions of the unit and subtraction of impeller volume.

In Metric Units

Choose a commercial unit at 0.37 kW per 43.8 l/sec of flow.

Power input of in-line blender:
  = 65.7/43.8 x 0.37 = 0.57 kW
  = 560 N m/sec or watts
Volume to which power is supplied:
  $= \dfrac{0.4 \text{ sec} \times 65.7 \text{ l/sec}}{10^3 \text{ l/m}^3}$
  = 0.026 m^3
Viscosity at 8°C:
  μ = 1.387 centipoises = 1.387 x 0.001 Pa sec or N sec/m^2 = 1.387 x 10^{-3} N sec/m^2

Therefore, the velocity gradient is:

$$G = \left(\frac{P}{\mu V}\right)^{\frac{1}{2}} = \left(\frac{560}{1.387 \times 10^{-3} \times 0.026}\right)^{\frac{1}{2}} = 3900 \text{ sec}^{-1}$$

Similarly, at a temperature of 18°C:

$$G = \left(\frac{560}{1.060 \times 10^{-3} \times 0.026}\right)^{\frac{1}{2}} = 4500 \text{ sec}^{-1}$$

Example 2

The water supply to a town is taken from a lake. The raw water is of reasonably uniform quality, and typical analyses for a year are shown in Table III. With design flows similar to Example 1, design the rapid mix unit.

Table III. Data for Design of Example 2 Lake Source

Occurrence, %	24	33	21	14	8
Raw Water Analyses					
Turbidity, TU	12	21	35	49	60
Alkalinity, mg/l as $CaCO_3$	36	29	34	31	29
Temperature, °C	14	19	22	18	20
Jar Tests					
Alum Dose,[a] mg/l as alum	36	28	31	27	29
Lime Dose, mg/l as $Ca(OH)_2$	17	13	14	12	12
pH	6.6	6.4	6.5	6.7	6.5

[a]Most jar tests did indicate a decrease in turbidity, which remained constant at higher alum doses, *i.e.*, possible restabilization was not indicated.

Design

The raw water turbidity remained less than 35 TU for 78% of the time and less than 49 TU 92% of the time. In this instance, direct filtration with or without contact coagulation should be seriously considered as a design alternative. However, analysis here will be for a conventional coagulation-sedimentation-filtration system.

The jar tests indicated alkalinity addition was needed and alum doses were in the confined range of 27 to 36 mg/l with little restabilization. *Hence, the major mechanism appears to be sweep coagulation.* A backmix reactor would be efficient, provide for alum precipitation, and optimize the initial formation of flocs. Using Equation 4, with a mean alum dose of 31 mg/l,

$$GT_{opt} = \frac{5.9 \times 10^6}{31^{1.46}} = 39200$$

It must be emphasized that this equation, based on experiments with one type of a colloid-coagulant system, gives order of magnitude quantities that must also be cross checked with values used in typical practice.

$$\text{For } G = 1000, \; T_{opt} = 39.2 \text{ sec}$$
$$\text{For } G = 700, \; T_{opt} = 55.7 \text{ sec}$$

Use a design value of 50 sec for the mean residence time and a G value of 700 sec⁻¹. As a check, *Water Treatment Plant Design*[17] suggests a mixing time greater than 40 sec for $G = 700$ sec⁻¹, and 20 sec for $G = 1000$ sec⁻¹. Hence, the values above are satisfactory.

In English Units

The volume of the rapid mix chamber for 1.5 mgd = 2.32 ft³/sec x 50 sec = 116 ft³. The actual dimensions of the chamber should be based on the raw water influent conduit (pipe or channel) dimensions. Always design for a minimum of two units.

Assuming the layout of the treatment plant is best for two units,

$$\text{Volume of each rapid mix chamber} = \frac{116}{2} = 58 \text{ ft}^3$$

Using square chamber of 4-ft side dimensions,

$$\text{Depth of chamber} = \frac{58}{4 \times 4} = 3.6 \text{ ft}$$

Design dimensions of each rapid mix tank = 4 ft x 4 ft x 4 ft. Use a flat-bladed mixer and feed chemicals at the level of the blades. The actual dimensions of the mixer are to be obtained from manufacturer's catalog.

The required motor horsepower can be computed from Equation 1. In this equation, the volume of each tank is 4 x 4 x 4 = 64 ft³. Since the design is for a minimum value of G (700 sec⁻¹), use the highest viscosity, which occurs at the lowest temperature of 14°C. This means that during summer conditions, a higher velocity gradient will be developed with the same power.

$$\text{Viscosity at } 14°C, \; \mu = 1.1748 \text{ centipoises} \times 2.088 \times 10^{-5} \text{ lb sec/ft}^2/$$
$$\text{centipoise} = 2.45 \times 10^{-5} \text{ lb sec/ft}^2$$

By rearranging Equation 1,

$$\text{Power required, } P = G^2 \mu V = 700^2 \times 2.45 \times 10^{-5} \times 64 \text{ ft lb/sec} =$$
$$768.3 \text{ ft lb/sec}$$

$$\text{Water hp} = \frac{768.3}{550} = 1.40 \text{ hp}$$

Assuming a motor efficiency of 80%,

Motor horsepower $= \dfrac{1.40}{0.8} = 1.75$ hp. Use standard manufactured unit.

Design horsepower of motor = 2 hp.

Equation 3 gives the variation of G with the rpm of rotation as $G \propto N^{3/2}$. To accommodate the range of G from 700 to 1000 sec^{-1},

Use a speed range $= \dfrac{N_1}{N_2} = \left(\dfrac{1000}{700}\right)^{2/3} = 1.27$

Design variable speed drive of mixer for a ratio of 1.3:1

In Metric Units

The volume of the rapid mix chamber for 65.7 l/sec

$$= \frac{65.7 \text{ l/sec} \times 50 \text{ sec}}{10^3 \text{ l/m}^3} = 3.29 \text{ m}^3$$

Using two units,
Volume of each chamber $= 3.29/2 \text{ m}^3$
$= 1.65 \text{ m}^3$

Using square chamber of 1.2 m side dimensions,
Depth of chamber $= \dfrac{1.65}{1.2 \times 1.2} = 1.15$ m

Design dimensions of each rapid mix tank = 1.2 m x 1.2 m x 1.2 m
Volume of each tank $= 1.2 \times 1.2 \times 1.2 = 1.728 \text{ m}^3$
Viscosity at lowest temperature of 14°C
$= 1.1748$ centipoises $= 1.1748 \times 10^{-3}$ Pa sec or N sec/m^2

Using rearranged Equation 1,

Power required, $P = G^2 \mu V,$
$= 700^2 \times (1.1748 \times 10^{-3}) \times 1.728$
$= 995$ watts or N m/sec

Motor power with 80% efficiency $= \dfrac{995}{0.8 \times 1000}$ kW
$= 1.24$ kW

Using standard manufactured unit,

Design power of motor   = 1.5 kW

The above design allows flexibility in operation for G values ranging from 700 to 1000 sec^{-1}. In this example the flexibility in operation is not of dominant importance since the raw water quality is reasonably consistent. However, for variability in raw water quality, such flexibility is essential for optimum operation of the plant.

Two important considerations in the design of the rapid mixer need to be noted as addenda.

1. The alum solution lines should be designed for feeding an alum feed concentration of 0.5% or less into the raw water. Several studies[6,12,20] have shown that weak solutions provide better coagulation than higher concentrations. The lime for alkalinity may be added upstream or downstream of the rapid mixer, although limited tests have indicated better performance when added downstream.
2. The siting of the rapid mix units must be as close as possible to the coagulant feed tanks to minimize clogging of chemical feed lines, especially for lime.

REFERENCES

1. O'Melia, C. R. "Coagulation and Flocculation," *Physicochemical Processes for Water Quality Control*, W. J. Weber, Ed. (New York: Wiley-Interscience, 1972).
2. Stumm, W. and C. R. O'Melia. "Stoichiometry of Coagulation," *J. Am. Water Works Assoc.* 60:514-539 (1968).
3. Stumm, W. and J. J. Morgan. "Chemical Aspects of Coagulation," *J. Am. Water Works Assoc.* 54:971-994 (1962).
4. Hahn, H. H. "Effects of Chemical Parameters Upon Rate of Coagulation," unpublished Ph.D. dissertation, Harvard University, Cambridge, Massachusetts (1968).
5. Hahn, H. H. and W. Stumm. "Kinetics of Coagulation with Hydrolyzed Al(111)," *J. Colloid Interface Sci.* 28:134-144 (1968).
6. Letterman, R. D., J. E. Quon and R. S. Gemmell. "Influence of Rapid-Mix Parameters on Flocculation," *J. Am. Water Works Assoc.* 65:716-722 (1973).
7. Hudson, H. E., Jr. and J. P. Wolfner. "Design of Mixing and Flocculating Basins," *J. Am. Water Works Assoc.* 59:1257-1267 (1967).
8. Vrale, L. and R. M. Jorden. "Rapid Mixing in Water Treatment," *J. Am. Water Works Assoc.* 63:52-58 (1971).
9. Stenquist, R. and W. Kaufman. "Initial Mixing in Coagulation Processes," Sanitary Engineering Research Laboratory, Report No. 72-2, University of California, Berkeley (1972).
10. Wilson, G. E. "Initial Mixing and Turbulent Flocculation," unpublished Ph.D. dissertation, University of California, Berkeley, California (1972).
11. Stumm, W. and J. J. Morgan. *Aquatic Chemistry* (New York: Wiley-Interscience, 1970), pp. 230-231.
12. Kawamura, S. "Coagulation Considerations," *J. Am. Water Works Assoc.* 65:417-423 (1973).
13. Camp, T. R. "Floc Volume Concentration," *J. Am. Water Works Assoc.* 60:656-673 (1968).
14. *Recommended Standards for Water Works* (Ten State Standards). Health Education Service, Albany, New York (1972).
15. Camp, T. R. and P. C. Stein. "Velocity Gradients and Internal Work in Fluid Motion," *J. Boston Soc. Civil Eng.* 30:219-237 (1943).
16. Rushton, J. H. "Mixing of Liquids in Chemical Processing," *Ind., Eng. Chem.* 44:2931-2936 (1952).

17. ASCE, AWWA and CSSE. *Water Treatment Plant Design* (New York: American Water Works Association, Inc., 1969).
18. Gemmell, R. S. "Mixing and Sedimentation," *Water Quality and Treatment*, American Water Works Association, Inc. (New York: McGraw-Hill Book Co., 1971).
19. Moffett, J. W. "The Chemistry of High Rate Water Treatment," *J. Am. Water Works Assoc.* 60:1255-1270 (1968).
20. Griffith, J. D. and R. Williams. "Applications of Jar Test Analysis at Phoenix, Arizona," *J. Am. Water Works Assoc.* 64:825-830 (1972).
21. Hudson, H. E., Jr. "Dynamics of Mixing and Flocculation," *Proceedings of the 18th Annual Public Water Supply Conference,* University of Illinois, Dept. of Civil Engineering, 41-54 (1976).
22. Kawamura, S. "Considerations on Improving Flocculation," *J. Am. Water Works Assoc.* 68:328-336 (1976).
23. Chow, V. T. *Open Channel Hydraulics* (New York: McGraw-Hill Book Co., 1959), pp. 412-414.

CHAPTER 9

SEDIMENTATION

J. Donald Walker, P.E.
Walker Engineering
841 North Lake Street
Aurora, Illinois 60506

INTRODUCTION

In the treatment of potable water, sedimentation is a customary step used after coagulation-flocculation and ahead of sand filtration. There are a few exceptions in direct filtration where the sedimentation step has been omitted, but, by and large, sedimentation is an important step in the delivery of water of the greatest clarity and lowest possible turbidity in the finished water. Clarity also aids filter operation by making possible reasonably long filter runs of low head loss increase without the danger of early breakthrough by minimizing mud balls and crack problems in the filter and by saving pumping energy.

Sedimentation can be effected in several ways. The considered choice depends a great deal on the size of the plant throughput. The use of automatic mechanical sludge removal may also have bearing on the choice of sedimentation geometry. Bear in mind that very small plants, even middle-sized plants, may typically run at higher rates than average demand by operating the water treatment plant only for several prime hours during the day and storing finished water for use during the remaining hours. This increases the size of the plant.

Clarification is typically performed: (a) in long rectangular basins with or without sludge scrapers, (b) in small square (or round) hopper-bottomed tanks, and (c) in tube groups or lamella high-rate separators. Another

149

variation is a modular combination system consisting of rapid mix, slow mix, and up-flow clarification in a "solids contact" system.

Sedimentation in clarification basins is a low head loss and low-energy operation that will ensure its perpetuation through the ensuing decades of high cost of energy.

CLARIFIER PERFORMANCE

All flocculated solids sedimentation performance is affected by several design and operating features, most of which can be improved by good design and selection of appropriate parameters.

Density currents and poor effluent weir placement too often have resulted in unsatisfactory performance. Pittsburgh, Pennsylvania, had exceptionally good instant coagulation blending and series flocculation for river water turbidity removal, but each 10-mgd sedimentation basin could only handle half of the design flow because of a poor effluent withdrawal system. Even at half rate, the basins produced an effluent with high turbidity, which resulted in filter runs of less than 6 hr. The flow in the long, rectangular basin was well clarified until it reached the overflow weir, which extended across the far end. There, the slurry blanket was boosted upward by the density current and joined the otherwise clear effluent. A weir trough system, well-distributed over most of the surface area with mechanically-controlled slurry removal, resulted in an effluent containing less than one turbidity unit (TU) with filter runs of 72 hr at low head loss. The improved basins were then able to accept twice the original design flow. Long filter runs are not necessarily a goal, because energy savings with shorter filter runs of very low head loss might be more desirable. But here, the comparison indicates the improvement obtained.

Ainsworth[1] abandoned a long, baffled rapid-mix channel with a detention time of 2.5 min, and converted the extreme far end of this channel into a two-second, high-intensity, air-agitation section. The alum feed was relocated and introduced at the "instant" blending section with a remarkable reduction of alum and an increase of the clarity of the sedimentation basin effluent. Table I shows the improvement.

Hudson[2] reported a 30% reduction in alum dose when the alum dosage point was changed from a poorly located, horizontal, perforated pipe diffuser above the raw-water channel to a section of turbulent flow just before a hydraulic jump in a Parshall flume. This relocation evidently produced improved dispersion blending of the coagulant before hydrolysis and polymerization were complete. There was also a substantial saving of coagulating chemical, and a significant decrease in clarifier effluent turbidity.

Table I. Comparison of Long and Short Rapid Mix at Santa Barbara

Parameter	Original Rapid Mix Channel	Converted Blending Section
Mixing Time, sec	150	2.5
Alum Dose, gpg	2	1 (or less)
Clarifier Turbidity, TU	3-5	0.5-1
Filter Runs, hr	30	100

The discussion to follow treats other factors affecting the performance of sedimentation basins and practical water plant design in large, medium and small plant sizes. Plants with a capacity of more than 30 mgd (1.3 m^3/sec) are herein considered "large," a capacity of about 5 mgd (0.22 m^3/sec) is "medium" and a plant serving a community of fewer than 3000 people and producing less than 0.5 mgd is considered to be "small."

LARGE PLANT DESIGN

General

Plants treating raw water flows of about 5 mgd in each basin are apt to have long rectangular sedimentation basins—typically, but not always, mechanically cleaned. These basins clarify coagulated floc-laden waters. The water may also contain additives such as clay or carbon, which increase the floc volume to be removed as underflow. In water softening plants, mechanical scrapers must be used to cope with the large volume of lime sludge.

In general, the most hydraulically stable clarifier is the long rectangular basin. Sometimes, circular basins, popularized by the maintenance-free center drive sludge collector, are used, especially on industrial clean water supplies. The traditional iron chain-and-drag underwater mechanisms are subject to wear and are far from maintenance-free.[3]

Among other new developments, there is a light, plastic 6-in. (15-cm) pintal drag chain that can eliminate nearly all of the underwater friction wear. There is also: (a) the practical, light-weight carriage collector, whcih removes bottom slurry through a suction header or by a blade, and (b) the floating sludge removal mechanism, which is easily towed back and forth along the length of the tank to remove slurry through a suction header and a siphon. Both of these have the advantage of no underwater rotating units or bearings, which contributes to low maintenance.

Design and Operation

Large plants are apt to run continuously at a constant rate, but they are often designed to operate at two (or even three) rates with the chemical feeders and ancillary equipment programmed for automatic operation at any of the rates. During seasons of lower water use, the plant may shut down temporarily for a few hours during the night.

The settling basin requirement is met by designing two or more units to act in parallel. This reduces the size of sludge cleaning equipment and also helps hydraulic stability by reducing the detrimental effects of wind on large surfaces. If circular clarifiers are used, it avoids using outsized basins of more than 150-ft (46-m) diameter. And finally, there is the advantage of flexibility for servicing desludging equipment.

In older water plants, many of the sedimentation basins had no sludge scrapers and required the arduous, thankless job of desludging by manually hosing about every four months. This long sludge detention time does lead to some putrescence and deterioration of water quality. Plain basins also accentuate the sludge disposal problem. It is important to have multiple basins so that some continue to operate while one is out of service for cleaning.

Mechanical scrapers should always be used for cold lime softening because of the voluminous lime sludge and its ability to exude moisture and become extremely thick and viscous when left in place for more than a day. Without mechanical desludging equipment, such basins would quickly fill with slurry.

Modern practice is to use mechanical desludging equipment in all sedimentation basins to effect continuous desludging. Mechanical desludging is a requirement for good sludge treatment.

When clarifying alum sludge, a circular scraper with corner sweeps is sometimes employed to desludge only in the first half (or third, where most of the sludge first settles) of a long rectangular basin, and the remaining plain floor is hosed once or twice a year. In these designs, a barrier wall 5-ft high (1.5-m) blocks sludge otherwise carried by density currents into the nonmechanical part. But one wonders whether saving of the cost of the second mechanism is worth the disadvantages.

More frequently, especially with lime softening sludge, the entire basin is served by a mechanical desludging mechanism and an end cross hopper is used for sludge blowdown. This cross hopper is a long trench containing either a transverse helicoid sludge screw driven by a gear motor or a mechanical chain drag collector. The hopper function may also be served by a transverse series of deep pyramidal hoppers with steep, *smooth* slopes that are never less than 60°.

Sludge blowdown from one or more hoppers is controlled by a metering pump or a programmed blowdown valve. Both are designed for intermittent operation to move all of the thick sludge underflow but not to "posthole" (or overdraw) at any time and break through the viscous bottom sludge layer, which would draw thin sludge and leave the thick sludge to pile up on the floor.

A few years ago, alum sludge piled up on the floor of a long rectangular clarifier at Contra Costa, California, and stopped the heavy-duty carriage scraper mechanism. Lack of sludge blowdown control was causing postholing in the cross hopper, and the movement of dense sludge nearly ceased. The simple cure lay in controlling the sludge blowdown cycle on a realistic, programmed, start-and-stop basis.

Long Rectangular Basins

These basins are the choice of most plants large enough to use them because they are hydraulically more stable and have less short-circuiting. Conversely, be particularly wary of center-feed clarifier units of more than 150-ft diameter. These outsized circular clarifiers are sometimes substituted for the superior long rectangular basins. The long rectangular basins also perform better than peripheral-feed, circular basins.

Design faults possible with modern rectangular basins lie mostly with: (a) the uniformity of admitting the influent across the entire width, and (b) effluent overflow drawoff. Other faults may be lack of programmed sludge underflow control and impractical side water depths.

Influent Diffusion

An influent cross weir discharging into a forebay against a solid target baffle is the least-effective means for good influent distribution. It usually leads to short-circuiting by jetstreaming along one or both walls.

Sometimes a false wall having open, checkered brickwork is built into the influent end to provide good lateral distribution. The spacing and net amount of opening must be designed correctly and for one flow rate. There must be enough head loss for good distribution but not enough to cause floc break-up. As a practical matter, the net velocity of flow through such a false wall should be about 3-5 ft/sec (1-1.5 m/sec) to produce a head loss of 3 or 4 in. of water column (WC). Fine tuning can be accomplished by adding loose half-bricks. These checkerwork brick false walls have been successfully used as an addition to upgrade basins with poor influent control. Such upgrading results in much better effluent clarity because there is less short-circuiting. Figure 1 shows a typical design.

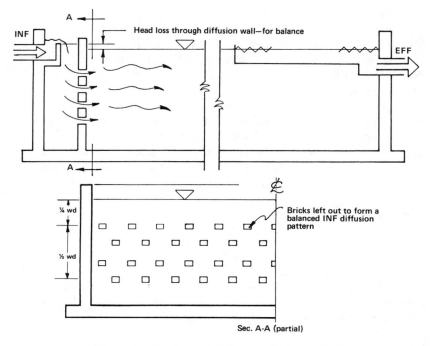

Figure 1. Checkerwork influent diffusion wall.

A new version of influent control is to use a false wall with the proper number and spacing of flared-entry, 4-in.- (10-cm-) diameter openings made of fiberglass reinforced polyester (FRP). This new form seems to be more workmanlike and is designed for about a 4-in. WC head loss to ensure uniformity of entry of the influent.

Another control method is to use a lateral distribution trough with ports placed on about 5-ft (1.5-m) centers to bring the flow near to the surface. Adjustable, biased gates cause these equalized flows to act tangentially across the surface with an energy dissipation of about 4-in. WC. These tangential flows encounter a baffle with a shelf at about middepth, which causes a horizontal rolling action. This rolling energy produces homogenous floc distribution and the displacement under the baffle is uniform and laminar. This energy-dissipating inlet, shown in Figure 2, not only affords good lateral distribution but also keeps the floc homogenized so that discrete sedimentation starts only below middepth. The result is slower density currents because of the reduced distance of floc fall.

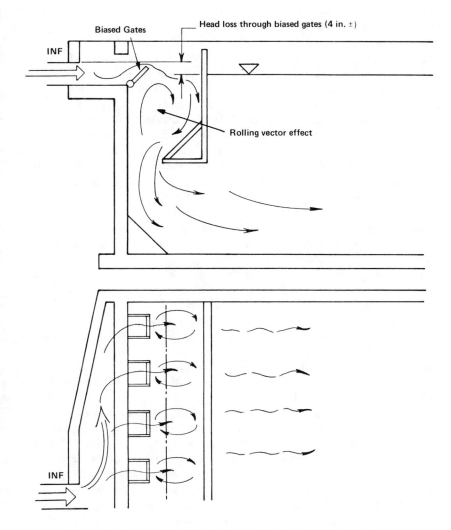

Figure 2. Lateral inlet distribution with adjustable biased gates.

Overflow Weir Troughs and Orifice Troughs

The overflow system may be either submerged orifice troughs or V-notched overflow weir troughs. Years ago someone deduced that overflow weirs broke up residual, coagulated floc. Now it is known that the local G-value or energy dissipation of both kinds of collectors is the same.

Oddly enough, the G-value in the top of the sand filter bed is greater than that of either the overflow weirs or the orifice system.

Typically, weir troughs are evenly distributed at the effluent surface and cover approximately the last third or half of the surface. At one time it was considered that the weir overflow rate should not exceed about 10 gal/min/lin ft (0.12 m^3/m min) of weir. Recent research[4,5] has established that overflow rates even as high as 75 gal/min/lin ft (0.93 m^3/m min) does not disturb the sludge blanket. Rampant density currents are responsible for forcing the sludge blanket up to and over the weir, no matter how low the linear rate is. However, a limitation of 10 gal/min/lin ft tends to require greater lengths of weir troughs, which should be spaced well away from the effluent end wall. It is the distribution of the weir troughs that is all-important for "working" a large portion of the basin surface. A high, up-welling, rise rate under a short or crowded trough system (even though it meets the regulatory requirement for linear rate) is not as good as a well-distributed weir trough system, even if the latter is linearly "overloaded."

The sedimentation of coagulated floc bears little resemblance to subsidence by Stokes law. The influent mass of discrete flocs join together as they rapidly settle into the forebay at the head end of the clarifier. This mass becomes hindered and soon forms a sludge blanket with a rather sharp interface between the clear water above and the hindered blanket on the floor. This blanket, undulating and drifting slowly toward the effluent end, remains mostly on the bottom except for sporadic blow-ups. There is a density gradation within the blanket, with the densest sludge on the bottom in the compression zone.

The clear water immediately above the sludge blanket interface is (except for disturbances) about as clear as the water overflowing the effluent collection system. Also, the clear water above the blanket is as clear near the influent end as it is anywhere along the blanket, even at the effluent end. Then, except for the rising and falling of some fragmented, discrete floc, there is no additional clarification between this sludge blanket interface and the surface overflow, which may suggest that this clear water depth is wasted. This, however, is a snare and a delusion! To a greater or lesser degree there are disturbances within this blanket that may fragment some floc masses and send them up into the clear zone. Also, there are major blowups from within the blanket that send "thunderheads" of floc seething and rolling toward the surface. Sometimes these phenomena are caused by: (a) jetstreaming, which is often the result of short-circuiting or hydraulic instability due to poor design of the influent displacement control, (b) excessive solids loading rate, (c) excessive depth of sludge blanket in a clarifier, or (d) other design or operation problems.

Water depth is important in providing freeboard above these blanket interface disturbances. It is also important in making possible the proven advantage of a well-distributed collection trough system by reducing stray currents.

Finger weir spacing is important on long, rectangular basins. Finger weirs are the long weirs that typically extend longitudinally from the effluent end toward the middle of the basin. If carriage or floating sludge cleaners are used, the longitudinal finger weirs are a "must."

If drag chain and flights are used, the weir troughs are typically distributed across the width of the tank. However, the weir length and distribution is comparable to that of the finger weirs, and most all that is said here about finger weirs applies also to cross troughs. Weir troughs are equally spaced to serve the centroid of the area they control. In general, they are spaced on about 15-ft (4.6-m) centers but never more than 20-ft (6.1-m) centers. There is a vast empirical background for such spacing and there is little advantage in placing them closer, particularly if closer spacing results in too small a cluster that would increase the average vertical rise rate. Long weir troughs are usually supported on concrete piers rising from the floor. The troughs are made of FRP or of coated steel with FRP serrated weir plates.

No weir or cross trough should ever be located against or even close to the end wall. But finger weirs can run to the end wall because overflow near the end is minute compared to the total weir length.

Rise Rate Under Overflow Troughs

A useful, empirical parameter for the average rise rate under the weir troughs is 1-1.5 gal/min/ft^2 (0.041-0.061 m^3/m^2) of surface covered by the the trough system. In general, the weir trough group should be distributed from the effluent end to about halfway toward the influent end. This is reasonable because the clear water just under this trough system flows from the effluent end toward the influent end and opposite to the general displacement, whereas the density current flows along the bottom from influent to the effluent end, as shown in Figure 3. Many successful sedimentation basins have even carried the finger weirs (or cross troughs) all the way up the basin to within about one-quarter of the distance to the influent end.

Water Depth

The side water depth (SWD) is important and should be at least 15-18 ft (4.6-5.5 m) for basins more than 100 ft (31 m) long or 70 ft (21 m) in diameter. The SWD can be as shallow as 10-12 ft (3.1-3.8 m)

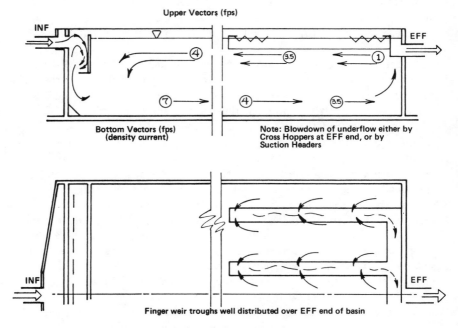

Figure 3. Long rectangular sedimentation basin vectors.

for small-to-medium basins but 8 ft (2.4 m) is too shallow for even very small basins.

Considering surface loading rates to be all important, it was once thought that multiple shallow trays with 3-6 ft (1-2 m) SWD would be advantageous. The tray-like clarifiers did not prove practical. During the late 1930s and 1940s, there were a great number of field trials for making shallow sedimentation basins work. All of these attempts failed. One in Tokyo uses a throughput of 90% through the top tray (open to the weir troughs) and 10% or less through the bottom tray. Tube clarifiers work very well, but in no way do they resemble the basins under discussion, and they are not examples of shallow sedimentation.

Water depth is a factor of safety against sludge fractions leaving the undulating sludge blanket interface, which (if disturbed) is all too ready to rise as a "thunderhead" and partially join the overflow. Sludge blankets are prone to being disturbed or fragmented by jetstreams and density currents, and spatial distance above this blanket is required to obtain a clear effluent reliably.

Sludge Underflow Control with Mechanical Desludging

One successful method of sludge underflow blowdown is an adjustable time cycle to start and stop either a sludge suction pump or an automatic blowdown valve. The temporary high velocities keep the sludge suction lines cleared of deposits and allow additional thickening in the accumulating hopper between blowdown cycles. The sludge discharge line should often be purged for 20 sec with a spurt of clear water at a velocity of 6 ft/sec (1.8 m/sec) following the blowdown cycle, especially for lime sludges.

The underflow discharge should just pace the accumulation of thickened sludge on the bottom. It should never be either so intense or so prolonged as to break through the thick sludge blanket in the hopper (or on the floor if suction headers are used). Breaking through the dense sludge layer leaves the dense sludge mass to pile up on the floor and pumps too thin a sludge.

Alum sludges are very slowly putrescent and the operator does not need to keep as close a check on the depth of this sludge blanket because a reasonable amount of time (8 hr or so) is not apt to cause putrescence. This is not a license to ignore the depth of sludge blanket and to invite putrescence with taste and odor. Even though they are thixotropic at first, these sludges can in time become fixed to the floor as anchor sludge, which resists mechanical cleaners. The depth of dense sludge on the floor should be measured at least weekly to keep track of sludge blowdown effectiveness.

The volume of lime precipitation sludges is apt to be increased by solids recycle, but putrescence from long dwell time is not a problem. However, the rapid accumulation of dense lime slurry is, and the sludge blanket has to be watched for orderly blowdown. Failure to monitor this underflow can stall the desludging mechanism with all its implications for trouble.

Circular Basins

Circular clarifiers, as compared with square ones, have a stellar reputation for mechanical reliability. Circular sweep mechanisms applied to square tanks (or to long rectangular basins) have corner sweeps that are used to hook sludge from the corners so that big fillets can be avoided. These corner sweeps are vulnerable to bearing trouble and occasionally they run up and off the top of the steel corner curb that guides the pantograph sweep arm (see Figure 4). This serious problem can be avoided by three remedies: (1) use an *18-in. high* (0.5-m) curb for ample leeway, (2) fine-tune the outboard wheel, which should always be large and should never be a skid, so that it takes a *slightly downward* path as it presses against the curb when it folds the pantograph, and (3) *pay close attention to the condition of these underwater pantograph support bearings and the wheel bearings.*

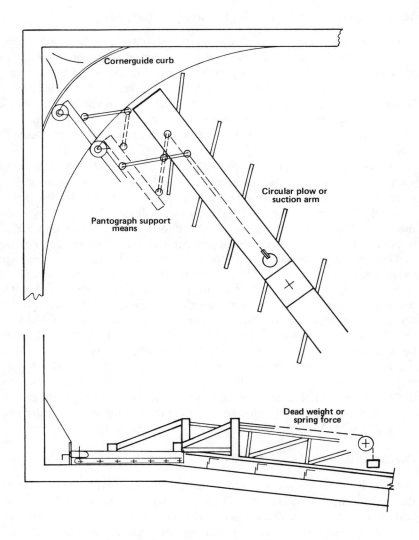

Figure 4. Pentagraph support mechanism.

Circular basins, if not larger than 125-150 ft (38-45 m) in diameter, can do an excellent job of clarification. But the design engineer must pay attention to a few essential details such as: (a) a good method of ensuring uniform radial distribution of the influent, (b) an extensive system of overflow weir troughs, and (c) an ample side water depth.

Inboard Circular Weir Troughs

Circular clarifiers employ weir troughs spaced well inboard to gain a better placement and greater overflow weir length. The inboard spacing tends to avoid the disturbance of small density currents that climb the peripheral wall, and it limits the magnitude of rise or up-welling velocity. The weir trough should be placed about 15-20% of the tank radius inboard from the peripheral wall. It is not effective design to hang this inboard trough at or near the peripheral wall as a matter of convenience. Such a design may as well be a simple peripheral overflow weir because the region of rise adjacent to the wall would be vastly overloaded anyway.

The design parameters for finger weirs or cross weir troughs also apply to inboard circular weir troughs. In square tanks, the inboard weir trough should cut across the corners rather than parallel the walls to avoid chronic rising "corner floc" that generally occurs in square basins due to the converging energy vectors toward the corners.

Circular weir troughs are designed as segmented straight lines that approximate a circle. Depending on the number of landings or corbels selected, the troughs are fabricated in straight sections 15-20 ft (4.6-6.1 m) long as shown in Figure 5. This is a practical method of manufacturing, and it also places most of the overflow weirs closer to the centroid of the surface. Each trough section, with its adjustable V-notch weir plates, should also have a 1.5-in. (3.8-cm) hole in the bottom for drainage when the basin is dewatered. This hole does not suck floc into the effluent and it takes many years to close or clog.

Influent Diffusion

Center-feed, radial-flow basins need more than casual design of the feedwell. Because of the overloading of the relatively small feedwell region (usually less than 2% of the area and 1% of the volume), there is little chance of stilling the influent feed. If a simple, nondescript circular feedwell extending a few feet into the water is used, the flow from this feedwell is apt to be a jetstream, often visible through surrounding clear water, as has been witnessed by too many operators.

Some method of creating 4-5 in. (10-13 cm) WC of headloss should instead be employed to subdivide the influent into uniform radial flow from the feedwell. One such design shown in Figure 6 uses an inner diffusion well with multiple biased gates that dissipate about 4 in. of WC in order to attain equal discharge. The divided, tangential flow streams into a surrounding feedwell with an area of about 4 or 5% of the tank surface. The feedwell skirt extends to middepth. The dissipating energy homogenizes the inner contents of this feedwell, and it imparts a rotating drift

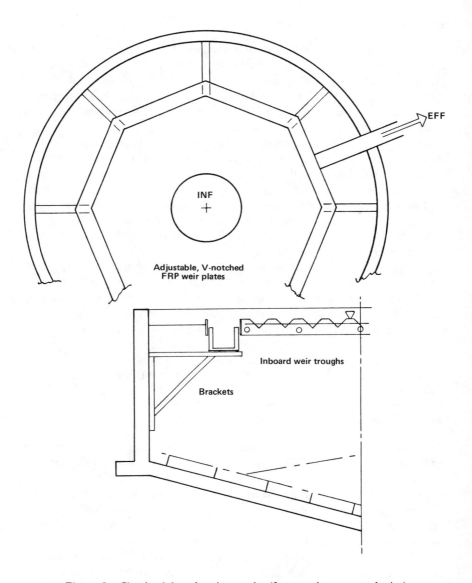

Figure 5. Circular inboard weir troughs (for round or square basins).

to the feedwell mass to ensure equal velocities from the bottom of the skirt. This diffusion system (or its equal) costs slightly more than a simple, undersized feedwell, but it is worth the cost because it does the job properly.

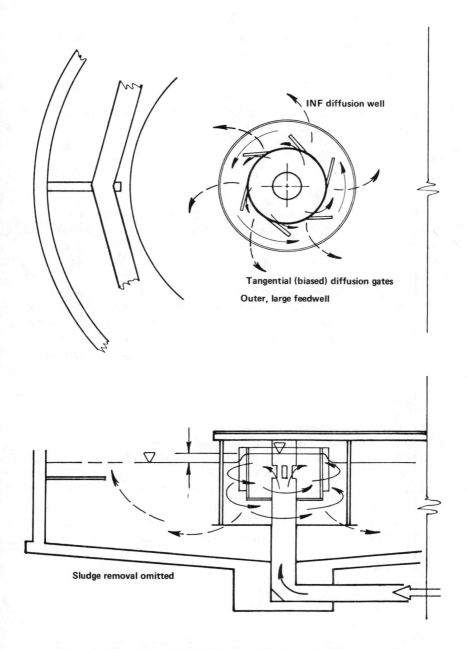

Figure 6. Flocculating feedwell (with multiple biased diffusion gates).

Power to Operate

The operating torque requirements for the circular sludge removal mechanisms are given in the Appendix of this chapter.

Other Considerations

The discussion in *Long Rectangular Basins* about sludge blanket control and water depth apply in the same way to circular tanks.

Design Parameters

Detention Period

Detention time is simply the time to fill the basin at the given rate of influent flow. Unfortunately, it is a poor parameter, and it often bears little resemblance to the actual displacement period because of short-circuiting through the basin. Some regulatory agencies have arbitrarily required water plant clarifier detention periods of as much as 6 hr. Some texts suggest 2-4 hr, which is not too unreasonable when compared to 6, but the spread would result in a great variation of basin size. For a given surface area, detention can be increased by increasing the SWD. However, beyond a reasonable water depth, there is nothing to be gained with a deeper basin. Sedimentation basins should not be designed on the basis of detention period.

Displacement Velocity

Another nebulous parameter is the average linear velocity or the displacement velocity through a long rectangular basin. Some texts and standards set a very large range of 0.5-6 ft/min (0.005-0.030 m/sec) for this velocity, but this is too wide a spread for practical design. This indeterminate parameter is meaningless in the sedimentation of floc-bearing waters because there is no plug-flow displacement through this basin— not with the contradictory bottom and surface currents. There is no corresponding parameter for radial flow velocity in circular basins.

Length-to-Width Ratio

The length-to-width ratios are meaningless by inspection. Large plants certainly must use relatively long tanks, and even so the aggregate width of all parallel tanks frequently exceeds the length. A rectangular basin handling a small flow is apt to be slender primarily for equipment selection. An interesting example is a successful floc sedimentation unit, which is about 150 ft wide and 30 ft long (46 m x 9 m). A carriage collector sucks up sludge along the long dimension, whereas the throughput is displaced across the short dimension, which results in an L/W ratio of less than 1. The L/W ratio is not a practical design parameter.

Surface Loading Rate

The surface overflow area is the most important parameter for basins clarifying flocculent solids. This has been stated by Camp[6] and others in many publications. The surface overflow rate for any given sedimentation tank could be determined by jar test studies wherein the best coagulant, optimum dosage, and the best flocculation are used. But usually the design engineer must rely on past empirical experience and estimate a safe basin overflow rate based on representative water analyses and on the contemplated coagulant use. There remain the problems of changing seasons and chancing water quality.

Table II gives a few safe overflow rates. Medium-sized plants should use rates 15-20% less to provide an additional factor of safety, and small-sized plants need even lower rates. These conservative rates assume good influent diffusion and well-spaced effluent overflow weirs. They apply to both circular and rectangular basins.

Table II. Safe Overflow Rates for Clarifiers

Application		Surface Loading Rate	
		gal/day/ft^2	m^2/m^3 day
Lime Softening	(low magnesium)	2000	81
	(high magnesium)	1600	65
Alum Coagulation	(turbidity removal)	1200	49
	(color removal)	900-1000	37-41

Solids Loading Rate

When clarifying flocculated water carrying 3,000 mg/l of solids or more, the solids loading rate is apt to be limiting when compared to the surface loading rate. Hence, the overflow rate needs to be limited to keep within bounds of solids loading on the floor of the clarifier. In waterworks practice, the basin influent rarely bears more than 1000 mg/l of flocculant solids. Thus, the solids loading rate is seldom limiting, but nevertheless it should not be ignored. Rectangular basins handle solids loading better than circular basins do. As a general average, long rectangular basins can handle up to 51 lb/ft^2 day of floor area (244 kgf/m^2) because much of the density current (which is a function of solids loading rate) is emitted from the hopper blowdown or the sludge suction pipe. A limit of about 30 lb/ft^2 day (150 kgf/m^2) probably applies to circular clarifiers.

Mechanical Desludging

Carriage Collectors

The traditional carriage collector is a heavy rolling bridge spanning the width of one or more long rectangular tanks. It is supported at both ends by double-flanged iron wheels or trucks that roll back and forth on T-rails fastened to the top of the walls of the sedimentation tank. On the heavier units these trucks may carry cog wheels that engage longitudinal cog racks attached to the two opposite tank walls. The cog racks prevent crabbing or slipping while heavy sludge is bladed to the end hopper. The use of rubber tires in lieu of flanged, iron wheels is to be discouraged in any location subject to frost, sleet or snow.

Power for the drive and hoists is usually supplied to a control center on the carriage bridge through a flexible power cable carried on a reel mounted on the carriage. Lighter carriages on shorter spans are driven by friction between the flanged wheels and the T-rails or by a stainless steel flexible cable wrapped around a powered sheave mounted at the end of the basin. The underwater part of the carriage collector is typically equipped with a deep transverse scraper blade to move either lime or alum sludge to the hopper. Carriage systems for basins shorter than about 100 ft (30 m) are sometimes powered by means of an overhead power cable that slides in loops along a stretched centerline cable.

Carriage collectors are relatively maintenance-free, and they are slowly displacing the traditional (and vulnerable) iron chain drag mechanism for moving settled slurry to the underflow blowdown system. Unfortunately, the heavy carriage units can only be installed in very large plants on a competitive basis. They are not cost-effective in medium-sized plants, and they are nearly out of the question for small plants.

Cost-effective types of light carriage collectors have been recently developed to compete with the traditional metallic chain-and-drag collectors, long a standard in almost every consulting engineer's computerized specifications. The new carriage collectors are competitive even in the small plants. They are particularly valuable for medium-sized plants with basin widths of less than about 40 ft (12 m). These units are structurally sound and they possess the advantages of having extremely low maintenance costs and no underwater bearings. They roll on flanged iron wheels and T-rails and usually are pulled by a stainless steel flexible cable with a motorized drive mounted on one end of the basin, although they are sometimes pulled by two side cables. The carriages are fitted with a blade for moving sludge into a hopper, and this is an advantage when the solids loading is very light as the blade adjusts to the sludge available. A blade is also advantageous when the sludge is voluminous or heavy, which is usual for lime sludge or primary sludge. A flexible cable can supply electricity to the bridge for power to the wheels, for raising the blade or for heat tape.

The floating bridge siphon unit, shown in Figure 7, has come into prevalent use during the last five years and it has established an enviable position in the market, particularly for its low maintenance cost where it replaces iron chain drags in long rectangular basins. Suction tubes and headers are coated aluminum, light in weight, and are supported from rugged, nonsinkable floats. The suction tubes are 9-10 in. (20-25 cm) in diameter. They discharge laterally across the float system and through a downleg or siphon into a longitudinal sludge trough running the length of the basin. The multiple-orifice suction removal of thickened sludge from the floor of the basin has proven to be sound engineering practice. Control of the dense sludge underflow by use of siphon technique has also proven to be sound.

Drag Chain-and-Flight Collectors

There are two general types of drag-and-flight collectors. The traditional unit of 50 years ago was a heavy iron chain and water-logged wooden flights. The modern version is light plastic chain and FRP flights.

Heavy iron chain has frequently been criticized in the literature[3] for lack of cost-effectiveness because of its early wear, high maintenance and replacement cost, and the out-of-service time of the basin when it is dewatered for repair of the collector. Friction between the many shoes and the rails, friction between chain and sprocket teeth, elongation of the chain due to pin wear, and the resulting poor fit between chain and sprocket teeth, which causes a decreasing pitch diameter, all tend to cause excessive operating torque and chain and sprocket destruction. It takes considerable maintenance (chain turning and repinning, sprocket turning) to keep one going.

Light, high-density plastic chain is relatively new, but it has been applied for some time in industrial and trash rack service. The 6-in. (15-cm) chain is strong, and, typically, the breaking strength is 2000 lb (0.9 tonne) whereas the service tension is usually under 400 lb (0.18 g) with a maximum of 800 lb (0.36 t). The light weight of the entire assembly and

Figure 7. Floating bridge siphon unit (courtesy Leopold/Clari-Vac Co.).

limited sprocket tooth wear, together with the low friction between plastic wearing shoes and rails, results in 30-50% savings in chain pull force as compared with iron chain. The light plastic chain, then, is practical and advantageous because excessive friction is removed, the chain pull is lessened, and the head and idler shafts and gear reducer drive can be reduced in size. There is little chain elongation, so the chain continues to fit the drive sprockets even after long service.

Solids Contact Basins

Solids contact systems are applied equally well to large or small plant design. Fresh precipitate for nucleating reactions is recycled at low rates in plants to remove turbidity and at large rates in lime softening plants to drive softening reactions to completion.

A solids contact basin consists of either round or square vertical outside walls with a rapid mix tank and a recycle system nested in the center surrounded by a frustoconical skirt, as shown in Figure 8. The annular section between the skirt and the outer wall is the clarifier. The skirt and the inner rapid mix parts are of steel. A bridge carries a variable-speed, 4:1 recirculation turbine drive and a motorized reducer drive for the bottom scraper. Thick slurry is scraped along the bottom to a central well from whence it is automatically discharged to waste by programmed intermittent sludge blowdown.

The annular clarifier section discharges the clarified overflow through multiple, radial weir troughs arranged to work the entire surface and to ensure vertical flow. This unit exhibits great hydraulic stability without the complication of density currents because the feed is uniformly introduced near the bottom under the large-diameter skirt. In lime softening or when too much recirculated solids are carried in suspension, there may occasionally be a solids-loading rate overload, with density currents resulting. This operating problem can be avoided by better underflow (sludge blowdown) control.

The same parameters as for separate flocculation basins apply to floc aggregation time under the frustoconical hood. The recycled solids may be of benefit, depending on circumstances, but do not shorten the time parameter. Flocculation is enhanced by the high recirculation rate (varying from 1:1 to 4:1), which accentuates early floc contacts in the rapid mix section during the formative micelle state. However, the recirculation and the resulting high G value in the rapid mix section does not eliminate the need for instant coagulant blending just ahead of this system if a trivalent coagulant is used for turbidity or color removal.

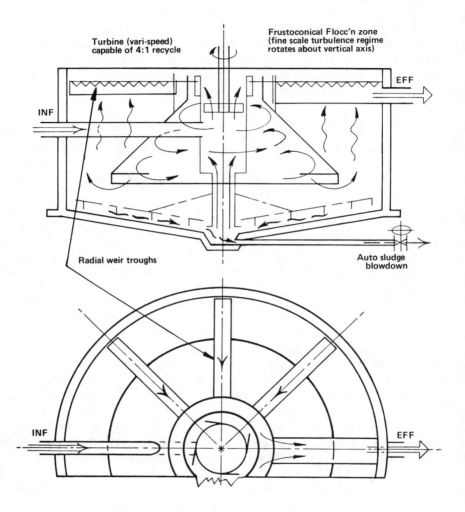

Figure 8. Solids-contact basin (upflow basin).

Subsequent displacement through the conical floc aggregation chamber has to be controlled carefully (by baffle design) to avoid uncontacted short-circuiting. Slow-mixing G values as great as 100 sec^{-1} can easily be accommodated in this chamber, and a variable-speed drive on the recirculation turbine can vary this at will.

Inlets

The influent into the annular clarification space is by radial displacement under the conical skirt. Careful design and installation technique must be employed to ensure even flow all around the skirt. Otherwise, localized flow and jetstreams initiate short-circuiting.

Overflow System

The radial trough overflow system provides good hydraulic stability and clarification. Radial weir troughs are typically placed so that their outboard ends are about 18-20 ft (5.5-6.1 m) apart. But good weir trough placement cannot effect good results all alone; the influent flow must also be uniform.

Advantages

The advantages of the modular solids contact system are: (a) size savings (since the detention time does not exceed 1.5-2 hr), and (b) savings in construction costs and mechanical piping. These combine to produce a plant of lower capital cost.

For excess lime softening, especially, these systems (through massive recirculation) yield a more stable water with a minimum of unstable hydroxyl alkalinity. This, in combination with the addition of about 2 mg/l of hexametaphosphate, often makes it possible to omit the expensive CO_2 feed.

Flocculation Clarifier

A cousin system of the solids contact basin consists of an open-bottom flocculation chamber nested inside an annular clarification section, as shown in Figure 9. Slow-speed, large-diameter axial flow turbines are usually used for slow mixing the contents. G values as high as 100 sec^{-1} can be used for slow mixing without disturbing the annular clarification section, but typical operation is at a lower G value of about 50 to 25 sec^{-1} by using variable-speed drives.

These systems do not have the hydraulic stability of a well-designed solids contact basin and are not recommended for water flocculation and clarification because they are subject to short-circuiting in the single-stage flocculation chamber. The flocculating clarifier has its greatest application with industrial wastes bearing coarse grit (that would pile up in a standard flocculation basin). It can also be used as a presedimentation basin where some coagulants are to be added. They are applicable to: (a) wastewater tertiary treatment, (b) primary high-lime coagulation and sedimentation, and (c) the second stage in lime-softening split-treatment flow schemes.

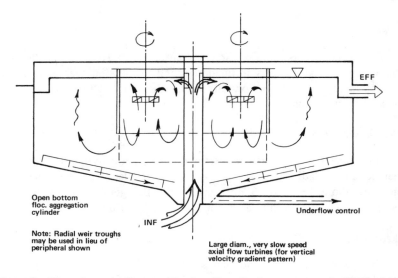

Figure 9. Flocculator-clarifier combination (for circular or square clarifier basins).

MEDIUM-SIZED PLANT DESIGN

Medium-sized plants could be characterized as those with daily average flows ranging from 2000-4000 gpm (130-250 liter/sec). Frequently, plants of this size and smaller are designed to operate for only two shifts per day (for lower demands they may operate for only one shift) to relieve the burdens of continuous operation. Basins designed to operate only part of the day cost somewhat more but they are able to handle unusual water demands, and they provide for community growth. The plant might even be sized to operate for one shift during most of the year and two shifts during the summer or other periods of high demand. An economic analysis should be made to determine the most advantageous plan for each community.

These plants can be automated to come on the line unattended in order to satisfy unusually high reservoir withdrawals. Or, if clear water storage is scant, the plant can also be automated to run during low demand periods at some fraction of the maximum flow—a two-speed plant to follow more or less the demand on, perhaps, a 15-hr basis.

The greater design flow rate resulting from an 8- to 16-hr operation affords duplication in most of the critical treatment units, which pays with better operation and maintenance conditions. Furthermore, a filter or sedimentation unit out of service is not as critical because the remaining active unit can meet the demand if the plant is temporarily operated longer.

Clarifiers

Plants within this intermediate flow category are apt to be served by two clarifiers, either circular or long rectangular types. A long rectangular clarifier might be desludged by: (a) plastic chain-and-flight drag conveyor, (b) floating bridge sludge suction unit, or (c) carriage collector (of the light class) carrying either a blade or a suction tube. A circular, center-feed clarifier would be desludged with a rotating plow system, either pier-supported or bridge-supported.

Sometimes square basins, with *very rugged* corner sweeps, are used in lieu of circular basins. The use of corner sweeps makes large corner fillets unnecessary. Corner fillets tend sporadically to dump sludge as an avalanche, which causes stirring and collector overload problems. On the other hand, corner fillets can be used instead of corner sweeps if the tank is very small, the slope is 60°, and the surface is very smooth (even coated with epoxy).

Design Parameters

Design considerations and parameters for both the long rectangular and the circular basins are, in general, the same as discussed in *Large Plant Design*. However, the surface loading rates in Table II should be downgraded about 15-20% in medium-sized plants. A realistic and conservative viewpoint is that operation may not be quite as competent or continuous as in the larger plants. There are likely to be fewer jar tests run, small feeder hoppers are more apt to run out of chemicals, and the operator is apt to have other municipal duties and so leave the plant to run unattended more often. Adding a couple of feet to the diameter of a basin costs very little and reduces the surface overflow rate. It provides a more comfortable margin of safety against indifferent operation and unknown future requirements.

SMALL PLANT DESIGN

This category could be divided into two subcategories: (1) 1000 population class with a daily average plant flow of about 80 gal/min (5.1 liter/sec), and (2) 3000 population (upper quartile of towns in many western plains regions) with about 230 gal/min (14.5 liter/sec) daily average. In both of these categories the plants might well be advantageously designed to operate only a nominal 8 hr/day, which in practice becomes 7 hr/day actual running time.

Sedimentation Basins

These basins are cleaned mechanically and hence the circular basin sizes suggested in Table III are given, in general, to the closest 2 ft. The rectangular basins have been selected for practical widths to accommodate typical, long-basin cleaners.

Table III. Clarifier Basin Criteria for Alum Coagulation

Treated Flow (gpm)	Operation Time (hr)	Surface-Loading Rate (gal/day/ft^2)	Surface Area (ft^2)	Clarifier Type Long Rectangular Width x Length (ft)	Clarifier Type Circular Diameter (ft)
1000 Population Equivalent					
80	24				
120	16 (semiauto)	500	346	1-15 x 24	1-22
274	7 (of 8 hr)	500	789	1-18 x 44 or 2-15 x 28	1-32 or 2-24
3000 Population Equivalent					
230	24				
345	16	500	994	1-18 x 56 or 2-15 x 34	1-36 or 2-26
789	7	500	2271	2-18 x 64	1-55 or 2-38

Table III shows basin sizes for both 16-hr and 7-hr water production. The 16-hr mode suggests that the plant may run on a semiautomatic basis, starting and stopping automatically, but assuming operator attendance during the first 8 hr of each day for maintenance, chemical handling, precautionary checking, testing and reports. Such a plant could also be programmed to start up at any time on water demand occasioned by fire or serious main break. Times of automatic operation could be posted in the police station.

The 7-hr mode is selected for operating the plant manually on one 8-hr shift a day. Manual start-up and shutdown take about an hour, which leaves seven hours for production.

The additional basin sizes required for these shorter periods of operation have to be balanced against manpower cost. For semiautomatic design, the cost and maintenance of automatic controls and instrumentation must be considered. However, this mode may be a good compromise as compared with the extra size and cost of basins for mixing, sedimentation and filtration.

Surface Loading Rate

Small plants are faced with more operator problems than large plants, and experience demonstrates that it is not worth while trying to design surface rates with parameters having little factor of safety. Also, in small tanks, a little additional diameter or length costs little extra to construct and yields considerable comfort to both the operator and the responsible design engineer. Because of the small size of the plant, surface rates for clarifiers have been further discounted to about 500 gal/day/ft^2 (20 m^3/m^2 day) to produce a safe, workable design free of future regrets. These rates are safe even for cold surface waters.

Lime floc clarification basins for softening can operate at a conservative surface rate of about 1100 gal/day/ft^2 for water containing mostly calcium hardness and at about 1000 gal/day/ft^2 (41 m^3/m^2 day) where more than about 20 mg/l of magnesium hardness is present. Magnesium in bicarbonate form acts as a good coagulant when it is present in water being raised to high pH for softening.

Circular or square basins lose their dynamic effectiveness when they have less than a 15-ft (4.6-m) diameter. In such a small vessel there can be too much short-circuiting induced by the short flow path from feedwell to the peripheral weir, and still smaller surface rates should be assigned to them.

Water Depth

The basins of Table III should have a side water depth of at least 10 ft if the basin is automatically desludged. If not, extra depth for sludge storage is required. An assumption of 6% density and six months between cleanings is reasonable.

Feedwell Design

Feedwell design to create stable radial displacement is as important in small tanks as in large ones. Unfortunately, there seems to be a disregard of this important function, and the smaller the tank, the greater the disregard. Small tanks, whether circular with radial flow, or rectangular with longitudinal flow, have a special problem: any imbalance in influent distribution may result in direct short-circuiting to the overflow weirs.

Effluent Weirs

Radial effluent weir troughs make an efficient overflow control system for circular tanks, and they may be used with all tanks larger than 20 ft (6.1 m) in diameter where skimming is not required. Tanks smaller than

20 ft in diameter can have a single peripheral overflow weir. Radial systems work the entire surface of the tank and tend to create a stable hydraulic vertical flow. (See *Solids Contact Basins.*)

Inboard weir troughs are also desirable. On small diameter tanks the weir trough position should come reasonably close to the centroid of the overflow surface to enhance floc settling and improve density currents.

Sludge Hoppers

Circular tanks smaller than 40 ft (12.2 m) in diameter have plow-type sludge scraper systems that are bridge-supported. The conical sludge underflow hopper is 2 or 3 ft (0.7 to 1 m) deep, lies on the centerline of the drive tube, and is spaded by stirrups that extend into the hopper. The sludge discharge line is usually a 6-in. ductile iron, glass-lined pipe that runs under the basin to the automatic blowdown or sludge pump pit at the side of the clarifier. As stated before, the desirability of automatic blowdown for underflow control cannot be overemphasized.

In small long rectangular basins, the underflow is drawn from pyramidal hoppers located at the *far end* of the settling basin, and automatic blowdown is effected much the same as with center hoppers in circular basins. The hopper location at the effluent end is for moving the sludge with (rather than against) the density current. Large tanks use mechanical cross collectors with automatic sludge blowdown from the accumulating hopper.

Solids Contact Basins

For plants in this size range, the solids contact system is well applied to save total construction cost and for ease of operation. A change in chemical dosage produces discernable results quickly (within 2-2.5 hr), so trial and error operation is simplified when treating a changeable surface supply.

Solids contact basins are ideal for softening because the reactions are helped by the variable-speed rapid mix and the high recycle rate of the lime slurry solids, which drive hydroxy reactions nearly to completion. This almost eliminates the need for carbon dioxide to reduce residual plating. Typically, some Calgon is then used for stability.

HIGH RATE SEDIMENTATION

In recent years, there has been an awakening interest in the modular, high-rate floc sedimentation units. Better effluents have been produced in several unsatisfactory industrial wastewater clarifiers by passing the basin effluent through short, tube settler modules. In these "retrofit" basins, the tube modules intercepted turbid effluent, reclarified the flow within the tubes, and thus produced an effluent with low suspended solids and turbidity.

This suggests that equivalent clarification can be achieved in a much smaller unit with very short detention times (10-15 min vs 120-180 min) for an ordinary clarifier and higher surface-loading rates. This is true, and the tube settlers are especially practical in small flow applications. Where the tendency to the formation of slime or calcite deposits is small, the tube or plate modules can do an excellent job of removing settleable solids.

Clarifier basins of good design that are not overloaded do produce clear effluent and, even though larger, they may be cheaper than smaller units with the high-rate tube or plate modules. But even so, the smaller plants may well be better served by a system of these high-rate tube or plate module clarifier units, both for construction expense and performance. These high-rate modules are as easy to operate as a typical clarifier basin. Chemical dosages, coagulation, and flocculation procedures and equipment are the same for ordinary clarifiers and clarifiers containing tube settlers.

Many high-rate settler designs are proprietary, and are generally variations of either inclined tubes of small dimension (roughly 2-in. or 5-cm diameter), or inclined lamellae (plates) spaced about 2 in. apart. The velocities are low in order to produce laminar flow at very low Reynolds numbers. After the flow enters the inclined narrow space, discrete particles need to fall only a very short distance to the lower plate or tube surface where the sludge quickly compacts. If the tubes of lamellae are steeply inclined, the sludge slides down the slick plastic surfaces to discharge at the bottom. Tubes that are only slightly inclined must be backwashed.

The surface loading rate is about two to perhaps three times the rate used for standard clarifiers and they achieve 90 to 95% removal of settleable solids. Of course, flocculated waters require application rates consistent with their different theoretical settling velocities. For example, if alum floc settles at 2 in./min (5 cm/min) in the laboratory, the corresponding idealized surface-loading rate is 1795 gal/day/ft² (73 m³/m² day). An 80-ft diameter x 15-ft SWD (24.3 m x 4.6 m) center-feed clarifier with a well-designed diffusing inlet and weir troughs well inboard would be designed for alum floc at a maximum surface rate of about 900 to perhaps 1000 gal/day/ft² (36.7 to 40.7 m³/m² day), which allows a margin of safety to compensate for hydraulic instability or short-circuiting. A basin with a poor inlet, or a well-designed basin for a smaller plant, would be designed for about 600 to 700 gal/day/ft² (24.4 to 28.5 m³/m² day). A tube clarifier could be safely designed at 90% of the theoretical subsidence rate, or 1600 gal/day/ft^2 (65.1 m^3/m^2 day). The required water depth could also be much less. Herein lies the charm and space-saving of the high-rate settler, but for large flows the cost is apt to be prohibitive because of the required provision for mechanically desludging from under the inclined tube modules. For smaller plants, hoppers can be used and these high-rate systems seem to be economical.

Only two of the most established types of high-rate settlers are discussed herein. For other references, consult Yao[7] and Culp *et al.*[8]

Lamellae Sedimentation

A typical commercial form of lamellae sedimentation is a downflow application wherein the settled sludge slides down smooth plastic plates, and travels in the same direction as the flow of water. The parallel plates, inclined at about 45°, are spaced 1-2 in. (2.5-5 cm) apart. The effluent is collected through a header system, which maintains equal flow through each of the spaces.

Laminar flow exists between the plates, and as the flow is equal through each section, there is great hydraulic stability. Each settleable solid particle flows in a very short trajectory inside the narrow sedimentation space and quickly joins the compacted sludge mass that is sliding to the bottom. There is no vertical or density current movement.

The transport of sludge down the inclined lamellae is claimed to be aided by the concurrent downflow. However, the displaced velocity is so low, especially near the deposited boundary layer, that the validity of this claim appears to be questionable. This seems particularly so since upflow designs have the same measure of success as do the concurrent, downflow designs.

Tube Sedimentation

There are several configurations of tube shapes and sizes (see Chapter 14). A typical commercial form is a system of square tubes, 2 in. x 2 in. (5 cm x 5 cm) in cross section inclined at 60° to the horizontal. For added strength, the inclination of each row of the tubes are alternated in the module. Usually the tubes are of smooth PVC or ABS plastic. In most applications the water flows upward and the sludge flows counter-currently downward.

At ordinary rates of application, the minuscule head loss through the tubes suffices to balance the throughput for good hydraulic response and the tube modules do an excellent job of clarification. The throughput moves upward through the tubes as a balanced laminar flow, and the sludge flow moves down the tube countercurrent to the throughput. The clarification results are comparable with those of lamellae. Typical designs are shown in Chapter 14.

Again, because of the superior hydraulic stability, the application rates in terms of gal/day/ft^2 of horizontal tube surface can be expected to be about two times (and perhaps three times) the surface rates applied to standard clarifier basins. The side water depth can also be less than for standard clarifier basins.

SLUDGE DISPOSAL

No clarifier design discussion is complete without a plan for the disposal of the unwanted underflow product—sludge. Traditionally, the sludge was disposed of simply by discharge into a nearby stream, but this is now forbidden.

In large plants, the best method of disposal is a complex problem and at present still somewhat of a dilemma. The cost of sludge disposal is sometimes apt to exceed, in both construction and operating cost, that associated with the treated water product. In lime softening plants, slurry blowdown is further thickened for disposal as cake and the filter wash water is frequently returned to the flocculation and clarifier sections. In a very large softening plant, the number of units tends to spread out this filter wash water return flow, which is intercepted in a holding tank and pumped back at a uniform rate of 24 hr/day. In a smaller plant, this does not work well because of the long periods between filter washings and sporadic and relatively great flow of wash water.

In small plants, clarifier alum sludge, filter backwash sludge, and pre-sedimentation sludge (if any) should, if at all possible, be ponded instead of dewatered by elaborate filter presses or centrifuges. When the ponds fill, perhaps they can be mucked out and hopefully the gunk can be hauled to land disposal. The overflow from these ponds is typically collected and slowly pumped back to the plant where it joins the raw water supply. At lime softening plants, the pond overflow, which is softened water, should always be salvaged by means of a small pump installed in a pit receiving the pond overflow. The salvaged water should always be pumped back very slowly (on a 20-hr/day basis) to avoid a deluge of recycle to the plant.

DESIGN OF SLUDGE HANDLING MECHANISMS

Long Rectangular Basins

The force required to move mechanical sludge collectors can be calculated according to the equation:

$$F = f W n = f A \qquad (1)$$

where F = the force in pounds in the cable or chain
       f = the load factor in lb/ft
       W = the width of the basin (length of header, blade or flight) in feet
       n = number of headers, blades or flights in contact with sludge
       A = total area in ft^2 of all blades in contact with sludge

(In metric units, F is in kg, f is in kg/m, and W is in m). The load factors are given in Table IV. These load factors have been gathered from empirical field testing and through years of observing collector overload conditions under a gamut of applications.

Table IV. Load Factors for Sludge Collectors

	Load Factor (f), lb/ft of width		
Type of Sludge	Suction Header Units	Bladed Carriage Units[a]	Plastic Chain Drag Units[b]
Alum or Iron	5	9	6
Lime (cold lime softening)	-[c]	22	13
Rethickened Lime	-	-	50

[a]For blades 12 in. deep. For deeper blades multiply f by the ratio, blade depth/ 12 in. Rail friction, if any, to be added. Carriage units frequently carry transverse suction headers for which suction header values should be used.

[b]For flights 6 in. deep. For deeper flights multiply f by the ratio, flight depth/6 in. Load factor includes sprocket and bearing friction, which are small for plastic chain, flights, and wearing shoes. For iron chain add 50% for friction (which tends to be a low estimate).

[c]Lime sludge should not be collected by suction headers.

The designer should realize that, for any given category of sludge, there is bound to be a wide variation in the resistance to move a scraper or fluidizing blade through that sludge. The extent of the resistance also depends a great deal on the dwell time and actual density of the sludge encountered. Also, to be practical, the resistances enumerated must lie in the upper quartile of empirical experience, but the "average" resistance may not be more than one-half of these values. Even so, continuous force to keep the mechanism going through high resistance situations should be provided. Thus, Equation 1, as applied to chain drag units, is modified to

$$F = f W [1 + 0.85 + (0.85)^2 + \cdots (0.85)^{n-1}] \qquad (2)$$

The upper layers of the sludge blanket have little effect. In alum sludges, settling near the forebay or influent end is hindered and dense sludge is carried by the density current to about the last 70% of the basin.

Example

Alum sludge in a basin 100 ft (30 m) long and 18 ft (5.5 m) wide is collected by a plastic chain drag unit with flights 6 in. (0.15 m) deep spaced at 10 ft (3 m). Compute the torque on a suitable sprocket to drive the unit.

The dense sludge is estimated to be confined to the last 70 ft of the basin; therefore, 7 flights carry the load. From Table IV and Equation 2,

$$F = 6 \times 18 \; [1 + 0.85 + (0.85)^2 + (0.85)^3 + (0.85)^4 + (0.85)^5 + (0.85)^6]$$
$$F = 6 \times 18 \times 4.53 \; = \; 489 \text{ lb (222 kg)}$$

A typical drive train would be an 11-tooth gear on the drive shaft and a 40-tooth sprocket with a pitch diameter of 2.77 ft (radius = 1.385 ft or 0.422 m) on the head shaft. The drive shaft torque is

$$T = 489 \times 1.385 \times 11/40 \; = \; 186 \text{ lb-ft (25 kgf m)}$$

This torque does not include friction.

If the drag chain unit were of iron with 6-in. (15 cm) iron pintal chain, reinforced wooden flights, and iron shoes bearing on iron rails, friction would have to be added. If 50% is added for friction, the torque is

$$T = 186 \times 1.50 \; = \; 279 \text{ lb-ft (39 kgf m)}$$

Circular Basins

A formula for the torque required to rotate a two-arm collector (wherein an arm is a unit extending from the pivot to the wall) is

$$T \; = \; \frac{D^2 f}{4} \tag{3}$$

where T = torque in lb-ft
   D = basin diameter in ft
   f = load factor for plastic chain drag units (Table IV)

The drive shaft torque must be reduced in proportion to the reduction gear system.

SUMMARY

Although sedimentation of flocculent solids to produce a clear effluent can be accomplished in several ways, the most tried, economical and easily operated unit is a well-designed clarifier basin. Such basins are typically equipped with a sludge scraper and underflow control system, and require only a very low head loss from influent to effluent.

Larger installations typically consist of long rectangular sedimentation basins, although sometimes circular center-feed basins are used. All good clarifier basins have special well-designed forebays or feedwells, and a well-distributed overflow weir trough system. Solids contact systems, with their combination of solids return, high-rate rapid mixing, floc aggregation and circumferential vertical rise clarification, also merit consideration, especially for lime softening.

Floc preparation before sedimentation should consist of instant blending, rapid mixing, and staged floc aggregation, and it has as much significance as any other feature of the plant on the clarification of the product.

The newer high-rate settlers such as inclined tube or plate modules warrant careful consideration, especially for small flows. They are also applicable for upgrading existing designed or overloaded clarifiers or poorly designed basins.

The design engineer may well select conservative surface-loading rate parameters, especially when dealing with small clarifiers serving small towns.

REFERENCES

1. Ainsworth, L. A., Superintendent of Water Treatment, Santa Barbara, California. Private discussion (1972).
2. Hudson, H. E., Jr. "Residence Times in Pretreatment," *J. Am. Water Works Assoc. Res.* (1975), p. 46.
3. Thwaytes, G. W. "Settling Tank Maintenance at Hyperion," *J. Water Poll. Control Fed.* (1962), pp. 1235-1243.
4. Crawford, G. "Toronto, Ashbridges Bay Plant," OWRD and Circa (1966-1968).
5. Flood, F. L. "Sedimentation Tanks," in *Seminar Papers on Wastewater Treatment and Disposal*, Boston Soc. of Civil Engineers (1961).
6. Camp, T. R. "Sedimentation and the Design of Settling Tanks," *Trans. Am. Soc. Civil Eng.* 111:895-936 (1946).
7. Yao, J. M. "Design of High-Rate Settler," American Society of Civil Engineers, EE5:621-637 (1973).
8. Culp, G. L., K. Hsiung and W. R. Conley. "Tube Clarification Process, Operating Experiences," American Society of Civil Engineers, SA5:829-849 (1959).

CHAPTER 10

SLUDGE DISPOSAL

Louis R. Howson, P.E.
Senior Partner
Alvord, Burdick & Howson, Engineers
Suite 1401, 20 N. Wacker Drive
Chicago, Illinois 60606

HISTORY

Sludge disposal is as old as public water supplies although it has only been recognized and designated as a "pollution" problem in recent years. The early public water supplies in America, in other than small communities, were usually taken from the nearest surface water sources. The selection of locations for many early American cities was based largely on the availability of surface water.

Initially, in rivers such as the Ohio, which normally carries relatively high turbidity, the water was pumped from the river to large reservoirs usually located on nearby hills. There it would be held for periods varying from a few days to as much as a month or more, during which the coarser materials in suspension would settle, and the water decanted from the top would be distributed to the consumers with no other treatment. Many such reservoirs are still in existence at Louisville, Cincinnati, Covington and elsewhere. They still serve a purpose, but today it is recognized their purifying ability is limited and totally inadequate for producing an acceptable water quality. Nevertheless, several important cities, including New York and Boston, still take their water supply from large impoundments which, through plain subsidence of long duration, produce water whose only treatment is chlorination before delivery to the distribution system without filtration.

Impounded supplies, whether large or small and whether or not they are followed by filtration, are sludge collectors, although not a part of the sludge disposal problem discussed here. In many places, however, the accumulation of sludge in impounding reservoirs has seriously reduced their storage capacity and thus its very presence has developed a water supply problem. Waterworks sludge disposal, as herein discussed, really started with rapid sand filtration about the turn of the present century.

A major development in the improvement of surface water supplies came through what are generally referred to as the "Louisville experiments." This was an extensive study made under the direction of Fuller[1] who assembled a staff of capable chemists and engineers to investigate a means of improving the quality of turbid waters. This work, which was the fore-runner of mechanical filtration, demonstrated that turbid river waters could be clarified using chemicals for the coagulation of turbidity, after which the water could be filtered through granular filters at relatively rapid rates to produce a clear, sparkling effluent of superior bacteriological quality.

Growing out of the Fuller experiments, mechanical filtration (called "rapid sand filtration" as contrasted to "slow sand" filters) made rapid strides and in the next few decades rapid sand water filtration expanded rapidly. With it came problems related to the disposal of the materials precipitated in the pretreatment (coagulating) basins and the solids removed by backwashing of the filters.

Initially, the problem of disposing of the wastes from water treatment plants was not considered important. Practically all of the filter wash water and most of the coagulation sludges were discharged back into the source from which the supply had been taken. The disposal of the heavier sludge from the settling process sometimes caused sludge banks detrimental to navigation. It was not until the 1960s that the waterworks industry recognized that filter wash water, too, was a pollutant and that no water treatment plant sludges should be returned to the streams.

In 1969 the American Water Works Association Research Foundation carefully selected a group of AWWA members with expertise in water treatment and the handling of wastes. Their reports[2,3,4,5] constitute a valuable contribution to better understanding of the problems arising from water treatment, and each of the four sections contains an extended bibliography. But while discussing many attempts at solutions, they are inconclusive as to practicability and costs.

In this chapter, the general problem is more specifically defined, and the most practicable solutions for water treatment plants with capacities of less than 10 mgd (0.44 m^3/sec) are given.

The character and amount of sludge produced in the purification of individual water supplies varies with the variations in quantity treated and the character of the individual water supplies. For illustration, the water supplies drawn from the Great Lakes normally have turbidities of less than

10 TU but due to the effects of winds, storms, etc., their turbidity sometimes reaches 100. Rivers are much more variable in character and turbidities of 2000 or more are fairly common while peak turbidities on such
rivers as the Rio Grande reach several thousand.

WHAT IS WATERWORKS SLUDGE?

Waterworks sludge is the suspended solid material (sand, silt, clay and
sometimes algae) present in the raw water plus the precipitated chemicals
added to facilitate separation of the suspended solids. Some of the chemicals may partially dissolve in the process, whereas others such as lime (used
largely in the softening process) combine with the hardness dissolved in the
water to produce large additional volumes of sludge. Usually, they have
no offensive odors.

Alum sludges are predominant in water treatment because alum is the
most frequently used coagulant. This sludge is colloidal, sometimes sticky,
and is difficult to thicken or dewater mechanically. Lime-softening sludge
is large in volume but comparatively easy to dewater. Each pound of lime
used in softening results in about 2-1/2 pounds of sludge. Polymers, powdered activated carbon, and coagulant aids sometimes add to the sludge
volume but are usually not determining factors in the solution of sludge
disposal problems.

The major part of alum sludge is created in the pretreatment (coagulation) process in which it is precipitated. Where large volumes of sludge
are produced, the removal is usually accomplished by mechanical means
and is continuous during the operation. Where the treatment is limited to
alum or polymers, the volume of sludge is relatively much smaller, and it
is usually stored in the coagulation basins from which it is removed periodically. Coagulation basins are usually cleaned twice a year when water of
moderate or low turbidity (such as from the Great Lakes or impounding
reservoirs) is treated, or they may be cleaned several times a year if the
water is more turbid (such as river waters). In plants of the sizes for
which this discussion is prepared, two to four times a year is the usual
period for sludge removal from the coagulation basins in treatment plants
(except softening plants).

The largest single item contributing to the volume of sludge produced
by coagulation is the turbidity or suspended solids of the raw water. Even
with waters from the Great Lakes having a turbidity usually less than 10
TU, about 50% of the sludge volume results from the turbidity.

There are approximately 4000 water treatment plants in the United
States. Hudson[6] reported that the quantity of solids from U.S. water treatment plant coagulation processes totaled about 1 million tons per year.

This figure indicates why recognition has been given in recent years to the importance of discharge of the waterworks wastewater sludges into the water courses without treatment, and the waterworks industry now recognizes the importance of the pollution factor in this by-product.

Data with respect to both volume and character of water treatment plant wastes are generally available only from larger plants that have adequate laboratory facilities and personnel. However, the data secured from the operations of large plants can usually be utilized in studying the criteria governing the design of smaller treatment plants by analyzing and expressing the data from the larger plants in terms of, for example, pounds of sludge per million gallons per TU turbidity. Most treatment plants, even the smaller ones, keep records of turbidity, but usually only the larger plants keep records of suspended solids.

A number of investigators have found that suspended solids in mg/l bear a substantially direct relationship to water turbidity in TU. Recently, Eckmann[7] showed this relationship, herein given in Figure 1. For Chicago, suspended solids in mg/l and turbidity in TU are interchangeable.

Waterworks sludges derive from two sources and are quite different in character: (a) sludge from coagulating basins, and (b) sludge removed by backwashing of the filters. The liquid volume of filter waste usually amounts to some 2 to 3% of the water filtered; it varies widely in character, ranging usually from 200 to 800 TU turbidity. The two types of sludge can be handled quite differently.

Filter backwash water occurs at frequent periods (once or twice a day), and the liquid volume is large (85 to 140 gal/ft^2 for 3.5 to 5.7 m^3/m^2) per wash. This waste is intermittent, of short duration (5 to 10 min), and is relatively low in solids content.

The volume of coagulation basin sludge usually amounts to less than 0.1% of the water filtered, but it has a high solids content.

CHARACTER OF SLUDGE

Plain-settled sludge, silt, sand and rock are relatively coarse. Plain-settling basins are now rarely included in new plants, but a few, built before coagulation and rapid sand filters, are still in service. Before coagulation at Louisville, Kentucky, about 12,000 tons of sludge per year (about 25% solids) is settled in basins which provide about 24-hr retention.

Coagulation basin sludge varies widely depending upon the character of the raw water, the type and quantity of chemical coagulants used and the character of the plant itself. This sludge normally has a solids content between 1 and 2% as drawn from the basins. In plants less than 10 mgd in size, there is rarely any treatment such as thickening. In large plants much research has been done in an effort to consolidate this sludge and reduce its water content. To date, most efforts have ended in failure and "thickening" of alum sludge cannot be economically accomplished.

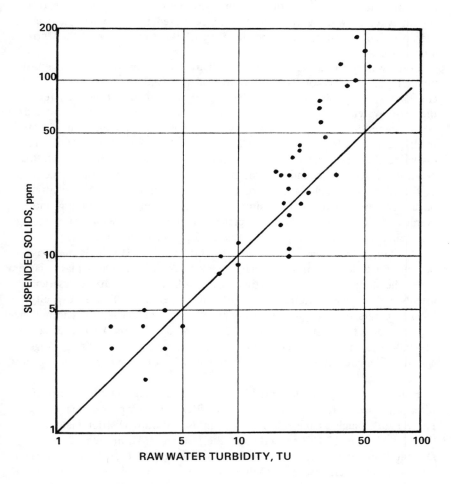

Figure 1. Comparison of turbidity with suspended solids on Lake Michigan water at Chicago.

METHODS OF SLUDGE DISPOSAL

The filter backwash sludge and the coagulation basin sludge are so different that different methods of disposing of them should be used.

Filter Backwash Sludge

Practically all new plants and, increasingly, by special construction a great number of the older plants are returning the filter wash water to the raw water intake and thus through the coagulation facilities and the filters. In this way, all waterworks sludge is concentrated in and collected from the coagulation basins. Separate handling of the two wastes is then no longer necessary.

Filter washing is usually accomplished by an upflow rise rate of approximately 15 gpm/ft^2 or 24 in./min vertical rise (0.61 m^3/m^2 min or 61 cm/min) which is about 6 times the rate of normal operation. To return the wash water directly to the raw water in a moderate size plant would not be practicable. Accordingly, it is becoming common practice in both small plants and large plants with a small number of large filter units to discharge the wash water into a detention reservoir. The capacity of this detention reservoir depends on the number of filters in the plant, the area of the filters (particularly the largest ones), the number of filter washings on a day having a maximum number of washes, and the maximum and average day's output of the filter plant. Obviously, the wash water storage reservoir should have a capacity in excess of a single wash.

Ordinarily, filter washes vary from 35,000 to 50,000 gallons per wash per million gallons of filter capacity being backwashed. On a mass diagram basis, the desirable capacity of the waste wash water equalizing reservoir usually falls between 8,000 to 16,000 gallons per mgd of rated capacity. Smaller plants require proportionally more retention capacity. The equalizing reservoir at Evanston, Illinois, (72 mgd or 3.2 m^3/sec) has a capacity of 19,400 gal/mgd. Wilmette, Illinois, has a capacity of 12,000 gal/mgd, and the Glencoe, Illinois, (2 mgd or 0.088 m^3/sec) reservoir has a capacity of 16,600 gal/mgd.

Extensive studies of the variable factors experienced in the return of a filter wash water to the influent of the filter plant have been made. Large plants, such as the 480-mgd (21 m^3/sec) South District Plant at Chicago with its 120 filter units and the 960-mgd (42 m^3/sec) Central District Plant at Chicago with its 96 filters, each of 10-mgd (0.44 m^3/sec) rated capacity, can discharge their wash water to the plant influent without interposing an equalizing basin between the filters and the plant inlet. The two Chicago plants have demonstrated that direct return can be accomplished without detriment to the plant operation.

Detailed studies show that plants with a design capacity of 100 mgd (4.4 m³/sec) or less require the construction of reservoirs to equalize the variations in the rates of the filter backwash water production, so wash water can be returned to the raw water intake at a uniform rate.

These retention reservoirs should have adequate, readily operable outlet control valves, to regulate the delivery of wash water to the influent at a rate no more than 10% of the filtration rate. The Green Bay, Wisconsin, plant, built in 1954, was one of the earliest designed to operate in this manner, and it has returned wastewater to the influent at as much as 16% without detriment to plant operation. Nevertheless, 10% is a reasonable basis for design.

The rate at which the wash water flows from the detention basin to the treatment plant inlet should be valve-controlled, so the rate of wash water return is substantially constant.

This method of handling the filter backwash water sludge problem has the following advantages:

1. All filtration plant wastewater sludge is collected in the coagulation basins.

2. Whereas filter backwash water is produced in large volume and relatively frequently, the sludge in the coagulation basins is accumulated over long periods and usually removed only two to four times per year.

3. Filter backwash water low in solids (turbidity) in the range from 200 to 800 TU is collected in the coagulation basin as coagulation sludge containing 1 to 3% solids.

4. In its passage through the coagulation basins, it combines with coagulated sludge whose solids content is much greater than that of the wash water and, therefore, much easier to handle.

5. The backwash water, amounting to 2 to 3% of the total pumpage, is retained in the system.

6. Withdrawal of sludge to the lagoons at intervals of several months allows time for drainage and evaporation in the lagoons. When the lagoon is needed again, the earlier deposited sludge should have dried to a moisture content of about 50%. A 12-in. (30-cm) layer of sludge delivered to the lagoon should be reduced to a depth of about 1 in. (2.5 cm) which indicates a long life expectancy for the lagoon.

7. The return of waste wash water to the coagulation system may produce some economy in the use of coagulant.

8. This method of recycling reduces the volume of waste wash water that eventually goes to the lagoons to a small proportion of the amount if handled separately. The coagulation basin is in effect a "thickener" for the filter backwash water.

If the backwash water is returned to the coagulation basin (the best practice for old as well as new plants), there is only one source of water plant sludge, the coagulation basins.

Sludge Lagoons

Lagoons function for both storage and drying, as they are used alternately for receiving and drying periods operated on an "apply and dry" rotation principle.

The combined sludge removed from the coagulation basins can readily be disposed of in lagoons, preferably on the treatment plant site, but, if necessary, the sludge can be readily pumped for relatively long distances to the lagoon site. Distance between filter plants and lagoons, while adding to the cost of a pipeline, is a minor cost item in total water filtration plant costs. At York, Pennsylvania, the sludge is pumped 2 miles to the lagoon and at Appleton, Wisconsin, the sludge lagoons are 2-1/2 miles from the softening plant. The Louisville (Kentucky) Water Company, which purifies and softens Ohio River water, has three lagoons located seven miles from the major treatment plant. No difficulty has been experienced due to distance.

Lagoons should have good drainage. That is usually best accomplished by constructing the lagoons by berms with the bottom of the lagoon at natural ground level. Preferably, sludge lagoons should not be built in excavated pits and never with bottoms below ground water level. Sludge should have drainage into the subsoil and its surface should be open to evaporation.

Waterworks sludge, particularly lime, does not consolidate under water, but it dries readily when exposed to air and by drainage through the soil. When lagoons are built above ground, the berms should be from 10 to 15 ft (3 to 5 m) high and far enough from property lines so that, if needed, their top elevation can be raised. This has been done by removing dried sludge from the interior of the lagoons for use as embankment material. Lagoon berms for larger plants or those with softening should be about 12 ft (4 m) wide at the top to facilitate the use of construction equipment.

Coagulation basin sludge is usually withdrawn from the basins to the lagoons in periods of low water use, such as April and October. Two or more lagoons should be provided for alternating use to allow between 6 months and 1 year for evaporation and seepage. Such a drying period reduces the sludge to the consistency of the soil at the site. Most treatment plants of 10 mgd (0.44 m^3/sec) capacity or less usually have only 2 to 4 coagulation basins so that cleaning affects operations as little as possible.

At Green Bay, the filter plant operated with a single, 2-acre (0.81 ha) lagoon for its first 9 years, during which it filtered approximately 30,000 million gallons of Lake Michigan water. At that time 2000 yd³ (1530 m³) of dried sludge was removed and the lagoon deepened. The lagoon has since operated another 15 years.

Sludge stored in lagoons in cold climate areas many benefit from freezing. Sewage sludge left on sand-drying beds over the winter dries to a porous, finely divided, somewhat granular material well-suited for direct application to lawns.

Softening Sludge

Softening plants produce a large volume of sludge. The volume may be approximately estimated by assuming that the dried sludge will, because of the combination of the lime with the calcium and magnesium in the water, be 2.5 times the volume and weight of the commercial lime used in the softening process (see also Figure 2). Where the water is from a turbid surface supply, the turbidity is often removed by coagulation before softening. The coagulated sludge and the softening sludge can both be disposed of in the lagoons. Because of the large volume of sludge produced by lime softening, withdrawal from the basins to the lagoons is more frequent. The number of lagoons, should, therefore, be increased so that when used in rotation, each lagoon has 6 months' rest and consolidation period. Howson[8] reached the following conclusions:

1. Land, when available for sludge disposal by lagooning, is usually the most economical method.

2. The cheapest means of transporting sludge from the treatment plant to the sludge disposal site is by pipeline.

3. Although sludges vary in their character, composition and volume, it may be assumed that on a dry basis there will be approximately 2.5 lb of sludge produced per pound of commercial lime used in the softening process.

4. Most softening plant sludges air dry in well-designed lagoons with supernatant removed to facilitate drying to about 50% moisture. New sludge should not be added to a lagoon until previous applications have dried.

5. When lime-softening sludge is reduced to 50% moisture content, it will occupy only about 16% as much volume as when applied at 90% moisture, or 8% as much volume as when applied at 95% moisture (see Figure 2).

6. Although data are meager, it is believed that sludge deposited in a lagoon where it is always submerged will not compact to less than about 70% water. In this condition it occupies twice as much storage space as when dried to 50%. The wet sludge is very difficult to handle.

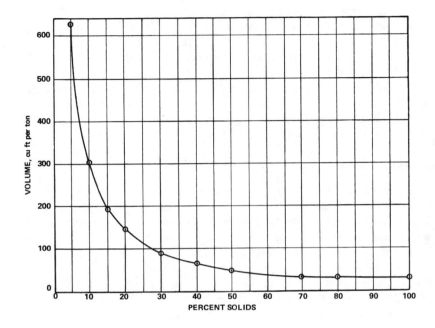

Figure 2. Relation between lime sludge volume and percent solids concentration.

7. When sludge is dried to 50% moisture, the total volume per year produced by a 1-mgd (0.044 m^3/sec) plant removing 100 mg/l hardness is about 2/3 ac-ft (820 m^3).

8. There should be at least three lagoons for lime-softening sludge, each capable of holding approximately one year's supply of wet sludge. Application of sludge to the lagoons should be rotated as consolidation to 50% is reached.

9. As a general figure, where sludge can be lagooned to 10-ft (3 m) depth, it is desirable to have 3 to 5 acres (1.2-2 ha) available for sludge beds for each mgd (0.044 m^3/sec) plant capacity in order to provide for the future.

10. Lagoons should be operated on a "fill and let dry" basis. Sludge should be applied to a depth of 3 to 5 ft (1-1.7 m), which usually shrinks to 6 to 10 in. (15-20 cm) depth when dried to 50% moisture.

11. Because of the desirability of applying as much as 3 ft (1 m) at a time, berms should be built 3 to 5 ft (1 to 1.7 m) above maximum sludge level, when filled. Berms should be broad enough (about 12 ft or 4 m) to permit operation of a drag line and to permit raising their top elevation.

12. Sludge dried to about 50% moisture is easy to handle and makes excellent material for berms. It dries more and compacts to become impervious.

13. In operating lagoons for sludge disposal, it is desirable to fill at one end and decant or otherwise withdraw supernatant at the other end. Relatively long and narrow lagoons are preferable.

14. It is desirable to keep the water level in lagoons as low as possible to facilitate compaction.

15. Lime sludge does not dewater properly under water. Compaction under water does not result in solids concentration of more than 30-40% solids. Lagoons should never be operated as ponds.

SUMMARY

There is no other method of sludge disposal that can successfully compete with lagoons in small treatment plants on the bases of economics or results. There have been numerous pilot plants using presses, centrifuges or other mechanical thickeners. The upper limit of solids concentration is usually about 30 or possibly 40% solids. The cost of installation usually exceeds that for lagoons and pipelines.

The partially mechanically dewatered sludge must be trucked to a disposal area with greater installation and operating costs than are incurred with lagoons. The water leaving the thickening equipment must usually be treated further for solids removal.

REFERENCES

1. Fuller, G. W. *The Purification of the Ohio River Water at Louisville, Kentucky.* (New York: D. Van Nostrand Co., 1898).
2. AWWA Research Foundation Report. "Disposal of Wastes from Water Treatment Plants—Part I," *J. Am. Water Works Assoc.* 61:541 (October, 1969).
3. AWWA Research Foundation Report. "Disposal of Wastes from Water Treatment Plants—Part II," *J. Am. Water Works Assoc.* 61:619 (November, 1969).
4. AWWA Research Foundation Report. "Disposal of Wastes from Water Treatment Plants—Part III," *J. Am. Water Works Assoc.* 61:681 (December, 1969).
5. AWWA Research Foundation Report. "Disposal of Wastes from Water Treatment Plants—Part IV," *J. Am. Water Works Assoc.* 62:63 (January, 1970).
6. Hudson, H. E., Jr. "How Serious is the Problem," Proc. Tenth Sanitary Engineering Conference, University of Illinois, Urbana, 1968.
7. Eckmann, D. E. Partner, Alvord, Burdick & Howson, Private communication.
8. Howson, L. R. "Lagoon Disposal of Lime Sludge," *J. Am. Water Works Assoc.* 53:1169 (September 1961).

CHAPTER 11

DESIGN OF FLOCCULATION SYSTEMS

A. Amirtharajah, Ph.D.
Associate Professor
Department of Civil Engineering and Engineering Mechanics
Montana State University
Bozeman, Montana 59717

INTRODUCTION

The flocculation stage in the water treatment process train is the aggregation or growth of the destabilized colloidal suspension. Flocculation follows, and in some instances overlaps, destabilization. Destabilization is essentially controlled by the chemistry of the process while flocculation is the transport step that results in collisions between destabilized colloidal particles, leading to the formation of flocs.

Two major mechanisms of flocculation are generally identified: (a) perikinesis, which is aggregation resulting from random thermal motion of fluid molecules and is significant for particles less than 1 to 2 μ, and (b) orthokinesis, which is induced by velocity gradients in the fluid. The latter is the predominant mechanism in water treatment. In addition, differential and fluctuating velocities can also cause contacts between particles leading to aggregation. This is a major mechanism in sludge blanket or solids contact clarifiers. A knowledge of the kinetics of the process of orthokinetic flocculation forms the basis for the design of the flocculators in a water treatment plant. The parameters that determine the rate at which aggregation occurs define the design dimensions of the process tank and its equipment.

This chapter summarizes the available kinetic models of the flocculation process and presents the design of mechanical flocculators and sludge blanket clarifiers, incorporating several ideas presented in recent literature. Mechanical flocculators and sludge blanket devices are intentionally identified as separate flocculation systems to stress their striking dissimilarities. Typical illustrative design calculations are included.

THEORETICAL CONCEPTS FOR DESIGN

Mechanical Horizontal Flow Flocculators

In 1917, von Smoluchowski showed that orthokinetic flocculation is characterized by the equation:

$$\frac{dN}{dt} = \frac{G}{6} n_1 n_2 (d_1 + d_2)^3 \tag{1}$$

in which dN/dt =   rate of collision between 1 and 2 particles
       G =   uniform velocity gradient
       n_1 and n_2 =   number density of 1 and 2 particles
       d_1 and d_2 =   diameters of 1 and 2 particles

Camp and Stein[1] presented an equation for estimating the mean velocity gradient, G, in terms of the power input, P, to the system as,

$$G = \left[\frac{P}{\mu V}\right]^{1/2} \tag{2}$$

in which V = volume of system and μ = dynamic viscosity of liquid. Camp[2] analyzed several flocculation basins and found satisfactory performance in basins that had the nondimensional parameter, Gt, with values in the range 2×10^4 and 2×10^5 and G values from 20 to 74 sec^{-1}.

For paddle-type mechanical flocculators, the power dissipated, P, in the liquid can be determined from the expression

$$P = \tfrac{1}{2} C_D A \rho v^3 \tag{3}$$

in which C_D =   coefficient of drag = 1.8 for flat plates
       v =   velocity of paddles relative to liquid ≈ 0.75 x velocity of paddle
       ρ =   density of liquid
       A =   area of paddle

If the paddle is such that significant velocity changes would occur along its length due to the distance from the shaft, then the expression needs to be integrated for an elemental area.[3]

Even for a constant G value, it is not possible to integrate Equation 1 analytically. However, the equation has been integrated numerically.[4,5]

The equation derived by Harris, Kaufman and Krone[4] from von Smolu-chowski's Equation 1 for a series of m identical continuous stirred-tank reactors is

$$\frac{n_1^o}{n_1^m} = (1 + K \phi G \frac{T}{m})\, m \tag{4}$$

in which n_1^o and n_1^m = number concentrations of primary particles in the influent and effluent from the mth reactor

$\quad\quad$ K $\quad\quad$ = constant

$\quad\quad$ ϕ $\quad\quad$ = floc volume fraction

$\quad\quad$ T $\quad\quad$ = overall residence time.

Several investigators[5-8] have confirmed two facts: (a) that there is a limiting floc size that bears an inverse relationship to the velocity gradient— that is, the smaller the velocity gradient, the bigger the limiting size, and (b) the flocs produced under higher velocity gradients, though smaller in size, are more dense. These facts can be partially rationalized by intro-ducing simultaneously the mechanisms of aggregation and break-up during flocculation.

Argaman and Kaufman[9] incorporated a break-up mode into the mech-anism of aggregation of flocs and developed

$$\frac{n_1^o}{n_1^m} = \frac{\left[1 + K_A G \frac{T}{m}\right]m}{[1 + K_B G^2 \frac{T}{m} \sum_{i=0}^{i=m-1} (1 + K_A G \frac{T}{m})^i\,]} \tag{5}$$

in which K_A = aggregation constant (dimensionless) and K_B = break-up constant (time^{-1}). The break-up or erosion of formed floc has been observed by several researchers.[5,10] In the above development, the floc volume fraction, ϕ, has been included in the constant, K_A. For clay tur-bidity and alum floc, the values of the constants are of the order of $K_A = 10^{4.3}$ and $K_B = 10^{7}$ sec. The variation of these parameters is shown in Figure 1, which may be used to design flocculators with alum as coagulant.

Two important conclusions can be drawn from the work of Argaman and Kaufman[9]: (a) there is a minimum time below which no flocculation occurs, whatever the value of G (*i.e.*, aggregations and break-up balance each other), and (b) compartmentalization significantly reduces the over-all detention time for the same degree of treatment. The second conclu-sion has been confirmed by several investigators and the recommended design practice is a minimum of three compartments.[2,11]

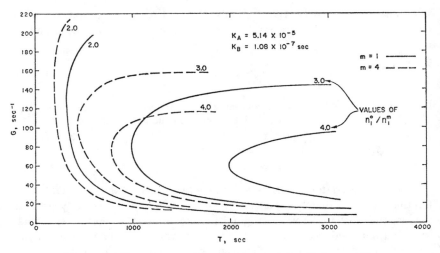

Figure 1. Flocculator performance curves (after Reference 9).

The concept that aggregation and break-up of flocculated particles occur simultaneously is generally accepted at the present time and most recent studies[12-15] have incorporated this concept. The mechanism of floc break-up in turbulent flocculation processes has been studied by Parker, Kaufman and Jenkins.[12] They also confirmed the validity of the form of Equation 5 for activated sludge floc.

Kao and Mason[16] studied nonadhering assemblages of spheres in shear fields and derived an approximate relationship for the aggregate radius, R_t, after a shearing time, t, as

$$R_0^3 - R_t^3 = k \, Gt \qquad (6)$$

in which R_0 = initial radius of assemblage and k = a constant depending on the flow field. The equation was derived by assuming that the number rate at which spheres are pulled off the periphery of the aggregate is proportional to the tensile stress generated on the surface of the aggregate. An analysis of the above equation implies that even at a constant value of G, the size of the aggregate radios R_t reduces with increasing time, t. This equation, based on fundamental considerations, lends support to Camp's[2,6] original use of the dimensionless product Gt as the predominant parameter of design. Most textbooks[3,17,18] recommend Gt values of 10^4 to 10^5 for design. Spielman and Quigley[15] have extended these studies by by using motion picture photography to show the break-up of cohesive aggregatives such as ferric hydroxide, and they indicate that erosion of aggregates occurs at the outermost tips of the elongated particles, which produces a steady stream of smaller particles.

Andreu-Villegas and Letterman[13] in a recent study (1976) indicated that optimum values exist for the flocculation time, T, and velocity gradient, G. Their bench-scale batch studies with alum floc and kaolin clay indicated that the flocculation period should be 20-30 min, since increases in time beyond this period do not significantly improve flocculation. The optimum value for G is given by the expression

$$(G^*)^{2.8} T = K \tag{7}$$

in which G^* = optimum velocity gradient, sec^{-1}
 T = flocculation time in minutes
 K = 4.9×10^5, 1.9×10^5 and 0.7×10^5 for alum concentrations of 10 mg/l, 25 mg/l and 50 mg/l, respectively.

These empirical results can be combined into an approximate single expression,

$$(G^*)^{2.8} T = \frac{44 \times 10^5}{C} \tag{8}$$

in which C = alum concentration in mg/l in the range 0-50 mg/l. The optimum value of G was defined as the velocity gradient that minimizes residual turbidity by flocculation and settling. The range of optimum G was between 20 and 50 sec^{-1}. The expression $(G^*)^{2.8}$ CT is similar to GCT proposed by others[18, 19] as the parameter for design.

In a simulation study, Ives and Bhole[14] confirmed Camp's ideas[2] that tapered flocculation with a diminishing velocity gradient was more efficient than uniform velocity gradient flocculation. The velocity gradients were reduced in the ratio 2:1 along the length of the flocculator. Other evidence[2,20] for improvement in flocculation by tapered velocity gradients from G = 150 sec^{-1} to 30 sec^{-1} has also been presented. Cockerham and Himmelblau[21] used Argaman and Kaufman's[9] kinetic equation (Equation 5) and performed a stochastic analysis of orthokinetic flocculation with a Monte Carlo simulation. They found that the overdesign factor (the overdesign necessary to cater to variation of the flow rates about the mean) was approximately 1.4 for all variations in flow rates. This result provides an objective means for providing safety factors in design of flocculators.

Flocculation Zone of Sludge Blanket Clarifiers

The flocculation zone of a sludge blanket clarifier functions as a fluidized bed with hydrodynamical similarities and dissimilarities to a horizontal flow flocculator. The interaction of the fluid liquid system has been shown to cause constantly fluctuating velocities amenable to analysis as a turbulent field. However, the actual fluid velocities are in the transitional range.[22,23] The similarity of a sludge blanket clarifier to a

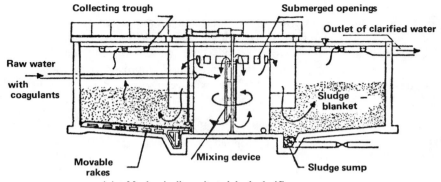

(a) Mechanically agitated bed clarifier.

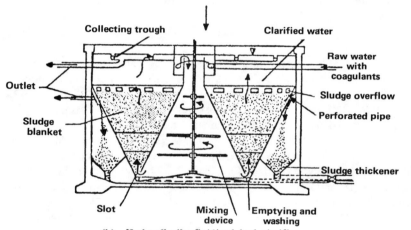

(b) Hydraulically fluidized bed clarifier.

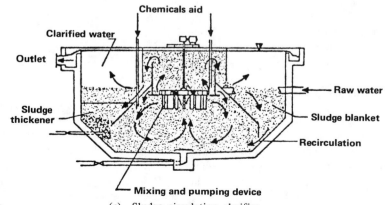

(c) Sludge circulation clarifier

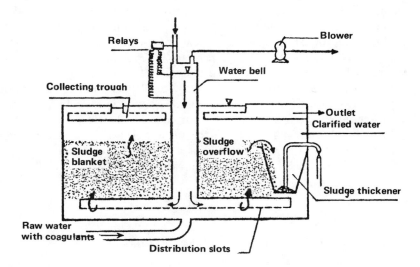

(d)   Unsteady discharge clarifier.

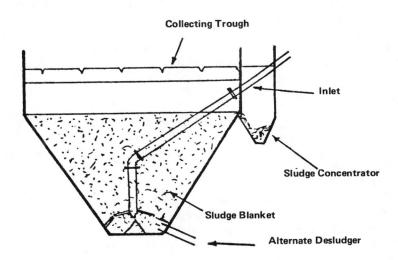

(e)   Hopper-bottom, hydraulically fluidized clarifier.

Figure 2.   Clarifiers with sludge blanket flocculation (after Reference 27).

horizontal flow flocculator lies in the fluid system functioning like a turbulent field. The predominant dissimilarity is in the high concentration of solids. Design approaches[6,24] have been commonly based on application of von Smoluchowski's Equation 1. However, Bond[25,26] Tesarik,[27] Ives,[19] and Brown and LaMotta[28] have sought design parameters related to the properties of a fluidized bed.[29]

In practice, several variations of the sludge blanket clarifier exist. Some include mechanical flocculation in a central chamber, others have no external power input. Designers often do not consider the flocculating effect of the sludge blanket when used in the vertical flow sedimentation section of a clarifier. Some typical sludge blanket clarifiers are shown in Figure 2.

Starting from von Smoluchowski's Equation 1, Hudson showed[6,24] that

$$\frac{n_T}{n_o} = e^{(-\eta G \phi T)/\pi} \tag{9}$$

in which n_T and n_o = number concentrations of particles in effluent and influent

η = collision efficiency factor

T = liquid residence time

Hudson's original development[6] was for horizontal flow flocculators, though several researchers[17-19] indicate the usefulness of the approach for sludge blanket clarifiers and suggest use of the parameter $G\phi T \approx 100$ for design of sludge blanket clarifiers.

Bond[25] theoretically developed an expression for the settling velocity of a suspension, v_s, in terms of the settling velocity of the individual particles of the suspension, v_p, as,

$$v_s = v_p (1 - f\phi^{2/3}) \tag{10}$$

in which f = shape factor = 2.78 for ferric and alum floc. The relation is similar to that derived by Rouse[30] in 1941 and a modification of which was used by Kalinske[31] for determining settling rates in solids-contact units.

Typical values for v_p developed using the above expression[26] are shown in Table I, in English and SI units. The term v_p decreases with a fall in temperature, but increases as the floc ages and turbidity accumulates. However, v_s/v_p is independent of temperature. Therefore, seasonal variations in temperature can be met by altering the throughput. Bond's design procedure[26] was to establish an applicable value for v_p based on typical values of temperature, coagulants used and, if possible, model tests. For design, the nominal upflow velocity, v, at the slurry separation line (which is the same as v_s), was equated to 0.5 v_p.

Table I. Settling Velocities for Floc Particles

Kind of Floc	Range of Settling Velocities v_p at 15°C	
	(ft/min)	(mm/s)
1. Fragile Floc, Color Removal with Alum	0.12-0.24	(0.61-1.22)
2. Medium Floc, Algae Removal with Alum	0.20-0.30	(1.02-1.52)
3. Strong Floc, Turbidity Removal with Alum	0.24-0.35	(1.22-1.78)
4. Strong Floc, Lime-Soda Softening	0.24-0.35	(1.22-1.78)
5. Crystalline Floc, Calcium Carbonate Granules	0.40-0.66	(2.03-3.35)

One of the important aspects of the above theory is that the floc volume fraction, ϕ, in the fluidized bed is dependent on the upflow velocity. Fluidization theory[22] establishes this fact with no doubt. The range of values for floc volume fraction for alum and lime floc is from 0.06 to 0.10. Using $v_s/v_p = 0.5$ and f = 2.78 in Equation 10 gives values of $\phi = 0.076$, which may be used for purposes of design.

Tesarik[27] and Brown and LaMotta[28] have shown that the sludge blanket behaves as a fluidized bed. The Richardson and Zaki[29] equation for its description is of the form,

$$\frac{v}{v_p} = \epsilon^n = (1 - \phi)^n \tag{11}$$

in which ϵ = porosity or fraction void and n = a coefficient. Tesarik found n = 4 for flocs formed with alum, alum and silica, and chlorinated ferrous sulfate. Brown and LaMotta found n = 1.75 and 3.74 for alum floc in different countries, n = 5.68 for iron floc, and n = 6.40 for alum with polyelectrolyte. However, these values of n were dependent on the assumed initial porosities, which were questionable.[32] It is not surprising that the value of n varies since fluidization literature[22,33] indicates that n is a function of particle Reynolds Number (i.e., particle size and density), and particle shape. These would change with the coagulants used, quality of raw water and conditions of initial mix. The assumed value for the initial porosity would also affect the results.[32]

Two important considerations for design were noted by these authors:

1. Using potential flow theory, Tesarik[27] showed that the minimum depth from the clear water collection launder to blanket level was equal to half the distance between the launders.
2. Brown and LaMotta[28] indicated that the time, t, used in parameters for evaluating the performance of a sludge blanket should be the solids residence time, t_s, given by:

$$t_s = \frac{sV}{s_b Q} \tag{12}$$

in which s and s_b = volumetric solids concentration in influent and blanket
V = volume of blanket
Q = flow rate

In horizontal flow flocculators, the solids residence time and liquid flow-through time are the same. O'Melia[17] also notes this distinction, but it is not commonly made in the literature[3,18,19] where the liquid flow-through time is used. Flocculation in a sludge blanket can be rationalized with either viewpoint; this concept needs further work for elucidation. The solids residence time concept seems more applicable since it provides a basis for rationalizing the contrasting flocculation in the sludge blanket.

Ives[19] presented a theoretical study of the operation of a sludge blanket clarifier by explaining its operation as orthokinetic flocculation in an infinite series of elemental layers. The theory was for a cone-shaped flocculator of 54° angle. Using Bond's Equation 10 and the power loss in a fluidized bed to determine the velocity gradient, he developed an expression for the flocculation criterion as,

$$\sum_{L_\ell}^{L_u} G_L \phi_L T_L = [\frac{(\gamma_s - \gamma)\pi}{4Q\mu f^{9/2}}]^{\frac{1}{2}} \int_{L_\ell}^{L_u} L [1 - (\frac{L_\ell}{L})^2]^{9/4} dL$$

$$= \frac{L_\ell^2}{2} [\frac{(\gamma_s - \gamma)\pi}{4Q\mu f^{9/2}}]^{\frac{1}{2}} F(\frac{L_u}{L_\ell}) \tag{13}$$

in which T_L = liquid detention time at level L
γ_s and γ = specific weights of floc and water
f = Bond's shape factor for flocs = 2.78
L_l and L_u = diameters of lower and upper ends of a conical frustrum of 54° angle (also equal to distances from vertex of cone).

The function $F(L_u/L_l)$ was integrated numerically and its values are shown in Figure 3.

Assuming a diameter for the size of floc particles and $\sum G_L \phi_L T_L = 100$, it was possible to use Equation 13 to design the dimensions of a conical sludge blanket clarifier. The typical range of values for $\sum G_L \phi_L T_L$ was from 60 to 120. An important point indicated in earlier studies[25,26] was that, due to the decreasing velocity in the expanding cone, the concentration of floc is highest at the blanket surface, and a slurry weir at the surface level provides effective control of the blanket. This was confirmed by Ives.[19]

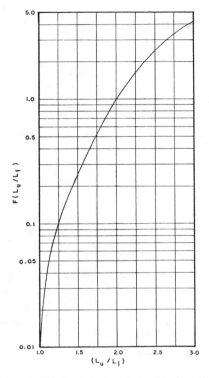

Figure 3. Graph of function, $F(L_u/L_l)$, for sludge blankets (after Reference 19).

The above review indicates that the functioning of horizontal flow flocculators is significantly better understood than that of sludge blanket clarifiers. While further theoretical and experimental work is urgently needed, the design of flocculation units cannot, obviously, be postponed until elegant and exact theories are developed. Hence, the later sections of this chapter illustrate procedures for design based on the above theories. At present, these theories are the best available for design although they are, admittedly, incomplete.

DESIGN ALTERNATIVES—MIXING DEVICES FOR FLOCCULATION

Three types of mechanical mixing devices for flocculators are in common use: (1) paddle or reel-type devices, (2) turbines and (3) axial flow propellers. Typical units and the spatial distribution of velocity gradients they produce are shown in Figures 4 and 5. Paddle or reel-type devices

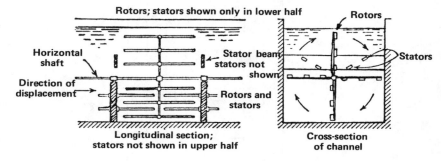

(a) Paddle type with rotors and stators.

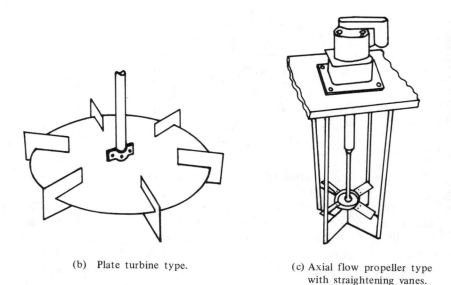

(b) Plate turbine type.

(c) Axial flow propeller type with straightening vanes.

Figure 4. Typical mechanical flocculators.

are mounted horizontally or vertically and rotate at low speeds, 2-15 rpm. Design is based on limiting the tip speed to 1-2 ft/sec (0.3 to 0.7 m/sec). Argaman and Kaufman[9] used a stake and stator device similar to reel-type units and found it superior to a turbine. A design of this type of flocculator is illustrated in Fair, Geyer and Okun.[3]

Turbines are commonly flat-bladed devices connected to a disc or radius arm. The plane of the flat blades is in the plane of the rotating

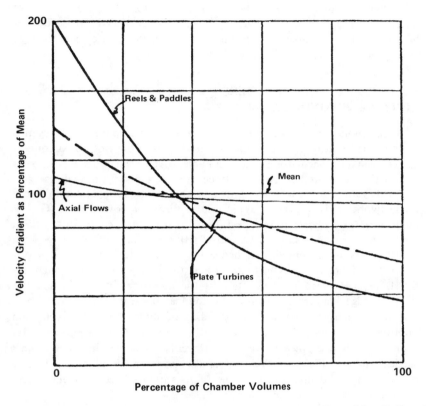

Figure 5. Distribution of velocity gradients in mechanical flocculators (after Reference 24).

shaft. The blades can be mounted horizontally or vertically and operate at 10-15 rpm. Several researchers[9, 24, 34, 35] have found this unit the least effective for flocculation. Walker[34] found that plate turbines were effective up to velocity gradients of 40 sec^{-1}, but produced high-velocity currents at G values greater than 45 sec^{-1}. His suggested design parameter was to limit the maximum peripheral velocity to 2 ft/sec for weak floc, and to a maximum of 4 ft/sec (1.2 m/sec) for strong floc.

The third type of device is an axial flow propeller shaped like a ship's screw. The pitched blades are inclined at 35° to the plane perpendicular to the axis of rotation with a large hub area. The unit may be installed vertically or horizontally. The vertical unit can be installed off center.[3] These are high-energy flocculation devices operating at 150-1500 rpm. With stationary straightening cross vanes installed ahead of the propeller, these devices can be used for G values up to 90 sec^{-1}. There is no limitation on tip speeds.[24,34] Hudson[24] and Walker[34] favor these devices because of their simplicity in installation and maintenance and because they produce uniform turbulence in the flocculator.

SLUDGE BLANKET CLARIFIERS—
APPLICABLE CONDITIONS

A commonly posed question with conflicting viewpoints is determination of the conditions under which a horizontal flow flocculator or a sludge blanket flocculator is most suitable. American practice has been to use the horizontal mechanical flocculator. A historical analysis shows that this evolution can be attributed to three reasons: (1) Camp's theoretical work[1,2] was directly applicable in terms of power input (G, GT requirements) to horizontal-type flocculators, (2) the control of variable power input made operation easy, and (3) many variants of sludge blanket clarifiers are patented, and hence their use is discouraged. It should be evident from the previous review and the names of authors such as Bond, Tesarik, Ives, and LaMotta that the literature[19,25,27,28] and the use of sludge blanket clarification is dominated by European practice. The writer believes that as understanding of sludge blanket flocculation improves this type of flocculator will become much more popular. Its capital cost is low since it combines flocculation and sedimentation in a single structure, the operating cost is also low because of minimal power requirements, and in addition, some of its variants have no mechanical equipment that needs maintenance.

The sludge blanket is the ideal flocculation system when it is used for lime softening or combined softening and alum coagulation. Under these conditions the solids contact provide ideal sites as nucleii for precipitation and growth of the calcium carbonate crystals.

Engineers often rationalize that the conventional horizontal flow flocculator is inadequate for treating low-turbidity waters due to insufficient contact opportunity and hence suppose that the sludge blanket unit is the preferable flocculation system. This analysis is only partially correct. It is true that if a heavy blanket is already in existence in the clarifier and the turbidity of the raw water becomes low, the flocculation efficiency is still high in a sludge blanket system. However, the system is dynamic, and trying to build a blanket that does not initially exist with a low-turbidity water is difficult because the microflocs tend to overflow into the weirs. Operational data[36] have shown that a medium-heavy turbid raw water (80-200 TU) that did not change in quality over several days produced excellent sludge blanket flocculation. In fact, a pilot plant[36] that had an unstable blanket (due to low turbidity in the raw water) was easily improved by adding turbidity from a soil suspension.

Under tropical conditions[36] the stability of the blanket declines during the afternoon hours. It has been hypothesized that absorption of the infra red rays of the sun causes this instability due to production of density currents at the surface of the blanket. However, this needs to be substantiated with definitive studies.

Another common criticism of sludge blanket clarifiers is that they are not suitable for stop-start operation. A plant-scale parallel study in Sri Lanka of intermittent and continuous operation of alum flocculation of a surface water in hopper-bottom type flocculators indicated no difference in clarified water quality. During initial start-up, accelerated formation of the blanket may be obtained by gradually increasing the flow rate while adding turbidity artificially. Nozzles to break up compacted sludge in start-up are also recommended.[26] Theoretically, it is possible to operate sludge blanket clarifiers at considerably higher loading rates than presently practiced.[19] The chief limitation is design of the inlet so as to prevent high velocity streaming of the water. Baffles or other arrangements at the inlet would assist considerably in this respect for hopper-bottom type clarifiers.

DATA NEEDED FOR DESIGN

The minimum data required for design of the flocculation unit are similar to those for the rapid mix unit indicated in Chapter 8. Procedures for data collection from detailed predesign studies are shown in Chapter 3. Using Equation 1 of Chapter 3 developed by Tekippe, it is possible to use the simple jar test apparatus to obtain values for some of the parameters detailed under theoretical concepts for design. It is strongly recommended that maximum predesign studies, within the constraints of time and money, be made with the actual raw water to be treated. The use of a pilot plant and the development of values for the required coefficients in Equation 5 (as done by Argaman[37]) are recommended for the design of mechanical flocculators.

ILLUSTRATIVE DESIGN PROBLEMS

Example 1

The water supply for a town is taken from a river with considerable variations in quality. The raw water analyses for a year are shown in Table II, with typical coagulant doses obtained from jar tests. The design

Table II. Data for Design in Example 1 River Source

Occurrence, %	11	34	28	13	9	3	2
Raw Water Analysis							
Turbidity, TU	28	45	81	126	140	180	>200
Alkalinity, mg/l as $CaCO_3$	86	132	92	84	93	77	82
Temperature, °C	13	10	8	18	18	16	15
Jar Tests							
Alum Dose,[a] mg/l as alum	18	27	36	28	41	31	36
pH	6.4	6.7	6.4	6.4	6.4	6.2	6.4

[a]Most jar tests indicated an optimum alum dose and possible charge reversal and restabilization.

flows for 24-hr operation of the coagulation-flocculation units are: average day = 1.0 mgd (43.8 liter/sec) and maximum day = 1.5 mgd (65.7 liter/sec). Design the flocculation unit for the treatment plant. The example is a continuation of that in Chapter 8 for which an in-line blender type of rapid mix unit was designed.

Optimum Velocity Gradients

Turbidity varies considerably over the year, with the higher ranges (> 126 TU) occurring 27% of the time. For operational flexibility with varying raw water quality, a mechanical horizontal flow flocculator is chosen. Alum is the only chemical needed since sufficient alkalinity is available in the raw water. The most common alum doses are 27-28 mg/l and 36 mg/l.

The maximum design day is 1.5 mgd for the coagulation-flocculation units. It should be noted that the plant output will be 3-8% less than 1.5 mgd to allow for sludge and backwash water losses. Designs usually consider these losses. The overdesign factor estimated by Cockerham and Himmelblau[21] to compensate for variations in flow rates is 1.4 for flocculation units. A conservative design approach will use 1.5 mgd for design flow and a factor of 1.4 on this flow for variations in flow rates. The present design is based on a flow rate of 1.5 mgd.

The advantages of compartmentalization are indisputable. Assume a four-compartment flocculator. Using Equation 7 and the common alum doses,

$$(G^*)^{2.8} T = \frac{44 \times 10^5}{27} \text{ to } \frac{44 \times 10^5}{36}$$

$$= 1.63 \times 10^5 \text{ to } 1.22 \times 10^5$$

in which T = flocculation time in minutes. Assume a total flocculation time of 30 min, since greater times[13] do not produce corresponding increases in efficiency.

$$(G^*)^{2.8} = \frac{1.63 \times 10^5}{30} \text{ to } \frac{1.22 \times 10^5}{30}$$

$$= 5433.3 \text{ to } 4066.7$$

$$G^* = 21.6 \text{ to } 19.5 \text{ sec}^{-1}$$

For T = 20 min, the corresponding range of G* values is 24.9 to 22.5 sec^{-1}.

For an optimum G value of 22 sec^{-1} and a flocculation time of 20 min, GT = 2.6 x 10^4. This falls within Camp's guidelines of GT = 10^4 to 10^5. Jar tests with these values of G and t should be run on the raw water to confirm the adequacy of the parameters chosen.

Using G = 22 sec^{-1}, T = 20 min = 1200 sec and m = 4 in Equation 5,

$$\frac{n_1^o}{n_1^4} = \frac{[1 + 10^{-4.3} \times 22 \times \frac{1200}{4}]^4}{[1 + 10^{-7} \times 22^2 \times \frac{1200}{4} \sum_{i=0}^{3} (1 + 10^{-4.3} \times 22 \times \frac{1200}{4})^i]}$$

$$= 2.87$$

Alternatively, this result can readily be read from Figure 1 for G = 22 sec^{-1}, T = 1200 sec and m = 4. The design is satisfactory. The curves on this figure indicate that increasing the G value at this same detention time would improve the efficiency of flocculation. Furthermore, Figure 1 shows that the point (G = 22, T = 1200) is not too close to the minimum flocculation time. It would be more conservative to increase the total detention time to 30 min, but the writer uses 20 min since compartmentalization and variable G values permit attainment of higher efficiencies. This is an engineering judgment decision. Ideally, the design should be based on the constants K_A and K_B developed for the actual water and flocculator characteristics of the prototype using a pilot plant. Argaman[37] illustrated the excellent correspondence between pilot-plant tests and plant-scale operation by this approach, obtaining values for K_A and K_B equal to 1.8 x 10^{-5} and 0.80 x 10^{-7}, respectively. These values are not very different from those determined by Argaman and Kaufman.[9]

In order to incorporate the advantages of tapered flocculation, the optimum design G of 22 sec^{-1} is used for the third compartment and the G values are tapered on either side to produce velocity gradients G_1 = 40 sec^{-1}, G_2 = 30 sec^{-1}, G_3 = 22 sec^{-1}, G_4 = 18 sec^{-1} in the four compartments, respectively.

Dimensions and Layout of Flocculator
in English Units

The detention time in each compartment of the flocculator is 5 min.

Design flow rate   = 1.5 mgd x 1.547 cfs/mgd = 2.32 cfs
Volume of each compartment = 2.32 x 5 x 60   = 696 ft^3

It is preferable to have the plan area as a square.

The depth of the flocculator depends on ground levels, the levels of adjacent units and the general layout and hydraulics of the plant.

Assume a required water depth of 9 ft.

$$\text{Side dimension of flocculator} = \sqrt{\frac{696}{9}} = 8.79 \text{ ft}$$

With a 2-ft freeboard allowance,

Design dimensions of each of
compartments   = 9 ft x 9 ft x 11 ft deep

If the water depth that can be accommodated at this site is only 7 ft, the dimensions of the compartments would be 10 ft x 10 ft x 9 ft. Designers usually use a horizontal flow layout as shown in Figure 6-A. A far better layout, assuming ground conditions permit, is Figure 6-B. It has several advantages: (a) it uses common wall construction, which reduces capital costs, (b) the reversal of flow from compartment 2 to 3 provides additional hydraulic baffle flocculation, (albeit at the sacrifice of some additional head loss), and most important, (c) with additional hydraulic conduits it provides for a parallel path for easy duplication and expansion, and (d) emergency cleaning is possible by shutting down one path temporarily with two compartments. Possible piping arrangements and layout for expansion are shown schematically in Figure 6-B.

Design of Stirrer

For operational flexibility, simplicity in construction and because the flocculator is small, axial flow type propellers are selected for mixing. Calculations for designing the mixer are based on Equations 1, 2 and 3 of Chapter 8.

For the first compartment, probable operation is at $G = 40 \text{ sec}^{-1}$. However, provision should be made to operate with a variable speed motor in the range of $G = 15$ to 75 sec^{-1}. Because G varies as the 1.5 power of speed, a 1:5 range of G needs a 1:3 range of speed. Use the highest viscosity, which occurs at the lowest temperature of 8°C.

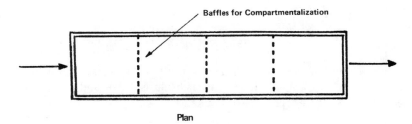

Plan

A. Commonly used layout.

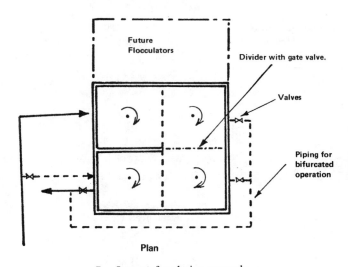

Plan

B. Layout for design example.

Figure 6. Alternative flocculator layouts.

Viscosity at $8°C, \mu$ = 1.3872 centipoises x 2.088 x 10^{-5} lb sec/ft^2/centipoise
= 2.896 x 10^{-5} lb sec/ft^2

Using the highest value of G for motor power and rearranging Equation 1 of Chapter 8, the power required is:

$$P = G^2 \mu V = 75^2 \times 2.896 \times 10^{-5} \times (9 \times 9 \times 9)$$
$$= 118.75 \text{ ft lb/sec}$$

Assuming a motor efficiency of 80%,

$$\text{Motor horse power} = \frac{118.75}{550 \times 0.8} = 0.27 \text{ hp}$$

Use a standard 1/3-hp motor.

Equation 2 of Chapter 8 is

$$P = \frac{k}{g} \rho N^3 D^5$$

For a three-bladed propeller impeller, $k = 1.0$ as indicated in Chapter 8. Assuming the impeller is 2 ft (*i.e.*, each blade is 1 ft), and power input to mixer at 80% efficiency,

$$(\frac{1}{3} \times 550 \times 0.8) = \frac{1.0}{32.2} \times 62.4 \times N^3 \times 2^5$$

$$N = 1.33 \text{ revolutions/sec} = 79.9 \text{ rpm}$$

Try a variable speed drive with a speed range of 3:1, say 27 to 81 rpm. The cheapest design would be to use a standard range of speeds from typical manufactured units. At 27 rpm,

$$P = \frac{1.0}{32.2} \times 62.4 \times \left(\frac{27}{60}\right)^3 \times 2^5 = 5.65 \text{ ft lb/sec}$$

hence,

$$G = \left[\frac{5.65}{2.896 \times 10^{-5} \times (9 \times 9 \times 9)}\right]^{1/2} = 16.4 \text{ sec}^{-1}$$

if it is assumed that all power input to mixer is transmitted to water. An alternate calculation approach to check the G value at 81 rpm is the use of Equation 3 of Chapter 8.

$$\frac{G_1}{G_2} = \left(\frac{N_1}{N_2}\right)^{3/2}$$

$$\frac{G_{81}}{16.4} = \left(\frac{81}{27}\right)^{3/2}$$

hence

$$G_{81} = 85.2 \text{ sec}^{-1}$$

Note that the power required at the higher G value must be provided for the motor. The identical motor and mixer should be provided in the second compartment.

Similar calculations indicate that with a motor design horsepower of 1/4 hp and a speed range of 72 to 24 rpm, the velocity gradients developed will range from 75 to 15 sec^{-1}. It should be noted that by tapered flocculation, the power of the motors in compartments 3 and 4 is less than the power of that in compartments 1 and 2. This type of design approach would provide for flexibility in operation and minimize power costs.

The details of the mixer arrangement would be three or four blades, each 1.0 ft long, pitched at 35° and installed on a 9-in.-diameter disc.

Vanes must be installed below the blades to straighten the flow. Plates 3 ft long and 9 in. deep arranged crosswise are adequate. The compartments must be separated from each other with a divider. An opening in the divider at floor level provides for each inlet. A gate valve can be incorporated for isolating streams. The velocity through the inlet should be limited to 0.5-2 ft/sec. It has been shown[38] that baffles with 3% openings produce relatively negligible short-circuiting. This would be an alternate arrangement for compartmentalization. However, this type of divider makes it more difficult for isolating chambers for cleaning purposes.

Design of Flocculator in Metric Units

Design flow rate = 1.5 x 43.8 liter/sec = 65.7 liter/sec

Detention time in each compartment of the flocculator is 5 min. Volume of each compartment = $(65.7 \times 5 \times 60 \text{ liter})/(10^3 \text{ liter/m}^3)$ = 19.71 m³. Assuming a water depth of 3.0 m, with a freeboard allowance of 0.5 m, the side dimension of the flocculator is $(19.71/3.0)^{1/2}$ = 2.56 m. Design dimensions of each of four compartments = 3 m x 3 m x 3.5 m

Viscosity at 8°C = 1.3872 centipoises = 1.3872 x 10^{-3} Nsec/m²

The first compartment is designed for a G range of 15 to 75 sec^{-1}.

$$\text{Power required} = G^2 \mu V$$
$$= 75^2 \times (1.38 \times 10^{-3}) \times (3 \times 3 \times 3)$$
$$= 209.6 \text{ watts or Nm/sec}$$

Assuming a motor efficiency of 80%,

$$\text{Motor power} = \frac{209.6}{0.8 \times 1000} \text{ kW}$$
$$= 0.26 \text{ kW.}$$

Design power of motor = 0.30 kW

Equation 2 of Chapter 8 is dimensionally correct, where the density, ρ, in English units is measured in lb. As the mass density is normally measured in SI units as kg/m³, the modified equation for SI units is,

$$P = k \rho N^3 D^5$$

with ρ = 1000 kg/m³ and k = 1.0 for a three-bladed propeller. Assume diameter of impeller is 0.60 m,

$$0.30 \times 0.8 \times 1000 = 1.0 \times 1000 \times N^3 \times (0.60)^5$$
$$N = 1.45 \text{ revolutions/sec}$$
$$= 87 \text{ rpm}$$

Use a variable speed drive with a speed range of 3:1, perhaps 30-90 rpm. Use standard manufactured unit. At 30 rpm,

$$P = 1.0 \times 1000 \times \left(\frac{30}{60}\right)^3 \times (0.60)^5 = 9.72 \text{ Nm/s}$$

$$G = \left[\frac{9.72}{1.3872 \times 10^{-3} \times 3 \times 3 \times 3}\right]^{1/2} = 16.1 \text{ sec}^{-1}$$

Using Equation 3 of Chapter 8,

$$\frac{G_1}{G_2} = \left(\frac{N_1}{N_2}\right)^{3/2}$$

$$\frac{G_{90}}{16.1} = \left(\frac{90}{30}\right)^{3/2}$$

hence,

$$G_{90} = 83.7 \text{ sec}^{-1}$$

Similar calculations indicate that a motor design power of 0.10 kW and a speed range of 75 to 25 rpm produce velocity gradients of 76 to 15 sec^{-1}. These would be the mixers for compartments 3 and 4.

The details of the mixer arrangement would be three or four blades, each 30 cm (0.3 m) long, pitched at $35°$ and installed on a 0.2 m diameter disc. Vanes are installed below the blades to straighten flow. They can be plates 1.0 m long and 0.2 m deep, arranged crosswise. The compartments must be separated from each other by a divider and an opening in the divider at flow level serving as an inlet. The velocity through the inlet should be limited to 0.1 to 0.5 m/sec.

Example 2

The water supply for a town with limited water resources is taken from a river and a ground water source. The mixed raw water has considerable seasonal variations in quality. Treatment requires removal of hardness, turbidity and color. Design a hopper-bottom type solids contact clarifier for a design flow of 1000 gpm (1.3 mgd) for 22-hr operation on the maximum day. (In metric units, the design flow = 63 liter/sec).

Analysis of the raw water (Table III) indicates that lime-soda ash softening is required during 53% of the year when large volumes of ground water are used, and alum coagulation is needed during other months. Jar tests indicate that softening the water with stoichiometric quantities of lime and soda ash reduces the total hardness to about 60-70 mg/l as $CaCO_3$, and the turbidity and color are reduced to values of 2-5 TU and 3-8 color units. Small amounts of sodium aluminate improve the magnesium hydroxide floc settling characteristics.

Table III. Data for Design in Example 2 Combined River Ground Water Source

Occurrence, %	10	31	26	17	9	5	2
Raw Water Analysis							
Turbidity, TU	30	80-140	40	60	180	140	240
Color, CU	20	60-140	40	50	140	130	150
Total Hardness, mg/l, as $CaCO_3$	180	80	120-250	120-250	40-60	40-60	-
Calcium Hardness, mg/l as $CaCO_3$	75	25	40-75	40-75	10-20	10-20	-
Magnesium Hardness, mg/l, as $CaCO_3$	105	55	80-175	80-175	30-40	30-40	-
Alkalinity, mg/l in $CaCO_3$	80	45	60-150	60-150	30-40	30-40	-
Temperature, °C	13	21	17	13	18	23	18

*Dimensions and Layout of Flocculator
in English Units*

Table I gives the range of settling velocities for different flocs at 15°C. Since color and turbidity occur together and 50% of the time the plant would soften the water, the controlling settling velocity for design is chosen to be 0.25-0.30 ft/min at 15°C. Assume a velocity of 0.28 ft/min for design. The actual design should be for 18-23°C, so the lowest temperature of 18°C should be used.

Assuming the flocs settle according to Stokes' law (for calculating temperature effects only),

$$v_p \propto \frac{1}{\mu}$$

therefore,

$$v_p \text{ at } 18°C = \frac{\mu_{15°C}}{\mu_{18°C}} \times 0.28 \text{ ft/min}$$

$$= \frac{1.145}{1.06} \times 0.28 \text{ ft/min} = 0.30 \text{ ft/min}$$

A conservative approach to design is to use the lowest settling velocity for floc and the lowest temperature. However, as we are already calculating for the average alum floc characteristic and as the plant would often soften water with very much stronger floc, the design can be based on the average floc and lowest temperature during alum flocculation.

Using Bond's criterion,[25] the design superficial velocity is

$$v_s = 0.5 \ v_p = 0.5 \times 0.30 \text{ ft/min} = 0.15 \text{ ft/min}$$

$$\text{Design flow} = 1000 \text{ gpm} = \frac{1000}{7.48} = 133.7 \text{ ft}^3/\text{min}$$

Therefore, the surface area required = 133.7/0.15 = 891 ft^2. Assuming at least two flocculators, the area of each flocculator = 446 ft^2. With square sides, the length of each side at slurry separation line = $\sqrt{446}$ = 21.1 ft.

Preliminary design dimensions of each of two flocculators = 22 ft x 22 ft. These dimensions should be checked with Ives'[19] approach given by Equation 13. The dimensions of the lower end of the pyramidal frustrum can be obtained by assuming the floc volume concentration, ϕ, equals zero at this level in Equation 10 as suggested by Ives.[19] Therefore,

$$v_s = v_p = \frac{Q}{L_1^2}$$

$$\text{Design flow for each flocculator} = Q = \frac{133.7}{2} = 66.9 \text{ ft}^3/\text{min}$$

$$L_1 = \frac{Q}{v_p} = \frac{66.9}{0.30} = 14.9 \text{ ft.} \quad \text{Use } L_1 = 15 \text{ ft.}$$

For a square pyramidal frustrum, the only change in Equation 13 is the omission of the constant, $(\pi/4)$. So the modified Equation 13 is

$$\sum_{L_1}^{L_u} G_L \phi_L T_L = \frac{L_1^2}{2} \left[\frac{\gamma_s - \gamma}{Q \mu f^{9/2}} \right]^{1/2} F\left(\frac{L_u}{L_1} \right)$$

$$\frac{L_u}{L_1} = \frac{22}{15} = 1.47$$

From Figure 3 for $L_u/L_1 = 1.47$,

$$F\left(\frac{L_u}{L_1} \right) = 0.22$$

Assuming a specific gravity of floc = 1.005,

$$(\gamma_s - \gamma) = (1.005 - 1.000) \ 62.4 \text{ lb/ft}^3 = 0.312 \text{ lb/ft}^3$$

$$\mu_{18°C} = 1.06 \text{ centipoises} \times 2.088 \times 10^{-5} \text{ lb sec/ft}^2/\text{centipoise}$$

$$= 2.21 \times 10^{-5} \text{ lb sec/ft}^2$$

$$Q = 66.9 \text{ ft}^3/\text{min} = 1.115 \text{ ft}^3/\text{sec}$$

Assuming f = 2.78 in Equation 13,

$$\frac{L_1^2}{2} \left[\frac{(\gamma_s - \gamma)}{Q \mu f^{9/2}} \right]^{1/2} F\left(\frac{L_u}{L_1}\right) = \frac{15^2}{2} \left[\frac{0.312}{1.115 \times 2.21 \times 10^{-5} \times 2.78^{9/2}} \right]^{1/2} \times 0.22$$

$$= 278$$

Therefore the parameter,

$$\Sigma G_L \phi_L T_L = 278.$$

Ives[19] suggests a range for this parameter of 60-120. Thus, the design is overly conservative on this basis. However, typical designs of hopper-bottom clarifiers without any analyses as above, but designed on conservative prior practice, are for upflow velocities of approximately 0.08-0.09 ft/min. The present design is at 0.15 ft/min based on Bond's criterion and so is about 1.8 times the design rate based on past practice. Bond[26] reported effective operational experience with sludge blanket flocculators at 2.6-2.8 times the design rate. Based on these experiences, the parameter $\Sigma G_L \phi_L T_L$ equal to 278 seems too conservative, so use a value of 120.

Assuming that $\Sigma G_L \phi_L T_L = 120$ and that all other parameters are unchanged, the Q that can be treated may be estimated from Equation 13 for $Q_1 = 66.9$ ft^3/min = 1.115 ft^3/sec.

$$\frac{(\Sigma G_L \phi_L T_L)_1}{(\Sigma G_L \phi_L T_L)_2} = \frac{Q_2^{1/2}}{Q_1^{1/2}}$$

$$\frac{278}{120} = \frac{Q_2^{1/2}}{(1.115)^{1/2}}$$

Therefore, $Q_2 = 2.46$ ft^3/sec = 2.21 times the design rate.

On the basis of the above analyses, it appears that the design is overly conservative. Assuming some settling tests can be done to confirm the range of settling velocities, the writer would prefer to design less conservatively. Increasing the design superficial velocity by 50% to 0.22 ft/min, and recalculating gives design dimensions of each flocculator in plan as 18 ft x 18 ft, the dimensions of lower end of blanket as 12 ft x 12 ft, and the parameter $\Sigma G_L \phi_L T_L$ as 202. The lower end dimensions are the theoretical values for sludge blanket operation. In practice, several reasons require that the lower dimensions be made smaller.

Since streaming velocities caused by the inlet often result in blanket instability, the pyramidal frustrum should be about 6 ft x 6 ft at the lower end, which increases the depth and the effective detention time. This provides sufficient room for maintenance operations, and yet prevents accumulation of sludge that may result in odors if the area is larger. The slope of the walls should be 55-60° from the vertical to cause sludge to slide continuously down the walls. The design depth of the blanket is 9 ft with wall slopes of 56°. A 2-ft freeboard can be allowed above the clarified water channels. The clear water depth above the blanket is determined later. The design dimensions of each of two flocculators are 18 ft x 18 ft at slurry level, 6 ft x 6 ft at bottom end, and 9 ft deep.

Slurry Weir Design

In a fluidized bed with decreasing upflow velocities, the concentration of floc particles is highest at the slurry separation line. This concentration may be determined using Equation 10 or Equation 11. For illustration, both models are presented.

Using Equation 10,

$$\frac{v_s}{v_p} = (1 - f\phi^{2/3})$$

$$0.5 = (1 - 2.78 \times \phi^{2/3})$$

Therefore, $\qquad \phi = 0.08$

Using Equation 11 with n = 4 as suggested by Tesarik,

$$\frac{v}{v_p} = (1 - \phi)^4$$

$$0.5 = (1 - \phi)^4$$

$$\phi = 0.16$$

The order of magnitude of the results is correct even though the difference is 100%. A value of $\phi = 0.12$ is used in the following calculations.

The slurry weir design is controlled by the floc produced during softening. For the worst water quality, the calcium hardness = 75 mg/l, magnesium hardness = 175 mg/l, and alkalinity = 150 mg/l. The bar diagram in Figure 7 gives the stoichiometric quantities of chemicals required and the precipitates produced.

$$
\begin{aligned}
\text{Total precipitate} &= (1.5 \times 2 + 3.0 + 2.0) \text{ meq/l } CaCO_3 \\
&\quad +(1.5 + 2.0) \text{ meq/l } Mg(OH)_2 \\
&= 8 \times 50 + 3.5 \times 29 \text{ mg/l of precipitate} \\
&= 501.5 \text{ mg/l}
\end{aligned}
$$

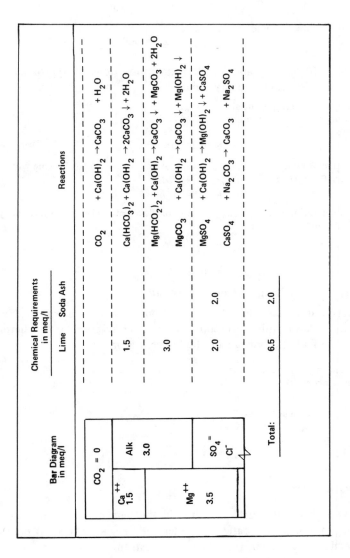

Figure 7. Chemical requirements for softening.

Assuming 60 mg/l of hardness remains in finished water, total dry solids removed

$$= 441.5 \text{ mg/l}$$
$$= 441.5 \times 8.34 \text{ lb/million gallons}$$
$$= 3682 \text{ lb/million gallons}$$

Volume of water treated in each flocculator $= \dfrac{1000}{2} \text{ gpm} = 500 \text{ gpm}$

Sludge with $CaCO_3$ and $Mg(OH)_2$ in proportions of approximately 4:1 has a density of 7.5 lb dry solids/ft^3.[26]

Volume of sludge produced per flocculator at unit solids fraction $= \dfrac{(3682 \times 500)}{10^6 \times 7.5} \text{ ft}^3/\text{min}$

$$= 0.245 \text{ ft}^3/\text{min}$$

At a floc volume fraction, $\phi = 0.12$, the volume of sludge flowing over the weir in each flocculator is
$$= 0.245/0.12$$
$$= 2.04 \text{ ft}^3/\text{min}$$

Assuming the weir runs along one side of flocculator (18 ft) with a slurry velocity of 3 ft/min[26] over weir,

$$\text{Head on weir} = \dfrac{2.04}{18 \times 3} \text{ ft}$$

$$= 0.038 \text{ ft}$$

Thus, the weir length is quite adequate.

The layout is such that the sludge from both flocculators is drawn into a concentrator and concentrated to a solids fraction of 0.3 before being drawn off at the bottom of the concentrator.

Volume of sludge drawn off from concentrator $= \dfrac{0.245 \times 2}{0.3} \text{ ft}^3/\text{min} = 12.2 \text{ gpm}$

Hence, percentage bleed-off in sludge $= \dfrac{12.2 \text{ gpm} \times 100\%}{1000 \text{ gpm}}$

$$= 1.2\%$$

Residence Times

As a check, the residence time in the blanket should be calculated, though this is not a basic parameter for design. The total depth of the pyramid to its vertex (of which the frustrum is a part) is 13.5 ft, and the frustrum is 9 ft deep.

Volume of blanket as pyramidal frustrum

$$= \frac{1}{3} [18^2 \times 13.5 - 6^2 \times 4.5]$$

$$= 1404 \text{ ft}^3.$$

Therefore, the hydraulic residence time

$$= \frac{1404}{66.9} = 21 \text{ min}$$

For the softening reaction,

Volumetric solids flow rate in influent $= 0.245 \text{ ft}^3/\text{min}$

The determination of the solids in the blanket is made by assuming an average floc volume fraction, $\phi = 0.06$. This is one-half the value at the slurry separation level. In actuality this would vary with the depth in the blanket, and it needs to be integrated from an elemental concentration as done in the discussion of the paper by Ives.[19] However, since the value for ϕ is questionable, an average would be adequate to determine the solids residence time.

Volumetric solids in blanket $= 1404 \times 0.06 = 84.2 \text{ ft}^3$

Using Equation 12,

Solids residence time, t_s $\quad = \dfrac{84.2}{0.245} = 344 \text{ min}$

$$= 5.73 \text{ hr}$$

$\Sigma G_L \phi_L T_L \qquad\qquad\qquad = 202 \text{ as calculated previously}$

Therefore,

$$\Sigma G_L \phi_L t_s \approx 202 \times \frac{344}{21} = 3309$$

This compares with the value of 3766 calculated by Brown and LaMotta.[28] This may possibly be the approximate value of the parameter for solids contact tank design. However, it needs further experimental confirmation for generalization.

Some Design Details

If the flocculators are designed for intermittent operation, a nozzle on the inlet pipe with a jet velocity of 2 to 3 ft/sec to break up compacted lime floc is a useful arrangement. The common difficulty in operating sludge blanket clarifiers at higher overflow rates has been due to blanket instability caused by streaming velocities from the inlet pipe. Some baffle arrangement to improve inlet hydraulics at the bottom of the pyramidal hopper is a useful design. Provision should be made for isolating and operating only one flocculator. This is necessary for maintenance and operation with bifurcated flows.

Four collecting launders should be provided to collect the clean water over the 18 ft. The two central launders should be spaced on 4 ft 6 in. centers, while the edge launders should be spaced 2 ft 3 in. from the wall. The minimum depth to the slurry level has to be one-half of 4 ft 6 in.; however, a depth of 3 ft 6 in. is provided.

By having tube settlers installed in the clear water zone, the flocculator performance may be considerably enhanced. This is an available operational alternative for flexibility or increased output. A schematic drawing of the flocculator is shown in Figure 8.

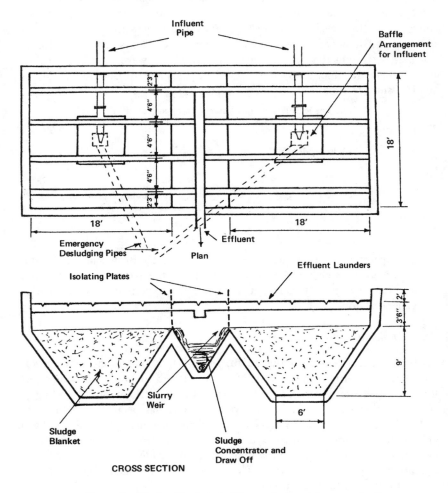

Figure 8. Sludge blanket clarifier as designed in example 2.

It should be evident from the analysis that the flocculator would probably function effectively at nearly twice the design flow. Hence, the writer strongly recommends that the hydraulics of the plant be essentially designed to carry flow rates 1.5 to 2 times the design rates. This general principle of overdesigning the hydraulics of a treatment plant is applicable for all the process units.

Dimensions and Layout of Flocculator in Metric Units

This section gives only the major calculations in metric units. For discussion of the various steps, see the calculations in English units.

From Table I the controlling settling velocity for design is 1.27-1.52 mm/sec. Use a design value of 1.42 mm/sec and correct for the operating temperature of 18°C. Assuming Stokes' law,

$$v_p \propto \frac{1}{\mu}$$

therefore

$$v_p \text{ at } 18°C = \frac{\mu_{15°C}}{\mu_{18°C}} \times 1.42 \text{ mm/sec}$$

$$= \frac{1.145}{1.06} \times 1.42 = 1.53 \text{ mm/sec}$$

Design superficial velocity $= 0.5 \, v_p = 0.77 \text{ mm/sec}$

The top and bottom dimensions of the pyramidal frustrum can be estimated on this superficial velocity and the settling velocity. Hence, using Equation 13, it is possible to estimate the parameter $\Sigma \, G_L \phi_L T_L$. An analysis of this parameter (see calculation with English units) indicates that the above design is too conservative. Therefore, the superficial velocity is increased by 50%.

Use a design superficial velocity $= 0.77 \times \dfrac{150}{100} \text{ mm/sec}$

$$= 1.16 \text{ mm/sec}$$

Design flow for each flocculator $= \dfrac{63}{2} \text{ liter/sec}$

$$= 31.5 \text{ liter/sec} = 31.5 \times 10^{-3} \text{ m}^3/\text{sec}$$

Surface area required $= \dfrac{31.5 \times 10^{-3}}{1.16 \times 10^{-3}} \text{ m}^2 = 27.2 \text{ m}^2$

With square sides, the length of each side at slurry separation line $= \sqrt{27.2} = 5.2 \text{ m}$

Use design dimensions: 5.5 m x 5.5 m.

The floc settling velocity $= 1.16 \times 2 = 2.32$ mm/sec

Therefore, lower dimensions of blanket L_1

$$= \left[\frac{31.5 \times 10^{-3} \ m^3/sec}{2.32 \times 10^{-3} \ m/sec} \right]^{\frac{1}{2}} = 3.7 \ m$$

Equation 13,

$$\sum_{L_1}^{L_u} G_L \phi_L T_L = \frac{L_1^2}{2} \left[\frac{(\rho_s - \rho) \ g}{Q \mu f^{9/2}} \right]^{1/2} F\left(\frac{L_u}{L_1}\right)$$

$$\frac{L_u}{L_1} = \frac{5.5}{3.7} = 1.5, \ \text{therefore from Figure 2,} \ F\left(\frac{L_u}{L_1}\right) = 0.25$$

$\rho = 1000 \ kg/m^3, \rho_s = 1.005 \times 1000 \ kg/m^3, \ g = 9.81 \ m/sec^2$

$f = 2.78 \ \mu_{18°C} = 1.06$ centipoises $= 1.06 \times 10^{-3} \ Nsec/m^2$

Therefore,

$$\frac{L_1^2}{2} \left[\frac{(\rho_s - \rho) \ g}{Q \mu f^{9/2}} \right]^{1/2} F\left(\frac{L_u}{L_1}\right) = \frac{3.7^2}{2} \left[\frac{(1005 - 1000) \ 9.81}{31.5 \times 10^{-3} \times 1.06 \times 10^{-3} \times 2.78^{9/2}} \right]^{1/2} \times 0.25$$

$$= 207$$

Therefore, $\sum G_L \phi_L T_L = 207$ and flocculation is more than adequate.

To increase detention time and provide for inlet streaming effects, increase depth to 2.8 m and use lower frustrum dimensions of 1.8 m x 1.8 m. A freeboard of 0.7 m may be allowed above the clarified water collecting channels.

Design dimensions of each of two flocculators

$= 5.5$ m x 5.5 m at slurry level, 1.8 m at bottom and 2.8 m deep.

Slurry Weir Design

Initial calculations are identical, as for English units.

Total dry solids removed $= 441.5$ mg/l

Volume of water treated in a flocculator $= 31.5$ liter/sec

Sludge with $CaCO_3$ and $Mg(OH)_2$ in proportions of approximately 4:1 has a density of 120.3 kg/m^3 dry solids.

Volume of sludge produced per flocculator at unit solids fraction

$$= \frac{441.5 \times 31.5 \times 10^{-6}}{120.3} \text{ m}^3/\text{s}$$

$$= 115.6 \times 10^{-6} \text{ m}^3/\text{s}$$

At a floc volume fraction, $\phi = 0.12$, volume of sludge flowing over weir per flocculator

$$= \frac{115.6 \times 10^{-6}}{0.12} \text{ m}^3/\text{s}$$

$$= 0.96 \times 10^{-3} \text{ m}^3/\text{s}$$

Assuming the weir runs along one side of flocculator (5.5 m), with a slurry velocity of 15.2 mm/sec over weir,

$$\text{Head on weir} = \frac{0.96 \times 10^{-3}}{5.5 \times 15.2 \times 10^{-3}} \text{ m} = 0.012 \text{ m}$$

Sludge is concentrated to solids fraction of 0.3

Volume of sludge drawn off from concentrator

$$= \frac{115.6 \times 10^{-6} \times 2}{0.3} \text{ m}^3/\text{sec}$$

$$= 0.768 \times 10^{-3} \text{ m}^3/\text{sec}$$

Hence percentage of bleed-off in sludge

$$= \frac{0.768 \times 10^{-3} \times 100\%}{2 \times 31.5 \times 10^{-3}} = 1.2\%$$

Residence Times

The total depth of the pyramid (with the frustrum dimensions as designed) is 4.16 m.

Volume of blanket as pyramidal frustrum

$$= \frac{1}{3} [5.5^2 \times 4.16 - 1.8^2 \times 1.36]$$

$$= 40.48 \text{ m}^3$$

Hydraulic residence time

$$= \frac{40.48}{31.5 \times 10^{-3}} \text{ sec} = 1285 \text{ sec} = 21.4 \text{ min.}$$

Solids residence time

$$= \frac{40.48 \times 0.06}{115.6 \times 10^{-6}} \text{ sec} = 21,010 \text{ sec} = 5.8 \text{ hr}$$

Therefore, $\Sigma \, G_L \phi_L t_s = 207 \times \dfrac{21,010}{1,285} = 3384$

Some Design Details

A nozzle on inlet pipe with a jet velocity 0.6-1.0 m/sec assists in breaking up sludge that settles due to intermittent operation.

Four collecting launders are provided over the 5.5 m. The two central launders are spaced at 1.4 m and the edge launders are spaced 0.65 m from the wall. A clear water depth of 1.0 m is provided.

REFERENCES

1. Camp, T. R., and P. C. Stein. "Velocity Gradients and Internal Work in Fluid Motion," *J. Boston Soc. Civil Eng.* 30:219-237 (1943).
2. Camp, T. R. "Flocculation and Flocculation Basins," *Trans. Am. Soc. Civil Eng.* 120:1-16 (1955).
3. Fair, G. M., J. C. Geyer and D. A. Okun. *Water and Wastewater Engineering, Volume 2* (New York: John Wiley and Sons, Inc., 1968).
4. Harris, H. S., W. J. Kaufman and R. B. Krone. "Orthokinetic Flocculation in Water Purification," *J. Environ. Eng. Div.* ASCE 92:95-111 (1966).
5. Fair, G. M., and R. S. Gemmell. "A Mathematical Model of Coagulation," *J. Colloid. Sci.* 19:360-372 (1964).
6. Hudson, H. E. "Physical Aspects of Flocculation," *J. Am. Water Works Assoc.* 57:885-892 (1965).
7. Michaels, A. S., and J. C. Bolger. "The Plastic Flow Behavior of Flocculated Kaolin Suspensions," *Ind. Eng. Chem. Fund.* 1:24-33 (1962).
8. Lagvankar, A. L., and R. S. Gemmell. "A Size Density Relationship for Flocs," *J. Am. Water Works Assoc.* 60:1040-1046 (1968).
9. Argaman, Y., and W. J. Kaufman. "Turbulence and Flocculation," *J. Environ. Eng. Div.* ASCE 96:223-241 (1970).
10. Thomas, D. G. "Turbulent Disruption of Flocs in Small Particle Size Suspension," *J. Am. Inst. Chem. Eng.* 10:517-523 (1964).
11. Hudson, H. E., and J. P. Wolfner. "Design of Mixing and Flocculation Basins," *J. Am. Water Works Assoc.* 59:1257-1267 (1967).
12. Parker, D. S., W. J. Kaufman and D. Jenkins. "Floc Breakup in Turbulent Flocculation Processes," *J. Environ. Eng. Div.* ASCE98: 78-99 (1972).
13. Andreu-Villegas, R., and R. D. Letterman. "Optimizing Flocculator Power Input," *J. Environ. Eng. Div.* ASCE 102:251-264 (1976).
14. Ives, K. J., and A. G. Bhole. "Theory of Flocculation for Continuous Flow System," *J. Environ. Eng. Div.* ASCE 99:17-34 (1973).
15. Spielman, L. A., and J. E. Quigley. "A Motion Picture Study of Floc Break-Up," presented at the Division of Environmental Chemistry, American Chemical Society National Meeting, New Orleans, Louisiana, March 20-27, 1977.
16. Kao, S. V., and S. G. Mason. "Dispersion of Particles by Shear," *Nature* 253:619-621 (1975).
17. O'Melia, C. R. "Coagulation and Flocculation," *Physicochemical Processes for Water Quality Control,* W. J. Weber, Ed. (New York: Wiley-Interscience, 1972).
18. Gemmell, R. S. "Mixing and Sedimentation," *Water Quality and Treatment,* American Water Works Association, Inc. (New York: McGraw-Hill Book Co., 1971).

19. Ives, K. J. "Theory of Operation of Sludge Blanket Clarifiers," *Proc. Inst. Civil Eng.* 39:243-260 (1968).
20. Kawamura, S. "Considerations on Improving Flocculation," *J. Am. Water Works Assoc.* 68:328-336 (1976).
21. Cockerham, P. W., and D. M. Himmelblau. "Stochastic Analysis of Orthokinetic Flocculation," *J. Environ. Eng. Div.* ASCE 100:279-294 (1974).
22. Amirtharajah, A. "Optimum Expansion of Sand Filters During Backwash," unpublished Ph.D. Thesis, Iowa State University, Ames, Iowa (1971).
23. Amirtharajah, A. "Optimum Backwashing of Sand Filters," presented at the ASCE National Environmental Engineering Conference, Nashville, Tennessee, July 13-15, 1977.
24. Hudson, H. E. "Dynamics of Mixing and Flocculation," *Proc. 18th Annual Water Supply Conf.*, Department of Civil Engineering, University of Illinois, Urbana, Illinois (1976), pp. 41-54.
25. Bond, A. W. "Behavior of Suspensions," *J. Environ. Eng. Div.* ASCE 86:57-85 (1960).
26. Bond, A. W. "Upflow Solids Contact Basin," *J. Environ. Eng. Div.* ASCE 87:73-99 (1961).
27. Tesarik, I. "Flow in Sludge Blanket Clarifiers," *J. Environ. Eng. Div.* ASCE 93:105-120 (1967).
28. Brown, J. C., and E. LaMotta. "Physical Behavior of Flocculent Suspensions in Upflow," ASCE 97:209-224 (1970).
29. Richardson, J. F., and W. N. Zaki. "Sedimentation and Fluidisation: Part I," *Trans. Inst. Chem. Eng.* 32:35-53 (1954).
30. Rouse, H. "Suspension of Sediment in Upward Flow," Iowa Studies in Engineering, Iowa Institute of Hydraulic Research, *Bulletin* 26: 14-22 (1941).
31. Kalinske, A. A. "Settling Rate of Suspensions in Solids-Contact Units," *Proc. Am. Soc. Civil Eng.* 79:1-8 (1953).
32. Parker, D. S. Discussion of "Physical Behavior of Flocculent Suspensions in Upflow," by J. C. Brown and E. LaMotta, *J. Environ. Eng. Div.* ASCE 97:793-795 (1971).
33. Amirtharajah, A. "Expansion of Graded Sand Filters During Backwashing," unpublished M. S. Thesis, Iowa State University, Ames, Iowa (1970).
34. Walker, J. D. "High Energy Flocculation," *J. Am. Water Works Assoc.* 60:1271-1279 (1968).
35. Walker, J. D. "High Energy Flocculation and Air and Water Backwashing," *J. Am. Water Works Assoc.* 60:321-330 (1968).
36. Amirtharajah, A. "Towns South of Colombo Water Treatment Plant," unpublished operating data, National Water Supply and Drainage Board, Ratmalana, Sri Lanka (1972).
37. Argaman, Y. A. "Pilot-Plant Studies of Flocculation," *J. Am. Water Works Assoc.* 63:775-777 (1971).
38. Kawamura, S. "Coagulation Considerations," *J. Am. Water Works Assoc.* 65:417-423 (1973).

CHAPTER 12

GRANULAR-MEDIA DEEP-BED FILTRATION

E. Robert Baumann, Ph.D., P.E.

Anson Marston Distinguished Professor of Engineering
Department of Civil Engineering
Iowa State University
Ames, Iowa   50010

INTRODUCTION

If water containing suspended solids is passed through a layer of porous media, some of the suspended and colloidal material is partially or completely removed. This process is called filtration and its efficiency and cost is a function of:

- the concentration and characteristics of the solids in suspension (particle size distribution, surface characteristics, organic vs inorganic),
- the characteristics of the porous media and other filtering aids used (particle size distribution, surface characteristics),
- the characteristics of the solids in solution in the water, and
- the characteristics of the filter and the method of filter operation.

The rate of flow through a filter at any time is directly proportional to the driving force and inversely proportional to the resistance of the filter media and the solids retained therein:

$$Q \propto \frac{\text{Driving Force}}{\text{Media Resistance}}$$

The driving force is measured by the water head (pressure drop or head loss) available to overcome the filter resistance. Three basic terms are used to describe the method of applying the water head required: vacuum filtration, pressure filtration and gravity filtration. In vacuum filters, the filter is located on the suction side of a pump and the pressure drop across the filter is limited to the suction lift that can be generated by the

231

pump, usually only 18-22 ft (5.5 - 6.7 m) of water. The filter itself is generally operated under a pressure of less than atmospheric pressure. In pressure filtration, the filter is located on the discharge side of the pump and the pressure drop across the filter can be any differential that can be generated by the characteristics of the pump; usually it is approximately 10-40 ft (3.0-12.2m) of water. The pressure within the filter is usually at a pressure greater than atmospheric.

In gravity filtration, the water flows through the filter under the force of gravity, the head available measured by the difference in water level above the filter media and in the clear well to which the filter effluent discharges. In other words, a gravity filter is a special type of pressure filter in which water is delivered to the filter and the water above the filter media is at atmospheric pressure. If the effluent water discharge elevation is higher than the filter media surface elevation, the pressure in the media is always greater than atmospheric pressure. If the effluent discharge location is below the media surface elevation, a condition of "negative-head" can develop in which the pressure within the filter media can be less than atmospheric pressure.

Filter media of many different types, particle size distributions and depths can be used in water filtration. The objective in the design of a filtration system is to determine which filter will provide the desired performance at *least cost*. All filters considered should be compared on their ability to provide *equivalent performance*, which may be defined as the production of the same quality of filtered water in the same time period from the same water source. In general, as the filter media size becomes larger, the depth of media required to produce a given quality of water becomes greater. This phenomenon gives rise to two distinct types of granular media filters:

- those in which the filter media is of such fine particle size (1-100 μm) and of such limited depth (3-5 mm) that the media is wasted after each filtration cycle, and
- those in which the filter media is of such coarse particle size (0.15-3 mm) and of such significant depth (0.5-6 or 8 ft or 0.15-1.8 or 2.4 m) that the same media is cleaned in place and put back into service.

The former type of filtration is properly referred to as "precoat" filtration, or more commonly "diatomite" filtration since diatomite filter media have been widely used for swimming pool and potable water filtration. The latter type of filtration is commonly referred to as granular-media deep-bed filtration, sand filtration or dual-media filtration. This chapter focuses on gravity, deep-bed filtration systems and their application to the production of potable water. Emphasis is placed on the principles and current advances in such applications and not on the details of gravity filter design, since these are covered in several recent text books.[1]

APPLICATIONS IN POTABLE WATER FILTRATION

Historical Review

Deep-bed granular-media filters find widespread use in potable water treatment for the removal of suspended solids in the following applications:

- removal of iron and manganese precipitates from ground water supplies,
- removal of softening precipitate carry-over from lime or lime-soda ash softening of ground or surface water supplies,
- removal of floc carry-over resulting from coagulation-flocculation-sedimentation of surface waters for the purpose of clarification of the water, and
- removal of microorganisms that are pathogenic to man and/or contribute to the chlorine demand of water, which can affect disinfection requirements.

Over the years, the role of filter has changed significantly as break-throughs have been developed in water treatment technology. The early English or slow-sand filter, dating from 1829, was operated at a rate of 2-4 mgad (million gallons per acre per day) (1.9-3.7 m^3/m^2 day) using about 3 ft (1 m) of ungraded sand media to filter *raw surface* water in runs that lasted for as long as 6-8 months. The filter was cleaned by draining the filter to below the sand surface and scraping off the dirty skin formed at the surface of the sand. This filter was designed specifically for the removal of suspended solids from surface water and was the *sole water treatment device used.*

After the Broad Street well epidemic in London in the early 1850s demonstrated that sewage contamination was related to human disease, there was a general international movement in developed nations to require all potable water to be filtered, generally using English-type slow-sand filters. In the 1860s and 1870s, Pasteur and others developed the germ theory of disease, and the primary role of the filter shifted from the need to remove suspended solids to the need to remove bacterial pathogens. The famous Hamburg, Attoona, Wandsbeck epidemic of cholera in Germany in the 1890s demonstrated conclusively the effectiveness of the slow-sand filter for the purpose. As a result, there was a general public clamor for the construction of water filtration plants in the United States, similar to current national interest in the construction of water pollution control plants.

The application of English-type filters in the U.S. was not as successful in clarifying the clay-bearing waters encountered in many parts of the country. Our high labor costs (even in the early 1900s) and the large filters required led to the commercial development of filters that were operated at higher rates. The classic work of Fuller[2] in Louisville,

Kentucky, in the late 1890s led to criteria for the design of new American-type or *rapid-sand* filters, which included:

- It was necessary chemically to pretreat water using alum or iron coagulants followed by flocculation and sedimentation of the water to reduce the solids load going to the filter.
- The filters should be operated at rates of 125 mgad, or 2 gpm/ft^2 (117 m^3/m^2 day), with filter runs normally 12-36 hr under a pressure differential of 8-10 ft (2.4-3.0 m) of water. The higher rates forced the solids deeper into the sand media and this, together with the significantly shorter runs as compared to English-type filters, required the development of new backwash methods.
- To reduce down-time for cleaning the filters, a reverse flow of filtered water *and* air was used to fluidize the filter media to wash the accumulated solids from the filter bed. American experience led to discontinuing the use of air for backwash as an unnecessary technique. British practice using these new rapid-sand filters expanded on the use of air for backwash. For either procedure, the backwash time was short (15-25 min) and led to a gradation of the filter media so that the media fines were at the filter surface at the end of backwashing.
- The filter media depths generally adopted were in the range of 3 ft (1 m), the same as in slow-sand filters. The media size adopted as the water works standard was 0.4-0.5 mm effective size, with a uniformity coefficient of about 1.6 to 1.7. (The effective size of granular media is that size below which 10% by weight of the media is "finer than.") The uniformity coefficient is the ratio of the 60% "finer than" size to the 10% "finer than" size. Figure 1 shows the size distributions of a typical coal and a typical sand media in use today. After backwashing, a typical 0.53-mm effective size sand would have a 0.3-0.4-mm sand layer at the media surface and a 1.0-1.2-mm sand at the bottom of the filter. Thus, once a suspended solid penetrated into the bed, it encountered larger and larger pores through which the water passed.

For the first half of this century, nearly all potable water filters without exception were designed on this basis. Thus, with this new technology, the fitler retained its role as the primary process protecting the quality of the filtered water, and chemical pretreatment was a secondary process designed to reduce the filter load to produce a filter that could be operated economically.

In the period between 1900 and 1915, the technology of disinfection of water first with hypochlorites and then with liquid chlorine was developed. With the general acceptance of chlorination of water as the *primary* process for protection of potable water against disease organisms, the filter was relegated to a secondary defensive role in removing bacteria, yet it retained its primary role in the removal of suspended solids.

After World War II, two criteria were recognized that required the termination of a filter run:

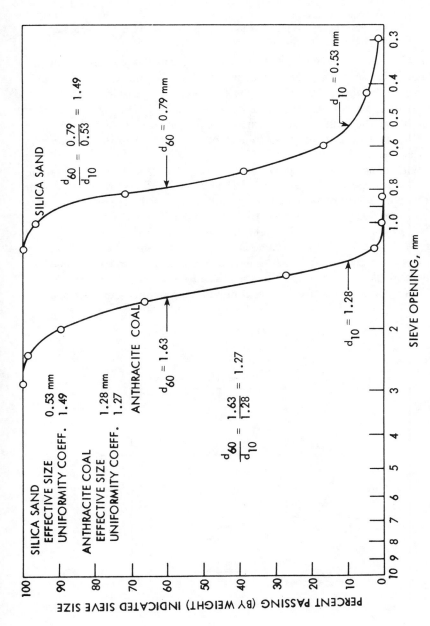

Figure 1. Size gradation curves for silica sand and anthracite coal filter media.

- either the suspended solids broke through the filter before the head available was used up, or
- the filter head available was used up before the solids broke through the full depth of the filter media.

In either event, the run would have to be terminated, in the former instance because of deterioration of effluent quality and in the latter, because of inability to maintain the filtration rate. Ideally, a filter should reach the point of exhaustion of its solids' holding capacity at the instant that all of the available head is used up, as shown in Figure 2. When this occurs, the filter is considered to be in an *operationally optimum* design condition.

With rapid population growths in the late 1940s and 1950s, filtration rates were increased to 2.5-4gpm/ft^2 (144-230 m^3/m^2 day) to increase plant production with existing equipment. As this happened, experience indicated that filter runs had to be terminated because all the available head was used. Thus, research in this period centered on increasing the hydraulic capacity of rapid-sand filters to accommodate higher rates with the same or longer filter runs. The research led to development of dual- and mixed-media filters, single-media upflow filters and single-media bi-flow filters. Figure 3 shows schematic diagrams of filter configurations used for granular-media

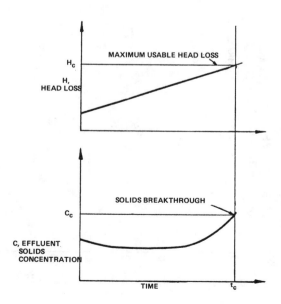

Figure 2. Critical head loss and critical effluent quality vs critical time of filtration.

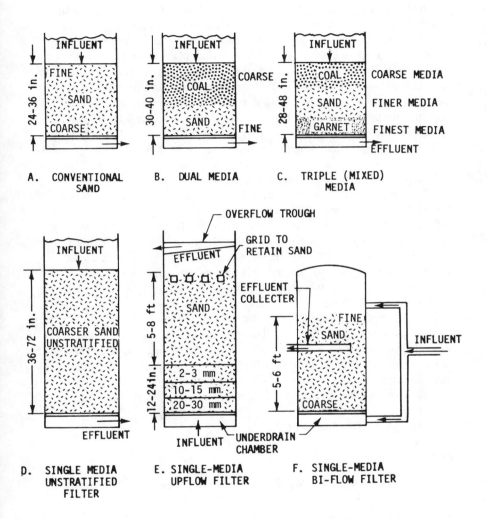

Figure 3. Schematic diagrams of filter configurations for granular-media filtration.

filtration. All of these filters when operated at rates above 3-4gpm/ft^2 (173-230 m^3/m^2 day) are referred to as high-rate deep-bed granular-media filters.

As filtration rates increased with the newer filter configurations, more and more filter runs had to be terminated because of solids breakthrough since the higher rates pushed the solids deeper and deeper into and through the bed. Thus, research efforts were shifted to solving this new problem, resulting in the development of new and improved technology in water coagulation, flocculation and sedimentation. With these improvements, water pretreatment became so successful that plant operators in the 1960s were able to report that the pretreated water going *to* the filters already met and exceeded the then-current drinking water standards. Thus, the developing technology had changed the role of the filter again. Now, it had lost its primary role in the protection of the water against pathogens to the process of disinfection and its primary role in the reduction of turbidity (proportional roughly to suspended solids) to chemical pretreatment processes. In fact, there was even discussion of whether filters had any role at all in potable water production.

With improved technology, there has been a corresponding lowering of the turbidity level permitted in drinking water. For example, the U.S. Public Health Service Drinking Water Standard (now the U.S. Environmental Protection Agency) became more restrictive, as shown in Table I. With the stricter turbidity requirements and the 1974-75 concern about potential carcinogenic particulates such as asbestos fibers in drinking water, the need for and role of filters in potable water quality has become widely accepted. Yet, with this increased role—now shared equally with chemical pretreatment and disinfection for the protection of water quality —there has developed a current tendency to load the pretreatment steps more heavily so as to put more burden on the filters. In other words, by developing each process closer to the limit of its potential performance in accordance with conservative application of existing technology, we can optimize the system by making optimum use of each of its parts.

Currently Accepted Technology

Plant Types

Currently, three main flow schemes are in common use for producing a low-turbidity potable water:

Table I. Drinking Water Standards

Time	Maximum Permissible Turbidity
Prior to 1962	10.0 TU
1962 to 1976	5.0 TU
1977	1.0 TU
American Water Works Association Goal	0.1 TU

1. Conventional treatment plants (raw water turbidities from 1 TU upwards)
    * Chemical pretreatment to *produce* a settleable suspended solid
        Rapid mix
        Flocculation
        Sedimentation
    * Polishing filters to remove residual flocs
        Polymer use to produce stronger flocs

2. Direct filtration plants (raw water turbidity to 50-60 TU)
    * Chemical pretreatment to produce a filterable suspended solid (10-20 mg/l alum or iron coagulant and polymer)
        Rapid mix
        Short-period flocculation
    * Filtration to remove coagulated-flocculated suspended solids

3. Contact coagulation-filtration plants (raw water turbidity to 75-100 TU)
    * Chemical pretreatment to produce a filterable suspended solid
        Rapid mix
        Short-period flocculation
        Up-flow, coarse media filtration
    * Filtration to remove coagulated-flocculated suspended solids carried over from the up-flow filter
        Polymer use to increase solids retention

In the use of chemicals such as polymers, it must be recognized that the type of chemical and the point of use of the chemical is important. For example, polymers can be used in many ways:

* as primary coagulants to make suspended solids more settleable,
* as filter aids added significantly ahead of the filter so that they interact with the suspended solids to make them more filterable,
* as filter aids added just ahead of the filter so that they interact primarily with the filter media to provide it with a higher attractive potential, and
* as a filter-media coating added before the start of a filter run to change the characteristics of the media.

This chapter, however, is not concerned directly with the design of the chemical pretreatment but with the design of the filter.

Types of Filters

In current practice, potable water filters are nearly all down-flow, dual-media or mixed-media, deep-bed filters operated at a high rate. In small plants, pressure filters (not fewer than two) are commonly used. The media is contained in a pressure tank and the water is pumped through the media so that it enters and leaves the filter under pressure. Frequently, the filter discharges directly into the distribution system. In large plants, gravity filters (not fewer than two to three) are used in which the filter is open at the top and the water flows through the filter by gravity and leaves the filter at atmospheric pressure.

NATURE OF DEEP-BED FILTRATION

Optimum Design

The optimum design of a filter requires consideration and specification of:

- media sizes and depths,
- the filtration rate,
- the pressure available for driving force, and
- the method of filter operation.

A mathematical model is needed to relate the combinations of these operating variables in order to produce combinations of filter designs that provide equivalent, operational-optimum performance. Obviously, the filter that provides this performance at least cost is the economically optimum filter to use.

Unfortunately, the satisfactory universal model for predicting filter performance is not yet available. However, models are being developed that can aid us significantly in optimum filter design. Now we must either:

- depend on the experience of others,
- conduct extensive pilot plant studies, or
- make an educated guess and build on our own experience.

The goal of economic optimum design requires the use of:

- a model for prediction of filtration results,
- data for predicting first and operating costs for the filters, and
- a computer program to relate the two.

A review of the factors affecting filter performance, therefore, is in order.

Solids Removal

When a filtration model is developed, different mechanisms that affect filter performance should be included. Filtration is comprised of three principal mechanisms: (a) transport, (b) attachment and (c) detachment. *Transport* mechanisms move a particle into and through a filter pore so that it comes very close to the surface of the filter media or existing deposits where *attachment* mechanisms serve to retain the suspended particle in contact with the media surface or with previously deposited solids. *Detachment* mechanisms result from the hydrodynamic forces of the flow acting in such a way that a certain part of the previously adhered particles, less strongly linked to the others, is detached from filter media or previous deposits and carried further into or through the filter.

Transport Mechanisms

Important transport mechanisms are screening, interception, inertial forces, sedimentation, diffusion and hydrodynamic forces. The suspended solids removal efficiency and the type of dominating transport mechanisms depend on the sizes of the particles and their size distribution in the filtering water. Yao *et al.*[3] found that there exists a size of suspended solids for which the removal efficiency is minimum. This critical suspended solids size is about 1 μm. For suspended solids larger than 1 μm, removal is enhanced by transport mechanisms of sedimentation and/or interception, *i.e.*, gravity forces. For suspended solids smaller than 1 μm, removal efficiency increases with decreasing particle size. Transport is made possible by the increasing effects of diffusion forces as particle size decreases. Thus, the effects of the applicable forces on particle transport are as shown in Figure 4.

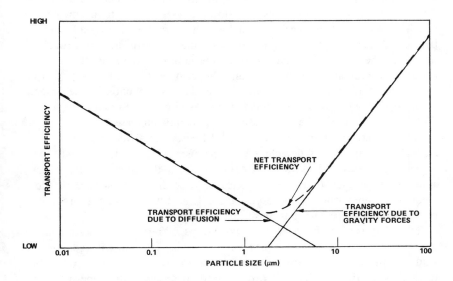

Figure 4. Relationship between transport efficiency and particle size. (After Reference 2.)

Attachment Mechanisms

Attachment of particles to the media surfaces has been generally attributed to physicochemical and molecular forces. Particles are often attached by molecular bridges, such as synthetic, natural or hydroxide polymers, which are specifically adsorbed to suitable sites. Yao et al.[3] observed that attachment can be improved by using the optimum polymer dose determined in the coagulation jar tests. The polymers probably enhance attachment in filtration by adsorption, producing charge neutralization and/or bridging. It is suggested that when conventional filters fail to produce efficient filtration, effective improvements can be made by altering the chemistry of the filter media, either by applying the optimum coagulant dose or by coating the media with a polymer or a metallic coagulant such as alum.[4,5]

Major attachment actions as suggested by various researchers include: straining, Van der Waals forces, electrokinetic interactions between surfaces and chemical forces, including molecular forces and surface tension.

Detachment Mechanisms

There is controversy concerning detachment of particles from filter media during filtration. Mintz[6,7] considers that deposits accumulated in a filter medium have an unequally strong structure. Under the action of hydrodynamic forces caused by the flow of water through the media, which increase with increasing head loss, this structure is partially destroyed. A certain portion of previously adhered particles, less strongly linked to the others, is detached from the grains. Consequently, as the deposits accumulate they become unstable and parts of them are torn away by the flow, to go back into suspension in the pores. The detachment phenomenon was also observed by Tuepker and Buescher[8] and Cleasby et al.[9] when the flow rate was suddenly increased.

Another group of researchers, Ives,[10] Lerk,[11] and Mackrle and Mackrle,[12] oppose this detachment mechanism, since as the interstitial velocity increases and as the surface available in the filter pores and the amount of divergence and convergence of flow diminish because of the deposits accumulating in the pores, there is a reduction in the probability of particles being brought to a surface for adherence.

The disagreement between the two groups of research workers concerning the role of detachment is not yet resolved. The ultimate goal of studying the physical nature of gravity filtration is to formulate a mathematical prediction model that can be used to describe the time-space variation and rate of accumulation of deposits within the filter.

Mathematical Prediction Models

All mathematical prediction models describing gravity filtration can be divided into two parts: one relating to the rate of clarification or the theory of suspension removal, the other relating to increase in head loss due to clogging in the filter pores.

Suspension Removal

One group of research workers has based their predictive models on the idealized assumptions that govern removal mechanisms. The models of Iwasaki,[13] Stein,[14] Mintz[7] and Ives[15] fall in this category. Another group of workers considers the transport-attachment-detachment mechanisms, in the course of filtration as a stochastic process, in which operations research techniques such as Queueing theory and Markov chains are applied in developing their models.[16-21] Still another group of workers has been interested in an empirical approach in which extensive experimental data are collected to develop empirical equations relating the media depth, size and gradation, filtration rate, and water characteristics to the filter effluent quality.[22-28]

In the development of the idealized model, several basic assumptions or equation modifications have been made.

1. The volume of the suspended solid removed from water, the specific deposit, σ, is equal to the volume of the solid deposited in the bed (Iwasaki[13]).
2. The removal of suspended solids with depth in a filter is directly proportional to the concentration of the solids times a proportional factor, λ, called an impediment modulus or filter coefficient (Iwasaki[13]).
3. The filter coefficient applicable to a given layer in the filter bed first increases with time of filtration as the amount of filter clogging increases and ultimately decreases as the capacity of the filter layer is reached (Stein[14]).
4. The porosity changes of the deposited solids affect the rate of solids removal through the depth of the filter bed (Ives[15]).
5. The efficiency of a filter depends on the surface area available for particle deposition and on the flow rate past such surfaces.

Thus, the equation of continuity, which takes into consideration the effects of porosity changes of the solids and equates solids deposited to solids removed, can be expressed as:

$$\frac{\partial C}{\partial 1} = \frac{1 - f}{v} \frac{\partial \sigma}{\partial t} \tag{1}$$

where σ = specific deposit, *i.e.*, the volume of particles deposited per unit volume of filter bed

t = the time from the beginning of the run

C = the suspension concentration
v = the flow rate
l = the distance from the top of the bed to the section under study
f = the porosity of the filter bed

The rate of removal of suspended solids as a function of solids concentration is as follows:

$$\frac{\partial C}{\partial l} = -\lambda C \tag{2}$$

where λ is the impediment modulus or filter coefficient. It is evident, however, that the filter coefficient is not a constant but varies with time during the run as follows:

$$\frac{\lambda}{\lambda_0} = (1 + \frac{\beta\sigma}{\epsilon})^y \ (1 - \frac{\sigma}{\epsilon})^z \ (1 - \frac{\sigma}{\sigma_u})^x \tag{3}$$

where λ_0 = the initial filter coefficient
ϵ = the initial porosity of the bed, *i.e.*, the porosity of the clean bed
β = a constant dependent on the packing of the grains
σ_u = the ultimate specific deposit
x, y, z = empirically derived factors

In 1970, Tchobanoglous[29] found that screening proved to be the principal removal mechanism operative in the direct filtration of settled sewage effluent without chemical pretreatment, and he proposed a modified first-order equation describing the rate of change of suspended solids concentration with filter depth.

The real problem involved in use of these models centers on the problems of determining the values of specific deposit appropriate for use in the filtration equation, and the variations in the filter coefficient, λ.

Head Loss

If filter media do clarify suspensions that flow through them, it follows that the pores of the media accumulate deposits that cause either a loss of media permeability or an increased flow resistance. The approach used most commonly to determine the head loss in a clogged filter has been to compute it with a modified form of the equations used to evaluate the clean water head loss. In all cases, the difficulty encountered in using these equations is that the media porosity must be estimated for various degrees of clogging. The complexity of this approach, unfortunately, makes most of these equations of little use. Nevertheless, the head loss development equations indicate first that, for a specific media and flow rate, the total head loss depends only on the volume of floc retained by the filter, and second that the time to reach a fixed head loss depends only on the volume of floc formed from the suspended solids in the raw water and the added chemicals.

Demonstration of Optimum Design

Both Ives[30] and Huang and Baumann[31] have demonstrated that existing fundamental and empirical models can be used to predict least-cost, optimum design combinations for single-media sand filters. Figures 5 and 6 show the results obtained when sand filters were designed to reduce 5.0 mg/l of iron in a well water to a maximum iron concentration of 0.3 mg/l at the end of the filter run. Figure 5 indicates that with 0.6-mm sand, the unit cost of filtration reaches an optimum when the filter run is about 34 hours long and the pressure drop across the filter is about 11.5 feet. Figure 6 indicates that the unit cost and terminal head both decrease with an increase in sand media size. In this situation, least-cost optimum design for the removal of 5 mg/l of iron was not reached since filtration costs are still decreasing with increase in sand size. Unfortunately, we cannot effectively backwash sand media much larger than 1.3 mm, using media fluidization techniques.

HYDRAULICS OF FLOW IN A FILTER

Design Considerations

The rate of flow through a filter is usually expressed in terms of flow rate per unit of area, or gpm/ft^2. In water filtration, filtration rates do not generally exceed a maximum rate of about 9 gpm/ft^2 (520 m^3/m^2 day). With the increased emphasis on wastewater filtration and the many recent papers describing such experiences, it is well to recognize the major differences between design considerations in water and wastewater applications. For example, consider the following:

	Water Filtration	Wastewater Filtration
Rate of Filtration	Adequate raw and treated water storage; therefore, filters operate at a constant rate. Peak rate set at peak future *daily* rate.	Wide variations of flow to filters. Peak rate at peak, wet weather *hourly* rate.
Suspended Solids	Low SS content with relatively constant characteristics, floc carryover, iron.	Much higher SS variations and characteristics. *Peak SS* accompany *peak flows.*
Filtrate Quality	Function of initial solids concentration (C_0), filtration rate, and media size and depth.	Significantly less dependent on filter characteristics and C_0.
Backwash	Sufficient experience to design adequate systems.	Less experience. Less effective due to "sticky" nature of biological SS.

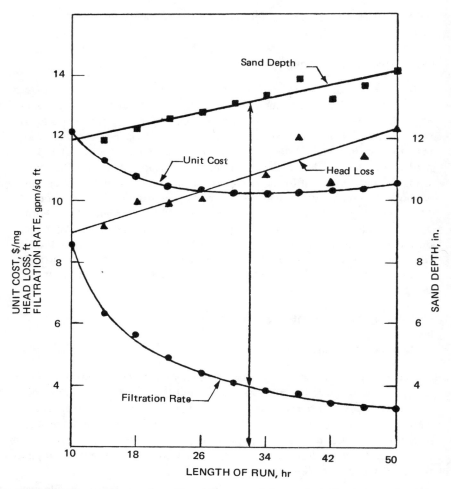

Figure 5. Effect of run length on filtration costs and the required filtration rate, head loss and sand depth. Sand size = 0.60 mm.

In view of the greater effect of C_o and filter characteristics on filtrate quality, the higher filtration rates currently used in wastewater filtration should *not* be transferred to water filtration.

The total head loss or available head on a filter is the difference in elevation between the water levels on the inlet and outlet side of the filter. For example, in Figure 7A, the available head loss is shown as the difference in elevation between the high water level above the media and the low level of water above the media controlled by the effluent weir.

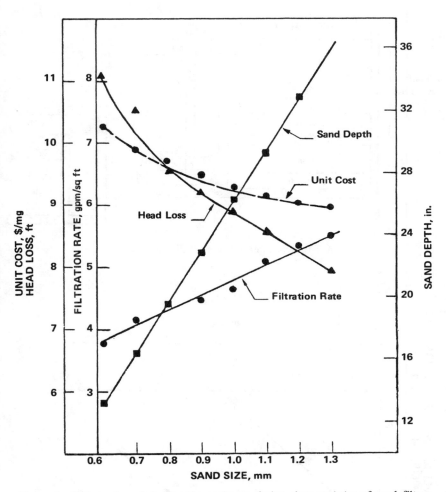

Figure 6. Effect of sand size on the optimum design characteristics of sand filters.

Head Loss in Media

In the grain size of interest in rapid granular-media filtration (0.4-2 mm), at the filtration rates of interest (2-8 gpm/ft² or 0.08-0.33 m³/m² min) with a water temperature between 0°C and 30°C (viscosity 0.018-0.008 poises, or 0.0018-0.0008 N sec/m²) laminar flow conditions exist. This means that the head loss through the media is directly proportional to the rate of flow. Figure 8 shows[32] the head loss as a function of filtration rate through a filter at the end of a run made at a rate of

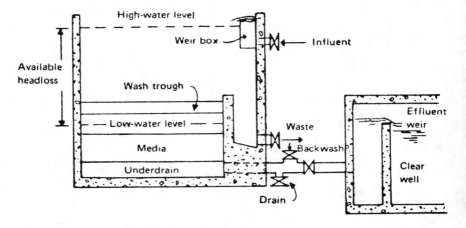

A. Typical filter and clear well arrangement.

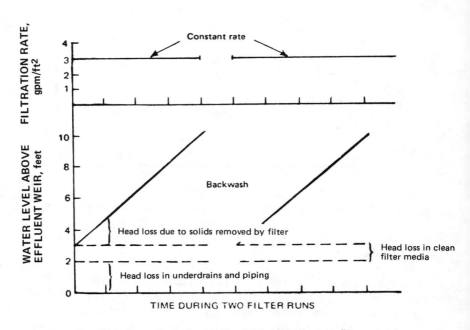

B. Filtration rate, water level and head loss in two filter runs.

Figure 7. Influent-flow-splitting filtration.

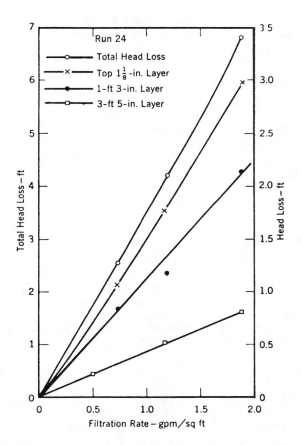

Figure 8. Head loss and filtreation rate.[32] After run at 2 gpm/ft^2, Ames filter influent water was used.

2 gpm/ft^2 (0.08 m^3/m^2 min). This figure demonstrates the existence of laminar flow conditions in filter media, an observation common to *all* water filtration applications in the usual range.

Head Loss in Underdrains

The flow through the filter underdrain system and effluent pipes is frequently under *turbulent* flow conditions. This means that the head loss is proportional to the *square* of flow rate. This is demonstrated effectively in Table II, which indicates that:

Table II. Head Losses in Filters

Item	Head, ft of Water		
Rate of Filtration, gpm/ft^2	2	4	6
Total Head Available	12.0	12.0	12.0
Head Loss Through Media	0.5	1.0^a	1.5
Head Loss Through Underdrain Piping System	1.0	4.0^b	9.0
Head Loss Available for Solids Accumulation	10.5	7.0^c	1.5

a $\dfrac{Q_2}{Q_1} = \dfrac{4}{2} = \dfrac{H_2}{H_1} = \therefore 2H_1 = H_2$

b $(\dfrac{Q_2}{Q_1})^2 = (\dfrac{4}{2})^2 = 4 = \dfrac{H_2}{H_1} \therefore 4H_1 = H_2$

c $12' - 1.0' - 4.0' = 7.0'$

- a major effect of increasing the rate on existing filters is to increase the head loss through the underdrain piping system faster than through the media, and
- the total head available for accumulation of suspended solids decreases rapidly with increase of rate.

Negative Head

As filtration progresses, the suspended solids removed are retained in the filter pores and the flow resistance increases which, in turn, decreases the pressure (driving force) available for maintaining the flow rate. The pressure distribution within the media of a typical gravity filter at various stages of a filter run is illustrated in Figure 9. Negative head (less than atmospheric pressure) can occur in a gravity filter when the summation of head loss from the sand surface downward exceeds the pressure available. The objection to negative head is the danger of forming air pockets in the zone of negative pressure. Air pockets reduce the effective filtering area, increase the local flow rate and head loss, and might result in serious degradation of filtrate quality. The filtration of aerated, iron-bearing water can cause air-binding of the filter due to the release of carbon dioxide or oxygen in the media. Thus, a part of the filter head loss attributable to iron is due to the accumulation of gas not iron.

In order to eliminate negative head in a gravity filter, it is *always* necessary to discharge the filter effluent water at a level *above* the sand media surface. This solution to the creation of negative head in filters is provided for the filters in both Figures 7 and 10. In order to keep the total

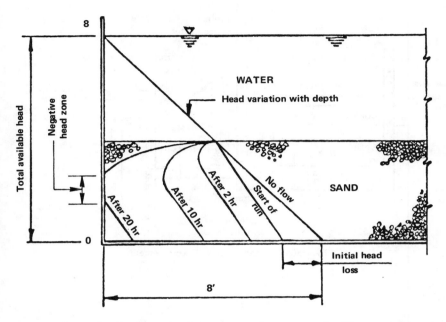

Figure 9. Head loss development at various stages of a filter run.

available head constant, the filter cell walls must be increased in height to make up for the distance to which the filter effluent discharge point is raised. Thus, this solution to negative head is applicable only in the design of new filters and not readily to the reconstruction of existing filters. Negative head conditions can also be eliminated by the use of pressure filters.

FILTERED WATER QUALITY CONSIDERATIONS

General Observations

At the beginning of a filter run, a short period of "bad" quality water is observed. The adjective "bad" is relative in that here it means that the suspended solids or turbidity level are higher than normal in the effluent. For example, Figure 11 shows that when a filter designed for iron removal at $C_o = 8$ mg/l was placed in service, the iron concentration in the effluent increased for the first 3 to 5 min (the approximate displacement time of the water in the media at the start of the run) and then improved over a period of about 1 hr. This phenomenon years ago led to the use of a filter-to-waste period when filters were first placed in service after backwashing.

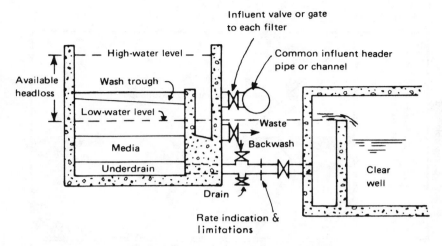

A. Typical filter and clear well arrangement.

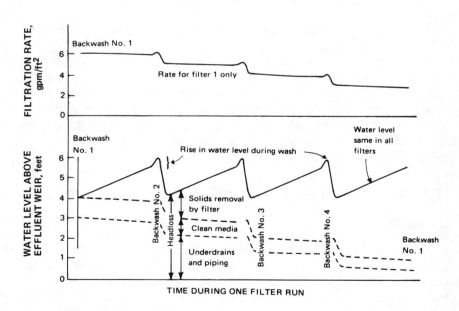

B. Filtration rate, head loss and water level in one filter run in a plant with four filters.

Figure 10. Variable declining-rate filtration.

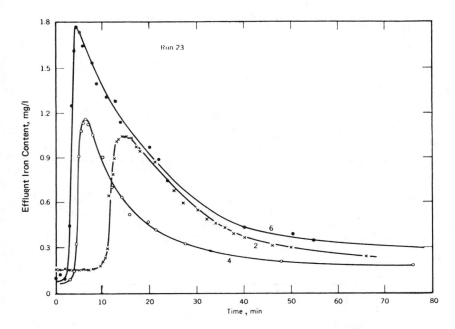

Figure 11. Initial effluent improvement. After water in filter has been displaced, turbidity drops to minimum. Numbers by curves indicate filtration rate in gpm/ft^2.

The initial period is usually followed by a long period of production of good water until later in the run when effluent quality may degrade again. The degradation may be gradual or sudden, and when the effluent quality exceeds a set level it is referred to as a solids *breakthrough*. Figure 12 shows typical effluent quality curves during this period as a function of filtration rate. Effluent quality may degrade at either low or high head losses. Runs should be terminated either on solids breakthrough *or* when the head loss limit is reached.

Effect of Filtration Rate

Low filtration rates do not ensure good quality water. With adequate chemical pretreatment and filter design, there is little difference in filtered water quality from filters operated at rates between 2 and 6 gpm/ft^2 (0.08 and 0.24 m^3/m^2 min) as shown by Figures 11 and 12. With poor pretreatment, even 2 gpm/ft^2 filtration rates can produce a bad water. Figure 13 demonstrates that the use of polyelectrolytes can improve filter effluent quality when added in very small doses to the filter influent. However,

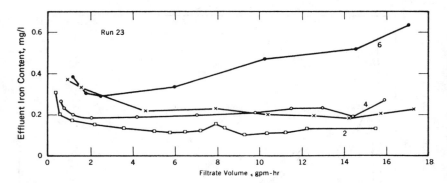

Figure 12.   Effluent iron content and filtrate volume. Observations for curves were made after initial improvement period. Numbers by curves indicated filtration rate. Curve along X marks indicates constant pressure. Rate of this run started at 6 gpm/ft^2 and decreased as head loss increased.

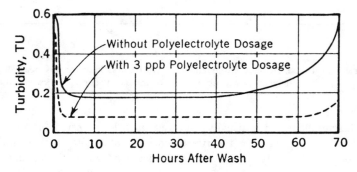

Figure 13.   Effluent turbidity with constant filtration rate.[8] The filtration rate was held constant at 3.5 gpm/ft^2.

such additions in this type of service cannot substitute for adequate coagulant dose in the pretreatment.

Effect of Filtration Rate Changes

Any rate increase or decrease on a dirty filter will cause some detriment to filtered water quality for a brief period thereafter (0.5-2 hr). The seriousness of the detriment depends on:[9]

> the magnitude of the rate change—the bigger the change the worse the effect;

- how suddenly the change is made—the more sudden the change, the worse the effect;
- the nature of the water and suspension being filtered—different waters have different sensitivities to rate change;
- other factors—use of polymer filter aids can reduce the magnitude of the effects observed.

Figure 14 shows the effect of a typical rate disturbance on effluent quality. Figure 15 shows the effect of the magnitude of the rate change and the

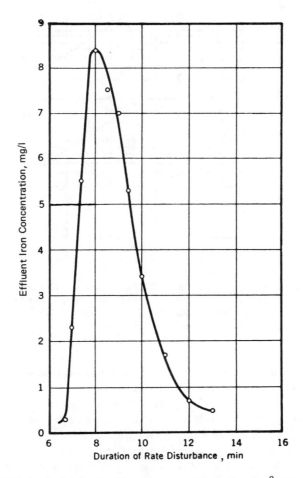

Figure 14. Effect of typical rate disturbance on effluent quality.[9] The iron concentration in the filter effluent builds up rapidly after the disturbance has been initiated. The curve represents Run 9a, which had an instantaneous rate change from 2 to 2.5 gpm/ft^2.

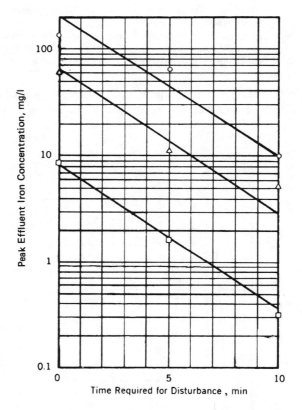

Figure 15. Peak concentration against disturbance time.[8] The curves indicate a first-order relationship between peak concentration and time required to make the disturbance. At a base rate of 2 gpm/ft^2, circles indicate 100% increase, triangles, 50% increase and squares, 25% increase.

duration of time over which the change occurs on the peak concentration of iron washed through a filter by the disturbance.

In discussion, the authors[9] suggested that rate disturbances such as those caused by rate controllers and those generated by taking filters in and out of service for backwashing should be avoided. This led to the authors' recommendation that constant-rate, effluent control filter design be abandoned to favor the use of influent flow-control or variable declining-rate filters to be described later in this chapter. These recommendations caused concern among filter control manufacturers and filter operators who doubted the magnitude of the disturbance effects. As a result,

Tuepker and Buescher[8] conducted studies at a plant-scale lime-softening plant to evaluate the effects of the rate disturbances.

Figures 16 and 17 show that they confirmed the fact that rate changes do produce a deterioration in effluent quality proportional to both the magnitude and rate of making a rate change. Again, the use of small doses of polymers as filter aids reduced but did not entirely eliminate the effects of the rate change. When the rate changes were applied gradually over a period of time, the extent of the quality deterioration was reduced, but not eliminated. With the use of polymers when the rate change was made over a period of 10 min, however, there were no observed effects of the rate changes (Figure 17).

Tuepker and Buescher[8] also evaluated the effectiveness of declining rate filtration. Figure 13 shows that with constant rate filtration at 3.5 gpm/ft^2, the water turbidity exceeded 0.2 unit after 46 hr of filtration without polymer addition and was below 0.1 unit for 60 hr with use of polymer. The results for declining-rate filtration are shown in Figure 18. Results equivalent to constant-rate filtration were achieved both with and without the use of polymer. Declining-rate filtration, however, eliminates the effects of rate disturbances and the need for rate controllers and complicated head loss gages.

These observations would imply the following:

- Effluent quality must be watched carefully. A continuous recording turbidimeter on each filter is desirable, and a good laboratory turbidimeter is the bare minimum needed for good operation.
- If filtered water quality is *not* good, why?
  - Poor pretreatment? Too much or too little coagulant?
  - Would a polymer help?
  - Are rate increases on dirty filters resulting from operations? Can they be avoided or made more gradually?

FILTER CONTROL METHODS

There are only two practical methods of operating filters and these differ primarily in the way that the driving force is applied across the filter. These methods are referred to as *constant-rate filtration* and *variable declining-rate filtration.*

In true *constant-pressure filtration*, the total available driving force is applied across the filter throughout a filter run. At the beginning of the filter run, the filter resistance is low and the rate of filtration is very high. (High driving force/low filter resistance = high rate of flow.) As the filter clogs with solids, filter resistance increases, and, because the driving force remains constant, the flow rate decreases. This method provides true declining-rate filtration.

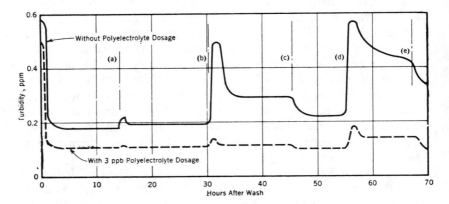

Figure 16. Effluent turbidity with rapid rate changes.[8] At (a) the rate was increased to 2.5 gpm/ft^2 within 10 sec. At (b) the rate was changed to 3.5 gpm/ft^2 within 10 sec. At (c) the rate was reduced to 2.5 gpm/ft^2 and at (d) was again increased to 3.5 gpm/ft^2 within 10 sec. At (e) the rate was decreased to 2.5 gpm/ft^2.

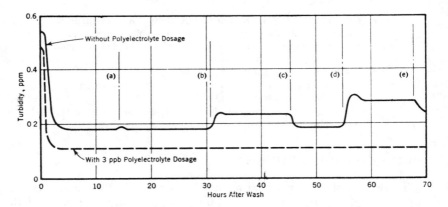

Figure 17. Effluent turbidity with gradual rate changes.[8] The rate changes were the same as in Figure 3, but the rate change was gradually accomplished over a 10-min period.

Constant-Rate and Constant-Water-Level Filtration

Current practice has tended to the use of *constant-rate* or *constant-water*-level filtration for gravity and/or pressure filters. A constant pressure

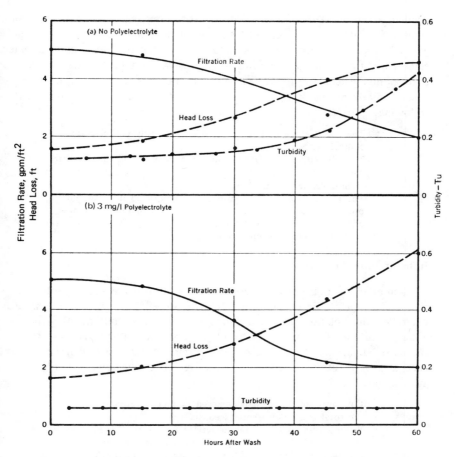

Figure 18. Filter effluent quality with declining-rate filtration.[8] At the beginning of the run, the rate was 5 gpm/ft^2; it declined to 2 gpm/ft^2 at the close of the run. (a) shows results without polyelectrolyte dosage; (b) shows results with 3 mg/l polyelectrolyte dosage.

is supplied across the filter system and the filtration rate or water level is then held constant by the action of a manually operated or automatic effluent flow control valve. At the beginning of the filter run, the filter is clean and has little resistance. If the full driving force were applied across the filter only, the flow rate would be very high. To maintain a constant flow rate or water level, some of the available driving force is consumed by an effluent flow control valve. At the start of the filter run, the flow control valve is nearly closed to provide the additional resistance needed to maintain the desired flow rate or water level. As filtration continues, the filter becomes clogged with solids and the flow control valve gradually opens. When the valve is fully open, the run must be terminated, since any further increase in filter resistance will not be balanced by a corresponding decrease in the resistance of the flow control valve.

The disadvantages of effluent-control, constant-rate filtration include the following:

- The initial and maintenance costs of the fairly complex rate control system are high.
- The filtered water quality is not as good using gravity granular media filters as that obtained using declining-rate filter operation in potable water filtration.[33,34]
- The rate or level control systems frequently do not function properly, causing sudden changes in rate, with the effect described previously on water quality. In many existing works, the control systems are nonfunctional.

Influent Flow Splitting

A number of alternative methods of flow control are coming into use that will supplant the effluent flow control valve[33] for gravity filters. For example, some plants have been constructed so as to split the flow nearly equally (influent flow splitting) to all the operating filters, usually by means of an influent weir box on each filter. A schematic diagram of such a gravity filter is shown in Figure 7A. The advantages of this system include:

- Constant-rate filtration is achieved without rate controllers if the total plant flow remains constant.
- When a filter is taken out of service for backwashing or returned to service after backwashing, the water level gradually rises or lowers in the operating filters until sufficient head is achieved to handle the flow. Thus, the rate changes are made slowly and smoothly without the abrupt changes associated with automatic or manual control equipment. Thus, there is little harmful effect to filtered water quality in potable water filtration.[8,9]
- The head loss for a particular filter is evidenced by the water level in the filter box. When the water reaches a desired maximum level (the desired terminal head loss), backwashing of that filter is required.
- The effluent control weir must be located above the sand to prevent accidental dewatering of the filter bed. This arrangement eliminates completely the possibility of negative head in the filter.

The only disadvantage of the influent flow splitting system is that additional depth of filter box is required because the effluent outlet is raised to its position above the filter media surface.

Variable Declining-Rate Filtration

Variable declining-rate operation is similar to influent flow splitting, and is the currently favored method of operation for gravity filters. Variable declining-rate operation achieves all the influent flow splitting advantages and some additional ones, without any of the disadvantages.

Figure 10A illustrates the desirable arrangement for new plants designed for variable declining-rate operation. Great similarity exists between Figure 7A and 10A with the principal differences being the location and type of influent arrangement and the provision of less available head loss.

The method of operation is similar to that described for influent flow splitting with the following exceptions. Figure 10B illustrates the water level variation and head loss variation typical to this operation. The filter influent enters below the wash trough level of the filters. When the water level in the filters is below the level of the wash trough, the installation operates as an influent flow splitting constant-rate filter. When the water level is above the level of the wash trough, the installation operates as a variable declining-rate filter. In general, the only time the filter water level is below the wash trough level is when all filters are backwashed in rapid sequence or after the total plant has been shut down, with no influent, so that the water level drops below the wash trough. In most cases, the clean filter head loss through the piping, media and underdrains ranges from 3-4 ft (0.9-1.2 m), and it keeps the actual low water level above the wash trough.

The water level is essentially the same in all operating filters at all times. This is achieved by providing a relatively large influent header pipe or channel to serve all of the filters and a relatively large influent valve or gate to each individual filter. Thus, head losses along the header or through the influent valve are small and do not restrict the flow to each filter. The header and influent valve are able to deliver whatever flow each individual filter is capable of taking at the moment. A flow-restricting orifice is recommended in the effluent pipe to prevent excessively high filtration rates when the filter is clean.

Each filter accepts at any time that proportion of the total flow that the common water level above all filters permits it to handle. As filtration continues, the flow through the dirtiest filter tends to decrease the most rapidly, causing the flow to redistribute itself automatically so that the cleaner filters pick up the capacity lost by the dirtier filters. The water level rises in the redistribution of flow to provide the additional head needed by the cleaner filters to pick up the decreased flow of the dirtier filters. The cleanest filter accepts the greatest flow increase in this redistribution. As the water level rises, it partly offsets the decreased flow through the dirtier filters and, as a result, the flow rate does not decrease as much or as rapidly as expected.

This method of operation causes a gradually declining rate toward the end of a filter run. Filter effluent quality is affected adversely by abrupt increases in the rate of flow, but here the rate increases occur slowly in the cleaner filters where they have the least effect on filter effluent quality.[8,9] Rate changes throughout the day in all of the filters due to changes in total plant flow, both increasing and decreasing, occur gradually and smoothly without any automatic control equipment.

The advantages of declining-rate operation over constant-rate operation are as follows:[33,34]

- For waters that show effluent degradation toward the end of the run, the method provides significantly better filter effluent quality than that obtained with constant-rate (or constant-water-level) filter operation.
- Less available head loss is needed compared with that required for constant-rate operation because the flow rate through the filter decreases toward the end of the filter run. The head loss in the underdrain and effluent piping system, therefore, decreases and becomes available to sustain the run for a longer period than would be possible under constant-rate operation with the same available head. Similarly, the head dissipated through the clogged portions of the filter media decreases linearly with the decreasing flow rate.

For the foregoing reasons, declining-rate filters are considered the most desirable type of gravity filter operation, unless the design terminal head loss is greater than 10 ft. Then, constant-level control or pressure filters may be a more economical choice. A bank of pressure filters can also operate using variable declining-rate filtration. However, any rate changes imposed on the plant cause sudden changes in filtration rates with pressure filters.

Some of the concerns and questions raised about variable declining-rate filtration are:

- It appears to be an uncontrolled system with little available operator manipulation. This is, in fact, an attribute that prevents operational abuse of the delicate solids removal mechanisms.
- If the rate-limiting device is sized for design year peak loads, it will permit higher than necessary filtration rates in the early plant life. This is true unless one limits the head loss utilized during the early plant life by using, for example, smaller orifices in the control section.
- What is the total available head loss to be provided? This is a difficult question, but no more difficult than it has been in the past for constant-rate filtration plants. It is best to be guided by past experience at the plant in question, or by pilot testing.

Surprisingly, the water level fluctuations in plants operating on this system are not as great as anticipated. Typical variations of 1.5-2 ft have been reported in potable water plants.[35,36]

GENERAL HEAD LOSS BEHAVIOR

Figure 19 shows several examples of head loss development during water filtration. Granular-media filters remove suspended solids in one of the following ways:

- by removal of the suspended solids at the surface by the finer media at the top of the filter, which forms a relatively thin layer of deposited solids at the surface,
- by depth removal of the suspended solids within the voids of the porous media—the better the distribution of the solids throughout the depth of the filter media, the better the use of the head available, and
- by a combination of surface removal and depth removal.

Most waters encountered in potable water filtration have nearly linear or straight line head loss vs time or filtration volume curves when operated at constant rate. This indicates no strong surface removal of solids, and production per run to a given terminal head loss decreases as filtration rate is increased. Figure 20 shows actual data from a run in which iron was being filtered. The very linear shape of the curves indicate depth removal of solids by adsorption to the surface of the filter media with more or less uniform blockage of the pore volume. Many waters produce head loss curves similar to those in Figure 19B, which, because of the slight curvature of the lines, indicate that some removal by straining takes place in the depth of the media, resulting in some pore blockage other than by adsorption to the media surface.

Some waters exhibit a strong surface removal of the solids being filtered because the finer media strains the solids at the filter surface. As a result, the head loss curve is exponential (concave upward) when operated at constant rate (Figure 19A). When this occurs, the media size at the surface must be increased or the filtration rate must be increased to drive the solids further into the filter to prevent such surface cake formation. Production per run can be increased at the high rates as shown by Figure 19C. Figure 21 shows results of runs filtering the clarifier overflow from a lime-soda ash softening plant.

Concave downward head loss vs time curves indicate weak removal and potential passage of turbidity in constant-rate filtration. However, in declining rate filtration, head loss curves for a given filter may be concave downward due to the decreasing filtration rate on that filter. This would not necessarily indicate potential passage of turbidity. Every operator of

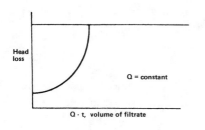

A. Surface removal of compressible solids.

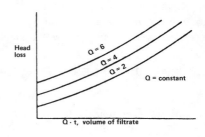

B. Depth removal of suspended solids.

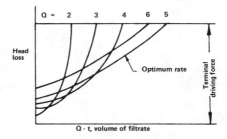

C. Depth removal of suspended solids with surface cake.

Figure 19. Head loss development during filtration.

a filtration plant should observe and plot filter head loss curves to show which type of head loss behavior exists. If the curves are concave upward, higher rates will increase production per run. If they are concave downward, there is a possibility of bad water. Pretreatment must be improved, and polymers should be considered as a filter aid.

FILTER RUN LENGTH

Since the capital cost of a filter is chiefly a function of the area of filter provided, the use of a high filtration rate is usually preferred. In general, the filter design should seek to maximize the net water production per square foot of filter consistent with filter operating feasibility. Useful relationships between net water production and run lengths obtained at different filtration rates are shown in Figure 22. Figure 22 is constructed

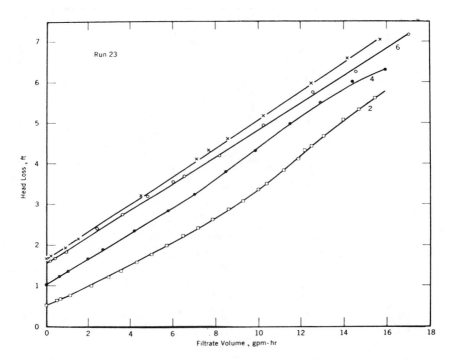

Figure 20. Total head loss and filtrate volume.[32] No optimum-rate tendency appears. Uncontrolled run (along X marks) started at 6 gpm/ft² and was allowed to decrease in rate as head loss increased.

for traditional potable water practice, and is drawn for the waste of all backwash water.

The data in Figure 22 were calculated assuming 30 min total down time per backwash to allow for draindown time, auxiliary air-scour time, actual backwash time and start-up time to reach normal rate. A total of 100 gal/ft² (4.1 m³/m² min) of wash water per backwash was used since it is typical of volumes adequate for most filtration situations.

The interpretation of this figure is relatively simple. At a filtration rate of 2 gpm/ft² (0.08 m³/m² min) and an infinite length of filter run, the net water production from the filter would be only 2 x 1440 min = 2880 gal/ft²/day and net water production would be the same as the filtration rate. However, when only 1.5-hr runs are obtained (which give 12 cycles/day with 0.5 hr for backwash), net water production would drop to a net production rate of only 0.67 gpm/ft². But we could shift to a filtration rate of 6 gpm/ft² and would have net water production of

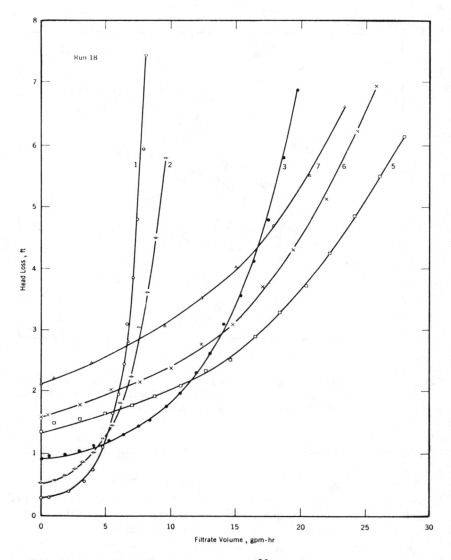

Figure 21. Total head loss and filtrate volume.[32] Ames filter influent was used while plant operated one well at 1100 gpm. In this run, optimum-rate tendency appeared at all flow rates.

at least 4.7 gpm/ft² even when the filter runs are as short as 3 hr. Normally, an increase in the filtration rate of this amount would not cause a decrease in run length from an infinite length run to one only 3 hr long.

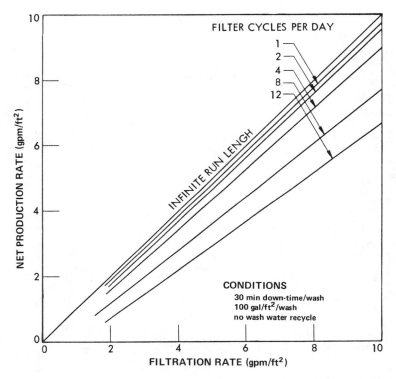

Figure 22. Effect of number of filter cycles per day on filtrate production with no recycle of backwash water.

In filter design, therefore, it should be remembered that long filter runs are not necessary to achieve a high net water production from a filter. In Figure 22, there is little to be gained in water production in going from a 24-hr run to an infinite length of run, and there is some danger. In iron removal, for example, the iron is oxidized to the insoluble state where it can be removed in the filter. If the filter runs are very long and bacterial action is present, the oxygen level in the filtered water can be depleted and the iron can be reduced back to a soluble form and can then leave in the filter effluent.

Very short runs, however, should be avoided. Because a filter needs to be taken out of service for cleaning, all filters are not always in service. If a battery of four filters or a four-cell declining-rate filter battery is used, the percentage of time all four filters are in service as a function of run length is shown in Figure 23. When run lengths are fewer than

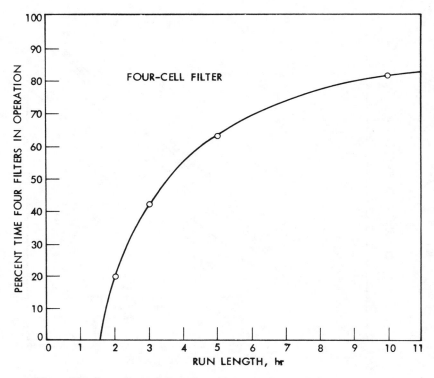

Figure 23. Percentage of time four filter cells in operation vs run length.

10-12 hr, one filter will be out of service more than 20% of the time. In filter design, this would mean that with minimum runs of 15-18 hr, the filter area could be based on the design filtration rate assuming all filters are in service. With runs of only 10-12 hr or less, the peak filtration rate should be reduced to allow for the down time of the filters. In general, the author prefers to use a filtration rate that under peak solids and flow conditions, allows the filter runs to be greater than, say, 15 hr and less than 24 hr. In general, runs longer than 30-36 hr are not desirable.

FILTER MEDIA

General

The quality of a filtered water is a function of both media size and media depth. The finer the media, the better the water quality produced, but the length of the filter run gets shorter. In general, the quality of filtered water is a function of the surface area of the media (which gets larger as the media gets smaller) and of the pore sizes between media particles (which get smaller as the media gets smaller). In general, the largest pore opening in a filter media is only about 15% of the media size. Thus, in practice, a filter should be able to remove *by straining* any particle whose diameter is greater than, say, 5% of the media size. Thus, a 0.5-mm sand (500μm) should be able to remove particles greater than 0.025 mm (25μm) in diameter by straining.

The granular filter media commonly used in potable water filtration include silica sand, garnet sand and anthracite coal. These media can be purchased in a broad range of effective sizes and uniformity coefficients. Table III is a partial list of media suppliers. Table IV shows the size range of media collected on various sieve sizes. The various media have specific gravities approximately as follows:

- anthracite coal: 1.35-1.75 for most U.S. anthracite; 1.6-1.75 for U.K.
- silica sand: 2.65
- garnet sand: 4-4.2

Selection of Filter Media

The selection of a filter media determines not only the filter effluent quality but also the filter backwash regime, and thus the backwash requirements become an integral part of the media decision.

For many years, the filter media selected for a deep-bed filter was usually sand. Because of the sorting process that goes on in the backwashing of a deep-bed filter, a bed of a natural material like sand becomes graded, with the fine-grained material at the top of the filter bed and the coarser-grained material at the bottom. The presence of the finest-grained material at the top of the filter bed where suspended solids first make contact with the filter favors rapid deposition near the surface of the filter. At flow rates less than the optimum, this unfavorable, though natural, gradation may promote the formation of a compressible mat of suspended solids on the surface that quickly uses up the available head.

This problem in particle size gradation within the filter bed can be corrected in a variety of ways. Ideally, a uniform size of media could be used. This ideal can be approached in a practical sense by using a material with a low uniformity coefficient. For example, we could specify that the media pass a U.S. Sieve No. 30 and be retained on a U.S. Sieve No. 40 (less than 0.60 mm and greater than 0.42 mm sand, Table IV).

Table III. List of Media Suppliers[a]

Anthracite Coal

1.	Carbonite Filter Corporation Box #1 Delano, Pennsylvania 18220	3.	Reading Anthracite Coal Co. 200 Mahantango Street P. O. Drawer F Pottsville, Pennsylvania 17901
2.	Shamokin Filler Company, Inc. Box 272 Shamokin, Pennsylvania 17872	4.	Glenn Alden Fuel Sales P. O. Box 568 Wilkes-Barre, Pennsylvania 18703

Silica Sand and Gravel

1.	Eau Clair Sand and Gravel Company 104 Gibson Street Eau Claire, Wisconsin 54701	6.	Cosby-Carmichael, Inc. Box 597 Selma, Alabama 36701
2.	Northern Gravel Company Box 307 Muscatine, Iowa 52761	7.	William Bird Sales Corporation Box 5442 Columbia, South Carolina
3.	Jesse S. Morie & Sons, Inc. Mauricetown, New Jersey 08329	8.	Southern Products & Silica Co. Box 38 Lilesville, North Carolina 28091
4.	The Parry Company 219 Church Street Chillicothe, Ohio 45601	9.	George F. Pettinos, Inc. 235 Bala Avenue Bala-Cynwyd, Pennsylvania 19004
5.	Monterey Sand Company, Inc. Box 928 Monterey, California	10.	New Jersey Pulverizing Company 205 West 34th Street New York, New York 10001

Garnet Sand

1.	Idaho Garnet Abrasive Company Box 1080 Kellogg, Idaho 83837

[a]This list is *not* a complete list of suppliers and is included for general information only.

Low uniformity coefficients (UC) are desired to achieve easier backwashing. This is especially true where fluidization of the media is required during backwashing, as with dual- and triple-media filters which require it because the entire media should be fluidized to achieve restratification. Therefore, the greater the UC (*i.e.*, less uniform size range), the larger the backwash rate required to fluidize the coarser grains thus provided. A UC of less than 1.3 is not generally practical because of the sieving capabilities of commerical suppliers. A UC of less than 1.5 can be obtained at a cost premium and is recommended.

Table IV. Size Range of Uni-Sized Media

Geometric Mean Size (mm)	Sieve Opening		U.S. Sieve No.	
	Passing (mm)	Retained (mm)	Passing	Retained
0.46	0.50	0.42	35	40
0.55	0.60	0.50	30	35
0.65	0.71	0.60	25	30
0.77	0.84	0.71	20	25
0.92	1.00	0.84	18	20
1.09	1.19	1.00	16	18
1.30	1.41	1.19	14	16
1.54	1.68	1.41	12	14
1.84	2.00	1.68	10	12

A UC of less than 1.5 has the advantage of ensuring that the coarser grain size in the media (such as the 90% finer size) is not excessively large which requires a large backwash rate. Sieve analyses of filter media usually plot linearly on either log-probability or arithmetic-probability paper. Advantage may be taken of the gradation caused by backwashing in a natural granular media if the filter is operated in an upflow mode because the coarsest part of the media is then exposed to the incoming flow and the finest part of the media is exposed to the effluent for final polishing of the effluent. Although the upflow mode of operation has been used in actual applications, it poses some technical problems in design and construction. These have prevented it from becoming a popular mode of operation in potable water filtration. Uplift of the media and the need for media restraint to prevent its displacement during filtration are the source of these problems. Approximately the same results may be achieved by layering one or more media, with the coarsest (but least dense) at the top.

Pitman and Conley[37] were among the first to apply this concept in the gradation of the media in a deep-bed filter by capping a sand filter with a layer of less dense but coarser granular material–anthracite coal. Subsequently, dual-media filters became common and multi-media filters, with as many as five layers of hydraulically separable materials, have been used to "reverse" the gradation of grain size in the filter media.

Dual Media

For dual-media filters, the sizes of the sand layer must be selected to be compatible with the coal that has been selected. The bottom sand (*e.g.*, the 90% finer size) should have approximately the same or a somewhat lower flow rate required for fluidization than the bottom coal to ensure that the entire bed fluidizes at the selected backwash rate.

To assist in the selection of the required backwash rate and to assess the compatibility question raised, empirical data on the minimum fluidization velocity of coal, sand and garnet sand at 25°C are presented in Table V. Empirical correction factors to be applied for other water temperatures are presented in Table VI. The temperature correction factors agree substantially with data presented by Camp.[38]

The effective size of the sand for a dual-media filter should be selected to achieve the goal of coarse-to-fine filtration without causing excessive media intermixing. If the coal density is in the typical range of 1.65-1.75 g/cm³, a ratio of about 90% finer coal size to the 10% finer sand size equal to about 3 results in a few inches of media intermixing at the interface.[38] A ratio of these sizes of 4 results in substantial media intermixing, whereas a ratio of 2 to 2.5 causes a sharp interface. Choosing media sizes to achieve a sharp interface means that the benefits of coarse-to-fine filtration are partly lost. Therefore, a size ratio of about 3 is recommended.

The use of Table V and the foregoing recommendation can be illustrated with an example. Assume a coal of 1.2-mm ES with a UC less than 1.5 (size range of 1.2-2.2 mm, 8-16 mesh range) has been selected. The sand should have an effective size about 0.7 mm to be one-third of the coarse coal size. A sand size range of 0.7-1.4 mm could be specified (14-25 mesh range), or one with an ES of 0.7 mm. The backwash rate for the coarse end of the coal (2.38 mm) is 30 gpm/ft² (1.22 m³/m² min) at 25°C and the coarse sand (1.4 mm) is 27 gpm/ft² (1.1 m³/m² min). Thus, they are compatible. If the peak operating temperature is expected to be 15°C, the required backwash rate would be 30 x 0.83 = 25 gpm/ft².

It should be noted that no harm would be done if the coarser sand grains were smaller than 1.4 mm. They would merely reach fluidization before the coarser coal grains. There is no danger of inversion of the coal and sand layers during backwashing or complete intermixing as there is with sand and garnet sand. The intermixing behavior of coal and sand, and sand and garnet sand has been experimentally demonstrated.[39]

The selection of the garnet sand in a triple-media filter would also have to be compatible with the silica sand in the same way that the silica sand is compatible with the coal. If the goal of filtered water turbidity is above about 0.2 TU, this author favors the use of dual-media filters. If the goal is less than 0.2 TU, then consideration of triple-media filters may be justified.

The particle size of the media selected for a given filter application depends upon the kind of suspended solids to be removed by the deep-

Table V. Minimum Fluidization Velocities for Various Uniform-Sized Media (Observed Empirically)[9]

Between U.S. Std. Sieves			Mean Size (mm)	Flow Rate to Achieve 10% Expansion at 25°C, gpm/ft^2		
Passing	mm	Retained		Coal	Sand	Garnet
7	2.830	8	2.59	37.0		
8	2.380	10	2.18	30.0		
10	2.000	12	1.84	24.0	41.0	
12	1.680	14	1.54	20.0	33.0	
14	1.410	16	1.30	15.7	27.0	49.0
16	1.190	18	1.09	12.5	21.0	40.0
18	1.000	20	0.92	9.9	16.4	32.0
20	0.841	25	0.78	8.4	12.6	27.0
25	0.707	30	0.65	7.0	9.0	22.0
30	0.595	35	0.55		6.3	18.0
35	0.500	40	0.46		5.4	13.7
40	0.420	45	0.38		4.0	11.3
50	0.297	60 (0.25 mm)	0.27			6.3
Specific Gravity				1.7	2.65	4.1

Table VI. Temperature Correction: Approximate Correction Factors to be Applied for Temperatures Other Than 25°C

Temperature (C°)	Multiply the 25°C Value by
30	1.09
25	1.00
20	0.91
15	0.83
10	0.75
5	0.68

bed filter. This must now be established by pilot plant studies, for as yet there is no rational basis for selecting a particle size or specifying an optimum flow rate. Experience in water filtration indicates that in dual-media filters, coal as coarse as 1.0-1.2 mm has been used effectively.

Specifications can be and are now written to provide a desirable, low-uniformity-coefficient filter media with a top coarse coal size of 1.0 mm by specifying that 90% of the coal shall pass a #14 sieve (1.41 mm) and that 80% shall be retained on a #18 sieve (1.00 mm). The 10% fines that pass the smaller sieve should be removed by skimming them off the surface of the filter media after several backwashings.

In dual-media filters, the two media must be sized for two considerations:

1. The 90% finer coal size and the 10% finer sand size must be related so as to control the degree of media intermixing.
2. The 90% finer coal size and the 90% finer sand size must also be related so that both media are fluidized to the same degree during backwashing at the same backwashing rate.

The coal specified above, for example, would require the use of a 0.45-mm sand to provide a coal/sand ratio of 3.0 and a uniformity coefficient of about 1.64 to provide a 90% finer sand that can be fluidized during backwash along with the (1.41 mm) 90% finer coal.

In addition to specifying the gradation of filter media used, the depth of media must be established. At present, there is no reasonable method (other than pilot plant operation) that can be used to determine the optimum depth of filter media. In general, this author prefers dual-media depth ranges as follows: coal, 18 to 24 inches; sand, 15 to 18 inches.

Robeck et al.[40] have presented data to indicate that the primary purpose of use of dual-media is the ability to lengthen filter runs significantly by better distribution of the solids load in the filter. With adequate pretreatment to produce a strong floc, Figure 24 indicates that the effluent quality is the same from sand (2 ft of 0.45-mm), coal (2 ft of 0.70-mm) and coal-sand (18 in. of 1.05-mm coal over 6 in. of 0.45-mm sand) filters, but the coal-sand filter produced a significantly longer filter run.[40] However, with a relatively weak floc due to inadequate pretreatment, the solids breakthrough was delayed 10-12 hr in the coal-sand media (Figure 25).

In summary, dual- and triple-media filters can be expected to:

- greatly increase run length for the use of the same sand size in a single-media filter, and
- produce a better quality water than can be obtained by a single-media filter under inadequate pretreatment conditions.

Media Backwashing

General

The principal problems in filter operation are associated with maintaining the filter bed in good condition. Inadequate cleaning leaves a thin layer of compressible dirt or floc around each grain of the media. Because pressure drop across the filter media increases during the subsequent filter run, the grains are squeezed together and cracks may form in the surface of the media, usually along the walls first.

The heavier deposits of solids near the surface of the media break into pieces during the backwash. These pieces, called mudballs, may not disintegrate during the backwash. If small enough and of low density, they

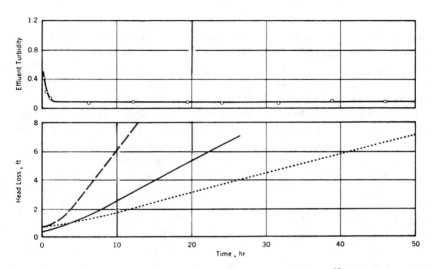

Figure 24. Effect of filter media on length of run with strong floc.[40] The data shown were obtained under the following operating conditions: raw-water turbidity, 30-45 units; alum dose, 7 ppm; activated silica, 20 mg/l; filtration rate, 2 gpm/ft²; settling tank effluent turbidity, 2 TU. In the upper graph, the curve is the same for all three filters. In the lower graph, the dashed curve is for sand; the solid, coal; and the dotted, coal and sand.

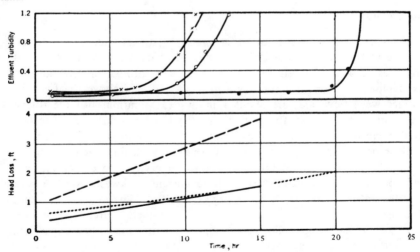

Figure 25. Effect of filter media on length of run with weak floc.[40] The data shown were obtained under the following operating conditions: raw-water turbidity, approximately 20 TU; alum dose, 100 mg/l; activated carbon, 2 mg/l; filtration rate, 2 gpm/ft²; settling tank effluent turbidity, 15 TU. In the upper graph, the curve determined by the X points is for coal; that by open circles, sand, and that by solid circles, coal and sand. In the lower graph, the dashed curve is for sand; the dotted, coal and sand; and the solid, coal.

float on the surface of the fluidized media. If larger or heavier, they may sink into the filter, to the bottom, or to the sand-coal interface in dual media filters. Ultimately, they must be broken up or removed from the filter because they will reduce filtration effectiveness or cause shorter filter runs by dissipating available head loss.

Potable water filter backwashing practice in the U.S. has used the high velocity wash with substantial bed expansion (20-50%). This method does not solve all problems with dirty filters, and it has created problems with shifting of the finer supporting gravel layers. The provision of a surface wash system has largely solved the problem of dirty filter media for potable water filters, but it has not solved the problem of shifting gravel. The growing use of wastewater filtration has further demonstrated the weakness of water fluidization backwash. Backwashing is substantially more difficult, and problems of agglomerates and filter cracks are prominent.

The problem of shifting gravel and the more difficult backwashing of a wastewater filter has stimulated renewed interest in the air-scour methods of auxiliary agitation, which has continued in use in European practice. There is also interest in the use of underdrain systems with fine strainers that do not require gravel, a system that was abandoned in the U.S. in the early 20th century due to clogging and corrosion problems.

Surface Wash

Evidence of the benefits of surface wash led to its wide adoption for potable water filters in the United States. Surface wash is introduced at pressures of 45 to 75 psig (310 to 510 kN/m^2) through orifices on a fixed piping grid or on a rotating arm located 1-2 in. above the fixed bed. Surface wash flow rates are about 1 in./min (2.5 cm/min) for the rotary type and 3-6 in./min (8-15 cm/min) for the fixed nozzle type. The desired operating sequence involves draining the filter to the wash trough level or below, applying the surface wash flow with no concurrent backwash flow for 1-2 min to break up surface layers on top of the media, and continuing the surface wash with concurrent backwash flow for several minutes until the backwash water begins to clear. The concurrent application may be at two rates, first a low rate to immerse barely the surface wash jets in the media followed by a period with normal bed expansion. The surface wash is then terminated and water fluidization backwash alone follows for 1-2 min to stratify the bed, a provision which is only important in dual- or triple-media filters.

Air Scour

Air scour consists of the distribution of air over the entire filter area at the bottom of the filter media so that it flows upward through the media. It is used in a number of fashions to improve the effectiveness of backwashing, and/or to permit the use of lower backwash water flow rates. The air may be used prior to the water backwash or concurrently with the water backwash. When used concurrently during backwash overflow, there is legitimate concern over loss of filter media to the overflow due to the violent agitation created by the air scour. When air is used alone, the water level is lowered 6-8 in. below the overflow level to prevent loss of filter media during the air scour.

Air scour may be introduced to the filter through a pipe system that is completely separate from the backwash water system, or it may be added through the use of a common system of nozzles (strainers) that distribute both the air and water, either sequentially or simultaneously. In either method of distribution, if the air is introduced below graded gravel supporting the filter media, there is concern over the movement of the finer gravel. This may occur as air is expelled by the water at the onset of the water wash, or especially by air and water used concurrently by intention or accident. This concern has led to the use in some filters of media-retaining strainers, which eliminate the need for graded support gravel in the filter. However, these strainers may clog with time, which causes decreased backwash flow capability or, possibly, structural failure of the underdrain system. Such failures have occurred.[41] This clogging may be due to fine sand or coal, which leaks through the strainer during down-flow filtration and later is lodged forcefully in the strainer slots during water backwash. The underdrain plenum below the strainers must be scrupulously cleaned before the strainers are installed to prevent construction dirt or debris from later clogging the strainers during backwash.

In view of the concerns expressed above and the renewed interest in air scour in the United States, a summary of European air-scour practice in potable water treatment is worthwhile because air scour has been used there since the beginning of rapid filtration.

Current potable water practice in the U.K. uses air first followed by water backwash. Plastic strainers with 3-mm slots are used in the underdrains, covered by two or more layers of gravel to support the media. Single-media sand filters are common, with a size range of 0.6-1.2 mm, but dual-media filters are being used more and more since about 1970. For single-media sand filters, the water wash rate is intended just to reach minimum fluidization velocity with only 1-2% bed expansion. Air is introduced for 3-5 min through the gravel layers at rates of 1-1.5 scfm/ft^2 (0.30-0.46 m^3/m^2 min), (sometimes up to 2 scfm/ft^2) followed by water at 12 in./min (7.5 gpm/ft^2 or 0.31 m^3/m^2 min). Problems with gravel movement have occurred, but only in a few cases. The absence of such problems must be attributed to the low water and air flow rates that are presumably not sufficient to move the fine gravel, and the fact that air and water are not used simultaneously. The preference for strainers with 3-mm slots and gravel is based on prior experience with clogging of strainers with 0.5-mm slots.

The renewed use of air scour in the United States has been patterned more after the British practice of using air scour alone first, followed by water backwash. U.S. air rates have been typically 3-5 scfm/ft^2 (0.91-1.5 m^3/m^2 min) for 3-5 min, and the subsequent water wash has been above fluidization velocity to expand and restratify the dual-media bed (typically 24-36 in./min or 15-22.5 gpm/ft^2 or 0.6 to 0.91 m^3/m^2 min). The filters are usually equipped with media-retaining underdrain strainers without graded gravel support for the media. Because of the fine sand media used in the dual-media beds (0.4-0.5 mm effective size), the strainer openings are very small (0.25-0.5 mm) and the strainer clogging problems and some underdrain failures have occurred therefrom. Because of this problem, a reconsideration of the U.S. air-scour design practice may be appropriate.

Backwashing Recommendations

In view of the difficulty of backwashing water filters, and the various filter media and backwash routines available, a research study was conducted at Iowa State University to compare the various alternatives as applied to wastewater filtration. Many of the conclusions of that study are applicable in potable water filter design. Some are important to the design of water filters and are, therefore, presented here:[42]

1. The cleaning of granular media filters by water backwash alone to fluidize the filter bed is inherently a weak cleaning method because particle collisions do not occur in a fluidized bed; thus abrasion between the filter grains is negligible.
2. Air scour followed by water fluidization backwash, and surface (and subsurface at the coal-sand interface) wash before and during water fluidization backwash proved to be comparable methods that can be applied to single-, dual- and triple-media filters. These two methods did not completely eliminate all dirty filter problems, but both auxiliaries reduced the problems to acceptable levels so that filter performance was not impaired.
3. For optimum water-only backwash, the media should be expanded to provide a void ratio of 0.7 at the surface of the filter media.
4. The use of graded gravel to support the filter media is not recommended where the simultaneous flow of air scour and backwash water can pass through the gravel by intention, or by accident, due to the danger of moving the gravel and thus upsetting the desired size stratification of the gravel.
5. Air scour is compatible with dual- or triple-media filters from the standpoint of minimal abrasive loss of the anthracite coal media.

However, the backwash routine must be concluded with a period of fluidization and bed expansion to restratify the media layers after the air scour.

The author suggests that the foregoing conclusions be used as design guides. In addition, the following design considerations concerning the backwashing provisions should also be considered.

First, consider the use of air scour as applied to dual- or triple-media filters backwashed with fluidization capability. For such a design:

1. Provide operational flexibility in the period of air scour between, let us say, 2 and 10 min so the operator could select the period he deems most appropriate.

2. If supporting gravel is not used, provide the capability for simultaneous air and water backwash. This technique requires provisions to allow for rapid draining of the filter to near the filter media surface, followed by the brief simultaneous air and water backwash until the water reaches within 6-8 in. of the wash troughs. The simultaneous wash is then stopped, and either air alone or water alone may be continued. The water rate during the simultaneous air-water wash should be below fluidization velocity to extent the time duration of that action to the maximum.

3. Provide a backwash water volume of at least 75-100 gal/ft^2 (3-4 m^3/m^2) of filter per wash. This is based on the observation that when backwashing at rates above the fluidization velocity for the media, the total wash water required for effective cleaning is about the same regardless of the backwash rate (about 75-100 gal/ft^2 of filter). This observation is for typical U.S. wash trough spacing with the trough edges about 3 ft (0.91 m) above the surface of the filter media. Larger spacing between troughs, or greater height of trough above the media, would increase the wash water requirements. No economy of total wash water use is achieved by adopting lower backwash rates (above fluidization) because the length of required backwash must be increased proportionately.

4. With dual- and triple-media filters, greater-than-normal bed expansion may be necessary to release dirt trapped deep in the bed. Optimum water backwash alone occurs at expansions of 40% for the typical sand beds and 25% for typical coal beds. A good guide would be to achieve 25% expansion of the coal bed and 20% expansion of the sand bed.

5. Appropriate media must be specified and provided so that the desired stratification of media layers occurs during backwash. Laboratory column studies are one simple way to be sure what happens with a given media. Interfacial mixing increases with higher rates and varies with the speed of closure of the backwash valve. Air-scour destratifies the bed and causes substantial media intermixing. Therefore, adequate time of water-only wash must be provided following an air-water wash to allow restratification of the bed and to wash out the residual air. The water valve closure should take place over a period long enough to restratify the entire bed.

6. The designer should remember that backwash rates required are controlled by the highest water temperature expected in the backwash water. The temperature correction diagram in Table VI can be used to convert the backwash rate requirement for different-sized media given at 25°C to any other water temperature.

REFERENCES

1. Weber, W. J., Jr., Ed. *Physiochemical Processes for Water Quality Control* (New York: Wiley-Interscience, 1972).
2. Fuller, George W. *The Purification of the Ohio River Water at Louisville, Kentucky* (New York: D. van Nostrand Co., 1898).
3. Yao, K. M., M. T. Habiban and C. R. O'Melia. "Water and Waste Water Filtration: Concepts and Application," *Environ. Sci. Technol.* 5:1105-1112 (1971).
4. Baumann, E. R. and C. S. Oulman. "Polyelectrolyte Coatings for Filter Media," *Filtration Separation* 7:682-690 (1970).
5. Bell, G. R. "Coagulant Coatings Open New Applications to Filter Aids," *Proc. Internat. Water Conf.* 129-133 (1961).
6. Mintz, D. M. "Kinetics of Filtration," *Dokl. Acad. Nauk. USSR* 78:12 (1951).
7. Mintz, D. M. "Modern Theory of Filtration," Special Subject No. 10, International Water Supply Congress and Exhibition, London (1966).
8. Tuepker, J. L. and C. A. Buescher, Jr. "Operation and Maintenance of Rapid Sand and Mixed Media Filters on a Lime Softening Plant," *J. Am. Water Works Assoc.* 60:1377-1388 (1968).
9. Cleasby, J. L., M. M. Williamson and E. R. Baumann. "Effect of Filtration Rate Changes on Quality," *J. Am. Water Works Assoc.* 55:869-878 (1963).
10. Ives, K. J. "Theory of Filtration," Special Subject No. 7, International Water Supply Congress and Exhibition, London (1969).
11. Lerk, C. F. "Some Aspects of the Deferrisation of Groundwater," Thesis, Technical University, The Netherlands (1965).
12. Mackrle, V. and S. Mackrle. "Adhesion in Filters," *J. San. Eng. Div., ASCE* 87:1732 (1961).
13. Iwasaki, T. "Some Notes on Sand Filtration," *J. Am. Water Works Assoc.* 29:1591-1597 (1937).
14. Stein, P. C. "A Study of the Theory of Rapid Filtration of Water Through Sand," Sc.D. Thesis, Massachusetts Institute of Technology (1940).
15. Ives, K. J. "Deep Filters," a paper presented to the 61st National Meeting, AIChE, Houston, Texas (1967).
16. Litwiniszyn, J. "Colmatage Considered as a Certain Stochastic Process," *Bulletin de L'Academie Polonaise Des Sciences Serie des Sciences Techniques* 11:81-85 (1963).
17. Litwiniszyn, J. "On Some Mathematical Models of the Suspension Flow in Porous Medium," *Chem. Eng. Sci.* 22:1315-1324 (1967).
18. Litwiniszyn, J. "The Phenomenon of Colmatage Considered in the Light of Markov Processes," *Bulletin de L'Academie Polonaise des Sciences Serie des Sciences Techniques* 16:183-188 (1968).
19. Litwiniszyn, J. "On a Certain Markov Model of Colmatage—Scouring Phenomena, I.," *Bulletin de L'Academie Polonaise des Sciences Serie des Sciences Techniques* 16:535-539 (1968)
20. Kraj, W. "Probabilistic Model of Colmatage and Scouring Phenomena," *Bulletin de L'Academie Polonaise des Sciences Serie des Sciences Techniques* 16:443-450 (1968).

21. Kraj, W. "The Changes in the Porosity Coefficient During the Process of Colmatage," *Bulletin de L'Academie Polonaise des Sciences Serie des Sciences Techniques* 18:239-243 (1970).
22. Hudson, H. E., Jr. "A Theory of the Functioning of Filters," *J. Am. Water Works Assoc.* 40:868-872 (1948).
23. Hudson, H. W., Jr. "Factors Affecting Filtration Rates," *J. Am. Water Works Assoc.* 48:1138-1154 (1956).
24. Gamet, M. B. and J. M. Rademacher. "Measuring Filter Performance," *Water Works Eng.* 112:117-118 (1959).
25. Hsiung, K. Y. and J. L. Cleasby. "Prediction of Filter Performance," *J. San. Eng. Div., ASCE* 94:1043-1069 (1968).
26. Cleasby, J. L. "Approaches to a Filtrability Index for Granular Filters," *J. Am. Water Works Assoc.* 61:372-381 (1969).
27. Hsiung, K. Y. "Filtrability Study on Secondary Effluent Filtration," *J. San. Eng. Div., ASCE* 98:505-513 (1972).
28. Conley, W. R. and K. Y. Hsiung. "Design and Application of Multi-media Filter," *J. Am. Water Works Assoc.* 61:97-101 (1969).
29. Tchobanoglous, G. "Filtration Techniques in Tertiary Treatment," *J. Water Poll. Control Fed.* 42:604-623 (1970).
30. Ives, K. J. "Optimization of Deep Bed Filtration," First Pacific Chemical Engineering Congress, Part I, Session 2, Separation Techniques, 99-107, Society of Chemical Engineers, Japan and AIChE, October 10-14, 1972.
31. Huang, J. Y. C. and E. R. Baumann. "Least Cost Sand Filter Design for Iron Removal," *J. San. Eng. Div., ASCE* SA2 97:171 (1971).
32. Cleasby, J. L. and E. R. Baumann. "Selection of Sand Filtration Rates," *J. Am. Water Works Assoc.* 54:579-602 (1962).
33. Cleasby, J. L. "Filter Rate Control Without Rate Controllers," *J. Am. Water Works Assoc.* 61(4):181-185 (1969).
34. Hudson, H. E., Jr. "Declining Rate Filtration," *J. Am. Water Works Assoc.* 51(11):1455 (1959).
35. Arboleda, Jr. "Hydraulic Control Systems of Constant and Declining Rate in Filtration," *J. Am. Water Works Assoc.* 66:87-94 (1974).
36. Cleasby, J. L. "New Ideas in Filter Control Systems," in Proceedings of Symposium "Procesos Modernos De Tratamiento De Aqua" XIII Congreso Interamericano De Ingenieria Sanitaria, published by the Pan American Health Organization (August, 1972).
37. Conley, W. R. and R. W. Pitman. "Test Program for Filter Evaluation at Hanford," *J. Am. Water Works Assoc.* 52(2):205-218 (1960).
38. Camp, T. F. "Discussion-Experience with Anthracite Filters," *J. Am. Water Works Assoc.* 53:1478-1483 (1961).
39. Cleasby, J. L. and C. F. Woods. "Intermixing of Dual Media and Multi-Media Granular Filters," *J. Am. Water Works Assoc.* 67(4):197-203 (1975).
40. Robeck, G. G., K. A. Dostal and R. L. Woodward. "Studies of Modifications in Water Filtration," *J. Am. Water Works Association* 56:198-213 (1964).
41. Cleasby, J. L., E. W. Stangl and G. A. Rice. "Developments in Backwashing Granular Filters," *J. Env. Eng. Div., ASCE* 101(EE5):713-727 (1975).
42. Cleasby, J. L. and E. R. Baumann. "Backwash of Granular Filters Used in Wastewater Filtration," Final Report, EPA Project R802140 (Jan., 1960).

CHAPTER 13

SUSPENDED SOLIDS REMOVAL

Robert T. O'Connell
Project Specialist
The Permutit Company, Inc.
E. 49 Midland Avenue
Paramus, New Jersey 07652

It has always been desirable to produce a potable water that has a minimal amount of suspended matter. Ideally, a water treatment plant effluent should contain less than one turbidity unit (TU). To attain this quality a water treatment plant must be equipped with the proper unit processes to remove suspended solids. These unit processes include chemical coagulation, flocculation, sedimentation, and filtration. This chapter will describe these unit processes and their practical application.

NATURE OF SUSPENDED SOLIDS IN WATER

All surface waters contain suspended matter in varying concentrations. These solids span a great range of sizes, shapes and materials. Naturally occurring suspended solids found in water include clays, organic matter, metal oxides, and minerals such as silica. These particles are normally too small to be removed by gravity settling. The diameter of most suspended solids ranges from $0.1\text{-}10^{-5}$ mm. Therefore, it is necessary to coagulate the suspended solids in order to ensure their settling within a reasonable time.

Before the suspended solids can amass, there are certain natural factors that must be taken into account. Almost all particles suspended in water possess a charge (predominantly negative). The cause of this charge can be any of the following:

- an imperfection in a crystal lattice (clay particles)
- the ionization of a surface group (the ability of this particle to remain in solution is pH-dependent)
- the abosrption of a specific ion on the particle surface.

Particles that depend upon surface charge for stability and do not possess bound water molecules on their surface are classified as hydrophobic (clays). These particles can be destabilized by an electrolyte. Conversely, suspended matter that has absorbed layers of water surrounding it is termed hydrophilic, and it is not easily coagulated.

CHEMICAL COAGULATION

Before the suspended matter in water can begin to amass, the stabilizing forces must be neutralized. The only useful method for destabilizing suspended particles is the addition of chemicals. The chemicals normally employed for this purpose are inorganic salts of iron or aluminum. The inroganic salts are electrolytes that bring about particle destabilization by one or both of the following mechanisms:

1. The repulsive, electrostatic charge surrounding the particles is reduced. This is accomplished by the absorption of the metal ion on the surface of the particle. The absorption of the counter-charged ions on the particle reduces the net particle charge thus allowing van der Walls forces to interact among the particles.
2. Gelatinous hydroxide complexes, formed by the reacting coagulant, enmesh the suspended solids. The acidic coagulants react with the natural alkalinity in the water to form the hydroxide products. If the water does not contain sufficient natural alkalinity to react with the coagulant, a source of alkalinity should be added in the form of lime, soda ash or caustic soda.

An alternative means of particle destabilization is the use of natural or synthetic polymers. There are long-chain, high molecular weight electrolytes with many active sites. One or more of these active sites adsorb onto the particle surface. At the other end of the polymer chain other active sites adsorb onto another particle, forming a "bridge" between the two suspended particles, which greatly improves settling characteristics of the particles.

Occasionally, it becomes necessary to add a component that aids the coagulant. These aids are used to coagulate more effectively the suspended solids (and/or increase floc size). These aids include weighting agents (bentonite clays, certain polymers), adsorbents (activated carbon), oxidants (chlorine, ozone, potassium permanganate) and polymers.

Some typical cases where an aid is required include:

- If the raw water suspended solids content is below 50 mg/1, it is probably necessary to add a weighting agent to obtain good removal. This weighting agent serves two purposes: it increases the probability of particle collisions, and it increases the weight of the floc particles formed.
- In the presence of organic matter, an oxidant may be required to achieve good coagulation.
- If an inorganic metal salt forms a light, dispersed floc, it is advisable to add a polymer to aid flocculation.

The use of a coagulant aid may also significantly reduce the coagulant dosage or expand the optimum pH range of the coagulant. These are economic considerations that must be evaluated when choosing chemicals for a water treatment plant.

SELECTION OF CHEMICALS

In the design of any water treatment plant for suspended solids removal, one of the most critical items is the proper selection of chemicals. Ideally, tests must be made on the raw water. This is the only way to determine the most effective combination and optimum dosages of chemicals. Obviously, this is not always possible for any number of reasons. It is necessary to have a detailed water analysis, and a year's sampling program discloses seasonal variations in the water quality, which may affect the chemicals necessary for proper coagulation. A good source for surface water quality data in each state is the U.S. Geological Survey, which distributes a compilation of water quality data from the major surface water sources within individual states.

Without the benefit of test work, provisions should be made for feeding a coagulant and one or more aids. The choice of the coagulant should take into account the price and availability of the chemicals under consideration. There are certain publications that list the manufacturers of the types of chemicals used in water treatment. Buyers guides are helpful, and market prices for the chemicals are available.

COAGULANTS

Alum

Alum [aluminum sulfate, $Al_2(SO_4)_3 \cdot 14 H_2O$] is undeniably the most widely used and accepted coagulant available. Coagulation is actually caused by hydroxide complexes of aluminum. The tri-valent aluminum ion is effective in reducing the net charge of the suspended matter. Alum is available as both a solid (containing approximately 17% aluminum as Al_2O_3) or a liquid (containing approximately 8.3% aluminum as Al_2O_3). Dry alum is sold in four commercial grades: lump, ground, rice and powdered. All four types are readily available in 100-lb bags or in bulk. Liquid alum is available in drums or in bulk (truck or tank cars).

A simplification of the alum reactions in water treatment are as follows:

$$Al_2(SO_4)_3 \cdot 14\ H_2O + 3\ Ca(HCO_3)_2\ \text{(natural alkalinity)} = \qquad (1)$$
$$2\ Al(OH)_3 + 3\ CaSO_4 + 6\ CO_2 + 14\ H_2O$$

$$Al_2(SO_4)_3 \cdot 14\ H_2O + 3\ Na_2CO_3\ \text{(soda ash)} = \qquad (2)$$
$$2\ Al(OH)_3 + 3\ Na_2SO_4 + 3\ CO_2 + 14\ H_2O$$

$$Al_2(SO_4)_3 \cdot 14\ H_2O + 3\ Ca(OH)_2\ \text{(lime)} = \qquad (3)$$
$$2\ Al(OH)_3 + 3\ CaSO_4 + 14\ H_2O$$

The most important variable in the proper use of alum as a coagulant is maintenance of the correct pH. The operational pH range for alum is 5.5 to 7.8. If test data to the contrary are lacking, it can be assumed that the optimum pH for turbidity removal is 6.8 and, for color removal, 5.6. With natural waters the pH values can be maintained by one of the following:

* Making use of the acidic nature of the alum to lower the pH to the desired value.
* If the dosage of alum required to maintain the ideal pH is too high to be economically acceptable, then the use of an acid to lower the pH may be necessary.
* If insufficient natural alkalinity is present in the raw water, the use of an alkali may be required to achieve the desired pH after a minimum dose of alum.

For estimating the chemical feed requirements, a minimum dosage of 35 mg/1 of alum should be considered for turbidity coagulation. Additional alum and/or acid may be necessary to lower the pH. An alkali may be needed to raise the pH. For color removal the minimum dose is 45 mg/1 of alum, plus either additional alum and/or acid to lower the pH to 5.6 or an alkali to raise the pH to 5.6.

There are two principal types of chemical feed systems employed for dry alum:

1. *Solution or batch type.* A specific amount of alum is added to a solution mixing tank (normally 100-lb bags are unloaded into the tank). Mechanical agitation is used to make up a maximum solution concentration of 12% alum by weight. Since it is important to control precisely the amount of alum added to the raw water,

a metering pump is normally used to control the feed. If the flow rate of the raw water is variable, the metering pump can be paced by an adjustable timer receiving a signal from either a mechanical meter with contacting head or an integrator receiving a signal from a differential pressure cell. With this system the necessary volume of solution can be fed to the raw water regardless of the raw water flow rate.

2. *Dry chemical or continuous type feed.* The chemical is metered out (either continuously or paced in proportion to flow) either volumetrically or gravimetrically from storage. The most popular type of dry chemical feeder for plants of under 10 mgd capacity is the screw type. The feed screw meters out a specific volume of alum that falls into a solution chamber equipped with a mechanical agitator and make-up water controlled by a float valve. The float valve is necessary to allow direct pumping from the solution chamber. Since the chemical is instantaneously metered, the solution concentration is immaterial. Hence a centrifugal pump can be used to deliver the solution. Variable speed drives on the screw motor ensure maximum flexibility of this system. It must be remembered that alum solutions are acidic and materials of construction for the chemical feed system components (pumps, agitator shafts, tanks) must be either made of or lined with corrosion resistant material.

Ferrous Sulfate

Copperas ($FeSO_4 \cdot 7 H_2O$) has a much wider effective pH range (4-11) than alum. Normally, the optimum pH for coagulation is above 8.0. Because of this, lime is fed in conjunction with ferrous sulfate to ensure the high pH. Ferrous sulfate is available in bulk and in 100-lb bags from a limited number of suppliers. Commercially available ferrous sulfate contains 55-58% $FeSO_4$. A simplification of the reactions of ferrous sulfate in water treatment are as follows:

$$FeSO_4 + 3 Ca(HCO_3)_2 = Fe(OH)_2 + CaSO_4 + 2CO_2 \qquad (4)$$

$$FeSO_4 + Ca(OH)_2 = Fe(OH)_2 + CaSO_4 \qquad (5)$$

Since ferrous hydrozide is soluble, the ferrous ion must be oxidized. Surface waters normally contain sufficient dissolved oxygen to accomplish this. If not, chlorine should be added.

$$4 Fe(OH)_2 + O_2 + 2H_2O = 4 Fe(OH)_3 \qquad (6)$$

For the purpose of estimating chemical feed requirements for suspended solids removed, a minimum dose of 50 mg/l of ferrous sulfate can be assumed. Sufficient lime should be fed to maintain the pH at about 9.3. Also, there should be provisions for feeding an oxidant to ensure the oxidation of the ferrous ions. The same feed systems used for alum can be used for ferrous sulfate. The maximum solution concentration is 5%. A minimum of five minutes detention time is required when mixing the solution to ensure proper dissolving of the chemical. Like alum, ferrous sulfate is acidic and care must be taken in selecting the materials of construction for the feed system.

Ferric Sulfate

The effective pH range of this coagulant, ($Fe_2(SO_4)_3 \cdot 7\ H_2O$), is the same (4-11) as that for ferrous sulfate. Ferric sulfate is available in 100-lb bags, drums, or bulk. The source of ferric sulfate is limited. The chemical should contain a minimum of 18% trivalent iron. As with ferrous sulfate, the optimum pH for coagulation is usually above 8.0. Therefore, lime (or other alkali) feed is required. The simplified ferric sulfate reactions in water treatment are as follows:

$$Fe_2(SO_4)_3 + 3\ Ca(HCO_3)_2 = 2\ Fe(OH)_3 + 3\ CaSO_4 + 6\ CO_2 \qquad (7)$$

$$Fe_2(SO_4)_3 + 3\ Ca(OH)_2 = 2\ Fe(OH)_3 + 3\ CaSO_4 \qquad (8)$$

For estimating chemical feed requirements a minimum dose of 25 mg/l should be considered for suspended solids removal. Sufficient lime, to maintain a pH of 9.3, would have to be added.

Again, the same feed systems that are used for alum can be used for ferric sulfate. Care must be taken to dissolve the ferric sulfate properly. A solution concentration of up to 25% is practical, but the addition of the make-up water must be carefully controlled. Also, a 30-min detention time is required in the mixing tank (with mechanical agitation) to ensure complete dissolving. Again, the acidic nature of the solution must be considered.

COAGULANT AIDS

Bentonite Clays

Clays are used in conjunction with an iron or aluminum salt in the following situations:

- raw water turbidities of less than 50 TU
- color removal
- treating waters with low mineral content.

In the first situation, the addition of the clay particles increases the probability of collision among the floc particles. This helps build larger, more readily settleable floc. In the second and third examples, the inorganic salts tend to form a light floc that does not settle rapidly. The clay combines with the floc particles and adds weight to them, which results in an increased settling rate. An added benefit of the clay is its ability to adsorb certain organics. For estimating purposes, a dose of 50 mg/1 of clay should be considered.

Again, the same basic types of feed systems that are used for alum can be used for clay. Care must be taken when mixing to ensure adequate wetting of the clay particles. Usually adding the clay to the vortex caused by the mechanical mixer is adequate. For tank sizing, a 5% slurry can be considered. Since the clay is noncorrosive, standard materials of construction can be used for either the suspension or dry chemical feed systems.

Powdered Activated Carbon

Activated carbon is often fed with a coagulant to help remove taste- and odor-causing compounds. A slurry of as much as 10% carbon by weight is possible. Mechanical agitation is necessary, with the precaution of feeding the carbon into a vortex. When using dry chemical feeds, a special wetting device is normally used in place of a mixing tank. Carbon is delivered by ejector. Since the amount of carbon to be fed depends upon the organics present in the raw water, the amount must be established by test. But for estimating purposes, a dose of 25 mg/1 of powdered carbon can be assumed.

Polymers

Organic polymers (polyelectrolytes) used in water treatment are high molecular weight, long-chain molecules. The polyelectrolytes are either natural (starches, polysaccharides) or synthetic (polyacrlimides). They are characterized by their charge (cationic, anionic or nonionic), which also dictates the circumstances in which the polymer is used. Cationic polyelectrolytes are used predominantly as primary coagulants. The polymers can be used either alone or in conjunction with an inorganic salt (*e.g.*, alum) to reduce the amount of inorganic salt required. Anionic polyelectrolytes most often are used as flocculant aids. When used in conjunction with an iron or aluminum salt, the anionic polyelectrolyte causes the formation of larger floc particles, which have an increased settling rate and increase the effluent clarity.

A great number of polymers are commercially available. All manufacturers supply information on the proper handling and feeding of their products. Polyelectrolytes are available in either solid or liquid form.

When using dry polymers, special care is needed in the preparation of the feed solution. Precise amounts of the polymer are added to a disperser funnel either manually or by use of a screw-type volumetric dry chemical feeder. The disperser funnel ensures adequate wetting and even distribution of the polymer. The disperser funnel discharges the wetted polymer directly into a mix tank equipped with a low-speed agitator, which is necessary to maintain a uniform solution. Aging time must be provided in the mix tank to hydrate fully the polymer molecules. The contents of the mix tank are then transferred to a second tank, which should be sized for no more than one day's storage to prevent deterioration of the polymer. Since polymers remain stable in solution, mixing is not required. From this second tank, the solution can be fed proportionately to its appropriate point of application.

The feed of liquid polymer is accomplished by transferring a portion of the concentrated liquid to a dilution tank. The polymer manufacturers can supply information on the proper concentration in the dilution tank. Agitation is required in the dilution tank to blend the concentrated liquid polymer and the dilution water uniformly. From the dilution tank, the polymer solution can be fed to the water treatment units.

SOLIDS CONTACT–SLUDGE BLANKET UNITS

Principles of Operation

The solids contact unit combines the following processes in a single unit: mixing, flocculation and solid-liquid separation. To accomplish this, the unit, shown in Figure 1, is divided by sloping baffles into two distinct zones: mixing and separation.

The unit is very efficient because chemical reactions occur faster and more completely in the presence of previously formed floc. Raw water and chemicals are introduced into the top of the unit. Mixing is provided to ensure adequate contact between the water, the chemicals and the previously formed floc. Throughout this mixing zone, a slow rolling agitation is applied to the slurry. This agitation results in the flocculation of the particles.

To control the flow between the mixing and separation zone, there is a port area, which is designed to produce uniform flow between zones. At this location, the downward flow in the mixing chamber is diverted to an upward flow in the separation zone. There is also recirculation of some of the sludge from the separation zone through the port area and into the mixing area, which helps to promote floc growth.

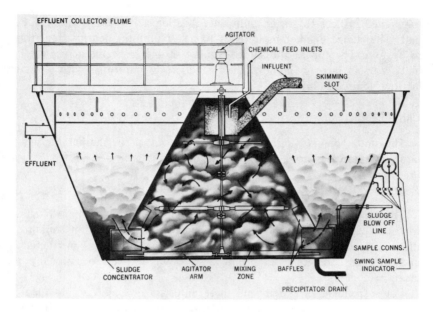

Figure 1. Sludge blanket solids contact clarifier.

The separation zone itself is designed with an ever-increasing upflow
area. This results in an ever-decreasing upflow velocity. There is a point
at which the upflow velocity of the water no longer supports the sludge
particles. At this point, a definite line of separation exists between the
sludge blanket and the clarified water, which has been filtered through
a bed of its own sludge. The floc particles left behind further combine
with other particles and grow larger. As a result they drop lower in
the sludge blanket. When the floc particle reaches the lowest part of
the separation zone, it is either recycled through the port area or it drops
into the sludge concentration section for blow-off. If the particle returns
to the mixing zone, the bottom agitator blade causes the particle to be
sent higher into this zone.

The sludge to be removed from the unit is allowed to concentrate in
a semiquiescent zone within the unit to reduce water losses. The sludge
is then wasted on a timer-controlled cycle to maintain a fixed balance
of solids in the unit. This balance is determined by the type of precipitate. The flow-through rate and the water temperature determine the
sludge blanket level. The concentration of the wasted sludge depends
greatly on the amount and type of chemicals fed.

After leaving the sludge blanket, the treated water continues to flow
upward to the collector system. The effluent is usually collected by

submerged orifices to ensure uniform collection across the surface of the unit. The sludge blanket itself tends to be of uniform density from top to bottom. The top of the sludge blanket should be stable across the unit. With an increase in flow, the blanket rises uniformly; with a decrease in flow, it drops uniformly.

Unit Classification

The sludge blanket units are classified according to whether the agitator orientation is horizontal or vertical.

Vertical Sludge Blanket Unit

The vertical unit (Figure 2) consists of an outer conical-shaped tank, with the widest section at the top. The inner conical section contains the mixing zone and the agitator. These units are usually constructed of steel on a concrete base, although the outer walls are occasionally made of concrete.

For suspended solids removal, these units are designed around the general parameters of 1.0 gpm/ft^2 (0.041 m^3/m^2 min) upflow rate at the surface and a detention time of 60 min. Design flow rates vary between 50 and 7500 gpm (3 to 473 l/sec). This corresponds to diameters ranging from 10-100 ft (3.3-33 m). Straight side wall heights vary from 10-14 ft (3.3 to 4.3 m). If longer detention times are required, or if more head is required for the effluent, additional side wall height can be added. This increase in the detention time does not significantly affect the upflow rate.

The agitator drive mechanism should have at least a 4:1 range. A walkway should provide access to the drive mechanism. Sample connections should be included to monitor the mixing and separation zones and the sludge concentrator.

Horizontal Sludge Blanket Unit

The outer chamber in the horizontal unit is rectangular and the outer walls are constructed of concrete. The inner mixing chamber is separated by an A-frame partition, which runs the full length of the unit (see Figure 3). The water and chemicals are introduced into a trough at the top of the mixing zone, thus assuring even distribution of water across the length of the unit. The water flows down into the mixing zone, where two horizontal agitators assure proper mixing, sludge recirculation, and no accumulation of sludge on the floor of the mixing zone.

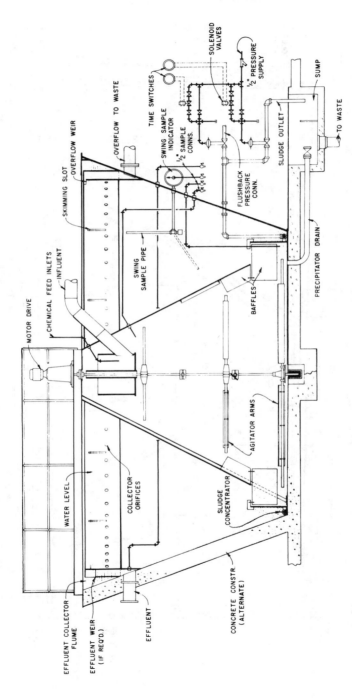

Figure 2. Vertical sludge blanket clarifier.

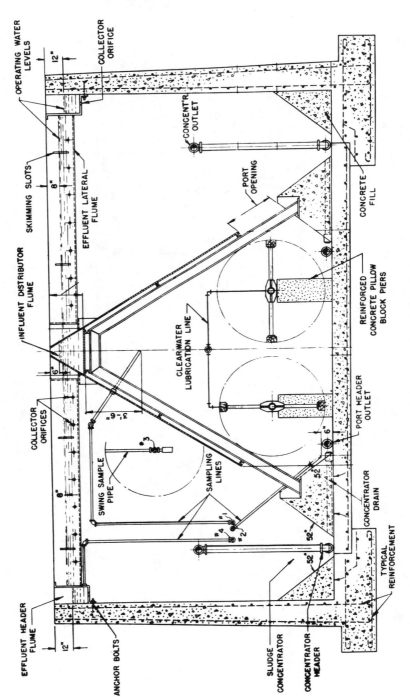

Figure 3. Horizontal sludge blanket clarifier.

The inlet flume, skirt plates (between the mixing and separation zones) and effluent flumes can be constructed of steel, wood or Fiberglass Reinforced Plastic (FRP). The popularity of the FRP is growing because of its ease of handling and low maintenance costs.

The typical size of these units ranges from 300-10,000 gpm or higher. The maximum upflow rate is 1.5 gpm/ft^2 and the minimum detention period is 60 min. Standard sizes are: 18.5 ft wide x 12 ft deep, 27 ft wide x 13 ft deep, 33.5 ft wide x 15 ft deep, 48 ft wide x 17 ft deep (5.6 x 36.5 m, 8.2 x 4 m, 10.2 x 4.6 m, and 14.6 x 5.2 m, respectively). The length of the unit is variable, to accommodate the flow. The detention time can be increased by raising the water depth of the unit.

With these units the agitator drive mechanism is located on one end of the unit and access must be provided. A walkway along the length of the unit should be provided to facilitate maintenance work.

Tube Settler Modules

Tube settlers, such as the Permutit Company's Chevron Shaped Tube Settler, offer both economic and performance advantages. The economic advantages are realized when considering that the allowable upflow rate for a given surface area can be increased significantly with the installation of tube settlers. For existing installations, plant capacity can be expanded for a lower capital expenditure. For new installations, equipment sizes and capital erection costs are reduced. A performance advantage in the optimization of effluent quality results from the short settling distance (4 in. or 10 cm) and the expanded effective settling area of 17.3 ft^2/ft^2 surface area.

Factors that should be considered when choosing a tube settler design include:

- laminar flow through the tube
- minimum settling distance by control of tube height
- continuous accumulated solids movement down the tube, without resuspension.

The chevron configuration provides all of the above. This shape has the highest perimeter of any common shape of the same area, and therefore, the lowest Reynolds number. The settling distance for particles entering the top of the tube is uniform. The 90° V-groove base provides optimum positive sludge flow. The tubes can be nested without loosing useful area. Tube modules can be shaped to fit almost any shape of basin with a minimum of waste and dead tube area. For alum flocs, the maximum particle upflow rate is approximately 2.5 gpm/ft^2 (0.1

m^3/m^2 min) of surface area. The type of suspended solids being coagulated and the chemicals fed affect the upflow rate, and all factors affecting upflow rate should be considered before determining the design rate.

FILTRATION

Since filtration is covered in other chapters, only one specific type of filter is discussed here.

Automatic Valveless Gravity Filter (AVGF)

The AVGF, shown in Figure 4, is well suited to filter the effluent from a solids contact unit. In the AVGF, coagulated and settled water from a constant level source, at least 21 ft (6.4 m) above grade, flows into the inlet pipe, which discharges into the upper part of the filter bed compartment, and filters down through the media. Water passes through plastic strainers at the bottom of the sand into the plenum chamber. The filtered water then rises up through the effluent pipe to an elevation of 14 ft (4.3 m) above the base. During the filter run the accumulated floc on the filter bed slowly builds a back pressure, which causes the

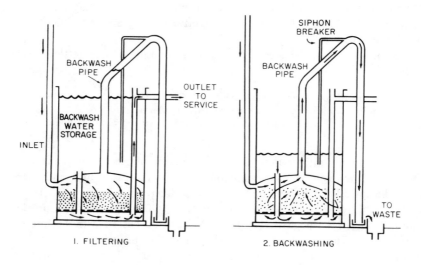

Figure 4. Operation of automatic valveless gravity filter.

water level in the backwash pipe to rise. When the pressure loss reaches a predetermined level, usually 4-5 ft (1.2 to 1.5 m) above the level of the filtered water outlet, a self-actuated primer system rapidly exhausts air from the backwash pipe and starts the syphoning action that backwashes the filter.

The AVGF has an integral backwash water storage chamber. When the unit backwashes, water from the storage chamber flows through ducts to the plenum chamber, through the strainers and backwashes the media. At the beginning of backwash the level in the storage chamber is highest. Therefore, the backwash starts at a high rate and diminishes gradually until the syphon is broken. During the backwash, the influent continues to flow to the filter chamber. The backwash cycle starts at 18-20 gpm/ft^2 (0.73 to 81 m^3/m^2 min) and decreases to 8-10 gpm/ft^2. The average rate is 13-15 gpm/ft^2 (0.53 to 0.61 m^3/m^2 min). The diminishing backwash rate ends when the backwash water level exposes a syphon breaker, which admits air to the backwash piping, breaks the syphon, and stops backwashing. At the end of the backwash cycle, the influent flow first fills the backwash storage chamber for the next back-wash. When filtering properly pretreated raw water at usual filter rates of 2-3 gpm/ft^2 (0.08 to 0.12 m^3/m^2 min) of filter bed area, filter runs of 2-3 days are usual.

When several AVGFs are installed, backwashing of more than one unit at a time is prevented by interlocking the top of the backwash pipe of each filter with the backwash storage tanks of the others. Interlocking is accomplished by running a pipe from the top of backwash pipes to a few inches below the backwash water surface of each other filter. While any one filter is backwashing, pipes of the other units are vented and cannot backwash. Thus, the sewer line need only be large enough to carry the backwash of a single filter.

The backwash rate is regulated by an adjustable restriction on the outlet and of the backwash pipe. This is set to give a 5-min backwash. An auxilliary pressure water supply to the backwash primer must be supplied to allow manual initiation of the backwash. The AVGE can be supplied with either a 24 in. (61 cm) deep bed of 0.45-0.60-mm silica sand (1.6 uniformity coefficient) or a 24 in. deep duel bed of anthracite and sand. The AVGF tank itself is about 15 ft (4.6 m) high. The elevation of the backwash pipe above the water level in the backwash storage tank determines the head loss that initiates the backwash. This is usually 5 ft (1.5 m) with the standard design. A turbidity breakthrough is unlikely because of the low head loss permitted.

BUDGETARY INFORMATION

Budget estimates (based on April 1977 prices) for capital equipment only, including solids contact units, gravity filters, chemical feeds and controls (FOB Points of Shipment) are given in Table I.

Table I. Cost (fob) of Permutit Filter Plants

Flow Rate	Estimated Budget Price
250 gpm	$ 90,000
500 gpm	140,000
1000 gpm	165,000
2000 gpm	240,000
5000 gpm	370,000

REFERENCES

1. Burnes and Roe, Inc. "Process Design Manual for Suspended Solids Removal," Technology Transfer, Environmental Protection Agency, EPA 625/1-75-003a (January 1975).
2. *Chemical Marketing Reporter.* (New York: Schnell Publishing Company).
3. American Water Works Association. *Water Quality and Treatment,* (New York:McGraw-Hill Book Company, 1971), Chapter 3.
4. Daniels, S. L. "A Survey of Flocculating Agents—Process Descriptions and Design Considerations," Presented at the 74th National Meeting of the A.I. Ch.E., New Orleans (1973).
5. Cities Service Company, Industrial Chemical Division. "Ferrifloc for Water and Waste Water Treatment," Atlanta, Georgia (1972).
6. "Precipitators for Clarification and Softening." Bulletin No. F220, The Permutit Company, Inc., Paramus, N. J. (1973).
7. George, G. S., and S. L. Bishop. "Application of Fiber-Glas-Reinforced Plastics at the Shoremont Water Treatment Plant," *J. Am. Water Works Assoc.,* 68 (5):223 (1976).
8. "Permutit Type V2B Chevron Shape Tube Settlers," Technical Bulletin No. 4422-01, The Permutit Company, Inc. Paramus, N.J. (1976).
9. U.S. Environmental Protection Agency. "National Interim Primary Drinking Water Standards," *Federal Register* 40:59566-59508 (1975).

CHAPTER 14

ADVANCED TECHNIQUES FOR
SUSPENDED SOLIDS REMOVAL

W. R. Conley, P. E.

Vice President
Research and Technical Services
Neptune Microfloc, Incorporated
1965 Airport Road
Corvallis, Oregon 97330

S. P. Hansen, P. E.

Senior Engineer
Culp-Wesher-Culp
3939 Cambridge Road
Shingle Springs, California 95682

INTRODUCTION

The suspended solids in water originate from a number of sources such as eroded soil, industrial waste, domestic waste, algae and various substances that, although not in suspension, are removed by coagulation, as are the suspended solids. All these materials can be removed by coagulation in the proper pH zone, with alum being the most common coagulant.

The single most important step in the removal of suspended solids from water is proper coagulation. Once the coagulation step is performed properly, operation of the plant is comparatively simple. However, because the raw water normally changes both seasonally and after every major storm, the operator must continuously adjust the chemical feed to cope with the constantly changing nautre of the raw water. If the water operator is lucky enough to have an impoundment or lake, changes in the water quality are less drastic, but nevertheless they do occur. The basic tools for determining proper coagulation are visual observation of the floc that is formed, the use of the jar test apparatus in the lab, and the pilot filter (Coagulant Control Center). An experienced operator can, by the use of one of these three methods, maintain control of his plant and produce a uniform quality filtered water regardless of the changing nature of the raw water supply.

High-rate sedimentation (using settling tubes) and high-rate filtration (using multimedia filters) have made it possible to decrease sharply the cost of new water treatment plants or to upgrade existing water treatment plants at minimum cost. A discussion of coagulation control, high-rate sedimentation, and high-rate filtration is the subject of this chapter.

COAGULATION

In order to achieve good coagulation, it is necessary to add the optimum amount of coagulant in the optimum pH zone. Too much or too little coagulant and too high or too low pH causes poor results. The coagulant and the pH adjustment chemical must be mixed quickly and thoroughly with the incoming raw water. The simplest and most practical way to do this is to inject the chemicals into the incoming raw water pipeline because turbulent flow in a pipeline is the cheapest and most effective way of distributing the chemicals into the raw water. If it is not possible to use this method, various quick-mixing devices are available. After the chemicals have been mixed with the water, coagulation is essentially instantaneous. The various suspended solids in the water, along with color and some dissolved substances, react with the alum to produce destabilized particles that can be flocculated and settled or, in some treatment plants, can be filtered directly without the flocculation and settling steps.

A device called the Coagulant Control Center can be used to determine whether coagulation is adequate in a very short time. After the chemicals have been added, a sample of the water is pumped to the Coagulant Control Center (see Figures 1 and 2). A portion of this water is passed through a small filter and the effluent from the filter is measured on a sensitive turbidimeter. The output from the turbidimeter is recorded. If the nature of the raw water changes or if something happens to the pH control or the coagulant feed, the effluent turbidity of the small filter increases and an alarm sounds. The operator can then determine the cause of the problem and make the necessary adjustments. The advantage of this system is that it detects any kind of abnormality in coagulation and alerts the operator in about 10-15 min so that the entire plant does not become upset. In addition, the results are continuously recorded so that the supervisor can determine if the operators are performing their job adequately. In addition to the turbidity from the small filter, the turbidity from the plant filters is also displayed, so the operator can manage his filter backwashing program efficiently. The first Coagulant Control Center was installed in 1962 and over 120 of these devices are now in operation throughout the United States.

Figure 1. Coagulant Control Center.

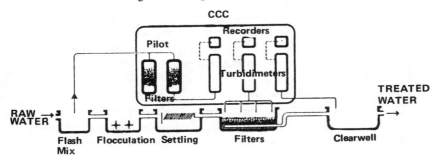

Figure 2. Schematic diagram of coagulant Control Center.

If the operator does not have a Coagulant Control Center, he normally relies on the visual appearance of floc leaving the flocculators to regulate his chemical dosage. A skilled operator can do an excellent job in this manner. However, the necessary skills for this are developed slowly and some people never achieve them. In a sense, it is an art that a few people simple cannot develop. The results are somewhat subjective and various operators may differ as to the proper coagulant dosage. The jar test in the laboratory is used in many filtration plants to supplement the visual observation method. The jar test, when run by an experienced operator, is extremely useful. However, in order to obtain good results the operator must have prepared accurately the solutions of alum and pH control chemical, he must be adept at pipetting solutions, and he must be a keen observer. These characteristics are not always well developed in plant operators. Consequently, the jar test apparatus is not used as often as it should be. The test has other disadvantages: it is not continuous, and the results are somewhat subjective.

More detailed discussions of this subject were presented by Conley,[1] and by Conley and Evers.[2]

FLOCCULATION

Once the proper coagulant has been mixed with water at the proper pH, flocculation is required if sedimentation is to be used. In a sludge blanket clarifier, the flocculation occurs in a slurry pool that is kept in suspension by gentle agitation for hydraulic detention times of 15-20 min. Flocculation in a slurry pool is extremely effective and would be more widely used except that this system is subject to upsets caused by flow and temperature variations or floc density changes. The skill required to operate sludge blanket clarifiers precludes their use in municipal practice.

The usual municipal water treatment plant has a horizontal flow basin with horizontal flocculators. Although this kind of flocculation is not as efficient as the sludge blanket type, it is comparatively easy to operate and is relatively insensitive to changes in the raw water characteristics and flow. Normally, this type of flocculation requires 15-30 min to achieve good results.

Whatever the method, flocculation must be sufficient to form a rapidly settling floc if the clarification equipment is to do its job. Just as flocculation cannot occur until there has been adequate coagulation, rapid settling cannot occur until there has been adequate flocculation.

SETTLING

The traditional settling basin consists of a rectangular tank with detention times of 2-6 hr. The first progress that was made to reduce the size of these tanks was introduction of the sludge blanket or upflow solids contact clarifiers. In these units, the total flocculation and settling time typically was reduced to 1 hr. This contrasts with the 15-30 min of flocculation time followed by 2-6 hr of settling in a conventional system. Reduction in size in the sludge blanket clarifier is achieved primarily because of the superior flocculation afforded by the slurry pool. Some observers compare the action in a sludge blanket clarifier to that in the filter. The small floc that is being formed contacts previously formed floc, adheres to it, and forms dense material several feet deep, which acts somewhat like a filter. By the time the water reaches the top of the slurry pool, all of the suspended solids are entrapped and there is a sharp interface between the slurry pool and the clear water above it. As long as the operator can maintain the slurry pool within a prescribed upper and lower level, and as long as conditions remain relatively stable, these units work remarkably well. However, if sludge quantity, flow rate, temperature, or density of the floc change, control of the slurry pool is difficult.

In the mid-1960s, rapid rate settling devices made of plastic were introduced to the water market. These devices, called tube settlers, used a principle of sedimentation first called to the attention of the industry by Hazen[3] and later by Camp.[4] The results of experimental work on these devices were reported by Culp and Hansen[5] in the late '60s. The tube settler is a variation of Camp's tray settler whereby the settling path is shortened so that the time of settling can be reduced. Two varieties of settling tubes were developed—the so-called horizontal tube, which consists of clusters of tubes with settling paths of 1-2 in (2.5 cm to 5.1 cm) (see Figure 3), and the upflow tube. Properly flocculated material will settle in horizontal tubes in less than 1 min. However, there must be space provided to hold the settled sludge. The actual settling time provided in the tubes proper is about 10 min. After the tubes are full, they are drained and backwashed at the same time as the filter. Total elapsed time in a plant using the horizontal tubes (for mixing, flocculation and sedimentation) is approximately 20-30 min. A typical flow diagram using horizontal tubes is shown in Figure 4.

Figure 3. Horizontal settling tube.

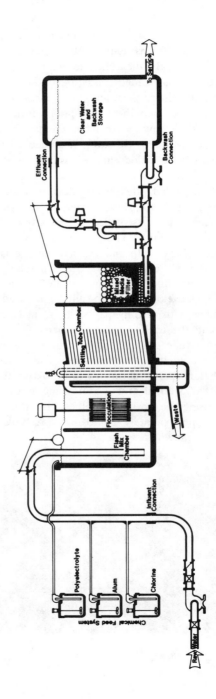

Figure 4. A typical flow diagram using horizontal tubes.

Another variety of settling tube is the upflow tube shown in Figure 5, which is placed in either conventional horizontal basins or in upflow basins to improve the sedimentation or to increase the rate of flow through these units. In the horizontal basin, the upflow tubes are ordinarily placed as shown in Figure 6. Generally, approximately one-third to two-thirds of the basin area is covered with tubes. In most applications in existing basins, it is not necessary to cover a greater area because of the much higher rise rates permitted with tube settlers. The front part of the basin is used as a stilling area so that the flow reaching the tubes is uniform. The design criteria recommended are typically 1-2 gpm/ft^2 (2.5-5 m/hr) across the total horizontal basin with 1.5-3 gpm/ft^2 (3.8-7.5 m/hr) through the tube part of the basin. For typical horizontal sedimentation basins, this requires a detention time of 1-3 hr. The use of these tubes to increase the flow rate through existing structures (and also for new plants) has been reported by Westerhof.[6]

Figure 5. Upflow settling tube.

The upflow tubes can also be used as shown in Figure 7 in sludge blanket clarifiers either to increase flow or to improve effluent quality. One positive factor for use of tubes in upflow clarifiers is that settling

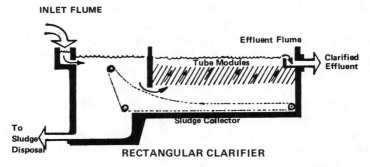

Figure 6. A typical flow diagram using upflow tubes.

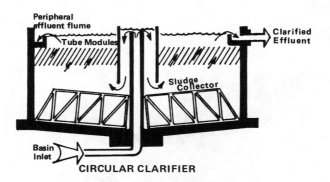

Figure 7. Use of upflow tubes in sludge blanket clarifiers.

tubes tend to stabilize the floc blanket as described by Hudson.[7]   Another
positive factor is that the tubes capture floc that would otherwise appear
in the settler effluent.   A factor weighing against the use of tubes in the
sludge blanket clarifiers is that, because time is shortened, there may be
an insufficient amount of time for adequate flocculation to be achieved.
This is especially true for wintertime operation when flocculation is slower.
However, by careful design, it is often possible to increase the flow through
sludge blanket clarifiers as reported by Eunpu[8] and by Livingston.[9]

The large surface of the tubes, especially on the edges at the top of
tubes, provides sites for the deposit of floc.   Cleaning is accomplished by
dropping the water level below the tubes and hosing the surface of the
tubes with water.   However, water jets can be installed to reduce the labor
required for cleaning.

Because time is shortened when tube settlers are used, velocities increase,
creating a danger of hydraulic disturbances which can interfere with settling.   In a horizontal basin, it is important that the water be introduced

uniformly into the basin with velocities not greater than about 1.5 fps (0.46 m/sec). The water from the flocculator to the settling basin must not cascade over a weir because it destroys the floc. The ideal distribution system is a baffle wall between the flocculator and the settling basin. A stilling zone should be provided between the baffle and the tube zone. In a normal setting basin it is recommended that not more than two-thirds of the horizontal basin be covered with settling tubes to provide a maximum stilling area ahead of the tubes. Installations wherein the entire basin area has been covered with tube modules have performed satisfactorily, however.

In a well-designed settling basin, the outlet conditions must not set up hydraulic disturbances. Good design requires provision for sufficient effluent weirs to produce uniform removal of the water without short-circuiting parts of the basin. The recommended design is to provide effluent weirs over the part of the basin that contains the settling tubes. The settling tubes should be located so that the bottom of the tubes is at least 4 ft (1.22 m) from the bottom of a rectangular clarifier. The top of the settling tubes should be located about 1-2 ft (0.3-0.6 m) below the top of the effluent weir. Weirs should be spaced so that the weir overflow rate is about 20,000-40,000 gpd/ft (250-500 m³/m day) of weir length. A ratio of tube module submergence depth to effluent weir trough spacing of 1:3 to 1:4 is recommended.

The maximum velocity at the inlet tube baffle should not exceed 3 ft/min (0.9 m/min). If sludge removal devices are used, the velocity of travel of the devices should not exceed 1 ft/min (0.3 m/min).

FILTRATION

The development of high-rate filtration [filter rates above 4 gpm/ft² (10 m/hr)] was made possible by the development of better filter media design, sensitive turbidimeters, and the use of polymers as filter conditioning agents. The conventional sand filter is not suitable for high-rate filtration because at high filtration rates, the head loss through a normally designed fine sand filter is unacceptable, and the filter runs are unduly short. If the designer attempts to overcome the head loss problem by making the sand coarse, he does this at the expense of the filtered water quality. For example, a 0.4-mm effective size sand does a good job in producing a good effluent quality at 5 gpm/ft² but the head loss is very high and the filter runs are very short. A 1.0-mm effective size sand solves the head loss problem, but the effluent quality is normally unacceptable. A coarse sand filter can be used for high-rate filtration when the applied settled water quality is so good that comparatively poor performance of the filter gives acceptable water.

If relatively coarse coal of 1-mm effective size is placed on top of fine sand of 0.45-mm effective size, there is a major reduction of head loss without much disadvantage in filtered water quality. Filters made of 18-24 in. (46-61 cm) of 1-mm coal and 6-12 in. (15-30 cm) of 0.45-mm sand are being used for high-rate filtration. These dual-media filters do a good job if enough polymer is added to ensure that the floc will not break through the filters. A third kind of filter made of the same sizes and depths of coal and sand as the foregoing filter is constructed by adding 1.5-4.5 in. (3.8-11.4 cm) of a very fine heavy mineral such as 0.25-mm effective size ilmenite, magnetite or garnet sand. A three-media filter does a superior job of water filtration and requires 5-10% less alum and about one-half to two-thirds as much polymer as a dual-media filter.[10] Figure 8 shows a typical design and Figure 9 shows a cutaway sketch.

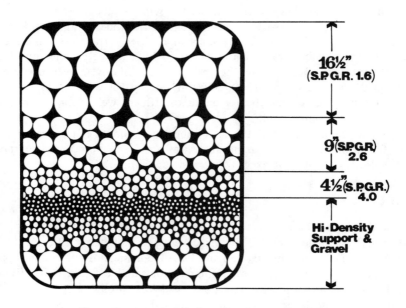

Figure 8. A typical design of a three-media filter.

The clean head loss through a high-rate three-media filter is given in Table I. Note that the head loss is in the range of 1-2 ft (0.3-0.6 m) of water at the normal high-rate filtration rate of 4-6 gpm/ft^2 (10-15 m/hr). Filter runs a high rate are normally long and the percentage of backwash is ordinarily about the same as for the conventional sand filter running at low rate. Plant data for a number of plants are presented in Table II.

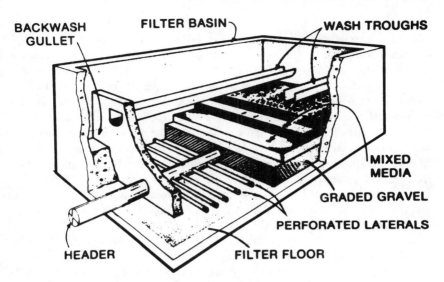

Figure 9. Cutaway view of mixed media gravity filter.

Table I. Clean Three-Media Filter Head Loss as a Function of Flow Rate

| Head Loss Media Only at 70°F | | Flow Rate | |
ft H$_2$O	mH$_2$O	gpm/ft^2	m^3/m^2 day
0.5	0.15	2	0.08
1.0	0.30	4	0.16
2.5	0.76	8	0.33
3.4	1.04	10	0.41

Table II. Typical Filter Runs and Percentage of Backwash Water at Various Filter Rates in Three-Media Filters

Plant	Filter Rate gpm/ft^2 (m^3/m^2 min)	Applied Floc TU	Run Length (Hr)	Percent Backwash Water
A	5 (0.20)	1	30	2
B	5 (0.20)	10	18	2
C	4 (0.16)	3	20	3
D	4 (0.16)	6	45	1
E	4 (0.16)	1	48	1
F	3 (0.12)	4	48	2
G	8 (0.33)	3	24	1

Note that correlation is poor between the length of run and percentage backwash water used. This occurs because some operators insist on removing all of the turbidity from the filters at each backwash while others do not. The water used ranges from as low as 75 gal/ft^2 (3.1 m^3/m^2) of media for each backwash to as high as twice that. Also note that there is poor correlation between the turbidity of the applied water to be filtered and the length of filter run. This shows that the nature of the material being filtered makes as much difference as the amount of material to be filtered. In many plants, the turbidity applied to the filter changes little from season to season and yet there are major changes in the length of filter runs. Runs varied from as little as 6 hr to as much as 50 hr throughout the year. Most of these differences in lengths of filter runs are caused by changes in the nature of the material to be filtered.

In some instances, it is possible and desirable to eliminate the flocculation and settling steps before filtration. If the raw water supply is relatively uncontaminated and contains relatively small amounts of turbidity and color, it may be possible to eliminate the settling and flocculation steps. Most of these so-called direct filtration plants are operating with water that contains fewer than 10 TU and fewer than 25 standard color units. Operating results have been good and there seems to be a trend in the U.S. toward more of this kind of plant. Capital costs of this kind of plant are usually about two-thirds of the cost of a normal filtration plant. Annual operating costs, including amortization, are significantly lower than for a conventional plant.

Another approach that has been fairly widely used to reduce the cost of filter plants has been the use of the contact basin. The contact basin, which usually is not preceded by flocculation, consists of a plain tank with about a 1-hr detention time. The design approach has been to use contact basins on waters that normally contain fewer than 200 standard units of turbidity or fewer than 100 units of color. The operating history has been outstanding and the capital costs are intermediate between a conventional plant and a direct filtration plant. The contact basins function as a reasonably good settling basin during periods of high turbidity, whereas at periods of low turbidity, the basins do very little work. Such plants serve the City of Richland, Washington, with water from the Columbia River, and the Clackamas Water District with water from the Clakamas River in Oregon.

SUMMARY

In order to achieve good suspended solids removal at low cost, it is first necessary to control coagulation precisely. In order to do this, a device called the Coagulant Control Center can be used to monitor coagulation conditions continuously and to alert the operator to any change in the raw water or to a failure in the chemical feed system. This same control center can also be used to monitor the turbidity of the final filtered water and to alert the operator if there is any abnormality in plant operation. After proper coagulation is achieved, some high quality raw waters can be filtered without flocculation and settling to produce the highest quality filtered water at minimum capital costs. However, most waters require some kind of settling in order to achieve satisfactory filter runs. Medium quality raw waters can use a contact basin, which is a simple 1-hr detention rectangular basin just ahead of filtration. The contact basin does not have flocculation or sludge removal equipment. Its function is to reduce the peak turbidity loads to the filters. For most raw water supplies, complete treatment is required. The treatment includes coagulation, flocculation for a sufficient period of time to form a fast settling floc, sedimentation and filtration.

High-rate treatment (high-rate tube settlers and high-rate mixed-media filters) can reduce capital construction costs significantly. High-rate tube settlers are of two types. The first uses a 1-in. (25 mm) settling path, and is called a horizontal tube settler. It is 39 in. (99 cm) long and is made from corrugated plastic sheets welded together. The detention time in this kind of settler is about 10 min. It is widely used in package treatment plants ranging in size from 10 gpm (0.61 l/sec) to 2 mgd (88 l/sec).

The second kind of rapid settling device is the upflow tube settler, which is installed in conventional type settling basins. By the use of upflow tubes the detention time in the settler can be reduced to about one-half to three-fourths of conventional detention time.

High-rate mixed-media filters are operated at 4-6 gpm/ft^2 (10-15 m/hr) or about twice the conventional design rate. Depending upon applied water characteristics and usage of filtered water, rates as high as 7-8 gpm/ft^2 (18-20 m/hr) are feasible at no sacrifice in quality.

A typical design consists of 16.5 in. (419 mm) of 1-mm coal, 9 in. (229 mm) of 0.45-mm sand, and 4.5 in. (114 mm) of 0.25-mm garnet. Several hundred mixed-media filters are in operation in the U.S. An essential ingredient in high-rate filtration is the use of traces of polymer (10-50 ppb), applied directly to the filter in order to improve the filtration properties of the floc.

The combination of a Coagulant Control Center to monitor coagulation, high-rate settling tubes, high-rate mixed-media filters and adequate polymer feed provides superior suspended solids removal at capital costs lower than those of conventional water treatment plants.

REFERENCES

1. Conley, W. R. "Integration of the Clarification Process," *J. Am. Water Works Assoc.* 57:1333 (October 1965).
2. Conley, W. R., and R. H. Evers. "Coagulation Control." *J. Am. Water Works Assoc.* 60:165 (February 1968).
3. Hazen, A. "On Sedimentation," *Trans Am. Soc. Civil Eng.* 53:45 (1904).
4. Camp, T. R. "Sedimentation and the Design of Settling Basins," *Trans Am. Soc. Civil Eng.* 111:895 (1946).
5. Hansen, S. P., and G. L. Culp. "Applying Shallow Depth Sedimentation Theory," *J. Am. Water Works Assoc.* 59:1134 (September 1967).
6. Westerhoff, Garret P. "Expansion Sparks New Design," *Water Wastes Eng.* June (1973).
7. Hudson, H. E. Jr. "Density Considerations in Sedimentation," *J. Am. Water Works Assoc.* 64:382 (June 1972).
8. Eunpu, Floyd F. "High Rate Filtration in Fairfax County, Virginia," *J. Am. Water Works Assoc.* 62:340 (June 1970).
9. Livingston, A. P. "High Rate Treatment Evaluated at Buffalo Pound Filtration Plant," *Water Sew. Works* June (1970).
10. Hsiung, A., W. R. Conley and S. P. Hansen. "The Effect of Media Selection on Filtration Performance," presented April 29, 1976, American Water Works Association Inaugural Meeting, Hawaii.

CHAPTER 15

PRECOAT FILTRATION

E. Robert Baumann, Ph.D., P.E.
Anson Marston Distinguished Professor of Engineering
Department of Civil Engineering
Iowa State University
Ames, Iowa 50010

INTRODUCTION

Precoat filtration is the usual term applied to the process of filtration employing a thin layer (3-5 mm) of filter media of such a fine particle size (1-100 μm) that the media is wasted after each filter run because it is so difficult to separate the solids removed from the filter media. Diatomaceous earth and perlite filter media, generally referred to as filter aids, are most commonly used as filter media in precoat filtration. The filtration process requires the use of a three-step operation:

- a precoating operation
- a filtering operation
- a backwashing operation.

Precoating Operation

In the precoating operation, a thin protective layer of filter aid is built up on a filter septum as a precoat (hence, the term "precoat filtration") by recirculating a slurry of the precoat filter aid through a septum (or septa) contained within the filter housing and back to the filter aid slurry tank (see Figure 1-A). The filter septum is a porous supporting structure capable of removing particles of filter media from the recirculating filter aid slurry by "straining" it out on the surface of the septum (see Figure 2). The septa used in precoat filtration vary according to

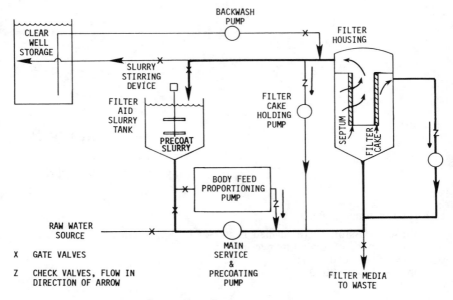

Figure 1-A.   Precoating operation.

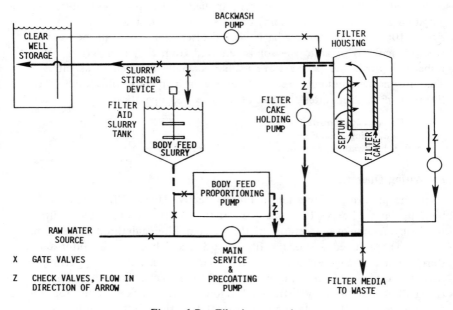

Figure 1-B.   Filtering operation.

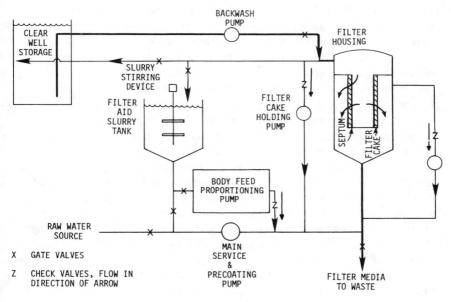

Figure 1-C. Backwashing operation.

Figure 1. Schematic diagram of precoat filtering system during (a) precoating, (b) filtering, and (c) backwashing operations.

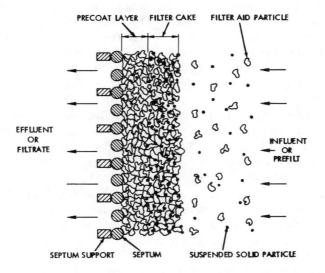

Figure 2. Cross section of a precoat filter.

shape (cylindrical or flat), the porous nature of the structure used to strain out the filter aid (well screen, porous screen, cloth, porous carbon tubes) and the manner by which the filter is backwashed.

In order to obtain a uniform thickness of precoat over the entire septum area, the upward velocity within the filter housing must be high enough to keep the largest filter aid particle in suspension until it reaches the uppermost level of the filter septa. A separate low-head, medium-flow slurry recirculation pump can be provided to operate only during the precoating operation. This pump draws suction from the raw-water section of the filter housing near the top level of the septa and returns the water to the inlet portion of the filter housing (see Figure 1-A). The sole purpose of the slurry recirculation pump is to keep the velocity of the precoat-bearing water moving upward in the housing around the septa high enough to provide a uniform precoat thickness on the septa. Without this slurry recirculation pump, the precoat layer at the bottom might be 10 cm thick and decrease in thickness to no precoat at all at the top. For good filtration, the precoat thickness must be uniform over the entire filter septa surface. Other methods of precoat application (such as a single pass of precoating water through a precoat pot containing the precoat filter aid and through the septa and to waste) can be and are used, but they sacrifice uniformity of precoat formation.

Filtering Operation

Once the precoat layer of filter aid is uniformly in place on the septa, the filtering operation can begin. The main service pump is adjusted to draw suction from the raw water source (the filter influent water) instead of from the precoat filter aid slurry tank, and the filter effluent (the clean water after filtration) is directed to the clear well storage tank instead of being returned to the precoat filter aid slurry tank (see Figure 1-B).

Once filtering begins, it should be apparent that the only filter medium in place is the 3.0- to 5.0-mm thick precoat filter cake, which results from precoating 0.1-0.15 lb of filter aid/ft^2 of septum surface. The size of the particles of filter aid in the precoat determine to a large degree the quality of the filtered water produced. Both diatomite and perlite filter media have mean particle diameters (4-30 μm) between 1 and 2 orders of magnitude smaller than the coal-sand media commonly used in granular-media, deep-bed filters (800-1500 μm). Precoat media are so fine that if filtration were begun through the precoat alone, an impermeable membrane of suspended solids (usually 1-50 μm in diameter) removed from the raw water would begin to form on the raw water side of the precoat.

This layer of suspended solids would in turn serve as the filtering medium, and the head loss through the compressible-impermeable layer of suspended solids would increase exponentially with the volume of filtrate (or the amount of solids collected on the precoat surface).

In order to eliminate the formation of a compressible-impermeable layer of suspended solids on the surface of the precoat, additional filter aid is added continuously to the raw water during the filtering operation as a body feed (hence, the use of the term "body feed filter aid"). The body feed filter aid (expressed in mg/l in the raw water) is usually fed in some direct proportion to the suspended solids level (in mg/l or in turbidity units) in the raw water. The ratio (C_S/C_F) of the suspended solids in the raw water, C_S, to the body feed filter aid concentration, C_F, controls both the thickness of the body feed "filter cake" layer that forms and its permeability and compressibility.

Body feed can be fed either dry, so that it enters the water passing through the main service pump, or in slurry form. Figure 1-B shows that the precoat slurry tank can also be used as a body feed slurry tank or for a separate precoat with body feed slurry tanks provided. Since diatomite and perlite filter aid particles can be reduced in size by violent mixing, slow-speed stirrers are usually used to keep the filter aid in suspension in the slurry mixing tanks. Various methods can be used to feed body feed slurry to the suction or discharge side of the main service pump to provide controllable, continuous body feed during filtration.

The body feed filter aid can be of the same grade or a grade different from the precoat. Thus, control of the precoat filter aid grade (particle size and surface characteristics), the body feed filter aid grade (particle size and surface characteristics), and rate of feed in proportion to the type and concentration of suspended solids to be removed by filtration can be adjusted to provide the desired filtered water quality and the desired hydraulic operating characteristics of the filter. The engineer can now optimize precoat filter design and operating conditions for:

- filtration rate, gpm/ft^2
- terminal head loss, ft of water across the filter septum-precoat-filter cake
- precoat grade and weight
- body feed grade and concentration.

This enables us to filter water containing a given type and concentration of suspended solids at least cost.

In granular-media, deep-bed filters, the medium size is designed to prevent surface cake formation and to encourage solids removal in the depth of the filter medium. Dual- and multi-media filters, upflow filters, and bi-flow filters are designed to encourage coarse to fine medium filtration to distribute the solids more uniformly in the depth of the filter bed. In precoat filtration, the filter medium is selected to encourage surface removal or straining of the suspended solids by the precoat medium, and body feed is added continuously to build up a filter cake that increases in thickness toward the raw water side of the filter. Depending on the suspended solids and body feed characteristics, filter cakes as much as 1-3 cm in thickness can be formed economically. This means that at the beginning of filtration the water passes through only 3-5 mm of filter aid, but near the end the water passes through as much as 2-3 cm of filter aid and filter aid plus suspended solids.

In normal operation of potable water precoat filtration plants, the filtering operation must be terminated or interrupted when the clear well storage tank (or elevated tanks floating on the line) are full. When filtration through a precoat filter is interrupted, the force (head loss through the septum-precoat filter cake) holding the cake on the septum no longer exists and the filter cake may fall off the septum and/or crack so as to form an unacceptable filter medium if filtration is resumed. To eliminate such an occurrence, *any interruption* to continuous flow through the filter medium requires either that filtration *not* be resumed until the filter is backwashed *and* reprecoated or that special provisions be made to keep the filter precoat and filter cake intact.

Figure 1-B shows how a separate filter cake holding system can be used for this purpose. A separate filtered water recirculation system, including a filter cake holding pump, is provided, which takes suction from the filtered (clean) water side of the filter housing and puts the filtered water back into the housing on the raw (dirty) water side of the filter housing. The filter cake holding pump operates *only* when the clear well or elevated tank is full and sends a signal back for the main service pump to be shut down. At least 10-15 sec before the main service pump shuts down, the filter cake holding pump comes on to cause a very low flow to pass through the cake when the main service pump stops. The holding pump flow (low flow and low pressure drop) is effective in preventing the filter cake from cracking and/or dropping completely off the filter septum. When the clear well or elevated storage system signals that more filtered water is needed, the filter cake holding pump system is shut off automatically *after* the main service pump has been in operation 10-15 sec and can resume the task of maintaining the integrity of the filter cake. When the pressure drop across the septum-precoat filter cake reaches the maximum terminal head provided for the system, the filter must be backwashed and reprecoated before filtration can resume.

Backwashing Operation

Once the terminal pressure drop across a precoat filter reaches the pre-determined terminal pressure drop, the filter must be backwashed and reprecoated before it is returned to service. Figure 1-C shows the filter system in a backwashing sequence. The purpose of the operation is to remove the entire precoat and filter cake from the septum and out of the filter housing as completely as possible in as short a down time and with as little backwash water as possible. Figure 1-C shows the use of a separate backwash pump that may be desirable in a potable water filtration plant; however, by selective location of the piping system it is often possible to use the main service pump as a backwashing pump. Some precoat filters are provided with special air-bump backwashing features, which are adequately described in the literature. After backwashing is completed (10-15 min of down time), the filter can be reprecoated (10-15 min) and returned to service. Thus, the normal out-of-service period at the end of a filtering cycle would be about 20-30 min.

Filter Types

There are two major types of precoat filters, depending on how the terminal pressure drop is obtained across the filter cake: vacuum and pressure filters. In the vacuum filter, the filter septum and its layers of precoat and filter cake are located in a filter housing open to the atmosphere on the suction side of the main service pump (see Figure 3). As a result, the total terminal pressure drop through the filter is limited to the total positive head of water above the filter plus the normal suction lift of the service pump. Although the theoretical suction lift of a pump is about 34 ft (101 kN/m^2) of water, the practical usable suction lift is only 20-24 ft of water. Thus, the maximum practical terminal pressure drop in a vacuum filter is only 20-24 ft of water, and the filter cake is normally under a pressure less than atmospheric pressure (operation under negative head conditions) except for a short period early in a filter run.

Since the vacuum filter housing is effectively an open tank, it can be constructed economically of a variety of materials and also can be back-washed manually or automatically with simple backwashing mechanisms. This type of filter is simpler and finds economical applications where the raw water is of relatively good quality, i.e., for the filtration of swimming pool water. Its major limitation lies in the relatively small terminal pressure drop that can be provided in filter design.

In a pressure filter, the filter septum and its layers of precoat and filter cake are located within a pressure filter housing on the discharge side of the main service pump (see Figure 3). Accordingly, the filter cake is

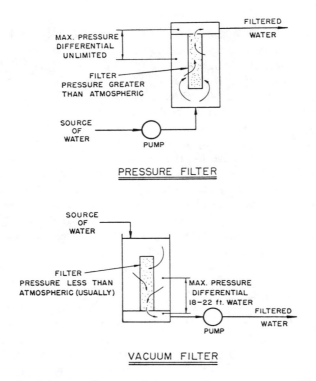

Figure 3. Schematic diagrams of vacuum and pressure precoat filters.

always maintained at a pressure greater than atmospheric pressure, and the terminal pressure across the filter is unlimited, since the pressure drop variable is determined by the characteristics of the main service pump. In practice, terminal pressure drops of 50-200 ft (150-300 kN/m^2) of water have been used successfully. The pressure filter housing must be constructed as a pressure vessel and thus is usually more expensive than a vacuum filter. Since the filter housing is enclosed, the backwashing operation is more difficult and must be more effective than with a vacuum filter because access to the filter septum is more difficult. The primary advantage of a pressure filter, and it is a significant advantage, is the ability to control the terminal pressure drop across the filter to the level found most economical in the optimization of the filter design.

Both vacuum and pressure precoat filters give identical results when operated at the same filtration rate to the same terminal pressure drop using the same precoat and body feed operating conditions. If dissolved gases are present (CO_2, O_2, H_2S, N_2) in the water to be filtered, these gases will tend to come out of solution under the negative head conditions in the vacuum filter. If the gas bubbles remain intact, they will block the filter cake, increase cake thickness, and increase the pressure drop across the filter in the same manner as the suspended solids the filters are supposed to remove. On the other hand, if the gas bubbles collapse, they may disrupt the uniformity of the filter cake and allow inadequately filtered water to pass through the cracks or openings formed, giving poor filtered water quality and a decreased or irregular rate of head loss build-up. Under such conditions, results differing from pressure filtration may be expected.

Figure 4 shows schematic sections through pressure- and vacuum-type precoat filters. Figure 5 shows typical municipal precoat filter installations at Massena, New York, and Lompoc, California.

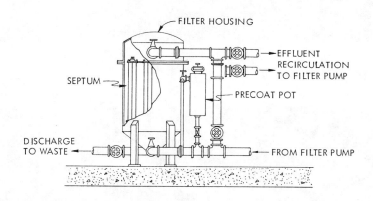

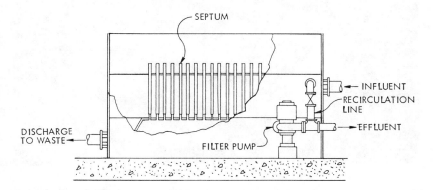

Figure 4. Schematic section through pressure- and vacuum-type precoat filters.

Figure 5. Typical municipal precoat filter installations: (a) pressure filters at Massena, New York, and (b) vacuum filters at Lompoc, California.

SELECTION AND TREATMENT OF
PRECOAT FILTER MEDIA

Filter Media Available

Figure 2 shows a cross section through a precoat filter. The clarity or the quality of the filtered water is a function of the type, grade (size distribution) and surface characteristics of the filter media used as a precoat. The hydraulic characteristics controlling the rate of head loss build-up in a filter run are a function of the ratio of the body feed to the suspended solids in the filter cake. Thus, you can control the filtered water quality and the hydraulic conditions of the filtering operation by controlling the size and surface treatment characteristics of the filter media used and the rate at which you use the filter media as body feed.

Precoat filter aids are produced by several manufacturers in several grades.[1,2] Diatomite filter aids are produced from mined deposits of fossil-like skeletons of microscopic water plants called diatoms. Processing the crude diatomaceous earth for use as a filter aid includes grinding, drying, flux-calcining and separation into different sized particles by air classification. Calcination affects the filtering properties of diatomite by changing the surface texture, agglomerating fines and converting clay minerals to aluminum silicate slag, which is largely eliminated in later processing steps. Figure 6 shows the particle size distributions and grade designations of diatomite filter aids produced by three different companies.[2]

Table I lists the trade names of several filter aid grades currently on the market that are approximately equivalent in performance. Filter aids are considered "equivalent" when they produce approximately the same flow rate and filtered solution clarity under the same operating conditions when filtering a standard sugar solution. Grades that produce the highest clarity also produce the lowest flow rate. In general, the high clarity filter aids are composed of very small particles of filter aid (mean size in the range of 3-6 μm), and high flow rate filter aids are composed of larger particles of media (mean size in the range of 20-40 μm). High clarity diatomite filter aids are cheaper, produce a better clarity of filtered water, and produce the highest rate of pressure drop increase across the filter and, therefore, shorter filter runs. The normal filter aid grades used in water filtration without use of media coatings or other chemical pretreatments produce Table I clarities between 936 and 960. The performance of diatomite as a filter aid depends on the unique physical structure of the diatom particles available in an almost infinite variety of shapes and sizes, which produce an extremely porous medium with numerous microscopic waterways and sieves that serve to trap impurities. The

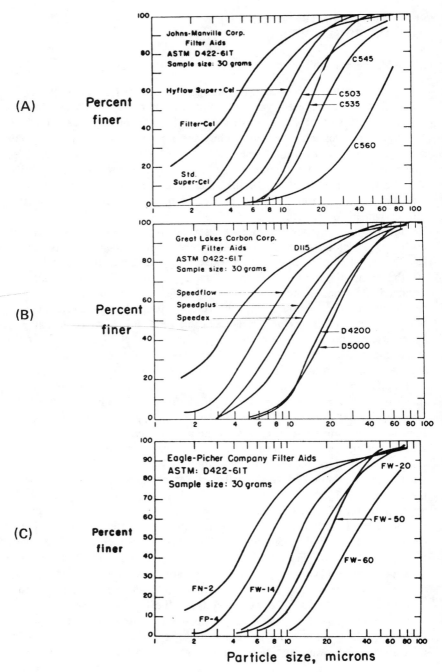

Figure 6. Particle size distributions of diatomite filter aids produced by three manufacturers.

Table I. Relative Rating of Diatomite Filter Aids[3]

Standard Ratios[a]				
Flow Rate	Clarity	Eagle-Picher	Johns-Manville	Dicalite
100	1000	Celatom FP-2	Filter Cel	215
125	1000	Celatom FW-2	Celite 505	Superaid, UF
200	995	Celatom FP-4	Standard Super-Cel	Speedflow
300	986	Celatom FW-6	Celite 512	Special Speed-flow, 231
400	983	Celatom FW-10	-	341
700	970	Celatom FW-12	Hyflo Super Cel	Speedplus, 689 CP-100
800	965	Celatom FW-14	-	375
950	963	Celatom FW-18	Celite 501	CP-5
1000	960	Celatom FW-20	Celite 503	Speedex, 757
1800	948	Celatom FW-40	-	-
2500	940	Celatom FW-50	Celite 535	4200, CP-8
3000	936	Celatom FW-60	Celite 545	4500
4500	930	Celatom FW-70	Celite 550	5000
5500	927	Celatom FW-80	Celite 560	-

[a]Based on bomb filter tests with $60°$ Brix raw sugar solution, $80°C$.

porosity of a clean diatomite filter precoat varies from 80-90% for the various grades of filter aid.

Several other materials have been used as filter aids. The most successful of these is perlite, a material obtained by processing volcanic ash. Perlitic rock is composed essentially of aluminum silicate and contains 3-5% water. When crushed and heated, the rock expands and fractures to produce a light porous material similar to diatomite in macroscopic appearance and hydraulic characteristics. As a filter aid, it is also available in different grades that vary in both particle size and specific gravity. In coarse particle sizes, bubbles of gas entrapped in the perlite particles that are not broken can produce "floaters." Thus, the presence of such material gives the perlite medium a range of effective specific gravities.

The main characteristics of filter media that affect their filtration properties include their particle size distributions, specific gravity, in-place bulk density and filtration resistance. The particle size distributions of the media affect cake porosity, the media surface area per unit of filter aid weight, and the size of the pores that are present. The specific gravity of the media affects the bulk-density of the filter media in place on the septum, usually measured in terms of pounds of media per cubic foot of precoat. The filter aid resistance, represented by ξ, is a measure of the hydraulic resistance of the clean filter media, for example, as it would be in the form of a precoat. Typical physical properties of several commercial perlite and diatomite filter aids are shown in Table II.[1]

Table II. Physical Properties of Several Commercial Perlite and Diatomite Filter Aids[1]

Filter Aid Designation	Effective Specific Gravity	In-Place Bulk Density (lb/ft^3)	ξ Index[a] (10^9 ft/lb)
272[b]	1.57	9.9	6.8
332	1.73	12.6	9.1
443	1.98	13.0	10.4
C-545[c]	2.30	19.7	1.8
C-535	2.32	19.9	1.9
C-503	2.30	19.9	3.1
HFC	2.30	20.7	5.2
FW-60[d]	2.22	19.3	1.1
FW-50	2.30	23.2	1.8
FW-20	2.28	20.7	2.5
4200[e]	2.27	23.5	1.8
Speedex	2.28	22.3	3.7

[a] Defined on page 350.
[b] Sil-Flo Corporation.
[c] Johns Manville Products Corp.
[d] Eagle-Picher Industries, Inc.
[e] Great Lakes Carbon Corp.

Selection of Filter Media

Every precoat filter requires the use of one or more grades of filter media for use as precoat and body feed. In addition, chemical pretreatment of the water and/or the filter aid can be used to enhance suspended particle retention or to improve the hydraulic characteristics of the resulting filter cake. In general, precoat filters are always operated as direct filtration processes without the use of chemical coagulation-flocculation-sedimentation preceding the filtration of the water.

Normally, the filter aid and/or filter aid treatments are selected that will produce the desired clarity of filtered water (or the desired removal of a specific suspended particle) with the lowest rate of head loss increase. The rate of head loss increase is measured by a filter cake resistance index, β.[4,5]

Since all grades of filter aid generally are not available simultaneously, Table I can be used to select representative grades for testing. The coarsest grade of filter aid that can provide the desired filtrate quality should be selected. The grade can be selected by filtering a sample of the suspension through a 10-15 mm thickness of each grade of filter aid in a Buchner funnel under a vacuum of 10-20 cm of mercury. Filter aid companies can provide "bomb" pressure filtration apparatus that can be used for the same purpose. If the desired filtrate quality can be reached with filter aid grades equivalent to a clarity rating of 960 or below, there is little need to use chemical pretreatment of the water or the filter media to improve particle retention and/or the hydraulic characteristics of the filter cake.

Sometimes, however, the particle to be removed cannot be retained by straining alone on such coarse filter media. For example, the filtration of Lake Superior water at Duluth was studied specifically to enhance the removal of asbestiform fibers from water.[3] The asbestiform particles present were of two main types:

- amphibole fibers: negative surface potential, size 0.1-0.3 μm in diameter, length 3 or more times diameter
- chrysotile fibers: positive surface potential, size 0.05-0.08 μm in diameter, length 3 or more times diameter.

Both asbestiform particles were present in about equal numbers, and both would readily pass through normal water filtration grades of filter aid. Figure 7 shows the zeta potential (related to surface charge) of the chrysotile particles as a function of the pH of the water in which they are suspended.[3] (The surface potential also varies from that shown as the mineral content of the water varies.)

As the mean particle size of filter aid gets smaller, the total surface area of a given weight of material increases. In order to remove very small particles of suspended solids such as asbestiform fibers, the particles must be *transported* through the fluid to the surface of the filter media, and then they must become *attached* to the surface of the filter media. Both particle transport and attachment are enhanced as filter media particle size is decreased. Diatomite filter aid normally has a negative surface potential, as shown in Figure 8. The surface potential will also vary from that shown as the mineral content of the water varies. Most suspended solids normally encountered in water also carry negative surface potentials. When such particles are filtered through negatively charged filter media, the transport forces have more difficulty in moving the particles to the media surface where attachment can take place. Thus, chemical pretreatment of the water or chemical pretreatment of the filter aid can be designed and used to change the surface potential either of the suspended solids or of the filter

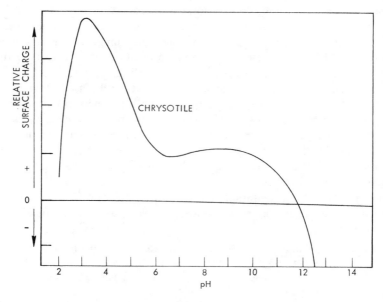

Figure 7. Surface charges on chrysotile asbestiform particles as a function of pH.[1]

medium. Preferably, one must be changed to carry a surface charge opposite to that of the other.

Since particles in water can be either positively or negatively charged (or a mixture of both), water or media treatments that are designed to enhance the removal of one species may interfere with the removal of the other species. In general, transport forces are a minimum when particles are 0.5-2.0 μm in size. Significantly smaller particles are transported effectively by diffusion forces; larger particles are transported effectively by gravity forces.

In precoat filtration, the prime filter medium consists of the precoat layer of filter aid. The addition of body feed filter aid throughout a filter run is designed primarily to maintain a desirable permeability of the cake formed (suspended solids removed by the filter plus the body feed filter aid). It does, however, also provide an increasing depth of filter medium as the run progresses and thus serves to increase both the area of media surface available for particle adsorption and the opportunities for transport of particles to the media surface.

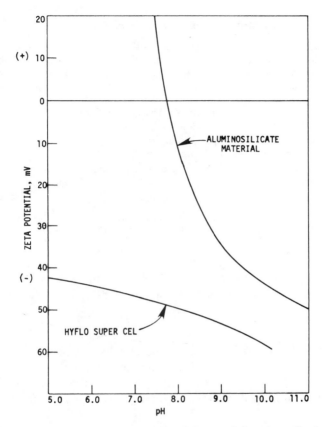

Figure 8. Surface potential of diatomite (Hyflo Super Cel) and an aluminosilicate material as a function of pH.

Improvement of Solids Removal

The removal of suspended material in a precoat filter can be improved using the following techniques:

1. Use with no chemical treatment of water or media

- Use of a finer grade of media as precoat to improve retention, but with a greater rate of head loss increase.
- Use of a greater depth of precoat, but at an increased cost for the filter aid.
- Use of two filter media, a fine medium for the precoat to improve solids retention and a coarser medium for the body feed to improve the hydraulic characteristics of the filter cake. The operation of the filter would be slightly more complicated by the use of two filter media.

2. Use of chemical pretreatments of filter media

- Application of an organic or inorganic coating to the filter media used as precoat to improve particle retention. In fact, it is theoretically possible to use two precoat layers, one negatively charged to enhance the removal of positively charged particles and one positively charged to enhance the removal of negatively charged particles. To the author's knowledge, however, this possibility has not yet been evaluated in practice.
- Application of an organic or inorganic coating to the filter media used as body feed to improve particle retention.
- Application of such a coating to both precoat and body feed filter aid.

3. Use of chemical treatment of the raw water

- Use of inorganic and/or organic coagulants to change the particle characteristics to improve their retention in the diatomite filter media.

If the desired filtrate quality cannot be obtained using filter aid grades coarser than clarity 960-963 grades (Table I), then Buchner funnel (or pilot plant) evaluations should be made of the use of coated filter media.

Filter Media Coatings

Among the materials that can be used to modify the surface of both suspended solids and precoat filter media are chemical coagulants such as alum[6] and organic polyelectrolytes.[7] These chemicals have been effective in attaching suspended particles to each other; they also are useful in attaching suspended solids to filter media. Both materials can be demonstrated to form a coating on filter media in such a way as to modify the surface charge of the media. Figure 8 indicates that the surface potential of diatomite filter aid is negative over a wide range of water pH. Baumann and Oulman[8] have demonstrated that diatomite in water can be coated with either alum (Figure 9) or Dow Chemical Co. cationic polymer C-31 (old designation Purifloc 601) (Figure 10) to change the surface potential from negative to positive as the dosage level (P/D, or grams of coating material per gram of filter media) is increased. The dosage level of coating material required has been demonstrated to be a function of the surface area of the filter media being coated, Figure 11. The surface charge imparted to the filter media is also a function of the pH of the water. Figure 12 indicates that the surface potential of Purifloc 601 coated diatomite changed from a positive to a negative surface potential as the water pH increased above a pH of about 7.0.

Baumann and Oulman[8] have hypothesized that, in coating a filter media, a coating chemical will form a single-layer coating or a multilayer

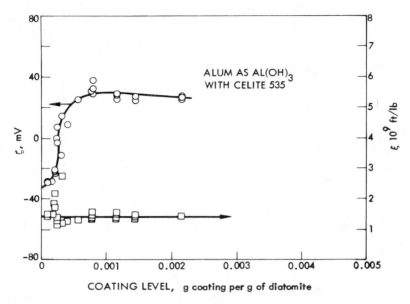

Figure 9. Zeta potential, ζ, and filter cake resistance, ξ, as a function of coating level for alum coated on Celite 535 at a water pH of about 6.0.

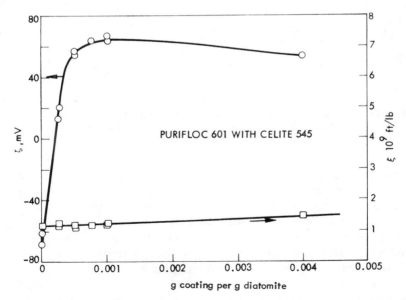

Figure 10. Zeta potential, ζ, and filter resistance, ξ, as a function of coating level for Purifloc 601 coated on Celite 545 at a water pH of about 6.0.

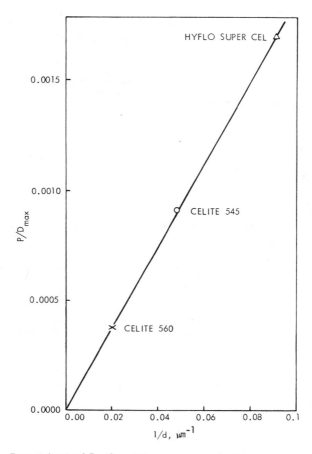

Figure 11.   Dosage level of Purifloc 601 polymer required to give a complete coating of diatomite filter media as a function of relative media surface area.

coating, or that bridging of the coating material between filter medium particles will occur. If a monolayer coating occurs, little change would be expected in the hydraulic resistance of the medium since the thickness of the absorbed coating would be insignificant compared to the dimensions of the hydraulic pathways through the medium. On the other hand, with either multilayer formation or bridging formation, significant effects of the coating on the hydraulic resistance of the filter medium could be expected. Theoretically, since only a monolayer of coating is needed to give the completely coated particle the surface characteristics of the coating material, monolayer adsorption appears to be the only

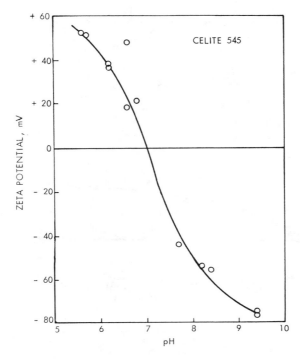

Figure 12. Effect of pH on the surface potential of diatomite coated with Dow polymer Purifloc 601.[9]

desirable coating since it would not significantly change the clean medium resistance when coated.

The resistance of a filter medium is measured by its ξ index, measured in feet per pound of medium. Baumann and Oulman[1] have developed a simple test procedure for determining the filtration resistance of precoat filter media which is used in military filter media specifications. Alum (Figure 9) and the cationic polymer Purifloc 601 (Figure 10) both apparently from monolayer coatings on filter media in such a way as *not* to change clean filter media resistance significantly. The alum coating remains positive over a broad pH range, but the Purifloc 601 polymer becomes negative above a pH of 7.0 (Figure 12). Other polymeric materials coat media in such a way as to demonstrate variable effects on filter media. Figure 13, for example, indicates that Jaguar 808 coated on C-545 does not change the medium resistance, and that it also does not change the medium surface potential from negative to positive.[8] Figure 14 indicates that coating Celite 545 with the polymer Reten 210 does change the surface potential

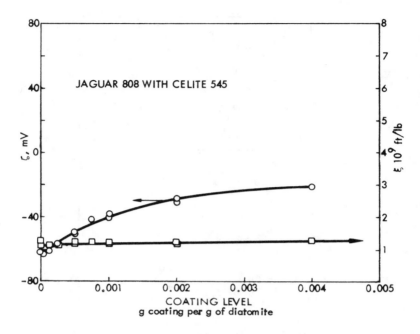

Figure 13. Zeta potential and filtration resistance of Jaguar 808 coated on Celite 545 at a water pH of about 6.0.[5]

from negative to positive and that the coating increases the resistance of the medium proportionally to the coating level.[8] This *may* indicate multi-layer formation of the coating on the medium. Figure 15 shows that, when Ucar Resin C-149 is coated on Celite 545, the medium surface potential can be changed from negative to positive and that the medium resistance increases exponentially with coating level, perhaps indicating bridging of the polymer *between* media particles.[8] Figure 16 shows that the use of a nonionic polymer, Separan NP10, for coating Celite 545 results in a significant increase in medium resistance but little beneficial change in the medium surface charge.

To be effective in improving particle retention in precoat filtration, filter media coatings should be able to provide a medium surface charge opposite to that of the particles to be removed at the water pH at which filtration takes place. Also, they should provide the surface charge without significantly changing the hydraulic properties of the filter medium. A study is currently under way at Iowa State University to determine the effects of coating level and water pH on both the medium surface potential and filtration resistance of a selective group of polymers when

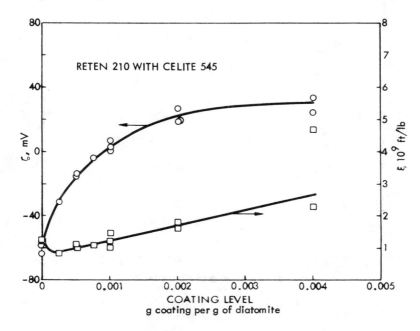

Figure 14. Zeta potential and filtration resistance of Reten 210 coated on Celite 545 at a water pH of about 6.0.[5]

coated on a "standard" diatomite filter medium. The coating of perlite medium gives the same results, generally, as the coating of diatomite filter medium. Figure 17 and 18 show the typical results obtained with a polymer that provides a positive zeta potential to filter media over a broad water pH range. Note, however, that the medium resistance increases with polymer coating level on the medium.

The coating of alum and/or polymers on filter media can be accomplished simply by adding the required weight of diatomite to a slowly mixing solution containing the required amount of chemical for about 10-15 min. With alum coatings, it has been demonstrated that it is possible to coat the medium, dry it, and ship factory-coated filter media. Normally, however, it is more economical to apply coatings in the field at their point of use.

The ability of alum-coated filter media to improve turbidity removal was first demonstrated by Bell.[10] He published curves, Figure 19, that show improvement in removal of turbidity, color and coliforms by coating a coarse filter medium (Celite 545) with 0.5-1.0% of $Al(OH)_3$. The coated coarse Celite filter aid gave significantly higher removal of

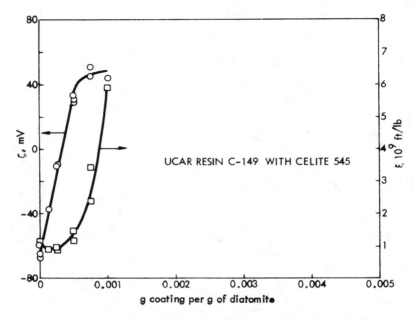

Figure 15. Zeta potential and filtration resistance of UCAR Resin C-149 coated on Celite 545 at a rotor pH of about 6.0.[5]

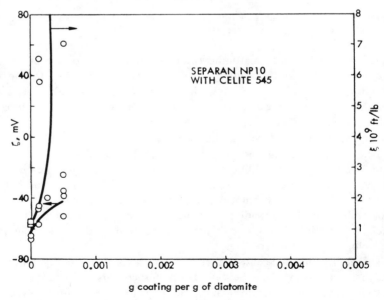

Figure 16. Zeta potential and filtration resistance of Separan NP10 coated on Celite 545 at a water pH of about 6.0.

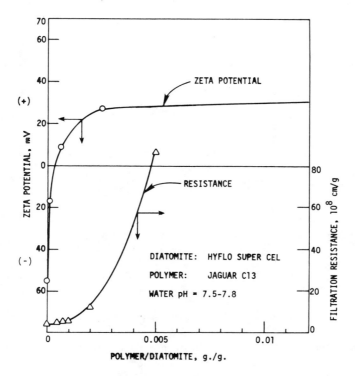

Figure 17. Surface potential and medium resistance as a function of coating level of Jaguar C-13 on diatomite at a pH of about 7.5-7.8.

turbidity, color and coliforms than did the use of the same material uncoated, yet the medium retained the hydraulic resistance of the same medium when uncoated. In fact, the coated Celite 545 removals were all significantly better than those obtained with a significantly finer medium, Hyflo Super Cel (Figure 6 and Table I).

Burns et al.[9] were the first to demonstrate that cationic polyelectrolytes like Dow Chemical Co. Purifloc 601 can also improve the adsorptive capacity of filter media. Figure 20 shows the capacity of Celite 560, a coarse filter aid that, when uncoated, permitted complete passage of a clay, but when coated with Purifloc 601 adsorbed the clay, which had a particle size less than 1 μm. When coated, the adsorption capacity follows the zeta potential curve, indicating that the adsorptive capacity is proportional to the amount of coating on the filter medium. Figure 21 shows the effect of the pH of the solution on both the zeta potential of the medium surface and the removal of clay on the coated filter medium. The data clearly suggest that positive surface potentials on the medium

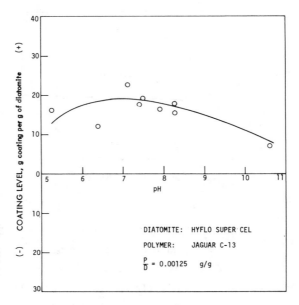

Figure 18. Surface potential and medium resistance as a function of water pH.

surface in the high basic pH range are needed to enhance adsorption of the clay in natural waters whose pH is basic.

The first significant application of the use of coated precoat filter media was in precoat filtration pilot plant studies designed for the removal of asbestiform fibers from Lake Superior water at Duluth, Minnesota. Both amphibole fibers and chrysotile fibers were present in about equal numbers. However, because of the larger size and total weight of the amphibole particles, the choice of media and chemical pretreatments was selected to remove the negatively charged amphibole particles. Since the Lake Superior water had a pH between 7 and 8, it was expected that alum-coated media might be successful in enhancing removals but that C-31* coated media would not because the surface charge on the C-31 coated medium would be negative at a water pH of 7 to 8. As there was a difference in the surface charge of the amphibole and chrysotile particles, it was expected that any treatment effective in amphibole fiber removal would not be effective in removal of the other asbestiform fiber.

The treatments designed and used for the removal of amphibole particles included: (a) use of different grades of filter aid (Celite 535, Celite 503, Hyflo Super Cel, Celite 512), (b) coating precoats and/or body feed

---

*Reported by Dow Chemical Co. in 1975 to be similar to their discontinued Purifloc 601 polymer.

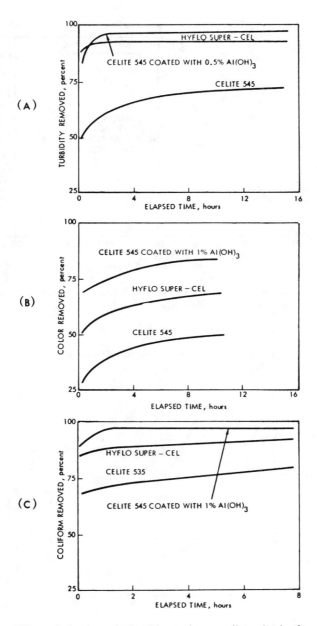

Figure 19. Effect of aluminum hydroxide coatings on diatomite in the removal of (a) turbidity, (b) color, and (c) coliforms from a river water by filtration.[9]

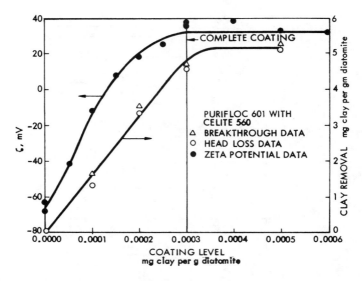

Figure 20.   Surface potential of Celite 560 coated with Dow polymer Purifloc 601 and its capacity for absorbing Panther Creek day.[11]

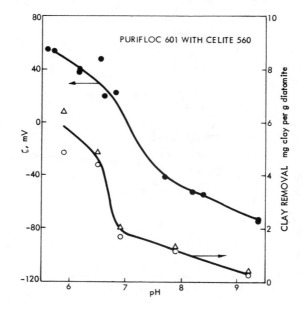

Figure 21.   Surface potential of C-545 coated with Purifloc 601 and removal of Panther Creek clay by Celite 560 coated with Purifloc 601 at various levels of pH.[11]

with alum, C-31 (a Dow cationic polymer), 573-C (a Cyanamid cationic polymer), A-23 (a Dow anionic polymer), Magnifloc 985 N (a Cyanamid nonionic polymer), and (c) coagulation of the water prior to filtration with Cat-Floc B (a Calgon Corporation cationic polymer).

The filter media listed above without coatings were not effective in reducing filtered water turbidity to the desired level.[3] Until 1962 the U.S. Public Health Service drinking water standards limited the turbidity of drinking water to a value less than 10 TU. In 1962 a new drinking water standard was issued in which the allowable drinking water turbidity was reduced to 5 TU. With the recent passage of the "Safe Drinking Water Act" by the U.S. Congress in December 1974, the Environmental Protection Agency has again revised the drinking water turbidity standard by requiring a drinking water turbidity less than 1.0 TU effective in June 1977. In 1968 the American Water Works Association adopted a turbidity "goal" (that water works should seek to achieve) of only 0.1 TU.[20] Therefore, all pilot tests conducted had a goal of producing filtered water with a turbidity less than 0.1 TU.

The tests with coated filter media indicated that use of alum-coated Celite 512 as a precoat and use of either uncoated Celite 512 or Hyflo Super Cel as body feed were able to produce the desired effluent turbidity and 80-90% amphibole fiber removal. Since Hyflo Super Cel is the coarser medium, however, its use as body feed provided better hydraulic characteristics in the filter cake. Use of alum-coated Hyflo Super Cel for both precoat and body feed provided effluent water turbidity of about 0.05 TU and 99-100% removal of amphibole fibers, but the hydraulic resistances of the resulting filter cake were higher than with other media and/or chemical pretreatments.

The most successful treatment was the use of alum-coated Celite 512 or Hyflo Super Cel as a precoat and the use of Hyflo Super Cel or a coarser medium as body feed together with a dosage of about 0.3-0.5 mg/l of Cat-Floc B as a coagulant added directly ahead of the filter. Such a treatment provided effluent water turbidity of 0.05-0.06 TU with 98-100% removal of amphibole fibers. (However, the removal of chrysotile asbestiform fibers was always, as expected, very low since the treatments were not designed specifically for their removal.) An economic analysis indicated that the latter treatment could be used in the design of a 30 mgd peak capacity precoat filtration plant that would produce an average of 20 mgd of filtered water in the year-around at a total cost of 7.81¢ per 1000 gal.

PRECOAT FILTER APPLICATIONS

Precoat filters can be used effectively for the filtration of water in swimming pools, for potable water filtration in military operations, for potable water filtration by municipalities, and for industrial water filtration. An AWWA Task Group in 1965 reported that about 88 precoat filtration plants had been built since 1953 for turbidity removal (73 plants), for iron and manganese removal from well supplies (13 plants), and for lime-soda-ash-softened water filtration (2 plants).[22] Several plants were subsequently abandoned.

Precoat filtration plants are normally selected for potable water filtration under the following conditions:

- when the precoat filtration plant is found to produce filtered water at lowest cost
- in small cities and towns (or private industries) where bonding capacity is limited and there is a desire to minimize the amount of invested capital because of the significantly lower first cost of a precoat filtration plant
- in systems with large seasonal variations in demand (summer resort areas)
- in systems needing emergency or standby service

Currently, the selection process must include a comparison between the total system costs based on use of granular-media deep-bed filters, and the use of precoat filters. In the U.S., the systems that are *currently comparable* and technically dependable and effective include the following:

- For raw water turbidity greater than about 100 TU
  1. Granular-media filtration plants
     Use of a complete conventional filtration plant using chemical pretreatment to produce a settleable suspended solid (rapid mix, flocculation, sedimentation) followed by granular-media deep-bed, dual- or multi-media filters.
  2. Precoat filtration plant
     Use of complete conventional filtration plant using chemical pretreatment to produce a settleable suspended solid (rapid mix, flocculation, sedimentation) followed by precoat filters.

  For raw water where turbidity and color are each less than 25 units, or color is low and maximum turbidity does not exceed 200, or the turbidity is low and color does not exceed 100 units [11]
  1. Granular–media filtration plant
     Use of a direct filtration plant with chemical pretreatment to produce a filterable suspended solid (rapid mix, short period or no flocculation) followed by granular-media, deep-bed, dual- or multi-media filters.

2. Precoat filtration plant

Use of precoat filters using plain or coated media and/or chemical pretreatment to produce a filterable suspended solid (rapid mix only).

In 1965, the AWWA Diatomite Filtration Task Group[12] cited cost figures indicating that, with waters containing low suspended solids, the cost of a precoat filtration plant would be 40-60% of the cost of a *conventional* rapid-sand filtration plant because fewer pretreatment facilities and significantly less building area would be needed. They found the total operating costs of many such plants to be comparable.

Since 1965, however, granular-media deep-bed filtration technology has improved to the point where direct filtration techniques are so advanced that we should not even consider use of a complete, conventional treatment system using deep-bed filters for filtration of water containing low levels of suspended solids. Direct filtration plants that are as effective can be built for the same purpose but with a capital cost saving "up to 30% under favorable conditions," together with a possible savings of "10 to 30% in chemical costs."[11] Direct filtration plants involve reduced operation and maintenance costs and produce less sludge which is more dense and more easily handled than a complete, conventional treatment plant. Over 100 direct filtration plants are currently in operation in the U.S.[13]

If precoat filters can be constructed at a "capital savings" of about 50% and a granular-media deep-bed "direct filtration" plants can be constructed at a "capital savings" of about 30% of the cost of a complete, conventional filtration plant, we should today (1977) expect to be able to build a precoat filtration plant with a capital cost of about 50% of the cost of a complete, conventional filtration plant and 80% of the cost of a direct filtration plant using deep-bed granular-media filters. It should be recognized, however, that cost comparisons are most valid when the "most economical" precoat filtration plant is compared with the "most economical" deep-bed filtration plant.

The AWWA Diatomite Filtration Committee reported in 1974 that there were then approximately 145 water plants using precoat filters for the production of a potable water. The only data of recent history on which to base a comparison of precoat filtration vs direct filtration deep-bed filter costs were collected in pilot-plant studies for filtration of Lake Superior water at Duluth for the removal of amphibole asbestiform fibers from water.[3] The lake water at Duluth has a history of water quality shown in Table III. After the pilot plant studies, the engineers made estimates of the cost of building and operating both a deep-bed direct filtration and a precoat filtration plant for producing 20 mgd of filtered

Table III. Percentage of Time Given Raw Water Turbidity is Equalled or Exceeded,
City of Duluth, Lakewood Pumping Station, 1952-1972

Raw Water Turbidity, TU	Percentage of Time Equal or Less Than
1.0	89.95
2.0	95.3
3.0	97.2
4.0	98.2
5.0	98.7
10.0	99.6
20.0	99.9
30.0	99.99
50.0	99.985

water of equivalent quality in a plant with a capacity of 30 mgd. The deep-bed plant used mixed-media filters operated at 5 gal/min/ft^2 and preceded by two-stage flash mixing followed by flocculation using 12-20 mg/l of alum and 0.05 mg/l of 985N*, a nonionic polymer.

Table IV summarizes the operating costs of the two plants. The first cost of the precoat filtration plant was 87.6% of the cost of the deep-bed filtration plant. The total operating cost of the precoat plant in this large installation (where economics of scale favor the deep-bed filter installation) was about 1¢/1000 gal greater than the cost of operating the deep-bed filters. In this installation, the deep-bed direct filtration plant was economically the more attractive alternative.

Table IV. Cost Comparison of a Precoat and a Mixed-Media Direct Filtration Plant
for Production of 30 mgd of Potable Water from Lake Superior at Duluth

	Direct Filtration Plant Mixed-Media (5 gal/min/ft^2)	Precoat Filtration Plant Producing	
		30 mgd	20 mgd
Total Annual Operating Cost			
Cost (¢/gal)	6.79	6.40	7.81
First Cost	4.49	3.93	5.90
Replacement Cost	0.57	0.50	0.75
Operation and Maintenance (Based on 20 mgd)	1.73	1.97	1.16
Plant Capital Cost ($)	$5,250,000	$4,600,000	

*American Cyanamid Company, Wayne, New Jersey.

PRECOAT FILTRATION THEORY

General

In order to optimize the design of a precoat filtration plant, it is necessary to find the proper combination of design variables (filtration rate, terminal pressure drop, and body feed rate) to produce the desired quality of filtered water at least cost. To accomplish such an analysis, it is necessary to have an appreciation of the operational effects of each of these variables.

The results of a typical series of precoat filter runs conducted at a constant rate of filtration but with varying amounts of body feed are shown in Figure 22. The total volume of filtrate is plotted against the total head loss in feet of water. With no body feed or insufficient amounts of body feed, the head loss increases more rapidly as the run continues, since the suspended solids are removed on the surface of the precoat filter medium and soon form a compressible, impermeable layer of solids that does not permit the ready passage of water. Under these conditions, the use of higher filtration rates would carry the suspended solids deeper into the precoat, and more water could be filtered before the impermeable surface layer is completed. Thus, more water might be filtered per run with higher filtration rates with no or insufficient body feed.

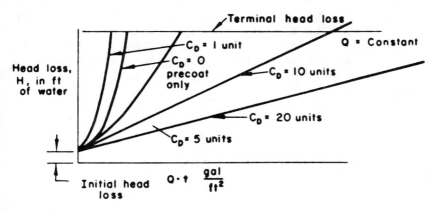

Figure 22. Pressure drop across a precoat filter as a function of body feed rate and volume of filtrate.

With further increases in the body feed, a body feed concentration is reached where the added diatomite results in an increased permeability of the filter cake. This serves to prevent the formation of an impermeable surface layer, and the rate of head loss build-up is decreased. With great enough body feed rates, filter aid may be the predominant suspended solid in the filter cake, and the resulting filter cake hydraulic characteristics may be more similar to those of the filter aid than to those of the suspended solids removed in the filter cake. Then, there is a straight line rate of head loss increase with volume of filtrate. With still higher rates of body feed, the slope of the straight line plot gets flatter and flatter. Eventually, however, still higher rates of body feed add so much filter aid unnecessarily that the slope of the relationship of head loss to filtrate volume again increases.

When the concentration of body feed is sufficient to produce a straight line relationship of head loss to volume of filtrate, the head loss through a given filter cake varies directly with the flow rate. If the flow rate is then halved, the head loss through the cake is halved. In fact, the ratio of the volume of water filtered at the lower filtration rate to the volume of water filtered at the higher filtration rate $(Q_1 t_1)/(Q_2 t_2)$, varies directly with the ratio of the higher unit flow rate to the lower unit filtration rate, Q_2/Q_1. In other words:

$$\frac{Q_1 t_1}{Q_2 t_2} \cong \frac{Q_2}{Q_1}$$

Under such conditions the volume of water filtered to any terminal head loss would be doubled by cutting the flow rate in half. If the volume of filtrate is doubled when the flow rate is cut in half, the filter run would have to last four times as long to produce that volume of filtrate. In fact, the ratio of the length of run at the lower filtration rate to the length of run at the higher filtration rate, t_1/t_2, varies directly with the square of the ratio of the higher filtration rate to the lower filtration rate, Q_2/Q_1. For example,

$$\frac{Q_1 \cdot t_1}{Q_2 \cdot t_2} \cong \frac{Q_2}{Q_1}$$

and

$$\frac{t_1}{t_2} \cong \left(\frac{Q_2}{Q_1}\right)^2$$

When the rate of body feed used is sufficient to produce a straight line relationship between head loss and volume of filtrate, the hydraulic characteristics of the cake enable the following:

1. The length of a filter run may be increased directly with an increase in the terminal head loss permitted across the filter cake at the end of the filter run.

2. The length of a filter run may be increased also by cutting the filtration rate; for example, cutting the filtration rate in half increases the length of run by a factor of 4.

3. The volume of filtrate produced per unit weight of precoat used may be increased either by increasing the terminal head loss or by reducing the filtration rate.

If the terminal head loss is doubled or the flow rate cut in half, the volume of filtrate per run is doubled and increases the economy of diatomite used, since the same weight of precoat filter medium produces twice the volume of filtrate.

In practice, the run length normally must exceed 24 hr to minimize labor costs involved in filter operation. The task of the engineer is to find the optimum combination (referred to earlier) of filtration rate, terminal pressure drop and body feed concentration to produce the filtered water at least cost.

Fundamental Equations

To minimize filtration costs, the engineer must complete a number of pilot-plant filtration tests to determine the hydraulic characteristics of filter cakes that result when body feed rates are sufficient to produce a straight-line head loss to volume-of-filtrate relationship. A recently proposed theory of diatomite filtration reduces the number of pilot plant tests that must be made. The total head loss, H, in a filter is equal to the head loss in the precoat layer, h_p, plus the head loss in the body feed layer, h_c (see Figure 23).

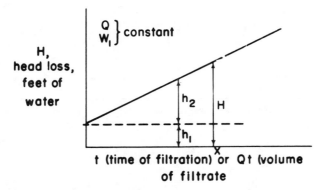

Figure 23. Pressure drop across a precoat filter as a function of volume of filtrate when the body feed rate is sufficient to provide a straight-line relationship.

Table V. Summary of Precoat Filtration Equations

	Equation	Equation Number
Any septum	$H_p = \dfrac{q\nu\xi w}{g}$	(1)
Cylindrical septum	$H_c = \dfrac{R_s\sigma}{\phi} \ln\left(1 + \dfrac{R_s\phi}{R_o{}^2}\right)$	(2)
	$L = R_o{}^2 + R_s\phi X - R_s$	(3)
Flat septum[a]	$H_c = \sigma X$	(4)
	$L = L_p + \dfrac{\phi X}{2}$	(5)
where:	$\sigma = \dfrac{q^2 \nu\beta C_F}{g}$	
	$\phi = \dfrac{2q\gamma_W C_F (10)^{-6}}{\gamma_p}$	
	$X = \dfrac{t - (1 - e^{-\delta t})}{\delta}$	
	$R_o = R_s + L_p$	
	$L_p = \dfrac{w}{\gamma_p}$	
	$\delta = \dfrac{Q}{V_f}$	
	$\beta = \dfrac{a_c\gamma_w (10)^{-6}}{\gamma_p}$	
	$\xi = \dfrac{a_p}{\gamma_p}$	

[a]Septum that does not exhibit increasing area effect.
[b]Dimensionless.

Table V (Continued)

Symbol	Meaning	Dimension
A_s	Septum area	$[L^2]$
a_c	Specific resistance of filter cake based on volume of filter media	$[L^{-2}]$
a_p	Specific resistance of precoat layer based on volume of filter media	$[L^{-2}]$
C_F	Body feed concentration, mg/l by weight	$[\text{- -}]^a$
C_S	Suspended solids concentration, mg/l by weight	$[\text{- -}]$
g	Gravity constant	$[LT^{-2}]$
H_c	Head loss through filter cake	$[L]$
H_p	Head loss through precoat layer	$[L]$
L_c	Thickness of filter cake	$[L]$
L_p	Thickness of precoat layer	$[L]$
L	$L_p + L_c$	$[L]$
Q	Flow rate	$[L^3 T^{-1}]$
q	Flow rate per unit septum area or filtration rate Q/A_s	$[LT^{-1}]$
R_o	Outer radius of precoated septum, $R_s + L_p$	$[1]$
R_s	Outer radius of septum	$[L]$
t	Elapsed time of filtration	$[T]$
V_f	Volume of filter housing	$[L^3]$
w	Precoat weight per unit septum area	$[FL^{-2}]$
X	Elapsed time corrected for initial dilution	$[T]$
β	Filter cake resistance index or β index	$[L^{-2}]$
γ_c	Bulk density of filter cake	$[FL^{-3}]$
γ_p	Bulk density of precoat layer	$[FL^{-3}]$
γ_w	Density of water	$[FL^{-3}]$
δ	Dilution rate, theoretically Q/V_f	$[T^{-1}]$
ν	Kinematic viscosity of influent	$[L^2 T^{-1}]$
ξ	Filter aid resistance index or ξ index	$[F^{-1}L]$
σ	Arbitrary group of terms	$[LT^{-1}]$
ϕ	Arbitrary group of terms	$[LT^{-1}]$

[a]Dimensionless.

Table V lists a summary of the precoat filtration equations used to represent the conditions shown in Figure 23.[4] The pressure drop through the precoat, which is assumed to remain constant through a filter run, is a function of the weight of precoat per square foot, the rate of filtration, the viscosity of the water and the filter aid resistance index, ξ. The filter aid resistance indices for several filter aids are listed in Table II. They may be determined readily with a simple apparatus and test procedure.[1]

The head loss through the body feed layer, which increases in thickness as filtration continues, must take into consideration the type of septa used. If flat or leaf-type septa are used, the filter surface area remains constant and cake thickness increases in direct proportion to the volume of filtrate. On the other hand, if cylindrical septa are used, the surface area of the cake becomes larger with time of filtration, and the cake thickness increase per unit volume of filtrate becomes smaller. These effects are, of course, more significant with smaller diameter (1 in. or 2.5 cm) than with larger diameter (3.5 in. or 8.9 cm) cylindrical septa.

Equation 1 in Table V can be used to calculate the pressure drop through the precoat on any septum, but Equations 2 and 4, respectively, must be used to calculate the head loss through cylindrical and flat septa. Both Equations 2 and 4 include a term σ. This term says that the head loss through the filter cake layer is a function of the square of the filtration rate and directly proportional to the water viscosity, the rate of body feed, and the resistance of the filter cake as measured by a cake resistance, β. The filter cake resistance index, β, is a function of the ratio of the suspended solids present in the water (measured by turbidity, or in units of mg/l) to the rate of body feed (measured in mg/l).

If filter runs are made under identical operating conditions, using flat septa (A) and cylindrical septa 3.5 in. (8.9 cm) in diameter (B) and 1 in. (2.5 cm) in diameter (C), the head loss time of filtration relationship would be as shown in Figure 24. When energy considerations are important, as they are today, this phenomenon related to the type of septa used in precoat filtration says that to conserve energy we should use small-diameter cylindrical septa in precoat filtration. However, it should be remembered that other economic and operating factors can influence the decision as to the type of septa to be used.

Optimum Filter Design

In order to optimize the design of a precoat filter, we must determine the combination of filtration design variables (filtration rate, terminal pressure drop and body feed rate) that provide the desired quality of filtered water at least cost. If the filtration rate and body feed

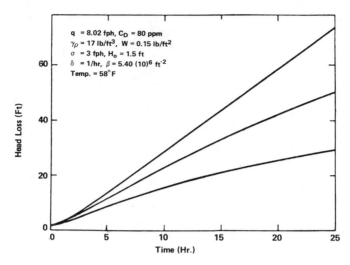

Figure 24. Theoretical head loss vs time of filtration curves using flat and cylindrical septa of different diameters. See Appendix for conversion of English to metric units.

concentration are held constant in a filter run, the effect of increasing the terminal pressure drop on filtration costs can be readily calculated as shown in Figure 25. These data were extracted from a run in which an iron-bearing water (7-8 mg/l) was filtered at 1 gal/min/ft² (0.041 m³/m³ min) with a 40 mg/l body feed.

The significance of this figure can be explained simply. The equipment cost is assumed constant since the cost of pressure filters is not affected significantly by the pressure drop across the filter in the range of 50-200 ft (15-61 m) of water. Yet, if we increase the pressure pumped against, the cost of power must increase (curve B) directly with the increase in pressure. The use of higher pressures provides longer runs and thus increases the volume of water that can be filtered through any given precoat. The total body feed used, a constant, is not affected by the terminal pressure drop. The total weight of filter aid used is, however, since it represents the sum of the precoat and body feed filter aid used. The higher the pressure drop and the longer the filter run, the lower the filter aid costs represented by curve C in Figure 25.

Under manual operation, a diatomite filter requires about 30 min service per day to check operations, mix body feed and precoat slurries and grease motors. In addition, about 30 min labor is required *each time* the filter is backwashed, reprecoated and a new run initiated. If runs are shorter than 6 hr, 24-hr attention is required. Even with automatic

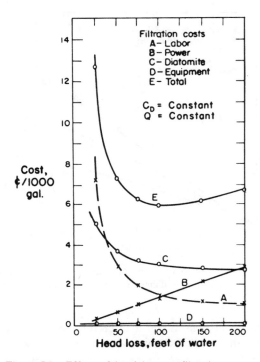

Figure 25. Effect of head loss on filtration costs.

filter operation (which might be more economical with shorter runs), more labor is required with the shorter runs to keep body feed and pre-coat slurry properly supplied. Curve A shows the typical effect of pressure drop across the filter on labor costs. The total filtration costs are represented by Curve E and equal the sum of the labor (A), power (B), filter media (C) and equipment (D) costs. In this example, they combine to indicate that the least cost of filtration at the *given filtration rate and body feed rate* would be with a terminal pressure drop of about 100-110 ft (30-34 m) of water.

In order to determine the least possible filtration cost, it is necessary to construct (or calculate) similar relationships for all practical values of filtration rate and all practical values of body feed rate. For example, the calculations might be made for all filtration rates from 0.3-5.0 gal/min/ft^2 in increments of 0.05 gal/min/ft^2 (written as 0.3/0.05/5.0) and pressure drops from 20 to 200 ft of water in increments of 1 ft (written 20/1/200). Since the possible combinations are many, it is impractical to make filter runs representative of each condition. Therefore, the results of such runs must be "simulated" using the basic filtration equations to reproduce head loss vs volume of filtrate curves from which calculations for data presentations similar to Figure 25 can be developed. All the calculations can be performed by a computer, provided that a "Program for Optimization of Plant Operation" (POPO) is developed and used. In order to simulate filter runs using the mathematical model, it is necessary to have a method of predicting values of the filter cake resistance index, β, for a wide range of values of both suspended solids concentration, C_S, and body feed concentration, C_F.

Determination of Filter Cake Resistance, β

In order to develop a method of predicting filter cake resistance,[5] β, for any given level of body feed and ratio of C_S/C_F in the filter cake, we must first develop a β-prediction equation. The β-index representing the filter cake resistance is a function of the concentration of suspended solids, C_S, the concentration of body feed, C_F, and of the clean filter aid (grade) resistance index, ξ. Many prediction equations are possible, but the following general form of the equation has been used in development of a POPO computer program:

$$\beta = 10^{b_1} (C_S/C_F)^{b_2} C_F^{b_3} \xi^{b_4}$$

in which b_1, b_2, b_3 and b_4 are exponents derived empirically. The ratio of C_S/C_F is the main variable affecting cake resistance. If C_S is relatively constant over a series of filter runs, then C_S/C_F and C_F would not be independent variables, and C_F would be dropped from the prediction equation ($b_3 = 0$). Similarly, if only one grade of filter aid is used in the development of the β-prediction equation, then the ξ term can be dropped ($b_4 = 0$).

Bridges[14] has reported that the equation

$$\beta = 10^{b_1} (C_S/C_F)^2$$

or

$$\beta = 10^{b_1} C_S^2 C_F^{-2}$$

is appropriate for suspensions that do not exhibit concentration effects. He recommends, however, that an equation of the type

$$\beta = 10^{b_1} (C_S/C_F)^{b_2}$$

be used for the analysis of pilot data, incorporating a range of values of both C_S and C_F.

In order to develop a suitable β-prediction equation, a series of pilot-plant filtration runs, in which the suspended solid level covers the range expected in normal plant operation and the body feed level provides a range of C_S/C_F ratios between about 0.005 and 1.0, should be developed.

It is important to recognize and avoid head loss vs time of filtration curves which become exponential and indicate that a compressible cake is formed. These may be expected to occur when the C_S/C_F ratio or T/C_S ratio reaches a value from 0.5 to 1.5. Bridges, for example, operated a precoat pilot plant (a military precoat filter with 15 gal/min or 0.95 liter/sec capacity) to filter raw water from the Missouri River at Council Bluffs, Iowa, using Hyflo Super Cel as filter aid.[14] He made runs with raw water turbidity values between 73 and 463 and body feed levels from about 40 to 480 to provide (C_S/C_F) ratios between 0.25 and 1.0. A preferred minimum number of filter runs would be 8 to 10, but β-prediction equations have been developed from the results of only 3 to 5 runs. β-prediction equations should *not* normally be used to predict cake resistances that fall outside the data used in the development of the equation.

The results of a filter run (mainly head loss as a function of time of filtration) are recorded along with all of the data required for the solution of the filtration equations. At Iowa State University, a computer program (BID, or the Beta Indices Determination) has been developed to calculate the β-index representative of the cake resistance during a filter run from the results of such a run.[14,15] Figure 26 shows the head loss vs time results from a filter run showing the actual head loss measured and the regression head loss time curve fit to the data by the BID computer program in determining the β-resistance index for the run, 193 x 10^{-6} ft^{-2}. The correlation coefficient, 99.975%, showing the excellence of fit of the curve to the data, is also reported.

The results of several similar runs can be plotted to show values of β as a function of the ratio of T/C_F, as shown in Figure 27.[14] A computer program (MAIDS) has been developed for Manipulation and Interpretation of Data Systems, which can accept data of the type in Figure 27 and provide the empirical values of the b_1 and b_2 exponents for the β-prediction equation. The equation representing the data in Figure 27 is

$$\beta = 10^{7.80} (T/C_F)^{2.43}$$

POPO, Program for Optimization of Plant Operation

Once a reliable β-prediction equation has been developed from pilot-plant data representative of conditions under which a precoat filter plant will be operated, the POPO computer program developed at Iowa State University can be used to find the optimum design characteristics of a precoat filtration plant. A separate computer run can be made to optimize plant design for filtration of *each* level of suspended solids to be encountered.

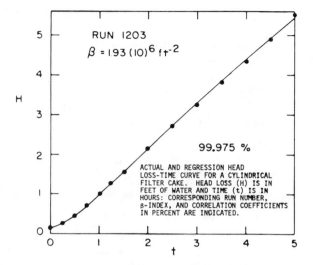

Figure 26. Determination of β-index for resistance of a filter cake.

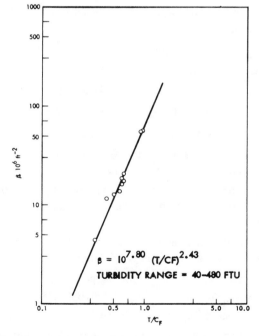

Figure 27. The relationship between β-resistance index and the ratio of T/CF for filtration of Missouri River water at Council Bluffs, Iowa, using Hyflo Super Cel.

The operation of POPO requires the input of only 20 items of data. Table VI shows a typical computer recording of the data that need to be input into the computer for optimization of plant design. These data were used in the interpretation of precoat filter design for removal of turbidity and asbestos fibers at Duluth, Minnesota, using alum-coated precoat and uncoated body feed together with a use of Catfloc B as a coagulant. In general, the data input of Table VI are self-explanatory, but some items, which deserve some additional comment, are explained here.

Item 1 The *design* capacity of the plant. All costs reported are for the filtration of this volume of water.

Item 2 The salvage value of the plant after its useful life.

Item 3 The wire-to-water energy conversion efficiency.

Item 4 The applicable interest rate (5.625% was the Federal Water Resources Council discount rate for making economic comparisons of alternative projects).

Item 5 The plant life, in years.

Item 6 The solids level or turbidity level for which the plant is optimized.

Item 7 The resistance of the filter aid used, ξ. E9 stands for 10^9.

Item 8 The critical water temperature, $°F$.

Item 9 The weight of precoat used, lb/ft^2 (Table II).

Item 10 The density of the grade of precoat used (Table II).

Item 11 Septum diameter (or flat side) in inches.

Item 12 The beta prediction equation exponent values b_1, b_2, b_3, b_4. The values of b_3 and b_4 are 0 in Table VI.

Item 13 The filtration range to be considered by the computer; here, filtration rates from 1.5 to 2.0 $gal/min/ft^2$ in increments of 0.05 $gal/min/ft^2$.

Item 14 The body feed range to be used: here, body feeds from 10 to 20 mg/l in increments of 1 mg/l.

Item 15 The terminal pressure drop range to be used: here, terminal heads from 120 to 170 ft in increments of 5 ft.

Item 16 The cost of filter aid delivered to the plant site.

Item 17 The first cost of the plant in $\$/ft^2$ as a function of the size of the plant. The data included are for pressure filter costs **in 1975** at Duluth.

Table VI. Typical POPO Input Data

0	JOB	X4.	ASHESTOS-TURBIDITY REMOVAL; DULUTH, MINNESOTA			
0			RUNS 73,74,75 HYFLOSUPERCEL BODY FEED + CATFLOC B			
0			PRESSURE D. F.	PLANT		
0			FINAL TURBIDITY: 0.05			
0						
1	DESIGN FLOW		30	MGD		
2	SALVAGE VALUE		0	PERCENT FIRST COST		
3	ENERGY CONVERSION		70	PERCENT		
4	INTEREST RATE		5.625	PERCENT		
5	PLANT LIFE		50	YEARS		
6	SOLIDS (CS)		2.5	MG/L (BY	TURBIDITY)	
7	XI INDEX		4.74E9	FT/LB		
8	TEMPERATURE		34	DEGREES F		
9	PRECOAT WEIGHT		0.15	LB/SF		
10	PRECOAT DENSITY		20.7	LB/CF		
11	SEPTUM DIAMETER		1.0	INCHES		
12	BETA PREDICTION		9.053/1.628/0/0			
13	UNIT FLOW RATE		1.5/0.05/2.0	GSFM		
14	BODY FEED		10/1/20	MG/L		
15	TERMINAL HEAD		120/5/170	FT		
16	DIATOMITE COST		134	$/TON		
17	FIRST COST		AREA	$/SF		
			1000	800.00		
			3472	545.51		
			6944	429.72		
			20832	309.17		
			41667	282.90		
			80000	270.00		
	*		100000	265.00		
18	POWER COST		1.627	CENTS/KWH		
19	LABOR COST		AREA	$/SF PER	MONTH	
			1000	4.00		
			3472	2.48		
			6944	1.66		
			20832	0.69		
			41667	0.48		
			80000	0.38		
	*		100000	0.34		
20	BACKWASH COST		10. 30	GAL/SF, MIN		
	BEGIN					

Under POPO control, the computer will perform the required optimization calculations and output:

1. The data entered into the computer.
2. A summary of the 10 best filter variable design combinations that produce the required filtered water at least cost. The 10 combinations

are reported when the cake resistance is 50, 75, 100, 125 and 150% of that estimated using the β-prediction equation (Item 2).

Table VII includes only the output data from the 75, 100 and 125% of predicated values. The first three columns give (for 100% predicated values), respectively, the *optimum* filtration rate (1.60 gal/min/ft^2), terminal head (135 ft of water), and body feed rate (13 mg/l). The remaining columns record, respectively, the actual β-resistance of the cake assumed, the run length in hours, the filter area required in square feet, the filter cake thickness in inches, and the total filtration cost, the first cost, the total operating cost, the labor and maintenance cost, the power cost, and the filter aid costs, all in $/mg. The last column reports the total monthly (30 days) operating costs. An engineer can now evaluate Table VII type data and determine the probable effects on economic design if the actual cake resistance is higher or lower than the value (100%) predicted by the β-prediction equation. In this example, the filter would be designed as follows:

Filtration rate: 1.60-1.65 gal/min/ft^2 (0.0652-0.0672 m^3/m^2 min)
Terminal pressure: 135 ft of water (41.1 m)
Body feed: 13 mg/l (when turbidity = 2.5 TU)
Filtration costs: 7.05¢/1000 gal ($18.63/m^3)

If we want to examine the effect of the suspended solids level on optimum plant design, all we have to do is change the Item 6 input data in several runs to reflect the range of solids levels for which optimum designs are required. Figure 28 shows the effect of solids level on the optimum design of precoat filters for the removal of a clay from a laboratory-spiked water. Figure 29 shows a similar diagram based on plant-scale tests conducted at Lompoc, California, which uses precoat filters for filtraton of a lime-softened water, where the *mean* turbidity level was only 7 TU. The effect of plant capacity on economical design can be evaluated using the POPO program by making several runs in which different plant capacities are entered as Item 1. Figure 30, for example, indicates that costs decrease as size increases up to 10 mgd for a lime-soda softening plant (Figure 29) but little economics of scale exists in building larger plants. Thus, it would be as cheap to build two 15 mgd plants as one 30 mgd plant.

Table VII. Typical POPO Output Data

				BETA INDICES = 75			PERCENT OF		PREDICTED VALUES					
1.70	125	12	3707	14.9	12767	0.16 *	67.8	28.7	39.1	16.3	9.2	16.3	13.6 *	61920
1.75	125	12	3707	14.0	12432	0.15 *	67.8	28.4	39.4	16.4	9.2	16.3	13.9 *	61927
1.70	120	12	3707	14.2	12792	0.15 *	67.3	28.7	39.1	16.3	8.8	16.2	14.0 *	61929
1.65	120	12	3707	16.1	13148	0.15 *	67.9	29.0	38.8	16.2	8.8	16.2	13.7 *	61831
1.65	125	11	4271	14.8	13161	0.15 *	67.8	29.0	38.9	16.3	9.2	16.3	13.3 *	61833
1.65	125	12	3707	15.8	13124	0.16 *	67.8	29.0	39.0	16.3	9.2	16.3	13.4 *	61833
1.70	125	11	4271	13.9	12805	0.15 *	67.8	28.7	39.1	16.3	9.2	16.3	13.5 *	61835
1.70	130	11	4271	14.5	12741	0.15 *	67.3	29.7	39.1	16.3	9.5	16.3	13.3 *	61935
1.75	135	12	3707	14.6	12409	0.16 *	67.8	28.3	39.5	16.3	9.5	16.3	13.6 *	61835
1.70	130	12	3707	15.5	12745	0.16 *	67.8	28.6	39.2	16.3	9.5	16.3	13.3 *	61834
				BETA INDICES = 100			PERCENT OF		PREDICTED VALUES					
1.60	135	13	4339	14.2	13594	0.15 *	70.5	29.4	41.1	16.2	9.9	16.2	14.9 *	64300
1.65	135	13	4339	13.4	13216	0.15 *	70.5	29.1	41.4	16.1	9.9	16.1	15.2 *	64311
1.65	140	13	4339	14.0	13193	0.15 *	70.5	29.1	41.4	16.1	10.2	16.1	14.2 *	64311
1.60	140	13	4339	14.9	13570	0.16 *	70.5	29.4	41.1	16.2	10.3	16.2	14.6 *	64312
1.55	135	13	4339	15.3	13996	0.16 *	70.5	29.5	40.8	16.2	9.9	16.2	14.7 *	64313
1.60	130	13	4339	12.7	13610	0.15 *	70.5	29.5	41.1	16.1	9.6	16.1	15.3 *	64314
1.60	140	12	4943	14.0	13607	0.15 *	70.5	29.4	41.1	16.2	10.3	16.2	14.6 *	64314
1.55	130	13	4339	14.7	14021	0.15 *	70.5	29.8	40.7	16.2	9.6	16.2	15.0 *	64315
1.55	135	12	4943	14.4	14034	0.15 *	70.5	29.8	40.7	16.2	9.9	16.2	14.6 *	64317
1.60	135	12	4943	13.5	13632	0.15 *	70.5	29.5	41.0	16.2	9.9	16.2	14.9 *	64320
				BETA INDICES = 125			PERCENT OF		PREDICTED VALUES					
1.55	145	14	4807	13.8	14066	0.15 *	72.8	29.9	42.9	16.2	10.7	16.2	16.1 *	66381
1.50	145	14	4807	14.7	14486	0.15 *	72.8	30.2	42.6	16.1	10.7	16.1	15.8 *	66386
1.50	145	13	5424	13.9	14526	0.15 *	72.3	30.3	42.5	16.1	10.7	16.1	15.7 *	66386
1.50	140	14	4807	14.2	14514	0.15 *	72.8	30.3	42.5	16.2	10.3	16.1	16.1 *	66387
1.55	150	13	4907	14.3	14047	0.16 *	72.8	29.9	43.0	16.2	11.0	16.2	15.8 *	66392
1.55	150	13	5424	13.5	14074	0.15 *	72.8	29.9	42.9	16.2	11.0	16.2	15.7 *	66393
1.50	150	14	4807	14.4	14503	0.15 *	72.8	30.2	42.6	16.1	11.0	16.1	15.4 *	66393
1.55	150	13	4807	13.2	14085	0.15 *	72.9	29.9	42.9	16.2	10.3	16.2	16.4 *	66304
1.55	145	13	5424	13.0	14098	0.15 *	72.8	29.9	42.9	16.2	10.7	16.2	16.0 *	66308
1.60	150	14	4807	17.4	13635	0.15 *	72.8	29.5	43.3	16.2	11.0	16.2	16.1 *	66401
				BETA INDICES = 150			PERCENT OF		PREDICTED VALUES					
1.50	155	15	5154	13.7	14534	0.15 *	74.8	30.4	44.5	16.1	11.4	16.1	17.0 *	68193
1.50	155	14	5769	13.0	14572	0.15 *	74.8	30.6	44.8	15.1	11.4	16.1	16.9 *	68193
1.50	150	15	5156	17.3	14550	0.15 *	74.9	30.3	44.5	16.1	11.6	16.1	17.3 *	68105
1.50	160	14	5769	13.5	14544	0.15 *	74.8	30.7	44.5	16.1	11.8	16.1	16.6 *	68199
1.50	152	14	5769	12.4	14597	0.15 *	74.8	30.4	44.4	16.2	11.6	16.2	17.2 *	68210

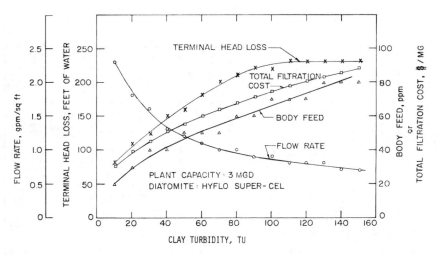

Figure 28. Optimum design characteristics of precoat filters for removal of Kentucky ball clay (kaolinite).

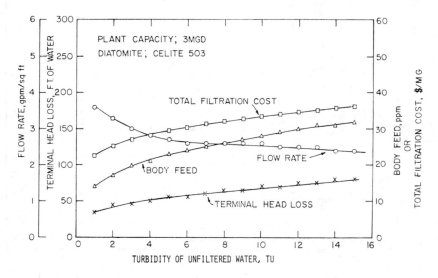

Figure 29. Optimum design characteristics of precoat filters for removal of lime-softening suspended solids.

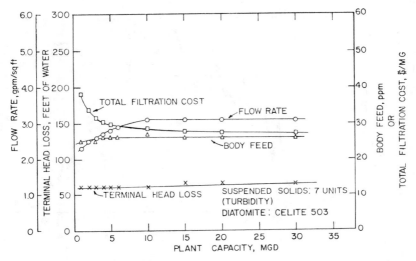

Figure 30. Effect of plant capacity on the optimum design characteristics of precoat filters for removal of lime-softening suspended solids.

APPROACH TO FILTER PLANT DESIGN

Determining Raw Water Characteristics

The AWWA Task Group of Diatomite Filtration has recommended that precoat filters be designed (optimized) for removal of the level of suspended solids or turbidity that is exeeeded only 10% of the time.[12] In the design of precoat filtration plants for removal of iron and/or manganese levels from ground waters, both the temperature and the iron and/or manganese levels are nearly uniform throughout the year. Such designs can be made effectively with relatively few POPO computer runs.

In the design of surface water filtration plants, on the other hand, both the level of water temperature and the water turbidity vary with time over rather wide limits. For example, graphs showing the variation in temperature and turbidity of weekly samples of water collected during 1968 and 1969 from the Des Moines River near Boone, Iowa, are shown in Figures 31 and 32.[16,17,18] Frequency distribution diagrams for these 104 weeks of data are also presented in these figures. During 1968 and 1969, the mean water temperatures were 12.9°C and 11.5°C, respectively. The mean turbidity was 30.5 TU in 1968 and 29.6 TU in 1969. If a precoat filter were designed to filter this water, the *design* temperature and turbidity should be selected to achieve a *minimum annual* cost of filtration. If the AWWA Task Group recommendation were followed,

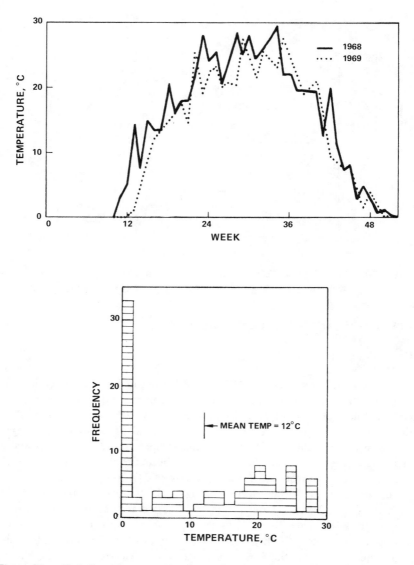

Figure 31. Variation in water temperature (and a frequency distribution diagram for same) observed in the Des Moines River at Boone, Iowa.

the plant would be optimized for a turbidity of 70 TU at a water temperature of 1°C.

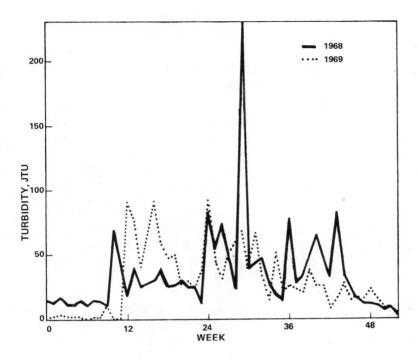

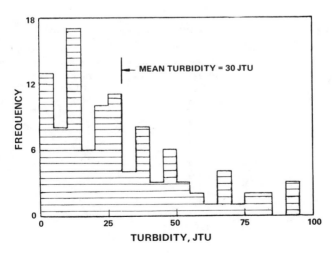

Figure 32. Variation in water turbidity (and a frequency distribution diagram for same) observed in the Des Moines River at Boone, Iowa.

Optimizing Plant Design

To demonstrate how the annual cost of filtering water from the Des Moines River is affected by the design water temperature and design turbidity, let us assume that the filter cake resistance prediction equation:

$$\beta = 10^{7.80} \, (T/C_F)^{2.43}$$

is applicable. This equation was developed by filtration of raw water in the Missouri River, Figure 27.[14] The head loss vs time curves using this water and Hyflo Super Cel filter aid became exponential when the (T/C_F) ratio was greater than 1.5. Therefore, the minimum value of C_F must be large enough that the maximum value of the ratio (T/C_F) does not exceed about 1.4. First, optimum design calculations were made using a design water turbidity equal to the mean 1968-1969 annual turbidity of 30 TU and with the design temperature varied from 0° to 30°C. The optimum design variable combinations for this temperature range are shown in Table VIII.[14] It is apparent in the results that temperature does not have a very large effect on the optimum design conditions. Therefore, a new series of optimum design calculations was made with a mean annual water temperature of 12°C and with the design turbidity varied from 5 to 100 TU. The optimum design variable combinations for this turbidity range are shown in Table IX. The results again indicate the very large effects the design turbidity has on the optimum design conditions. Thus, Table IX confirms the effects already observed in Figures 28 and 29.

Table VIII. Effect of Design Temperature on Optimum Precoat Filter Design Conditions (Influent Turbidity = 30 TU)[a]

Temperature		Optimum Filtration Rate	Optimum Terminal Head Loss	Optimum Body Feed Rate
(°C)	(°F)	(gal/min/ft²)	(ft)	(mg/l)
0	32	0.9	120	40
5	41	0.9	120	35
10	50	0.9	110	35
15	59	1.0	110	35
20	68	1.0	105	35
25	77	1.0	105	30
30	86	1.0	100	30

[a]See Appendix for metric conversion factors.

Table IX lists the optimum design conditions for a plant designed for continuous operation only at the one turbidity listed. An actual plant

Table IX. Effect of Design Turbidity on Optimum Design Conditions
(Water Temperature = 12°C)[a]

Design Turbidity (TU)	Optimum Filtration Rate (gal/min/ft²)	Optimum Terminal Head Loss (ft)	Optimum Body Feed Rate (mg/l)
5	1.9	45	10
10	1.6	60	20
20	1.2	90	30
30	1.0	115	35
40	0.9	140	40
50	0.8	150 (maximum	50
60	0.7	150 permissible	55
70	0.6	150 head loss)	60
80	0.6	150	65
90	0.5	150	70
100	0.5	150	80

[a]See Appendix for metric conversion factors.

must be designed using only *one set* of these optimum conditions and, in actual operation, is operated at turbidity levels both below and above that used in optimizing the design. In such a plant, the filtration rate and terminal head loss is fixed in the design, and only the body feed rate can then be changed when the water turbidity changes in order to reoptimize the plant operation. For example, if the plant design turbidity were 30 TU at a water temperature of 12°C, the filter would be designed to operate at 1.0 gal/min/ft² to a terminal head of 115 ft. If these values are input into the computer along with other levels of turbidity, we can determine the body feed rate that should now be used with each turbidity to minimize operating costs, Table X. As would be expected, both the body feed requirements and filter operating costs increase with the level of raw water turbidity.

It should be noted that if the plant is optimized to operate at a turbidity of 30 TU, operation with any turbidity level above or below this value results in operating costs higher than those that would result if the plant had been optimized to operate at the new turbidity levels. Figure 33 shows the cost of operating a precoat filtration plant if it could be optimized for each level of turbidity (the lower curve in the figure). If the plant is optimized for use with a water turbidity of 30 TU (Table X) or 50 TU, the actual operating costs when operated at other turbidities are also shown in Figure 33. The plant should be designed for that turbidity level that will produce the lowest *annual* cost of filtration.

Table X. Optimum Operating Conditions at Various Influent Turbidities
(Design Turbidity = 30 TU and Design Temperature = 12°C)

Influent Turbidity (TU)	Optimum Operating Body Feed Rate (mg/l)	Unit Cost ($/MG)
5	5	38.5
10	10	42.3
20	25	50.0
30	35	58.0
40	50	66.5
50	60	75.2
60	75	84.4
70	90	93.9
80	105	103.8
90	120	114.1
100	135	124.7

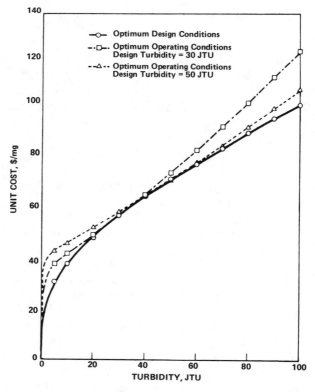

Figure 33. Effects of the design turbidity level on filter operating costs at other turbidity levels.

For example, let us suppose that our plant must produce 1 mgd of filtered water from the Des Moines River. If the plant design turbidity level were 30 TU, the annual plant operating costs can be calculated as shown in Table XI.[14] Operated with a turbidity of 30 TU, the plant operating costs would be $58.00/MG. Over a two-year period, the plant, operating with turbidity levels from 5-100 TU, would have a *mean annual* operating cost of $60.50/MG. If similar analyses are made at other turbidity levels and the *mean annual* operating costs determined, we can plot the *mean annual operating* cost as a function of the plant design turbidity level, Figure 34. The results shown indicate that the optimum plant design should be based on a turbidity level of about 35-40 TU. These results suggest that a precoat filter plant should be designed closer to the mean annual water turbidity or the suspended solids level rather than for that value which is exceeded only 10% of the time.

Table XI. Example Calculation of the Annual Cost of Filtration
(Design Turbidity = 30 TU)

(1) Influent Turbidity (TU)	(2) Number of Weeks	(3) Optimum Operating Cost ($/MG)	(4) (2) x (3) (Weeks x $/MG)
5	13	38.5	500.0
10	8	42.3	338.4
15	17	46.3	787.1
20	6	50.0	300.0
25	10	54.1	541.0
30	11	58.0	638.0
35	4	62.4	249.6
40	8	66.5	532.0
45	3	71.0	213.0
50	6	75.2	451.2
55	3	80.1	240.3
60	2	84.4	168.8
65	1	89.2	89.2
70	4	93.9	375.6
75	1	99.1	99.1
80	2	103.8	207.6
85	2	109.0	218.0
90	0	114.1	0.0
95	3	119.2	357.6
100	0	124.7	0.0

$$\Sigma = 104 \text{ weeks} \qquad \Sigma = 6307.0$$
$$= 2 \text{ yr}$$

6307.0 weeks ($/MG) x 7 MG/week ÷ 2 yr = $22,075/yr
$22,075/yr: 365 MG/yr = $60.5 per MG

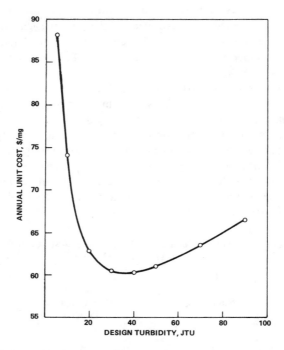

Figure 34.   Relationship between mean annual operating cost and plant design turbidity.

In the optimization study at Duluth, however, the economic analysis indicated that annual filtration costs were less when the plant was designed for removal of 1.0 TU of turbidity than it was when optimized for turbidity levels of 1.9 or 5.8 TU. See Table III for an analysis of water turbidity levels in Lake Superior water. With 20% increases in the cost of filter aid, the best turbidity for filter design shifted to the 1.9 TU turbidity level.

BASIS OF COST DATA

The cost figures in this chapter were collected over a long period (1960-1975) and should not be accepted as representative of actual filtration costs today. Precoat filtration costs today must be calculated and optimized on the basis of today's costs. Precoat filters have one disadvantage in that a larger portion of filtration costs, as compared to deep-bed filters, are in the variable operating areas which are subject to significant inflationary costs. For example, deep-bed filters involve terminal pressure drops of 10-15 ft (3-4.6 m) of water whereas precoat pressure filters involve terminal pressure drops of 50-150 ft (15-46 m). Thus, they are more subject to changes in energy costs. Similarly, about 20-30% of precoat filtration costs are associated with use of filter aid, which is mined and processed *predominantly* on the West Coast and shipped to the plant site. Thus, these costs are subject to inflationary costs involving both processing and transportation. Therefore, filtration costs included in this chapter should not be used as representative of today's conditions.

REFERENCES

1. Baumann, E. R., and C. S. Oulman. "Determination of the Effective Specific Gravity, Bulk Density, and Filter Cake Resistance of Precoat Filter Aids," Final Report on Filter Aid Specifications, Part IV, ERI-269, Iowa State University, Engineering Research Institute, Ames Iowa (August 1968).
2. Dillingham, J. H., and E. R. Baumann. "Hydraulic and Particle Size Characteristics of Some Diatomite Filter Aids," *J. Am. Water Works Assoc.* 56(6):793-808 (1964).
3. Baumann, E. R. "Diatomite Filters for Asbestiform Fiber Removal From Water," Proc., 95th Annual Am. Water Works Assoc. Conf., Paper No. 10-2C (1975), pp. 1-45.
4. Dillingham, J. H., J. L. Cleasby and E. R. Baumann. "Diatomite Filtration Equations for Various Septa," *J. San. Eng. Div. Proc., ASCE*, SAI:41-55 (1967).
5. Dillingham, J. H., J. L. Cleasby and E. R. Baumann. "Prediction of Diatomite Filter Cake Resistance," *J. San. Eng. Div., Proc. ASCE*, SAI:57-76 (1967).
6. Cummins, A. B. "Electropositive Composition and Process of Making the Same," U.S. Patent 2,036,258, (April 7, 1936).
7. Guebert, K. W., *et al.* "Coated Filter Aids," U.S. Patent 3,352,424 (November 14, 1967).
8. Baumann, E. R., and C. S. Oulman. "Polyelectrolyte Coatings for Filter Media," *Filtration Separation* (7)6:682-690 (1970).
9. Burns, D. E., E. R. Baumann and C. S. Oulman. "Particulate Removal on Coated Filter Media," *J. Am. Water Works Assoc.* 62:121-126 (1970).
10. Bell, G. R. "Coagulant Coatings Open New Applications to Filter Aids," Proc., Int. Water Conf., Eng. Soc. Western Penn. (1961), pp. 129-133.
11. Culp, R. L. "Direct Filtration," Proc., Direct Filtration Seminar, California Am. Water Works Assoc. Section Forum (Fall 1976), pp. 1-15.
12. Task Group Report (E. R. Baumann, Chairman). "Diatomite Filters for Municipal Use," *J. Am. Water Works Assoc.* 57(2):157-180 (1965).
13. Letterman, R. D., and G. S. Logsdon. "Survey of Direct Filtration Practice-Preliminary Report," Proc. Am. Water Works Assoc. Annual Conference (June 1976).

14. Bridges, H. R. "Design Requirements of Precoat Filters for Water Filtration," Ph.D. Thesis, Iowa State University Library, Ames, Iowa (1970).
15. Dillingham, J. H., J. L. Cleasby and E. R. Baumann. "Optimum Design and Operation of Diatomite Filtration Plants," *J. Am. Water Works Assoc.* 58(6):657-672 (1966).
16. Baumann, E. R. "Preimpoundment Water Quality Study Saylorville Reservoir, Des Moines River, Iowa," Annual Report ERI-558, Iowa State University, Engineering Research Institute, Ames, Iowa (August 1969).
17. Baumann, E. R., and M. D. Dougal. "Preimpoundment Water Quality Study, Saylorville Reservoir, Des Moines River, Iowa," Annual Report ERI-346, Iowa State University, Engineering Research Institute, Ames, Iowa (August 1968).
18. Baumann, E. R., and S. Kelman. "Preimpoundment Water Quality Study, Saylorville Reservoir, Des Moines River, Iowa," Annual Report ERI-80700, Iowa State University, Engineering Research Institute, Ames, Iowa (August 1970).

SELECTED READING LIST

1. Baumann, E. R. "Diatomite Filters for Municipal Installations," *J. Am. Water Works Assoc.* 49(2):174-186 (1957).
2. Baumann, E. R., and R. L. LaFrenz. "Optimum Economical Design for Municipal Diatomite Filtration Plants," *J. Am. Water Works Assoc.* 55(1):48-58 (1963).
3. "Quality Goals for Potable Water," Statement of Policy, *J. Am. Water Works Assoc.* 60(12):1317-1322 (1968).
4. Robinson, J. H., *et al.* "Direct Filtration of Lake Superior Water for Asbestiform Solids Removal," Proc., 95th Annual Am. Water Works Assoc. Conf., Paper No. 10-2a (1975), pp. 1-22.

CHAPTER 16

DESIGN AND APPLICATION OF POTABLE
WATER CHLORINATION SYSTEMS

Geo. Clifford White, C.E., M.E.

Consulting Engineer
556 Spruce Street
San Francisco, California 94118

INTRODUCTION

This chapter is meant to provide the designer with the necessary information to design a chlorination system and also to delineate the objectives
of chlorination and its many uses in water treatment. Since all of this
material is covered in great detail by White[1] in *Handbook of Chlorination*,
the following text is arranged in a modified outline form to guide the
busy designer to the most productive reading, to reference the pages for
each subject discussed and to provide warnings for pitfalls. As there are
frequent references to the book by White, a shorthand notation is used
wherein HC 301-304 means pages 301 to 304 in the *Handbook of Chlorination*. All figure numbers in the text (except Figure 1) refer to figures
in the *Handbook of Chlorination*. The text to follow also contains updated material and prices that are not available in the *Handbook of
Chlorination*. The costs given are, of course, approximate only. They
are based on a 1977 price index.

OBJECTIVES OF CHLORINATION

The primary objectives of the chlorination process are: (1) disinfection,
and taste and odor control of the finished water as it leaves the plant,

(2) taste and odor control in the distribution system, (3) prevention of the growth of algae and other microorganisms that might interfere with coagulation and flocculation, (4) elimination of slime growths and mud balls in filter media and the prevention of possible build-up of anaerobic bacteria in the filter media, (5) destruction of hydrogen sulfide and control of sulfurous taste and odor in the finished water, (6) removal of iron and manganese, and (7) organic color bleaching.

Disinfection

The various aspects of disinfection efficiency by chlorine are discussed in depth in HC 301-306 and shown in Figures 6.5 to 6.7. The most important factor in disinfection by chlorine is the chlorine concentration-time envelope.

Taste and Odor Control

The appropriate and judicious application of chlorine for this purpose is one of the most significant uses of chlorine in the production of potable water. These considerations are extensively discussed in HC 306-314.

Distribution Systems and Transmission Lines

The problem of delivering a palatable and safe water to the consumer does not end as the water leaves the treatment plant, pumping station or well discharge. There are a multitude of difficulties in maintaining water quality in transmission conduits and distribution systems. The problems of water quality control in distribution systems are of two types. Probably the most prevalent is complaints about taste, odor and dirty water. The other is deterioration of bacteriological quality.

The question of the proper solution to these problems is one of the most controversial subjects of modern water treatment practice. There seems to be more proponents of free residual chlorine as the proper treatment than those favoring chloramine residuals. There are also those who favor either ammonia-induced breakpoint or ammonia-controlled free residual chlorination as well as those who favor chlorine dioxide. In view of the recent controversy over chloro-organic compound formation and whether they are deleterious to health, the use of postammoniation in conjunction with dechlorination is certain to become more popular than before. The nature of the distribution system problem is discussed in HC 314-324 in great detail; it describes the nature of the problem, the bacteriology of closed water systems, the control of bio-fouling supplemented by case histories of various but common types of distribution system problems. These situations are important to the designer because the chlorination system design must have the flexibility to cope with distribution system problems.

While the designer is not necessarily concerned with the details of implementing a clean-up program, it is worthwhile to understand such a program. Because the problems of water quality degradation in distribution systems are legion, it is important for the designer to be familiar with this subject, which includes the restoration of transmission line capacity and how to cope with and determine the magnitude of such a problem.

Other Uses of Chlorine

Detailed discussions of other uses of chlorine such as an aid to coagulation and filtration, control of algae growths, preservation of filter media, hydrogen sulfide control and removal, iron and manganese removal, and color removal are found in HC 324-336.

Sulfide Control

The designer should be aware of the problems associated with hydrogen sulfide removal. This situation requires such supplementary safety factors in the design of chlorination facilities as: mixing, contact time, automatic control, dechlorination and rechlorination.

Iron and Manganese Removal

The removal of iron and manganese is just as significant. If a designer is confronted with a water containing iron at concentrations between 0.3 and 0.5 mg/l, he should be familiar with the discussion in HC 331-335. Furthermore, the mere presence of a significant amount of iron in a water supply suggests immediately the presence of manganese.

Manganese removal is a most important facet of water treatment. The use of chlorine and sequestering agents to deal with manganese in concentrations up to 0.1 mg/l is described in HC 333-335.

Color Removal

Color removal by chlorine is described in HC 336.

HOW TO ACHIEVE THE OBJECTIVES OF CHLORINATION

Current Practice

Modern chlorination of potable water falls into the following classifications (HC 296): plain chlorination, free residual chlorination, free residual chlorination followed by dechlorination, chloramination, prechlorination, postchlorination, and rechlorination. The practical significance of these practices is discussed in great detail in HC 297-301; Figures 6.3 and 6.4 remain today as classic descriptions.

It is indeed pertinent that the serious and dedicated designer and/or operator familiarize himself with these various definitions of current practices. It is also important that they are aware of the EPA findings[2] in 1976 delineating the detection of objectionable compounds resulting from the chlorination process. The compound of primary concern is the presence of chloroform, which is a known carcinogen. It has been definitely proven that in *some* waters the formation of chloroform, one of the trihalomethanes (THM), is enhanced by free residual chlorination. The waters most susceptible to chloroform formation by free residual chlorination seemed to be those of rather poor quality. This aspect is being investigated by the EPA and it is possible that the precursors in these waters that promote the formation of THM as a result of free residual chlorination can be removed in a pretreatment process, thereby eliminating the problem. Various aspects of this phenomenon were presented at a 1977 AWWA "Disinfection Seminar," for which proceedings are available.[3]

Several investigations currently underway at various treatment plants in the U.S. are directed at the manipulation of chlorine application that might result in the reduction of chloroform formation. Others are for the removal of the precursors that tend to enhance chloroform production. There are experiments with chlorine dioxide, chloramines (chlorine plus ammonia addition) and ozone because these oxidants do not promote the formation of THM. A report released by the EPA[4] in 1976 is worthy of review.

HC 336-342 contains a pertinent description of pre- and postchlorination practices beginning about 1938 when the breakpoint phenomenon was first recognized by Griffin[5] and others. The serious designer or dedicated operator should read this information carefully because it summarizes practically all of the chlorine control problems normally encountered by a potable water treatment plant operator.

The chlorination facility described in HC 337 and illustrated by Figure 6.9 demonstrates the modern approach to chlorine application for potable water. This system of chlorination followed by dechlorination to a specified residual entering the distribution system allows the operator a wide selection of chlorine dosages to produce a taste-free water and at the same time provides the operator with continuous automatic chlorine residual control of the treated water entering the distribution system to the specified residual of the operator's choice. Actual operator experiences are described in HC 339-342.

Dechlorination

The use of sulfur dioxide and other sulfur-bearing compounds that produce the sulfite ion in aqueous solution are the most popular agents for total dechlorination. The reaction is one of immediate stoichiometric relationship with no side reactions (see HC 343-349).

Sulfur Dioxide

This gas-liquid vapor product is described in chemical and practical operating terms in HC 344-347.

Activated Carbon

Carbon reacts with chlorine to effect total dechlorination. The various arrangements for dechlorination using activated carbon are discussed in HC 347-348.

Ammonia Nitrogen

Categorizing ammonia (NH_4OH) as a dechlorinating agent is somewhat misleading. The correct terminology should be described as the use of ammonia nitrogen for the conversion of free chlorine residuals to combined chlorine residuals, as stated in HC 348-349. The use of ammonia nitrogen for this purpose to produce a better tasting water has been demonstrated over and over again. The most significant achievement in the annals of water treatment using ammonia for dechlorination as well as for taste and odor control has been documented by Williams.[6-8]

Chlorination of High pH Waters

The treatment of these waters is a specialized chemical treatment system unlike any other raw water treatment situations. The definition implies that these waters are treated for the removal of calcium and magnesium hardness. The softening process must be controlled at pH levels at 10 or higher where the reaction of free chlorine with nitrogenous compounds is extremely slow. Temperature is also a factor because as the temperature drops the chlorine-ammonia reaction slows down.

The importance of these reactions is discussed in HC 350.

The Ammonia-Chlorine Process

This subject, discussed in HC 351-354, is of considerable importance because it now relates to a system that *does not* promote the formation of trihalomethanes, particularly when used as a pretreatment process.

Miscellaneous Applications of Chlorine

A great many uses of chlorine in the normal day-to-day production of potable water are generally overlooked.

Sterilization of Water Mains and Storage Tanks

These facilities are most conveniently disinfected by the use of portable water chlorination equipment. The use of chlorine for this purpose is described in HC 358-360. An AWWA committee[9] has described the use of hypochlorite solutions for these applications. The amounts of chlorine and hypochlorite tablets required for various pipe sizes are given in HC 360.

Table 6.4, HC 360, delineates the chlorine required to produce a chlorine concentration of 50 mg/l in 100 ft of pipe by pipe diameter. Table 6.5 delineates the number of 5 g hypochlorite tablets (3.75 g available chlorine per tablet) required to produce a chlorine dose of 50 mg/l in various lengths of pipe diameters from 2-12 in.

Other Uses

Chlorine is used for preventing biological growths in desalination facilities, ornamental water fountains and reflectory pools, and for restoring wells and injection wells. These uses are described in HC 354-358.

HYPOCHLORINATORS

Potable water chlorination systems can be designed with either hypochlorination or gas chlorinators.

Hypochlorite vs Gas

A good rule of thumb for deciding whether the situation calls for a hypochlorinator or a gas chlorinator is described in HC 648. However, lithium hypochlorite, mentioned in HC 648, should never be used for treating potable water. Lithium compounds for ingestion are now subject to medical prescription.

Importance of Contact Time and Mixing

The use of hypochlorite solutions in water treatment is generally limited to small water supplies. Optimum design of a hypochlorite facility for these supplies is rarely attained. All conventional equipment for metering liquid hypochlorite consists of positive displacement pumps, which means that the chlorine solution is injected in slugs. This requires greater attention to proper mixing and adequate contact time than is required for gas equipment because the latter injects a continuous stream of chlorine solution. One example of a poor system is the typical water pump and pressure system in which hypochlorite is injected into the pump suction and the pressure tank is considered as the chlorine contact chamber. Since the pressure tank "rides" on the pump discharge, the water short-circuits directly from the inlet to the outlet of the tank, which is, therefore, only a "wide" place in the line. These installations are common for roadside rest areas, outlying motels and restaurants, resorts, summer camps and bus stops.

Sufficient additional contact time should be provided to smooth out the slug-type chlorine injection, and the equipment should be selected so that for any given flow, the pumping speed should be fast enough to minimize this phenomenon. The reader should become familiar with the chlorine residual-contact time envelope described in HC 301-306.

Pumped Supplies

For the design and selection of equipment for this type of supply, see HC 649, 650, and Figures 12.6, 12.7 and 12.8 in HC 650-652.

Gravity Systems

These systems present the most difficult problems in selecting the proper equipment and its arrangement. HC 650 to 659 are devoted to these kinds of systems.

Gravity Systems with Power Available

Sometimes when power is available, some of the problems of the gravity system can be avoided. One solution is illustrated by Figure 12.44, HC 659, whereby the entire flow through a storage tank is chlorinated (including the overflow). This permits the use of a simple manually adjusted electrically operated diaphragm pump. Figure 12.15, HC 661, illustrates the use of an automatic valve in a storage tank to start and stop the hypochlorinator and treat only the water flowing into the tank.

Gravity Filter Systems

HC 660-662 tells why it is imperative that a flow-paced unit must be used on any filter system that "rides on the line." Sample calculations are given to guide the designer and operator.

Maintenance

The maintenance of these (and similar) systems using hypochlorite require the periodic flushing of the hypochlorinator with muriatic acid (HCl) to remove the white, scaly deposit resulting from the interaction of hypochlorite with the hardness in the make-up water as well as the deposition of the inherent impurities in the hypochlorite. These deposits interfere with the efficiency of the internal valve action, cause errors in pumping rates, and obscure the suction sight glass. When the sight glass becomes so coated with a deposit that its operation is obscured, it is time to flush the apparatus with muriatic acid.

Equipment Costs

The least expensive hypochlorinators are those in the category of 20 gal/day (76 liter/day) pumping rate. Converting this to a hypochlorite solution of 2% strength calculates to 3.34 lb Cl/day (1.52 kg Cl/day). This is well within the range of most hypochlorite applications where electric power is available and the use of manual or automatic start and stop control is acceptable. These units cost about $300 to $350 fob including solution tank accessories. Local contractors should be consulted for installation costs.

The next species of equipment is the more sophisticated, electrically driven diaphragm pump capable of providing either flow-proportional control or both flow-proportional and residual control. These pumps have a basic manual control arrangement cost of about $650 to $700. An electric or pneumatic adjustable stroke-length positioner for flow proportional control costs another $700 to $750. These pumps are also available with SCR drive whereby a flow or residual signal controls the speed of the pumping cycle. The SCR drive feature costs about $700, so for a compound loop-control hypochlorinator arrangement, the basic equipment cost would be $700 + $750 + $700 = $2150, not including the water flow measuring device or the chlorine residual analyzer.

The flow proportional, *i.e.,* meter-paced, hydraulic-operated hypochlorinator comes equipped with various sizes of water meters for flow-pacing control. A unit paced by a 2-in. (5-cm) disc meter costs about $2000 complete with meter. One paced by a 4-in. (10-cm) crest meter costs $3200. These units are popular and most effective as reliable chlorination systems, but it is imperative that the water supply to the hydraulic ram portion of the unit be free of debris such as fallen leaves and/or pine needles and with at least a 10 to 12 psi (0.7-0.84 kg/cm^2) pressure available at the unit.

GAS CHLORINATORS

System Description

In general, any chlorine gas installation consists essentially of three parts: chlorine supply system, metering and control system, and the injector system. As an installation of a simple chlorinator for a small supply progresses in sophistication from start and stop control to flow-paced control to flow-paced plus residual control and these grow in size to require evaporators, then tank cars and finally storage tanks, an additional array of equipment for dosage control, monitoring of process, alarms and special safety devices must all be considered part of a gas chlorinator description. All of these are described below.

Chlorine Supply System

Chlorine is packaged in special steel containers of various sizes, as follows:

- 100- and 150-lb cylinders
- ton containers
- single-unit tank cars
- multiple-unit tank cars containing 15 one-ton cylinders
- tank trucks of 15-17 tons capacity.

100- and 150-lb Cylinders

This size of cylinder serves the majority of chlorinator installations in the U.S. and Canada. Great care must be taken in the design of the housing for these cylinders as well as the requirement of scales and the optimum number of cylinders to be connected at one time. This topic is covered in great detail in HC 40 to 45.

Automatic Cylinder Switchover

The capability of automatically switching from an empty to a full cylinder is described in HC 45 and illustrated by Figure 2.4. This capability is available for both 100- and 150-lb cylinders and ton containers when the latter are operating on the gas phase. It is not possible to switch automatically containers of any kind by pressure differential *when operating on the liquid phase.*

Ton Containers

The use of ton containers requires much more careful design considerations than do the smaller containers. These containers allow the withdrawal of either gas or liquid. Their weight requires special handling equipment and their use in the gas phase readily subjects an installation to the phenomenon of reliquefaction, which burdens the operator with a variety of equipment problems.

Details on the layout and operation of ton containers using either the gas or liquid phase are described in HC 46-57. These pages include numerous illustrations of importance, particularly: (a) Figure 2.10, which shows monorail height location calculations, (b) Figure 2.11, which shows the proper location of a chlorine gas filter in a gas withdrawal system, (c) Figure 2.12, which shows the appropriate spacing of equipment and location of expansion tanks in a liquid withdrawal system, (d) Figure 2.14, which shows the details of a liquid expansion tank approved by the Chlorine Institute,* and (e) Figure 2.15, which shows the details of a chlorine storage and handling facility for a chlorinator capacity of 24,000 lb/day (10,800 kg/day).

Single-Unit Tank Cars

The details of tank car construction, siding location and other pertinent details are given in HC 55-60. The rules for tank cars are entirely different from ton containers so these pages should be read carefully.

---

*342 Madison Avenue, New York, New York 10017.

Stationary Storage

Whenever it becomes necessary to utilize a tank car for a supply system, the use of stationary storage becomes a consideration. Whatever the decision may be, if a stationary tank is to be used, it must conform to the design of a Chlorine Institute-approved tank car. In other words, the tank dome should be exactly as is shown in Figure 2.17, HC 59. Moreover, when this design is followed, the tank car emergency safety kit can also be used for the stationary tank. In recent years a variety of chlorine packagers have been attempting to promote tank truck deliveries to market their chlorine. They offer on-site storage tanks as a part of their service. *These packagers must be made to provide storage tanks that have a "dome" as provided by all chlorine tank cars.* The details of stationary storage design are described in HC 61-65.

Selection and Size of Containers.

See HC 65-66.

Designer's Check List for the Chlorine Supply System

See HC 66.

Material of Construction

See HC 67-68.

Flow of Chlorine in Pipes

Chlorine Gas and Liquid Under Pressure

The hydraulics of chlorine flow in the supply system become a consideration only in large installations where it becomes necessary to transport the chlorine over long distances and in large quantities. The hydraulics of liquid flow and gaseous flow are two entirely different phenomena. The subject of liquid chlorine flow is discussed in HC 69-71, and gaseous chlorine flow is discussed in HC 71-76. The flow of gaseous chlorine can become critical if the velocity exceeds 35 ft/sec (11 m/sec) as described in HC 71-72. The hydraulics of gaseous flow are fully covered by the usual classical approach using Reynolds number to establish the friction factor and the Fanning formula for circular pipes (HC 77).

Flow of Chlorine Gas under a Vacuum

This is a most important consideration when using a remote injector on large installations. When considering remote injectors, it is pertinent to consult page 5.305 of the Wallace & Tiernan Division Pennwalt Corporation[10] Chlorination Catalog #1. This shows the recommended injector vacuum pipe sizing for remote injector applications for chlorine feed rates up to 8000 lb/day (3600 kg/day) and distances varying from 1800 ft to 2500 ft (550 to 760 m).

An explicit approach to the design of injector vacuum lines is developed in HC 76-80. However, after these pages were written, White has decided, based on field experience, that all long injector vacuum lines should be designed on the basis of chlorine density at 25 in. Hg (85 kN/m^2) vacuum, 20°C (temperature of the gas), and a maximum head loss of 1.5-2.0 in. Hg (5.1-6.8 kN/m^2) regardless of pipe line length between the chlorinator and the injector. (This also applies to dechlorination systems using sulfur dioxide.) This differs from the HC text, which recommends an optimum vacuum level of 6 in. Hg (20 kN/m^2).

Chlorine Solution Lines

See HC 80-81 under this heading for guidelines and details of design.

The Injector System

The injector is the heart of any chlorination system. If there is not enough hydraulic power, whether it be static head or a pumped supply, the system will not operate.

All of the details concerning the hydraulic considerations of this part of the facility are detailed in HC 81-89. The primary consideration of any injector system is based upon the limiting strength of the chlorine solution discharge, which must not exceed 3500 mg/l (see Figure 2.29). Concentrations beyond this strength result in off-gassing at the point of application and gas binding in the chlorine solution lines. This reduces their carrying capacity and thus upsets the entire hydraulic design.

Figure 2.30 (HC 83) illustrates a typical situation of injector system hydraulics, Figure 2.31 (HC 85) illustrates the use of a booster pump, and Figure 2.32 (HC 86) shows what can happen if there were a negative head at the chlorine diffuser in an open channel. The latter condition must be avoided because a negative head at the diffuser results in the break-out of molecular chlorine from the chlorinator injector discharge. This results in what is described as chlorine "fuming" at the point of application. The designer should always be certain that there is at least a 2-psi (13.8 kN/m^2) pressure in the discharge piping immediately downstream from the injector (see HC 86).

Booster Pumps

Injector water booster pumps are generally of the turbine type that require an adjustable pressure by-pass assembly as shown in Figure 2.33 (HC 87). The other type of pump is the centrifugal pump. The selection of one vs the other is discussed in detail in HC 87-89; the reasons are important.

An extremely unusual situation sometimes occurs where it is not possible to provide an injector water pump of reasonable capacity to overcome the pressure at the point of chlorine application. These situations require the use of a special chlorine solution corrosion-resistant pump to transfer the chlorine solution from the injector discharge to the point of application. This is discussed in HC 88 and illustrated by Figure 2.34.

Designer's Check List for Injection System

See Items 1-6 inclusive, HC 89.

Chlorinators

HC 89-92 and Figure 2.26 describe conventional chlorination equipment and the fundamental theory of their operation.

Chlorination Stations

A wide variety of chlorination stations are described in Chapter 2. Beginning in HC 92, the prime considerations in the design and layout of these installations are listed.

Individual Deep Well Stations

This type of installation is widely used for gas chlorinators. HC 93-95 and Figure 2.37 describe the best way to achieve a reliable system. This system eliminates the use of a booster pump shown in Figure 2.38, but it is not always possible to install the system as shown in Figure 2.37.

Multiple-Well or Multiple-Pump Stations

Figure 2.39, HC 95, shows the possible variations of a chlorination station, which is required to compensate automatically for a variety of start-and-stop pumping conditions.

Relay Stations and Transmission Lines

Very often it is necessary to supplement chlorination of a large distribution system to prevent degradation of water quality in the far reaches of the system. This often occurs in long transmission lines. The situation is discussed in HC 95, and a typical chlorination station capable of dealing with the situation is shown in Figure 2.40.

Reservoir Outlets

Chlorination of reservoir outlets presents one of the most difficult control problems for a chlorination station. Such a problem occurs where the outflow may change to inflow and this immediately defines a situation where the flow into and out of the reservoir reaches zero. So for any given outflow or inflow, the flow-sensing devices controlling the chlorination equipment must have an infinite range. Since this is not possible, there are special arrangements of equipment and control devices that can solve the problems raised. These situations are described in HC 98-102 and illustrated by Figure 1 (page 387) and Figures 2.42 and 2.43 in HC.

Reverse Flows-Reservoir Systems

In HC 102, this subject is separated into two parts: predictable reversal and unpredictable reversal of flows. Reverse flows are most prevalent in small systems designed by engineers who were primarily interested in hydraulics and not water treatment. These systems are legion in the western part of the U.S. Flow reversals in reservoirs and the solution to the problem are described in HC 102 and illustrated by Figures 2.42 and 2.43.

Automatic Switchover of Auxiliary or Standby Equipment

Chlorination must be not only effective but continuous, so provision should be made not only for standby equipment but also for auxiliary equipment to take care of increased loads. Automatic switchover of the chlorine supply has already been described. Figure 2.44, HC 106, shows an arrangement for automatic starting and stopping of an auxiliary chlorinator.

Design of Chlorination Equipment

HC 105 explains precisely how to size a chlorinator, with a 10-mgd plant as an example. Other sizes such as those of 1-mgd plant are relative in nature. Smaller plants are likely to be designed for the use of hypochlorite.

Points of Application

The philosophy behind the various points of application used in modern filter plants begins in HC 105. In view of the recent concern about the formation of trihalomethanes (THM) by the use of free residual chlorination in some waters, the prechlorination point may have to be moved to a different location. Some thoughts have been expressed about eliminating prechlorination, which would be a serious mistake because of the many benefits of prechlorination.

Dechlorination

In HC 107 a rule of thumb is given on sizing equipment for dechlorination, which is a most important tool in the chlorination process. Fortunately, the same equipment and accessories used for chlorine can be used for dechlorination by sulfur dioxide or its derivatives.

Postammoniation

The proper sizing of ammonia equipment to convert all the chlorine residual to chloramines is mentioned in HC 107. This is a common practice in high pH waters. Ammoniation as a substitute for prechlorination will become more popular as a pretreatment because chloramines do not contribute to the formation of THM.

Ammonia metering equipment is available both in direct feed or solution feed. The latter is better for automatic control. However, solution-feed ammoniators require that the injector water be softened to a maximum total hardness of 25 mg/l.

Injector Requirements

The various hydraulic requirements and situations that may arise are also discussed in HC 107 with sample calculations that explain the various situations. Careful attention must be given to the provision of the injector water supply as this is the heart of the system. Solution pipe sizing is important because of the gross effect of too much friction loss. Therefore, equivalent lengths of these lines must be carefully calculated. Injector requirements depend upon four factors: (a) chlorine feed rate, (b) water pressure available, (c) water flow available, and (d) back pressure (hydraulic gradient) at the discharge of the injector. The last factor represents the friction loss in the chlorine solution line plus the pressure at the point of application.

Methods of Control

1. Open-loop flow proportional. This is a basic chlorinator control system and is described in HC 108.

2. Closed-loop flow proportional. This concept utilizes a chlorine residual signal superimposed on a flow signal. The flow signal proceeds to a multiplier via a ratio station and the dosage signal proceeds to the same multiplier via a controller that has proportional band and reset adjustments. The signal multiplier then puts out a single signal that adjusts the chlorine feed rate according to the two variables.

3. Compound-loop control. This is different from the previous method in that two separate signals are sent independently to the chlorinator. The various compound-loop arrangements are illustrated in Figure 2.41 and are described in HC 109. There is one exception. Since HC 109 was written and Figure 2.46 drawn, White has found that the *flow signal* should always go to the *chlorine orifice positioner.* Furthermore, a ratio station with a wheel-operated range of 0.4-4.0 should always be installed in the flow signal circuit as shown in Figure 1 of this text. The residual signal should *never* go to the chlorine orifice positioner because it results in an overly sluggish response. (This arrangement is shown in Figure 2.46, HC 111.) Figure 2.45, HC 110, illustrates a compound-loop system with an ideal loop time. The one missing item here is a ratio station on the flow signal line, which should be added as shown in Figure 1.

4. Direct residual control. Figure 2.49, HC 115, shows how residual control can be accomplished without the use of a flow signal. The most responsive arrangement is to send the residual signal to the vacuum regulating valve and to utilize the manual positioning of the metering orifice to set the proper dosage. Recent experience with these systems bears out the practicality of this method of control. However, it is not practical in any system where there are sudden changes in flows due to pumps starting and stopping. This system operates well whenever the flow variations follow the general diurnal pattern of a gravity system.

5. Cascade systems. These are variations of the closed-loop control concept. They are illustrated by Figures 2.50 and 2.51, HC 118 and 119, and described in HC 112-116. Figure 2.52 shows a highly sophisticated system used mostly in wastewater treatment.

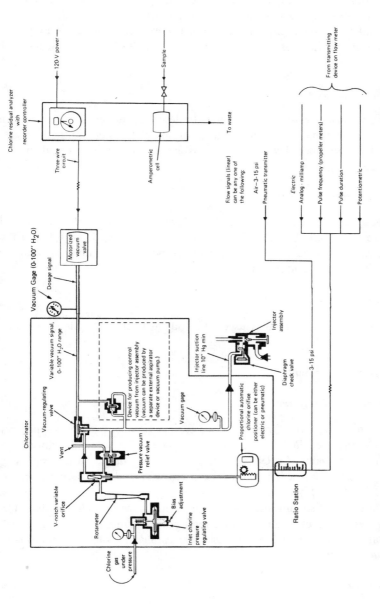

Figure 1. Schematic compound-loop control system. Both the chlorine orifice positioner and the motorized vacuum valve can be manually adjusted. (Courtesy Wallace & Tiernan Division, Pennwalt Corporation, and Van Nostrand Reinhold Publishing Co.)

6. Open-loop with superimposed dosage control. See HC 116 for a description of this simple but most helpful type of control. It is very popular with systems that do not have nighttime attendants.

Of all the systems illustrated, White believes the compound-loop control system is the most popular because of its flexibility and accuracy of control.

Housing Requirements

The importance of the requirements described under this heading cannot be overemphasized. The subjects of space, heating, ventilation, gas mask location, special vents and drains are all discussed in detail in HC 117-123.

Electrical Requirements

These requirements for chlorinators, evaporators and chlorine residual analyzers are described in HC 123-126. Figure 2.53, HC 124-125, shows a typical arrangement of a chlorinator, evaporator and analyzer with appropriate accessories and alarms.

Air Requirements

If a pneumatic system is used, the four general air requirements are covered in HC 126.

Accessory Equipment

Chlorine Supply System

The most important pieces of accessory equipment in the supply system are discussed in HC 126. These include: the external chlorine pressure reducing valve (CPRV) and the chlorine gas filter. Whenever an external CPRV is used, it should be preceded by a filter (see Figures 2.11 and 2.13, HC 50 and 52).

The designer should also provide enough gauges so that the operator can easily determine cylinder pressure, reduced pressure and pressure in isolated sections of the supply line. *Never use bushings anywhere in chlorine pressure piping.*

Evaporators

When the gas withdrawal rate from any group of containers exceeds 400 lb/day per cylinder connected, the system has to be converted to liquid withdrawal with evaporators preceding the chlorinators. A comprehensive discussion on evaporators is found in HC 127.

Chlorinator Alarms

In addition to the specific chlorinator alarms described in HC 128, the designer should consider an additional supply pressure switch in the chlorine header system if a handling system involves more than 3-ton cylinders.

Chlorine Flow Recorders

An important accessory for any residual control installation, these are described in HC 128. A chlorine flow recorder allows the operator to analyze a malfunction in the chlorination system, particularly during a time when it was unattended.

Chlorine Residual Analyzers

This accessory device, which has now become just as important as the chlorinator itself, is discussed comprehensively in HC 128-132. The history of its development, theory of operation and practical application are thoroughly described. The details of an amperometric cell are shown in Figure 2.54, and the methods of calibrating this instrument are described in HC 170.

Diffusers

Several types of diffusers are illustrated in HC 134-137. The importance of any chlorine diffuser is its ability to disperse the chlorine solution as quickly as possible. In closed pipes this means injection at the center of the pipe. Open channels are another matter, but good mixing by the diffuser itself cannot be expected.

Mixing

Various types of mixing devices are discussed in HC 136-139. In water treatment practice, mixing is not nearly as critical a consideration as it is in wastewater or water reclamation practice because the importance of proper mixing is a function of the relative quality of water treated. The poorer the quality, the more important mixing becomes.

The hydraulic jump in an open channel is an excellent mixing device in potable water chlorination. Open-channel devices are illustrated in Figures 2.62 and 2.63.

The turbine type mixer (Figure 2.64) is applicable in certain situations where velocities are too low to produce turbulent flow or where residual control is critical due to wide variations in flow which extends loop time beyond practical limitations.

Sampling

Sampling (HC 139) is as important as mixing for residual control measurement systems. Collecting a sample from a pipeline should be by a tube which draws from the middle third of the diameter of the pipe and not at the side of the pipe. Mixing is insufficient at the pipe wall and erratic readings result, even at 30 to 40 pipe diameters downstream from the point of application. Although the analyzer needs a sample flow of only one gal/min, it is better to select this from a larger sample flow. Hence, it is desirable to install the injector water pumps so that both the injector water and sample water requirements are combined. All sample lines should be designed so they can be purged of slime growths.

Contact Time

Adequate residence time for a chlorine residual in a potable water is readily achieved both in long transmission lines and treatment plant processes, which include clearwells and reservoirs. This is not true in small systems where the water is pumped directly into the distribution system. Regardless of the system details, contact times of less than 10 minutes are less than desirable, so the designer is cautioned to examine carefully the provision for proper chlorine residual residence time. Contact time requirements vary with the raw water quality of the water to be treated. Contact time should be consistent with those required for enteric virus inactivation.

Safety Equipment

It is difficult to say which is the most important safety device. First, there should be something to alert the operator that there is a malfunction of some sort. For safety considerations, a chlorine leak is the most hazardous situation, so a leak detector could be the most important piece of safety equipment. Chlorine leak detectors are described in HC 132, but since that was written there have been significant improvements in chlorine leak detectors, mostly toward simplicity. Therefore, Wallace & Tiernan, Fischer & Porter and Capitol Controls have developed new, reliable, inexpensive chlorine leak detectors. These are imperative for all chlorinator installations.

Equally important is the gas mask. Once the operator knows there is a leak, he can prepare himself by donning an appropriate type of breathing apparatus. These are described in HC 133.

For leaking chlorine containers, the operator should have an emergency kit for the type of container used, and each and every chlorinator station should have access to one of these kits (described in HC 133-134) for immediate use. The care and monitoring of chlorine leaks by operators is discussed in detail in HC 172 ff.

Alarms

HC 141 presents a check list of the most widely used alarm functions of a chlorination station. The mechanics of sounding these alarms has already been described. Figure 2.53 (HC 124) shows a typical alarm schematic. The possible alarm functions are:

1. high-low chlorine vacuum (vacuum switch on chlorinator indicates loss of either chlorine supply or injector water pressure),
2. evaporators
   a. low water temperature
   b. low water level
   c. high water temperature
   d. heaters on (indicating light)
3. low chlorine supply pressure, either gas or liquid
4. high pressure in liquid chlorine line
5. low and/or high chlorine residual
6. chlorine flow, low or high
7. low injector water pressure
8. chlorine leak.

Equipment Costs

Gas chlorinators are usually priced according to size and the capability of some type of proportional control. The price break for sizes usually occurs at 100 lb/day, 500 lb/day, 2000 lb/day and 8000 lb/day.

Manual and/or Start-Stop Automatic Control

The most popular model for this service is the ubiquitous wall-mounted chlorinator. It is rugged and dependable. There are variations of this type of unit that fall into the category of cylinder-mounted units. These types, regardless of manufacturer, are not as good an investment as the wall-mounted units. The following pricing is for wall-mounted units only and does not include installation.

A wall-mounted unit up to 500 lb/day capacity costs about $1800 fob. This unit can be used in any down-the-well or well-discharge application or in any pumped surface water supply. When a booster pump is required, add $250 fob for low-pressure applications and $600 fob when the pressure at the point of application is 60 psi or greater.

Cylinder-mounted units are slightly less in capital cost than the wall-mounted units. These units are available in various sizes up to 500 lb/day. The cylinder-mounted units are merchandized through distributors; therefore they are subject to a flexible "discount" schedule. Consequently, a cylinder-mounted unit for any given capacity can be purchased for about $150-250 less than comparable wall-mounted units. However, the designer is advised to confirm the accessories and services expected with this species of chlorinators with the manufacturer. There are other factors such as performance and maintenance characteristics favoring wall-mounted units over the cylinder-mounted units; these should be investigated by the designer.

Proportional Automatic Control

When considering automatic control, another species of chlorinator is required as compared to the manual control unit. A comparable 500 lb/day unit capable of providing proportional control is available in the manual mode at about $2600 fob. To provide flow-pacing control, add $1100, and for compound-loop control, add another $200 to $300.

Residual Control

Whenever compound-loop or direct residual control is required, a chlorine residual analyzer is required. Analyzers cost about $4500. To this price it is necessary to add the control functions required to complete the compound-loop system; this is another $100 to $1200.

Accessory Items

Every sophisticated chlorine gas installation requires a certain number of pressure alarm switches. These cost $200 to $250 each.

Every chlorinator installation should have a hi-lo vacuum alarm. This costs about $250 and provides the operator with the information that either the chlorine supply has failed or the injector water supply system is insufficient.

Evaporators

These units are required on those systems that must deliver more than 1500 lb/day (680 kg/day) chlorine. Evaporators come in 6000 lb/day (2.7 ton/day) and 8000 lb/day (3.6 ton/day) sizes. The cost for this equipment with all the required accessories is about $5600 to $6000.

Safety Equipment

There are three pieces of safety equipment which must be considered as part of a chlorination installation. One is a chlorine leak detector at $750. Another is the emergency container repair kits for various sizes of chlorine containers. Allow $550 to $650 for these units regardless of the size of the container. The third and probably most important is oxygen or air-breathing apparatus. Allow $600 for each of these units.

Contractor's Costs

All of the above figures are for equipment only. When equipment is to to be purchased and installed by a contractor, there are additional considerations.

The contractor will add to the equipment costs described above the man-hour labor cost for installing the equipment plus the miscellaneous piping and accessories to complete the installation, plus any additional rigging costs such as cranes or other equipment to complete the installation. To these costs the contractor will add about 25% for overhead and profit. Overhead includes rent, taxes, office expenses and managerial time with the vendor and engineers obtaining approval and working out the details of the installation. Chlorination equipment, like other chemical-feed equipment and instrumentation, has a high labor factor. A great part of this is due to problems during start-up and calibration and process acceptance by the client. To put a percentage factor for these costs on to equipment costs is not realistic. The designer is advised to consult his local contractor sources for each job to determine probable installed costs.

OTHER DISINFECTANTS

It is beyond the scope of this chapter to deal with other possible alternatives to the chlorination-dechlorination process. However, a second book by Geo. Clifford White titled *Disinfection of Wastewater and Water for Reuse* will be available from Van Nostrand Reinhold about July 1978. This book includes separate chapters on ozone and chlorine dioxide, and another chapter covers bromine, on-site generation of bromine, bromine chloride, iodine and UV radiation complete with comparative costs and illustrations. It also includes an entire chapter on hypochlorination and on-site generation of chlorine.

Chlorine Dioxide

Chlorine dioxide has been a long-neglected oxidant in the U.S. for a variety of reasons, mostly its cost and uncertain methods of on-site generation. However, it has been used to advantage in western Europe to aid in taste and odor control of low-quality waters. In the past few years (1974-1977), there has been a greatly renewed interest in the use of ClO_2 to replace prechlorination in France, Germany and Italy. The reason for this is twofold: (a) chlorine dioxide does not contribute to the formation of THM and (b) a French company, CIFEC,* claims to have on-site generating apparatus that produces a pure chlorine dioxide solution, thereby eliminating technical problems of quantifying residual measurements.

Very little is known about the superiority of chlorine dioxide over chlorine as an overall disinfectant and much work needs to be done in this country. CIFEC claims some 80 ClO_2 on-site generating installations, so it would be reasonable to expect considerable information from them by the end of 1978.

Three things we do know about ClO_2 are: (a) it is specific for the destruction of phenols, (b) residuals persist longer than chlorine, and (c) it does not react with ammonia nitrogen in concentrations normally found in potable waters or wastewaters.

Readers interested in chlorine dioxide should read HC 596-624.

Ozone

Ozone is undoubtedly the most powerful oxidant used in water treatment practice. It has never caught on in the U.S., Canada or Great Britain, mostly because much higher quality water is found in these areas. Ozone has been used in western Europe with great success as a super-polishing agent for the treatment of low-grade waters heavily contaminated by sewage and industrial wastes. If it were not for ozone, some of the large cities in western Europe that derive their water from the Seine, Rhine and other heavily polluted rivers would not be able to produce palatable water.

---

*Cie Industrielle de Filtration et d'Equipment Chimique S.A., 10, Avenue de la Porte-Molitor, F75016 Paris, France.

Ozone is excellent for taste, odor and color removal. However, nearly all plants using ozone also find it necessary to use chlorine because of its residual qualities. Ozone residuals last for only a few minutes. Ozone is expensive to generate and the equipment needs continuous maintenance. However, it has a definite place in water treatment practices as a polishing agent.

Readers are advised to consult HC 676-713 because it reviews other methods of disinfection.

REFERENCES

1. White, G. C. *Handbook of Chlorination* (New York: Von Nostrand Reinhold Company, 1972).
2. Symons, J. M., *et al.* "National Organics Reconnaissance Survey for Halogenated Organics," *J. Am. Water Works Assoc.* 67:634 (1975).
3. AWWA. *Disinfection Seminar Proceedings,* Annual Conference AWWA, Anaheim, California (May 8, 1977).
4. Symons, J. M. "Interim Treatment Guide for the Control of Chloroform and Other Trihalomethanes," Municipal Environmental Research Lab, EPA Report, Cincinnati, Ohio (June 1976).
5. Griffin, A. E. "Observations on Breakpoint Chlorination," *J. Am. Water Works Assoc.* 32:1187 (1940).
6. Williams, D. B. "The Organic Nitrogen Problem," *J. Am. Water Works Assoc.* 43:837 (1951).
7. Williams, D. B. "Control of Free Residual Chlorine by Ammoniation," *J. Am. Water Works Assoc.* 55:1195 (1963).
8. Williams, D. B. "Elimination of Nitrogen Trichloride in Dechlorination Practices," *J. Am. Water Works Assoc.* 58:248 (1966).
9. AWWA Committee, 8360 D Report. "Disinfecting Water Mains," *J. Am. Water Works Assoc.* 60:1085 (1958).
10. *Chlorination Catalog 1. Rev. 7-71.* Wallace and Tiernan Division, Pennwalt Corporation (1971).

CHAPTER 17

PUMP STATIONS

James R. Wright, P. E.
Project Engineer
Black & Veatch Consulting Engineers
P.O. Box 8405
Kansas City, Missouri 64114

A pump station is a fundamental auxiliary to most water treatment plants. With very few exceptions, water must be pumped either from the source of supply to the treatment works, or from the treatment works into the system, or both. These pumping functions are generally integrated into the overall water treatment plant design. Thus, pumping facilities must be consistent with the treatment plant design concepts.

GENERAL CONSIDERATIONS

Structural and Architectural

The pump station structure should be consistent with overall plant architecture and present a pleasing appearance. Design effort is required to accomplish this objective, but good results more than justify the effort. The structure should be of a substantial type because pump stations usually must serve a long useful life.

Equipment Layout and Arrangement

The layout and arrangement of equipment should facilitate operation and maintenance. Fundamentally this means adequate space, lighting and accessory equipment. These items are expensive, but an initial cost that

397

contributes to good maintenance throughout the life of the equipment is a sound investment.

TYPICAL PUMPING APPLICATIONS

The basic types of pumping associated with most water treatment plants are low-service and high-service pumping. The only exceptions are those few fortunate enough to be situated topographically so that pumping is not required. Low-service pumping conveys untreated water from its source through the treatment works. High-service pumping conveys treated water into the distribution system.

In plants that must provide treated water storage on the suction side of the high-service pumps at an elevation higher than the filter effluent, pumps must be provided to transfer the treated water from the filter effluent to the treated water storage. Transfer pumping might also apply to pumpage between any incremental treatment steps.

Filters may be backwashed by direct pumping from a clearwell located at an elevation below the filters, or from elevated storage that is replenished by separate wash water pumps, transfer pumps, or high-service pumps. Many distribution systems serve areas in which the topography requires more than one pressure plane, or zone. The interzone pumping required to move water from the lower to the high pressure zone is referred to as booster pumping.

There are many auxiliary and secondary pumping activities within a typical water treatment plant such as chemical feed, seal water, cooling water, and basin drain, but these are beyond the scope of this chapter, which considers only water supply and distribution systems.

Figure 1 is a schematic of a water treatment plant and a three pressure zone distribution system. Examples of all the types of pumping listed above are shown on the schematic. The low-service pumps deliver the water from the lake source to the pretreatment units. Flow is by gravity from the pretreatment units through the filters. The filter effluent is pumped by the transfer pumps to the ground-level clearwells. Filter backwash is by gravity from the clearwell. The high-service pumps deliver water to the first pressure zone, which has ground-level storage at overflow elevation 1165. Booster pumping station Number 1 delivers water from Zone 1 to Zone 2, which has an elevated storage tank at overflow elevation 1315. Booster pumping station Number 2 delivers water from Zone 2 to Zone 3, which operates at a hydraulic gradient of 1465. Schematics of the four types of pumping stations depicted in Figure 1 are shown in Figures 2, 3 and 4. A complete discussion can be found later in the chapter.

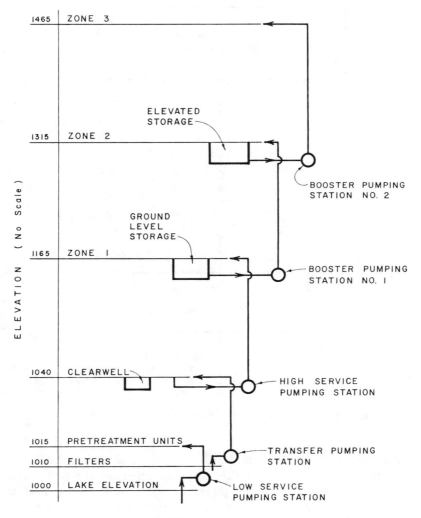

Figure 1. Water system schematic.

DESIGN CONDITIONS

Determination of Design Flows.

Perhaps the most perplexing question facing an inexperienced designer when approaching a pump station design is "where to start?" In almost any hydraulic engineering problem, once some of the factors are determined,

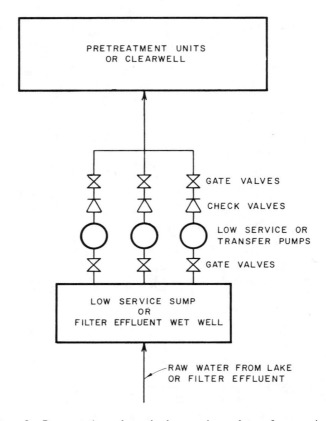

Figure 2. Pump station schematic; low-service and transfer pumping.

the rest fall into a mathematically fixed relationship. The difficult first step, then, often becomes: which factors to determine and how. Design flows, an obvious first step, are generally determined from statistical data guided by judgment. Using data and judgment, maximum, average and minimum daily flow rates can be predicted.

One reasonable approach to water treatment plant design is to provide treatment plant capacity to meet the maximum daily flow rate for the design period, and supply peak hourly demands from storage. If a plant is designed on this concept, the low-service and transfer pumping capacity should be equal to the maximum daily flow rate with an additional flow to supply plant operations such as backwash. Average and minimum daily pumping rates may be provided by use of multiple pumping units, variable speed units, throttling or a combination of these.

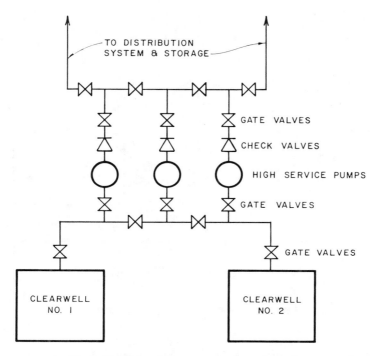

Figure 3. High-service pump station schematic.

The concept of firm capacity is basic to good design. Depending on the degree to which uninterrupted service is critical, judgment can be made about the standby capacity required. Generally, in small- and medium-sized water systems, providing firm design capacity with the largest unit out of service gives a reasonable degree of reliability.

Head Capacity Curves

In the following discussion, terms and definitions follow the Hydraulic Institute Standards.[1] A pumping station should have capacity to meet the design maximum flow. Consideration must also be given to average and minimum flows. It is, therefore, very important for the designer to think of a pump station in terms of a range of flow conditions rather than simply of a rated capacity.

The inherent relationship between a water treatment plant and the distribution system it serves cannot be ignored. For example, if the high-service pump suction storage on the distribution system storage is small,

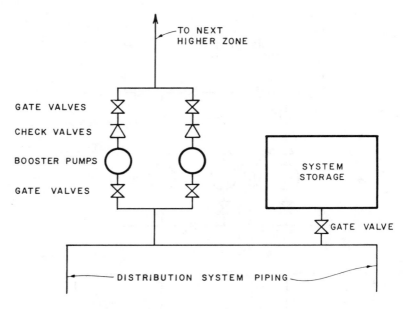

TO NEXT
HIGHER ZONE

GATE VALVES

CHECK VALVES

BOOSTER PUMPS

GATE VALVES

SYSTEM
STORAGE

GATE VALVE

DISTRIBUTION SYSTEM PIPING

Figure 4. Booster pump station schematic.

the treatment works must be designed to meet all conditions of demand
on the system. On the other hand, if sufficient storage is provided, the
treatment works can be operated at a predetermined constant rate and
system demand variations can be met by the fluctuating volume of water
in the system storage. Most systems operate somewhere between the two
extremes mentioned. This chapter is based on the previously stated
assumption that treatment works capacity is equal to the maximum daily
demand, and short-period extremes are being met from storage.

From a knowledge of the details of the distribution system and
standard hydraulic analyses, system curves can be prepared. A minimum
of four should be considered. The curves should be plotted starting with
the maximum and minimum static head. For low-service pumps discharging into a basin with a fixed effluent weir elevation, the two starting
points would be determined by the maximum and minimum suction water
level. In a similar manner these two starting points can be determined
for the other pumping applications of a water treatment plant by considering maximum and minimum static head conditions.

The range of friction losses for the design life of the pump station
should also be considered. Four system curves are shown in Figure 5,
with the maximum and minimum friction losses represented by a
Williams-Hazen coefficient of C = 100 and C = 140, respectively.

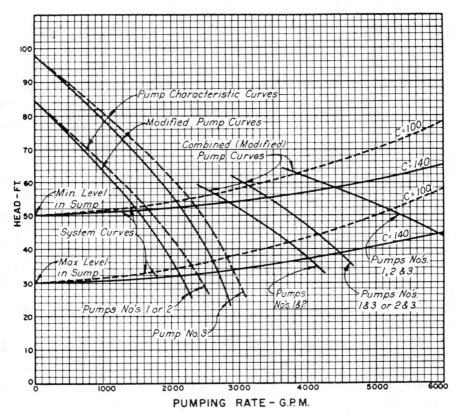

Figure 5. Typical head-capacity curves.

Keeping in mind the previously discussed concept of the range of flow conditions the pump station must meet, the number and capacities of pumps can be selected. When the pump selection is made, the pump characteristic curves can be superimposed on the system curves.

If two or more pumps discharge into a common header, it is usually advantageous to omit the head losses in individual suction and discharge lines from the system head-capacity curves.[2] This is advisable because the pumping capacity of each unit varies depending upon the units in operation. In order to obtain a true picture of the output from a multiple pump installation, *it is better to deduct the individual suction and discharge losses from the pump characteristics curve.* This provides a modified curve, which represents pump performance *at the point of connection to the discharge header.* Multiple pump performance can be determined by adding the capacity for points of equal head from the modified curve.

Figure 5 shows the pump characteristic curves, modified pump curves, and combined modified pump curves for multiple-pump operation. Intersection of the modified individual and combined pump curves with the system curves shows total discharge capacity for each of the several possible pumping combinations.

When the tentative pump selections have been made to meet head and capacity requirements, several factors should be verified. The pump efficiency should be near its maximum at the average operating condition. The power requirement at minimum operating head should also be considered to ensure that it is not excessive with respect to the other operating conditions. Other items are discussed under "Miscellaneous Considerations."

PUMP SELECTION

Classification of Pumps

A logical starting point for pump selection is a review of the types of pumping equipment available. Figure 6 is a Classification of Pumps from the Hydraulic Institute Standards.[1]

The positive displacement pump is the oldest type. Some were used before the time of Christ. The Archimedian Screw was first used for pumping irrigation water in Egypt about 200 BC.

With the advent of steam power, the reciprocating pump became widely used. Many years ago the steam-powered piston pump was the only type for municipal water supply systems. The City of Memphis, Tennessee, has only recently phased out the last of their steam-powered piston pumps. In Montana, the history of the Monster Pump of the Big Hole River Water Works of the Butte Water Company is well known. That installation, put into operation in 1902, was designated an American Water Landmark by the American Water Works Association a few years ago.

While the reciprocating pump was ideally suited for steam drive, the universal development of electric power led to the increasing use of centrifugal pumps. The development of the electric motor permitted use of much lighter and cheaper centrifugal pumps, which give steady flow at uniform pressures. It also provides the greatest possible flexibility, developing a specific head-capacity relationship at any design speed. Delivery can also be controlled by throttling. Centrifugal pumps have been built in sizes ranging from a few gpm to over 600,000 gpm

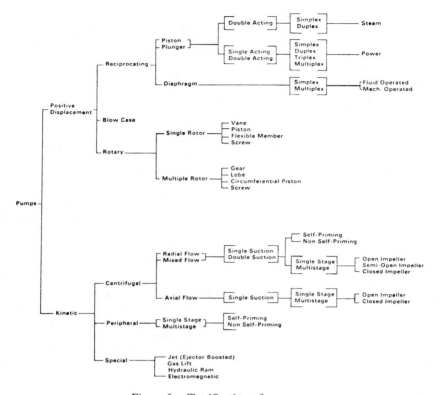

Figure 6. Classification of pumps.

(38,000 l/sec) and with head ranges of a few feet to over 6,000 psi (42,000 kN/m²).

Centrifugal Pumps

Only centrifugal pumps are discussed herein because they are the most commonly used for water works pumping. As shown in Figure 6, centrifugal pumps are classified as radial flow, mixed flow and axial flow. The radial-flow pump is one in which pressure is developed principally by the action of centrifugal force. The mixed-flow pump is one in which the head is developed partly by centrifugal force and partly by the lift of the impeller vanes on the liquid. An axial-flow pump, sometimes called a propeller pump, develops most of its head by the propelling or lifting action of the impeller vanes on the liquid.

Table I. Relative Characteristics of Centrifugal Pumps[2]

Characteristics	Axial Flow	Mixed Flow	Radial Flow
Usual Capacity Range	Above 10,000 gpm	Above 5,000 gpm	Any
Head Range	0 to 40 feet	25 to 100 feet	Any
Shutoff Head Above Rated Head (maximum efficiency point)	About 200%	160%	120 to 140%
Horsepower Characteristic	Decreases with capacity	Flat	Increases with capacity
Suction Lift	Usually requires submergence	Usually requires submergence (lift limited)	Usually not over 15 feet
Specific Speed	8,000 to 16,000	4,200 to 9,000	Below 4200— single suction Below 6000— double suction
Service	Used where space and cost are considerations and load factor is low	Used where load factor is high and for service where trash and other solid material are encountered	Used where load factor is high and for high efficiency and ease of maintenance.

Relative Characteristics of Centrifugal Pumps

Table I compares the relative characteristics of the three types of centrifugal pumps. This table provides a good general guide for the preliminary selection of pump type for a given service.

Specific speed is defined as that speed in revolutions per minute (rpm) at which a given impeller would operate if reduced proportionately in size to deliver a capacity of one gpm against a total head of one foot. This number is used to classify pump types and relates physical shape to performance characteristics. The mathematical definition of specific speed is:

$$N_s = \frac{RQ^{\frac{1}{2}}}{H^{\frac{3}{4}}} \tag{1}$$

where R is rpm, Q is gpm, and H is head in ft.

Referring to Figures 1 through 4 and Table I, selection of a pump for the various pumping units required might be made as follows. Axial-flow pumps might be selected for the low-service pumps. They could take suction directly from the low-service sump thus avoiding the cost of a separate dry pit and placing the motors above flood level. In warm climates they might be exposed on the top of the low-service sump with weatherproof motors and controls. In colder climates, however, a superstructure to house the pumps is required. As can be seen from Figure 1, the required pumping head is low and within the normal head range as shown in Table I and the service requirements are well-suited to raw water pumping.

Mixed-flow pumps might be selected for the transfer pumps since the head requirements are low and the space-saving features are important as these pumps are ordinarily installed within or adjacent to the filter building, which is expensive construction. The pump curve is fairly steep (shutoff head about 160% of rated head) and the horsepower curve is flat, which provides a certain automatic flexibility, as the relative levels in the suction sump and clearwell vary.

The high-service and booster pumps would logically be radial-flow pumps because of the required discharge heads. Ease of maintenance is especially important in the booster stations, which are usually remote from the treatment works.

Controls

Pumps can be controlled by level, pressure or flow. Level controls can start and stop pumps at predetermined water levels in storage reservoirs or treatment basins. Controls for variable-speed pumps can be used to vary the pump discharge to maintain a predetermined level, pressure or flow.

Flow control is used to meet a fluctuating demand by varying the speed of the pumping unit. Sometimes flow control is provided by throttling. However, it is usually more economical to vary the speed of the pump. Speed variation can be provided by magnetic couplings, hydraulic couplings, wound rotor motors and liquid rheostats, and variable frequency and voltage controllers. Speed adjustments can be manual or automatic.

Emergency controls are essential to good pump station design. Low-level shutoff should be provided in sumps and low-pressure shutoff in suction lines to prevent damage to pumping units by running them dry. High-level alarms should be provided in basins or reservoirs to indicate malfunctions and abnormal conditions.

A control scheme for the water system shown on Figure 1 might be as follows. The low-service and transfer pumps could be variable-speed units manually controlled. Assuming a reasonably large clearwell volume, however, sufficient flexibility could be provided with multiple constant-speed units as shown in Figure 2. The plant operator could start and stop low-service and transfer pumps as desired by observing clearwell level.

Low-service pumping is not automatically controlled because of the difficulty of automatic chemical-dosage control. The plant operator normally predetermines a treatment rate and then monitors treated water storage level and makes rate changes accordingly. The clearwell should have controls to shut the filters down at approaching high level and to sound an alarm that alerts the operator so that he can decrease or stop low-service pumpage. Obviously, the low-service and transfer pump capacities and filter rates must be coordinated. The low-service and transfer pumps and the treatment plant should be protected by appropriate low- and high-level shutoff.

Assuming the Zone 1 storage volume is reasonably large, the high-service pumps could be operated by level control in the reservoir. Varying demand could be met by multiple constant-speed units or by variable-speed units. If the Zone 1 storage volume were relatively small, it might be preferable to provide flow control for the high-service pumps by a flow meter in the discharge line that would start and stop constant-speed pumps and/or increase and decrease the speed of variable-speed pumps.

Booster pumping station Number 1 would operate very much like the high-service pumping station. The simplest control, however, would be by level control from the Zone 2 elevated storage. Booster pumping station Number 2 would probably best be controlled by flow. Since there is no storage on Zone 3, the entire flow range must be met by the pumps, which would probably require variable-speed units. The system could be controlled by pressure, but such systems are extremely sensitive and, unless the flow is very uniform, pressure control would probably not prove satisfactory.

The designer should always review his system to try to envision all possible emergencies and malfunctions and provide alarms and safety controls to prevent this occurrence from resulting in damage or loss.

Drives

The universal availability of electric power has led to the use of electric motors for most pumping units. However, gasoline, gas and diesel units are still used where the reliability of the electric service is questionable, or where special considerations dictate.

The choice between vertical- and horizontal-drive motors depends on the design conditions of the particular application. Horizontal pumps are usually easier to maintain and lower in first cost, but vertical pumps require less space and the motors can be located above flood potential in dry pit applications. Therefore, the designer's selection of vertical or horizontal drives requires careful consideration of many factors.

Miscellaneous Considerations

There are several miscellaneous items that should be considered in pump station design.

Cavitation

In the handling of incompressible liquids, all pumps have a tendency to cavitate when the total pressure energy in the pump inlet regions reaches the vapor pressure of the liquid being pumped. Each liquid has its own vapor pressure, which varies with its temperature. For example, the vapor pressure of water is 0.6 ft of water column at 60°F. Water boils when the vapor pressure is equal to the ambient pressure. Thus, whenever the absolute pressure in a pump and piping system is less than the vapor pressure, the water boils.

Normally, in a well-designed pump and piping system the lowest pressure will occur in the pump impeller inlet. If the absolute pressure in the inlet approaches the vapor pressure, the liquid boils, and the bubbles or cavities occupy part of the water passages, thus reducing the pump capacity. But, more important, the collapse of the cavities as they move into higher-pressure regions in the pump is accompanied by an exchange of energy, which causes noise, vibration and disintegration of waterway surfaces. Therefore, when making a pump selection, it is necessary to avoid cavitation. The pump must be located so there is sufficient pressure above the vapor pressure in the pump inlet to assure freedom from cavitation.

Manufacturer's pump performance charts show the net positive suction head required (NPSH). To avoid cavitation problems, careful attention must be paid to this requirement in designing pump suction or sub-mergence conditions. If there is any doubt after reviewing the performance charts, the manufacturer should be consulted.

Sump design

Care must be exercised in the design of sumps to avoid localized velocities that might cause vortex formation. Vortices can reduce capacity and cause noise, vibration and possible damage to the pump. The Hydraulic Institute Standards[1] contain recommendations on sump design.

Piping

Suction and discharge piping should be sized so that velocities are not excessive. Velocities of 5 ft/sec (1.5 m/sec) in suction piping and 8 ft/sec

(2.4 m/sec) in discharge piping are reasonable maximums. Piping should have sufficient flexibility and be adequately supported so that no stresses are transmitted to the pump. Expansion joints or couplings that do not provide an axially restrained connection should not be used between the pump and a point of anchorage in the piping. Such an installation causes the hydraulic reaction of the pump to be carried by the pump, pump base and anchor bolts. This force could be of a magnitude to govern the structural design of the pump, and could make the construction economically infeasible.

Valves

Valves should be installed in the suction and discharge of the pumps to permit isolating the pump for maintenance and to control flow. The type of check valve required depends on the piping system into which the pump discharges. Ordinary swing check valves may be adequate but in many systems power-operated stop-check valves are needed to control surge or water hammer. If the pump is to be started and stopped against closed valves to control transient surge pressures, power-operated butterfly or ball valves interlocked with the pump start-stop controls can be used. Auxiliary power in addition to the primary power source is needed for such installation in the event of power failure.

Arrangement

Figures 2, 3 and 4 illustrate several significant features of piping and valve arrangements. For example, Figure 2 shows an arrangement that permits isolation of any one of the three pumps for maintenance by closing the gate valves on the suction and discharge sides of the pumps. The check valves in the discharge lines from each pump prevent backflow through an idle pump.

Figure 3 illustrates several examples of design to facilitate operation and maintenance. Providing two clearwells with gate valves for isolation permits draining either of them for cleaning, repair, or maintenance. A suction header is provided so that all of the pumps may take suction from either or both of the clearwells. Valves permit isolation of any of the pumps for removal or repair, and check valves prevent backflow through an idle pump. The pumps discharge into a header that delivers water into the distribution system at two points. This permits isolation in the event of a line break in the distribution system (assuming proper valving within the distribution system) without disrupting service except in the immediate area of the line break.

Figure 4 depicts many of the previously discussed principles. The storage is arranged so that the reservoir can be isolated for repair or maintenance without disrupting pumping station service. When the reservoir is in service, the distribution system piping serves as pump suction when the system storage is being drawn down. The pump suction and discharge lines are valved and headered so that service can be provided by either pump while the other is out of service.

SPECIFICATIONS

All of the design effort that goes into a pump station is for naught unless it results in a physical installation that satisfies the design conditions. Practically all of the water treatment plant pumping equipment in the United States is furnished by competitive bidding, whether for public or private water systems. Therefore, it is imperative that the specifications upon whcih the bidders base their prices convey adequate information to meet the design requirements. A discussion of the minimum specifications follows.

1. Service and Installation Conditions. The intended service and general conditions of the installation should be clearly stated.

2. Performance and Design Requirements. The most significant aspect of the specification to the pump manufacturer is the performance and design requirements. The information should include at least the following:

- rated total head
- capacity at rated head
- maximum total head
- minimum capacity at maximum head
- normal operating head range
- minimum head for continuous operation
- maximum pump speed
- maximum brake horsepower for any head above minimum for continuous operation
- minimum efficiency, stated "for pump only" or "for pump and driver"
- available net positive suction head conditions
- pump rotation, if appropriate
- nominal size of suction and/or discharge flanges
- pump test pressure
- for vertical pumps: nominal column size and column and shaft dimensions

In addition to the foregoing applicable items, specifications for variable speed units should also include:

- minimum capacity
- electric motor type
- type of adjustable speed drive
- type of speed controller

A considerable portion of the data under this heading comes from the head-capacity curves discussed previously, which reemphasizes the importance of their careful preparation.

3. Materials and Construction. The cost of the manufactured pump is affected most by this portion of the specification. Accordingly, great care should be exercised in its preparation.

One of the most vexing problems that faces a designer-specifier is having to judge equality of competitive equipment items. In order to avoid the myriad of potential problems in this regard, the specifier should know what he wants, specify it explicitly, and accept nothing of lesser quality.

The specifier should indicate the type of construction desired for the basic pump components such as the casing, impeller, shaft, shaft sleeves, wearing rings, stuffing box, bearings, driver-to-pump coupling, and any pertinent accessory items. In addition, the materials of construction should also be indicated because the cost of the finished product is very sensitive to the materials required. For example, if conditions require stainless steel shafting instead of carbon steel, the effect on the cost of the unit is obvious and significant.

Naming certain manufacturers and model numbers as an example of type and quality desired is a reasonable and often useful procedure in preparing specifications; however, it must be realized that this is not a substitute for the explicit indication of the items discussed above. In the competitive free enterprise system of this country, each manufacturer seeks his own methods of finding a manufacturing and sales advantage. This is the basis on which our system works, but the economic facts are that there is usually relatively little cost difference in items of equivalent quality. The problems of a designer usually arise as a result of compromising quality, and the best defense against this problem is a clear and explicit specification.

4. Drives. Complete details of the drives should be given, including the speed controllers, if applicable. The data should include service factor, temperature rise, type of construction, and details of the power source. Information regarding the speed controller should include the type required, component details and dimensional limitations, if any.

OSHA requirements regarding noise levels should be reviewed and, if appropriate, a decibel level specified.

5. Tests. The required shop tests should be indicated. These tests should be performed and certified prior to shipment of the units. As a minimum, these tests should include capacity, power requirement and efficiency at critical heads. For very large pumps or unusual conditions it may be desirable to specify witness tests. This means that the engineer and owner must be notified sufficiently in advance of the shop testing so that they may make arrangements to be present and witness the shop testing. The necessity or desirability of witness testing is a matter of judgment to be determined by the designer. If appropriate, a field testing requirement can also be included to verify performance in the system.

6. Installation. Proper installation of pumping units is essential to a successfully operating pump station. Leveling, alignment, piping connections and grouting can all have a very significant effect on pump operation and useful life of moving parts. The Hydraulic Institute Standards[1] are a good source of information on pump installation. Adequate inspection during construction is vital to proper pump installation.

COST DATA

Figure 7 presents cost data for water pumping stations. These data are presented as a general guideline only. Each pump station has its unique features, each of which must be considered in estimating its cost. Different types of pumping stations obviously cost different amounts. For example, a wetwell type pump station would have a much greater structural cost than a pipeline booster station. See Chapter 30 also.

Curve A on Figure 7 presents cost approximations for pump stations that have relatively simple structures. Curve B presents cost approximations for pump stations with relatively complex structures.

The cost estimates include the structure, pumping equipment, piping, electrical equipment and controls and ordinary appurtenances. The cost excludes rock excavation, dewatering the site, and other nonstandard conditions. Obviously many details can significantly affect the pump station cost: the type and degree of instrumentation and control, esthetic considerations and the type of structure. The cost data in Figure 7 should be used with caution and be tempered by the judgment of the designer.

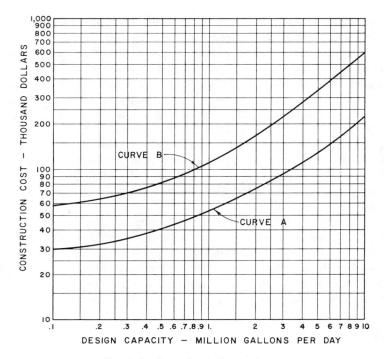

Figure 7. Cost of pumping stations.

ACKNOWLEDGMENT

Grateful appreciation is expressed to Mr. Henry H. Benjes, Engineering Manager, Civil-Environmental Division, Black & Veatch, and to Mr. R. R. Langteau, pump application specialist, Civil-Environmental Division, Black & Veatch, for assistance, guidance and data used in preparation of this chapter.

REFERENCES

1. *Standards of Hydraulic Institute,* 13th ed. (Cleveland, Ohio: Hydraulic Institute, 1975).
2. Benjes, H. H. "Sewage Pumping," *J. San. Eng. Div., Proc. Am. Soc. Civil Eng.* Proc. Paper 1665 (June 1958).

CHAPTER 18

PACKAGE WATER TREATMENT PLANTS

S. P. Hansen

Senior Engineer
Culp-Wesner-Culp
3939 Cambridge Road
Shingle Springs, California 95682

W. R. Conley, P.E.

Vice President
Research and Technical Services
Neptune Microfloc, Incorporated
1965 Airport Road
Corvallis, Oregon 97330

INTRODUCTION

Package water treatment plants have played an important role in the development of safe and adequate water supplies for many small U.S. communites. In the coming years more extensive applications of package equipment will be made, stimulated by rising construction costs that preclude the use of custom-designed treatment facilities. The package plant is factory built in a more efficient atmosphere, where costs can be more closely controlled which results in a more economical facility. Depending upon the package plant design, minimal on-site assembly and installation work is required, which further amplifies the economies realized with the package plant concept.

Package water treatment plants have been manufactured for many years. Treatment concepts employed in these package plants have, by and large, paralleled standard recognized water treatment plant design criteria. Approximately ten years ago the authors' company reviewed the package plant market and identified a need for an improved product of greater simplicity and better reliability. Design criteria were developed under the premise that they must be so simple that "Mrs Jones, the Girl Scout leader," could operate the equipment and be assured that a quality product would be produced at all times.

Technological advances (tube settling and mixed–media filtration) are utilized to produce a very compact package water treatment unit. With shallow-depth tube settlers, highly efficient clarification at detention times ranging from 10-15 min can be realized. Such a system has an obvious size-reduction effect on the clarification portion of the plant. Extensive research work, which led to the development and subsequent practical application of the shallow–depth tube settling concept, has been presented in detail elsewhere.[1,2,3]

Since filtration is one of the most important processes in water treatment, careful attention was directed to selecting a filter system that would provide a highly clarified, safe, and palatable finished product. The authors' company had previously developed a mixed-media filter utilizing anthracite coal, silica sand and garnet sand that had the established ability to operate safely at filter rates of 5 gpm/ft^2 (0.204 m^3/m^2/min) which produced filtered water of extremely high clarity. In-depth discussions of mixed-media filtration theory and practical applications can be found in the literature.[4,5] The established advantages of mixed-media made it an ideal selection for use in a package treatment plant where final product quality assurance is a requirement.

Design criteria developed from bench scale and pilot tests were used to construct a 20 gpm (1.26 l/sec) experimental unit. Extensive testing and equipment component qualifications established that the preliminary design criteria were sound. In-depth description of this work can also be found in the literature.[2] Several small design changes were made in the prototype to incorporate improvements uncovered in experimental unit testing. The prototype was then subjected to extensive testing on a wide variety of water supplies. Upon the completion of this testing, which was found to be highly successful, the equipment design was frozen and incorporated into production units. Designs were developed for units ranging in capacity from 14,000 gpd (0.61 l/sec) to 2.000,000 gpd (88.33 l/sec). The smaller units were named "WATER BOY," and the larger units were named "Modular Aquarius." Since the introduction of these package water treatment plants in 1965, over 250 units have been installed.

TREATMENT PROCESS DESCRIPTION

Both package treatment plants utilize a complete treatment process consisting of chemical coagulation, variable-speed mechanical flocculation, shallow-depth tube settling, and mixed-media filtration. WATER BOY package plants range in size from 10-100 gpm (0.61-6.08 l/sec) and have a flow process diagram as illustrated in Figure 1. Figure 2 is an illustration of the Modular Aquarius flow diagram, which uses the same process but consists of two modules ranging in capacity from 200-1400 gpm (12.2-88.3 l/sec).

In both systems raw water either flows by gravity or is pumped to the treatment plant. Influent flow is set at a fixed but adjustable rate with the control system designed to start and stop the treatment plant based on storage tank level, which reflects system demand. Coagulant and disinfecting chemicals are added upstream of the influent control valve. In-line mixing from the point of application to the plant flocculator is sufficient to disperse treatment chemicals completely with the incoming raw water.

The chemically treated water then passes into a mechanical flocculator designed to form a readily settleable floc rapidly. Flocculated water is then uniformly distributed over the inlet face of a bank of tube settlers. Tube settlers having an overflow rate based upon equivalent settling surface area of less than 150 gpd/ft^2 (9.17 m^3/m^2/day) provide for complete and effective removal of the suspended flocculated material.

After passing through the tube settlers, the highly clarified water is distributed to the gravity, mixed-media filters. The filters are designed with a pipe lateral underdrain system and operate at a constant rate. In the small plants, flow rate control is accomplished through the use of a low head filter effluent transfer pump, which discharges through a float-operated valve. The larger plants use a pneumatic signal from the filter to position the filter effluent rate control valve. In both systems, there is no change in filter rate throughout the entire filtration cycle once the plant raw water flow rate has been set. When the filter rate is controlled with a valve that reacts to filter level, the flow out of the filters is always precisely balanced to incoming raw water flow. Changes in filter rate are accomplished slowly, which prevents potential turbidity breakthrough conditions commonly caused by filter surging.

Filter backwash is initiated automatically once a preset filter head loss is reached. Manual backwash can be initiated at the discretion of the plant operator. During backwash, settled material accumulated in the tube settlers is automatically removed from the unit. By coordinating sludge removal from the tube settlers with backwashing of the filter, the need for operator attention for sludge removal from a clarifier has been eliminated. The advantage of this system over conventional package plant clarifiers is that no operator judgment is required to determine how often and how much sludge must be wasted.

Positive cleaning of the tube settlers is provided by the free-falling water surface, which scours the settling tubes clean. After the compartment

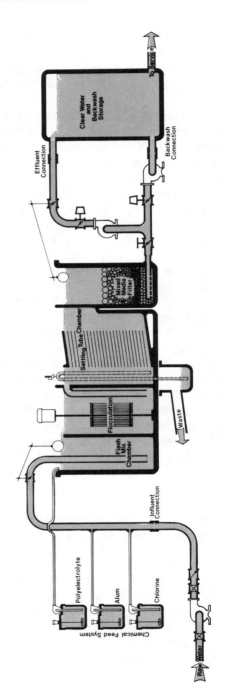

Figure 1. WATER BOY process flow diagram.

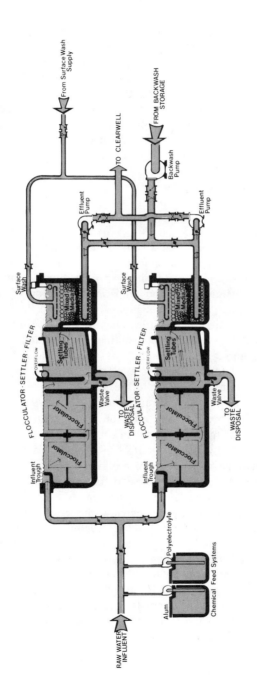

Figure 2. Modular Aquarius process flow diagram.

containing the tube settlers is completely drained, any remaining material is flushed out by the backwash water, which passes to waste through the tube-settling compartment. The draining technique does not result in any additional water usage over that required only for backwash of the filter. Prior to the end of backwash, the tube settling compartment waste valve is closed and the last portion of the backwash water, which is very low in suspended solids, is used to refill the tube-settling compartment. Upon completion of the backwash cycle, which is controlled by a cam timer programmer, the treatment plant is automatically returned to service.

PROCESS DESIGN DETAILS

Coagulation

For removal of turbidity and color from surface water, an inorganic coagulant, of which the most common is aluminum sulfate, is required to precipitate the suspended material. Depending upon plant size, alum is either prepared by dissolving and preparing a solution of dry aluminum sulfate or it is purchased in bulk liquid form. Bulk liquid alum becomes economically attractive for facilities ranging upward from 0.5-0.75 mgd (22 to 33 l/sec). If units are not equipped with flash mixing and pipeline mixing is employed, the alum must be added at a sufficient distance before the flocculator to ensure complete dissolution and mixing with the incoming raw water by pipeline turbulence.

In many surface waters the raw water alkalinity is insufficient to maintain the proper pH control and must be supplemented with artificial alkalinity. Soda ash, lime, sodium bicarbonate and liquid caustic have been used successfully for pH control in package plants. The choice of a particular material depends upon proximity to supplies, plant size and user preference. Soda ash is preferred for small plants; caustic is often an economical choice in larger package plants.

In package plants employing high rate processes for settling and filtration, a polyelectrolyte coagulant aid is usually added to the raw water to ensure production of a readily settling floc and a highly filterable floc as well. The polyelectrolyte dosages vary depending upon the chemical and physical characteristics of the raw water supply. Dosages can range from zero to as much as 0.5 mg/l when applied as a coagulant aid.

On units larger than 200 gpm (12 l/sec) provisions are also made to feed polymer prior to the filters to supplement polymer feed as a coagulant aid. Dosages range from 0.01-0.05 mg/l as a general rule.

Flocculation

In a package plant that utilizes high-rate sedimentation and filtration techniques, the flocculation process is extremely important. The short residence in the tube settlers does not allow time for additional flocculation to occur by particle contact. Early tube-settling studies showed that in order to realize optimum settling efficiencies the floc must be well formed and settle at a rate of at least 0.25-0.5 in./min. Close attention is given to the design of flocculation facilities in both small and large plants to satisfy these requirements. All models utilize tapered agitation mechanical flocculation. The 10-gpm (0.61 l/sec) through the 100-gpm (6.1 l/sec) units use fixed-speed flocculators. The larger plants use variable-speed drive units to provide an infinite range of flocculation intensities and the flocculators are divided into two compartments to minimize short-circuiting. The flocculators are designed to provide mean temporal velocity gradients "G" values ranging from 20 sec^{-1}-75 sec^{-1}. Flocculation detention times range from approximately 8 min in the smallest unit to 20 min in the 2 mgd (88 l/sec) unit.

Tube Settling

The key to the compactness of package plants is the low detention time, highly efficient tube-settling process. The compact tube settler allows development of a 2-mgd (88 l/sec) package plant that is truck transportable. The principles of shallow-depth sedimentaiton, utilizing 1-in. (2.5 cm) deep, 39-in. (99 cm) long tube settlers, allow for a clarification system to be designed with detention times ranging from 10-15 min. - By using tube settlers it is possible to satisfy the clarification requirements in 1/6 to 1/12 of the time required with conventional settling. Although the tube-settling detention time is extremely short, the overflow rate related to available settling surface area provided by the tube settlers is less than 150 gpd/ft^2. When the theory of shallow-depth sedimentation is reviewed, it becomes quite clear that the removal of settleable materials is a function of the overflow rate and settling depth and is independent of the detention time. Hence it follows that the extremely low tube-settling overflow rate should be consistent with a very highly clarified settled water.

Tube settlers used in these package plants are in modular form, such as illustrated in Figure 3. The modules are oriented so that the tube passageways are inclined at approximately 7.5° from the horizontal in the direction of flow. The inclination facilitates removal of collected sludge during draining of the basin containing the tube modules. The falling water surface scours the sludge deposits from the tubes and carries them

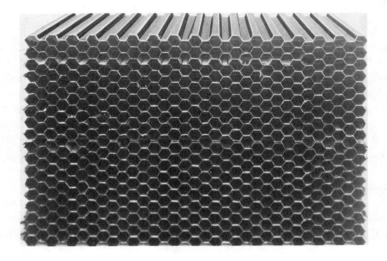

Figure 3. Horizontal settling tube.

to waste. During operation, the majority of the settleable solids is removed in the tube settlers and the rest is removed in the mixed-media filter. During the course of a filtration cycle, material collects within the tube settlers and builds up to the point where an increasing amount of material carries through to the filter which eventually initiates a backwash.

The original design requirements for the tube-settling system were based upon results from small-scale laboratory studies. With these studies there was no way to predict the efficiency level that might be realized on a full scale treatment plant. The goal was to develop a short detention time clarification system that would reduce peak turbidities to a level that could be handled economically by the mixed-media filter. It was found with both experimental and prototype treatment units that the tube-settling system outperformed initial expectations. Flocculated water turbidities of 10-20 TU and higher were routinely reduced to 1-2 TU after passage through the tube settlers.

In actuality, the design basis for the 7.5° tube settlers is related to available sludge storage volume and is not determined by flow-through velocities. Test results at tube flow rates in excess of 2.5 times design values yielded excellent turbidity removal.[1] These plants are designed with a performance-matched tube settler/filter unit wherein the useable sludge storage volume of the tube settler corresponds with the maximum available filtering head. In other words, the entire tube-settling volume is essentially utilized for sludge storage at the same time that the filter reaches the backwash initiation point.

Filtration

Both small and large package plants utilize a 30-in. deep mixed-media filter bed consisting of 18 in. (46.2 cm) of 1.0-1.2-mm effective size anthracite coal, 9 in. (23 cm) of 0.45-0.55-mm effective size silica sand, and 3 in. (8 cm) of 0.25-0.35-mm effective size garnet sand designed so there is a uniform gradation from coarse to fine in the direction of filtration. In all units the design filtration rate is 5 gpm/ft² $(0.20 \text{ m}^3/\text{m}^2/\text{min.})$. The particular filter-bed design coupled with this design flow rate provides an optimum balance between filtered water quality and filter run length.

One particular advantage of the mixed-media filter that makes it ideal for application in package plants is that a good quality water can be produced at even less than optimum chemical treatment conditions. With a coarser dual-media filter, slight departures from the optimum coagulant dosage result in increased filtered water turbidity. Mixed-media, by virtue of using fine garnet in the base of the filter, provides better stability against premature breakthrough or flow surging than the commonly used dual-media filter. The stability aspects allow for start/stop operation of the equipment without any deterioration in finished water quality.

EQUIPMENT APPLICATION

It is exceedingly important that a package plant be properly applied for it to perform successfully. The manufacturer's detailed application guidelines must be followed closely to ensure proper application. Routinely performed laboratory analyses of representative raw water samples also provide critical information, ensuring a proper application. In general, the majority of package plant applications have been on surface supplies where turbidity and color removal are the principle treatment requirements.

In any given application the characteristics of the raw water dictate the allowable rating for the particular package unit. Such parameters as raw water temperature, the amount of suspended matter and color content influence plant ratings. Both small and large package plants accommodate raw water turbidities up to 400 TU and/or alum dosages of 50 mg/l at water temperatures down to 50° F. At higher turbidity levels and/or alum dosages and at temperatures below 50°, it is suggested that the design capacity of the treatment plant be downrated. On the other hand, if the quality of the raw water is extremely good, consideration can be given to operating the treatment plants above their nominal design capacity.

There are also numerous ground water treatment applications of package units for iron and manganese removal.

Treatment Experiences

It is extremely difficult to acquire complete and accurate operating data for many small water treatment systems. (Similar problems occur with larger installations.) It is axiomatic that a manufacturer never hears about facilities that are operating properly, but is quick to learn of one that performs unsatisfactorily.

A considerable amount of operating data was accumulated from a 20-gpm (1.3 l/sec) trailer-mounted WATER BOY during the summer of 1966.[1] Table I gives the raw water characteristics at the various test locations. Table II is a summary of operating results. The data show that there were no difficulties in efficiently and economically producing a high-quality water with the package plant on all supplies studied.

Recently, operating information on package treatment plants was updated by a survey of nine existing facilities in Texas and Kansas during February of 1976. A summary of operating results is presented in Table III. With one particular exception, all treatment plants produced high-clarity filtered water with a turbidity of less than 1 TU. Most of the facilities were operated in a conscientious and dedicated manner and were well maintained. Several appeared to have had little or no maintenance since installation and, although disreputable in appearance, they were producing a filtered water of considerably less turbidity than the proposed Safe Water Drinking Act guideline of 1 TU. All the operators were highly complimentary of the equipment and were impressed by its reliability and simplicity. It is indeed fortunate that the treatment process and equipment employed in the WATER BOY and Aquarius is very forgiving. Even with minimal operator attention and maintenance the units yield excellent performance.

The City of Estacada, Oregon, installed a 1-mgd (88 l/sec) package plant in 1974. Raw water is obtained from the Clackamas River, where turbidities range from 3-5 TU in summer to 10-80 TU in the winter. When the plant was visited on 26 August 1975 the following chemicals were used: alum—11 mg/l; soda ash—7 mg/l; polyelectrolyte—0.03 mg/l; and chlorine—3-4 lb/day. Filtered water turbidities ranged from 0.1-0.4 TU over the previous filter run. The filters were backwashed every three days using 9,000-11,000 gal (34-42 m^3) per backwash. To produce the 750,000 gal (2840 m^3) per cycle, approximately 1.5% of production was used for backwash.

Effluent Quality Assurance

A major concern with unattended package plants (generally operated by a person with minimal skills and process familiarity) is that the quality of the finished product may not always meet acceptable standards. For example, a change in raw water conditions requires a coagulant dosage change to continue producing a high-clarity filtered water. Failure to replenish treatment chemicals or a chemical-feed pump failure could also result in high turbidity filtered water. If these conditions occur when the operator is not present, the plant could produce a substandard water quality. With the unattended package plant it is imperative that some assurance be provided to produce quality water at all times.

A turbidity control system has been developed that is designed to monitor the finished water turbidity continuously and, in the event that the finished water turbidity exceeds a preset maximum limit, shut down the plant and initiate an alarm. Briefly, the turbidity controller carries out the following functions:

1. continuously monitors the filter effluent turbidity and indicates or records effluent turbidity,
2. initiates a backwash cycle automatically if a preset high turbidity condition is exceeded,
3. returns the treatment plant to service after completion of backwash cycle and monitors filter effluent clarity for 15 min, and
4. shuts the treatment plant down automatically and initiates an alarm if the turbidity has not dropped below the preset high turbidity level after 15 min.

A 100-gpm (6.1 liter/sec) plant equipped with the turbidity controller is illustrated in Figure 4. With this feature it is difficult to produce an inferior filtered water quality. The treatment plant must either produce a quality water or produce none at all. The authors are of the strong opinion that turbidity control systems are mandatory on unattended package plants, especially if there is any possibility of human contamination of the raw water source. This equipment has been used extensively over the last ten years and has an outstanding service record. The small additional cost of this feature is more than offset by the quality assurance given to the final finished water.

Another quality control device that has been used on several occasions is a continuous chlorine residual monitor. A drop in the filtered water chlorine residual is immediately measured by the equipment and an alarm is initiated. In particular localities where a moderately to heavily polluted water supply must be used as a water source, this equipment would certainly be recommended.

Table I. Characteristics of Raw Water Supplies Treated with Package Plants

Test Site	Colored Water Newport, Ore.	Columbia River Rainier, Ore.	Willamette River Portland, Ore.	Well Supply Corvallis, Ore.	Muddy Creek Corvallis, Ore.
Turbidity (TU)	10 - 20	4-10	2.5 - 7.5	0.2	30
Color (CU)	110 -130	20	25 - 35	5 - 7	30
pH	7.6	7.4	6.1 - 6.3	6.3	6.9
Alkalinity (mg/l)	85	87	35	70	40
Hardness (mg/l)	17	70	15	105	10
Iron (mg/l)	4.7	0	1	4.0	10
Manganese (mg/l)	0	0	0	1.3	-
Threshold Odor	1	1	4	0	-
Temperature ($^\circ$F)	66	65	64	55	48

Table II. Test Results—Treatment of Raw Water (Shown in Table I)

Test Site	Colored Water Newport, Ore.	Columbia River Rainier, Ore.	Willamette River Portland, Ore.	Well Supply Corvallis, Ore.	Muddy Creek Corvallis, Ore.
Alum (mg/l)	30	13	25	0	38
Polyelectrolyte (mg/l)	0.3	0.15	0.1	0.1	0.45
Chlorine (mg/l)	1	3.0	2.2	8.1	2.0
Carbon (mg/l)	0	0	10	0	0
Permanganate (mg/l)	0	0	0	1.0	0
Turbidity (TU)	0.15	0.11	0.08	0.1	0.12
Color (CU)	5	0	0	3	5

Iron (mg/l)	0	0.1	0.1	0.15	0
Manganese (mg/l)	0	0	0	0.1	0
Threshold Odor	0	0	1	0	0
Chlorine Residual (mg/l)	1.0	0.6	0.5	0.1	0.3
Average Length of Filter Run Before Backwash (hr)[a]	22	18	26	20	20
Percentage Backwash Water	1.9	2.3	1.7	2.1	2.1

[a]Backwash at 8 ft of filter head loss.

Table III. Aquarius and Water Boy Plant Operational Data

Plant Location	Design Capacity of Plant (mgd)	Turbidity (TU)			Filter Runs (hr)
		Raw	Settled	Filtered	
Azle, Texas	2.0	15	4	0.2	24
Roma, Texas	0.4	6	3	0.1	240
Kemp, Texas	0.8	75	3	0.4	12
Alamo, Texas	2.0	5	4	0.5	48
Cedar Park, Texas	2.0	2	1	0.3	168
Trinidad, Texas	0.14	11	5	2.0	168
Latham, Kansas	0.01	68	12	0.5	72
Baldwin City, Kansas	0.14	40	20	0.3	12
Ozawkie, Kansas	0.14	300	9	0.3	96

Figure 4. Turbidity controller in a package plant.

INSTALLATION CONSIDERATIONS

Package plants are designed specifically to simplify and reduce the cost of installation. The WATER BOY unit is a completely factory-built unit and requires only: (a) connection of power to the load center on the unit, (b) installation of raw water, finished water and backwash water piping, and (c) placement of the filter media. A typical installation is illustrated in Figure 5.

The Modular Aquarius tankage is completely factory-built and painted but requires some on-site assemblage of influent and effluent piping and valves. A typical Modular Aquarius is depicted in Figure 6. An outside installation of an 0.5-mgd (22.1 liter/sec) Modular Aquarius in Figure 7 shows the compactness of the equipment.

Figure 5. Typical WATER BOY installation.

The simplicity of installation is reflected in the very economical construction costs of a number of package plant installations (see Table IV). The overall cost of a treatment facility utilizing a package plant obviously varies depending upon the size of the treatment unit, architectural design, the materials used in the treatment plant structure, complexity of raw water intake structures, design of filtered water storage facilities, and many other factors too numerous to mention. In general, the unit capital costs amortized over 20 years for the package units with capacities from 10-100 gpm (0.61-16 liter/sec) range from $0.40-0.15/1000 gal of treated water capacity.

Unit costs obviously decrease with the larger systems. Amortized capital costs for installations utilizing the Modular Aquarius, with a capacity of 200-1400 gpm (12-88 liter/sec) would range from $0.05/1000 gal of capacity to $0.10/1000 gal (3785 l) of capacity.

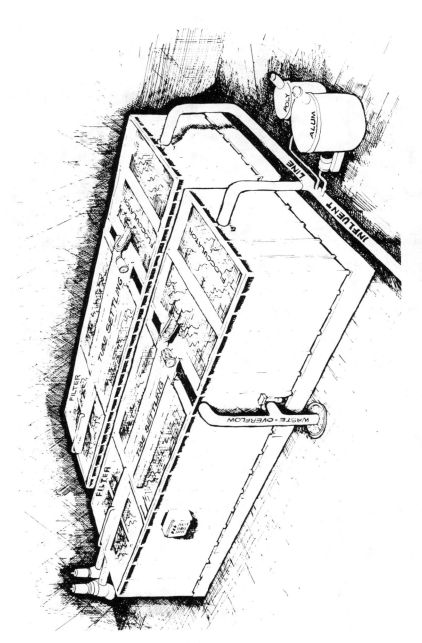

Figure 6. Typical Modular Aquarius installation.

Figure 7. Outdoor installation of Modular Aquarius.

SUMMARY

The package water treatment plant has become firmly established as a practical and economical solution for providing treated water for small communities. The problems of treatment plant process reliability have been overcome through the introduction of better techniques for flocculation, clarification and filtration. Variable-intensity mechanical flocculation, high-efficiency tube-type sedimentation, and mixed-media filtration have been successfully incorporated into package plants that provide safe finished water of high clarity. Features are available to monitor filtered water turbidity continuously to prevent introduction of unsatisfactorily treated water into the distribution system at any time.

Construction costs of treatment facilities employing package plants are less than conventionally designed facilities over the entire range of treatment capacities. Economics of operation can be realized because the plants are automatic and are designed for virtually unattended operation. Both the WATER BOY and the Modular Aquarius package plants have an outstanding record of performance during the decade since the first installations were placed in service. In consideration of: (a) the economics of construction, (b) operation and maintenance, and (c) reliability of performance, package plants should always be carefully considered for flows less than 5 mgd.

Table IV. Typical Installation Costs of Package Plants

City	Water Flow, gpm	Water Source	Treatment Purpose	Installation Facilities Include	Total Cost, $ (Approx)	Date	Adjusted 1977 Cost, $ Package Plant Only, fob	Adjusted 1977 Cost, $ Total Contract, Installed
Timber, Oregon	60 100	Surface	Turbidity, Color	1. Intake structure 2. Treatment plant **WATER BOY WB-82** 3. Backwash storage/clearwell 5. High-service pumps	45,000	1971	22,000	68,000
Oceanside, Oregon	100	Creek	Turbidity, Color	1. Intake structure 2. Raw water pumps 3. **WATER BOY WB-113** 4. 45' x 32' concrete block building 5. Below-grade clearwell 6. High-service pumps	202,000	1976	30,000	234,000
Gardiner, Montana	350	Yellowstone River	Turbidity	1. Intake structure 2. Concrete block building 3. Aquarius AQ-70 4. 100,000-gal elevated steel tank	129,000	1968	73,000	201,000
Mabank, Texas	540	Surface	Turbidity, Color	1. Intake structure 2. Aquarius AQ-112 3. Control building 4. Steel, ground-level water storage tank	180,000	1969	85,000	289,000

Estacada, Oregon	720	Clackamas River	Turbidity, Color	1. River intake structure 2. Aquarius AQ-150 3. Modular steel building 4. High-service pump 5. Steel, ground-level water storage tank	300,000	1973	104,000	405,000
Chinook, Montana	1440	Milk River	Turbidity	1. River intake pump station 2. Aquarius AQ-300 3. Below-grade clearwell 4. High-service pump 5. 40' x 124' prefab steel building for chem storage, chlorine room, lab, office	675,000	1976	171,000	742,000

REFERENCES

1. Camp, T. R. "Sedimentation and the Design of Settling Tanks," *Trans. Am. Soc. Civ. Eng.* 111:895 (1946).
2. Hansen, S. P., and G. L. Culp. "Applying Shallow Depth Sedimentation Theory," *J. Am. Water Works Assoc.* 57(9):1134 (1967).
3. Culp, G. L., S. P. Hansen and G. Richardson. "High-Rate Sedimentation in Water Treatment Works," *J. Am. Water Works Assoc.* 60(6):61 (1968).
4. Conley, W. R., and Kou-ying Hsiung. "Design and Application of Multi-Media Filters," *J. Am. Water Works Assoc.* 61(2):97 (1969).
5. Conley, W. R. "Integration of the Clarification Process," *J. Am. Water Works Assoc.* 57(10):1333 (1965).
6. Engelbach, R. L. "New Water Supply System for Gardiner," *Western City Magazine,* September 1968.

CHAPTER 19

PACKAGE WATER TREATMENT EQUIPMENT

Robert T. O'Connell

Project Specialist
The Permutit Company, Inc.
E. 49 Midland Avenue
Paramus, New Jersey 07652

Package equipment has come to mean a preassembled, modular treatment unit. These units should be factory-prepared to the greatest extent possible. Ideally, the package unit should arrive ready to: (a) be connected to the raw and finished water lines, (b) have utilities connected, (c) finish paint only and (d) be started in operation.

Upflow clarifiers and circular steel-shell gravity filters lend themselves readily to the package equipment concept. The main restriction on the size of equipment that can be factory assembled is road clearance. There are certain dimensional limitations that must be met, and these vary from state to state.

Package-type water treatment equipment for suspended solids removal are discussed in this chapter and examples of treatment systems are detailed.

PACKAGE SOLIDS CONTACT CLARIFIERS, SLUDGE BLANKET TYPE

The sludge blanket type unit, with a horizontally mounted agitator, is well suited for the preassembled package unit concept. These units are rectangular and in smaller sizes both the outer tank and the internal parts can readily be fabricated in steel or Fiberglass Reinforced Plastic (FRP). These units require no field welding. The steel units, sand blasted and primed on the interior and exterior, need only touch-up and finish coating in the field. The FRP units require neither coating nor preparation. The agitator assembly, internal sludge removal piping, sample piping, and influent and effluent flumes are factory-installed.

The surface area needed for proper operation is a function of the upflow rate chosen. The height of these units is fixed, but the width and length can be varied to obtain the necessary upflow surface area. To maximize upflow rate, or operating efficiency for a fixed surface area, tube settlers can be utilized. The use of tube settlers has expanded the nominal capacity of package units. Today, flows in excess of 1 mgd (0.04 m^3/sec) can be treated in a preassembled unit.

Typical of package sludge blanket clarifiers is the Hull Design Precipitator (The Permutit Co., Inc., Paramus, New Jersey). The height of this units is 10'6" (3.2 m). The width and length vary as shown in Table I. When the unit is received in the field, the contractor must place it on its pad and install the agitator drive motor and pipe to the inlet connection and away from the effluent and overflow connections. Power must be supplied to the agitator drive motor and, where required, for the flow measuring/chemical feed pulse transmitting equipment. The exterior sludge blowoff and pressurized water flush, piping, and valves must be assembled. Pressurized water or air must be supplied to the sludge blowoff, flushing water inlet valve operators, and (if necessary) to the inlet flow control valve. If tube settlers are to be used, the tube settler modules would have to be field-installed on factory-installed supports.

FILTERS

Steel-shell cylindrical gravity filters can be shipped partially preassembled (*i.e.*, no field welding required on the filter tank) in a number of sizes ranging from 3 ft to a practical upper limit of 12 ft diameter. The diameter is limited by state highway regulations. Occasionally, units as large as 15 or 16 ft in diameter have been shipped preassembled.

The two types of filters that are discussed herein are: (a) a syphon-actuated automatic backwashing filter (Automatic Valveless Gravity Filter, The Permutit Company, Inc., Paramus, New Jersey) with self-contained backwash water storage, and (b) an open tank, conventional style, steel-shell gravity filter with manual or automatic valve nests.

The Automatic Valveless Gravity Filter

The Automatic Valveless Gravity Filter (AVGF) is a cylindrical steel shell divided vertically into three compartments. The upper compartment is the backwash water storage chamber, the middle is the filtering chamber, and the lower is the plenum (collection chamber). The compartments are

Table I. Sludge Blanket Clarifier Sizes

Normal Flow Rate, gpm(m³/hr)	Flow Rate with Tube Settlers, gpm(m³/hr)	Size W x L, ft (m)
Steel Construction (See Figure 1)		
44 (10.0)	60 (13.6)	8' x 5' (2.44 x 1.52)
88 (20.0)	120 (27.2)	8' x 10' (2.44 x 3.05)
132 (30.0)	180 (40.9)	8' x 15' (2.44 x 4.57)
175 (39.7)	240 (54.5)	8' x 20' (2.44 x 6.10)
220 (50.0)	300 (68.1)	8' x 25' (2.44 x 7.62)
250 (56.8)	350 (79.5)	9' x 25' (2.74 x 7.62)
280 (63.6)	400 (90.8)	10' x 25' (3.05 x 7.62)
310 (70.4)	450 (102.2)	11' x 25' (3.35 x 7.62)
340 (77.2)	500 (113.5)	12' x 25' (3.66 x 7.62)
460 (104.4)	680 (154.4)	12' x 34' (3.66 x 10.36)
540 (122.6)	800 (181.6)	12' x 40' (3.66 x 12.19)
620 (140.7)	920 (208.8)	12' x 46' (3.66 x 14.02)
FRP Construction		
242 (55.0)	400 (90.8)	11' x 20'-4" (3.35 x 6.20)
308 (69.9)	510 (115.8)	11' x 26'-8" (3.35 x 7.82)
358 (81.3)	545 (135.1)	11' x 29'-8" (3.35 x 9.04)
452 (96.5)	705 (160.0)	11' x 35'-0" (3.35 x 10.67)
475 (107.8)	790 (179.3)	11' x 39'-0" (3.35 x 11.89)

separated by horizontal steel plates. When required, the backwash water storage and the filtering compartment can be physically separated to meet health department regulations. The filtering and plenum compartments are separated by a false bottom, steel strainer plate. The plenum and backwash water storage compartments are connected by riser pipes.

A separate inlet head tank and supports are included to ensure that the required inlet head is maintained. A syphon loop pipe system is provided to initiate backwash. The effluent connection is near the top of the filter tank. The height of the head tank is 21 ft (6.4 m) and the height of the filter tank is 15 ft (4.6 m). (For a more detailed description of the operation of this filter, see Chapter 13.)

When these units are delivered to the field, the contractor must: (a) place the unit on its pad, (b) prepare and paint the surface of the steel shell, (c) pipe up to the inlet connection, (d) install the effluent manifold and the backwash syphon loop and (e) pipe away from the effluent connection. The strainers must be mounted on the false bottom strainer plate, and the filter media must be placed in the filtering compartment. The unit would require a pressurized water supply to facilitate manual initiation of the backwash. The unit would then be ready for service.

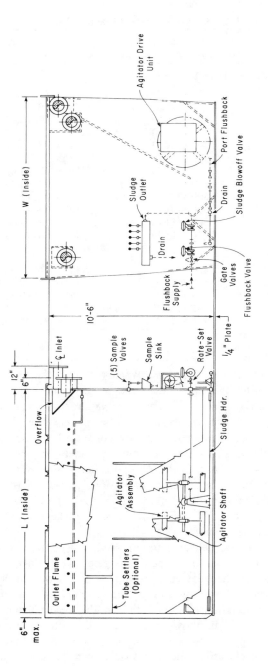

Figure 1. Typical package solids contact unit.

Steel Shell, Conventional Style Gravity Filters

This filter also has a steel cylindrical shell, and is divided into two compartments. The upper compartment is the filtration chamber separated by a steel strainer plate (false bottom) from the plenum chamber. The filter is provided with a steel wash water collection trough, to ensure uniform collection of the backwash water. The tank has two main connections. One is located in the filtration compartment for raw water influent and backwash water effluent, and the other is located in the plenum compartment for filtered water effluent, rewash water effluent and backwash water influent.

Installation requirements are similar to those described for Valveless Gravity Filters. When the unit is delivered to the field, it must be placed on a pad and painted. The strainers must be mounted on the strainer plate, the media must be loaded in the filtration compartment, and the face piping must be mounted on the tank connections. The valves are mounted on the face piping. If they are automatic valves, pressurized air or water must be connected to the valve operators. Piping would have to be installed to integrate the individual filter into the system. The unit would then be ready for service.

SYSTEM DEVELOPMENT

Referring to either of the two flow diagrams (Figures 2 and 4) and the corresponding plot plans (Figure 3 and 5), the raw water flows first through a measuring device, which is either:

- a mechanical flow meter with a pulse transmitter to pace the chemical feed timers, or
- a primary element (*i.e.*, orifice plate), a differential pressure transmitter to sense the difference in pressure across the primary element and to signal the square root extractor, which converts the differential pressure signal to a linear flow signal. This signals both the flow recorder and the flow integrator. The integrator is equipped with a pulse transmitter to pace the chemical feed.

The raw water then flows through the throttling valve. The float-actuated throttling valve maintains a predetermined level in the solids contact unit. The level in the solids contact unit sets the flow rate to the filters.

The chemically dosed water is introduced into the inlet flume, which is provided with orifices to assure even distribution into the mixing compartment. The agitator ensures contact between the incoming floc particles and the previously formed floc particles. The action of the agitator causes sludge, which has settled to the bottom of the mixing zone, to be forced upward. Thus, a portion of the sludge is recycled and continuously mixed

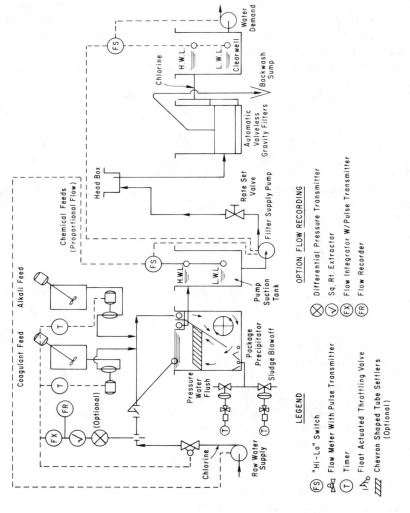

Figure 2. Typical flow diagram package precipitator-automatic valveless gravity filter.

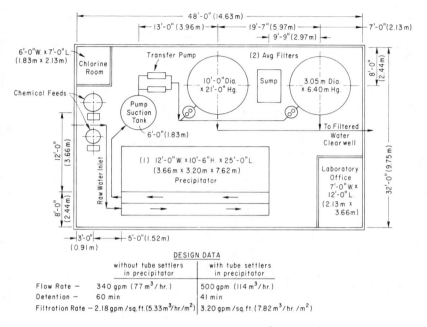

Figure 3. Typical plot plan 0.5 mgd (79 m³/hr) treatment plant.

with the incoming freshly formed floc particles. The mixture passes through a restricted port area, which evenly distributes the flow into the solid/liquid separation zone.

This separation (sludge filter) zone is isolated from the mixing zone by watertight baffles, which also keep the mechanical agitator from disturbing the sludge blanket. The slope of the baffle creates an ever-increasing up-flow area. The units are designed so that a plane is reached where the upflow rate can no longer support the sludge blanket. At this plane, the water separates from the sludge blanket and continues to flow upward until it is collected through the submerged orifices of the effluent flume.

A separation zone is also equipped with a sludge concentrator that collects the solids. The solids, which disengage from the sludge blanket, are those that have grown so heavy that the settling rate has surpassed the upflow rate. In the concentrator, the solids are allowed to compact and are hydraulically removed from the unit. This desludging operation and the alternating pressurized water backflush is controlled on a time cycle.

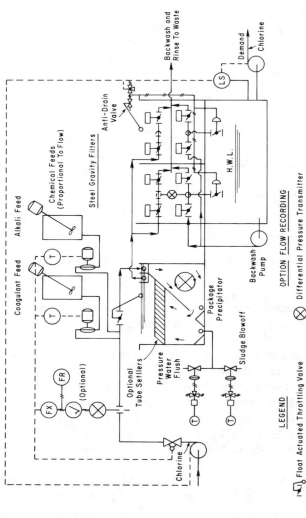

Figure 4. Typical flow diagram package precipitator-conventional steel gravity filter.

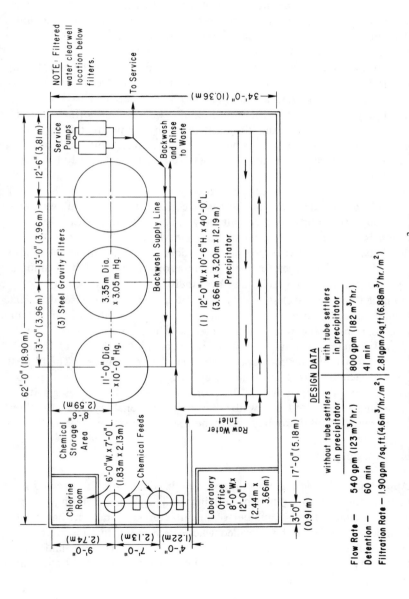

Figure 5. Typical plot plan 0.75-mgd (118-m³/hr) treatment plant.

By using tube settlers in the sludge blanket/clarified water separation zone, the flow rate in the solids contact unit can be increased. The increased upflow rate forces the sludge blanket into the tube settlers, which have approximately a 4-inch (10 cm) settling distance. Because the tube settlers are on a 60° angle (which is greater than the angle of repose of water treatment sludges), the tubes are self-cleaning.

Depending on the type of filters chosen, the flow from the solids contact unit is directed either to a pump suction tank or directly to the gravity filters. When using a filter that has a self-contained backwash water supply (e.g., the AVGF), the difference between the effluent head of the solids contact unit and the head required to operate this type of filter must be considered, and either the solids contact unit must be raised or a pump suction tank and transfer pump must be included (see Figures 2 and 3).

If a conventional style, steel-shell gravity filter is chosen, the height of the filter should match that of the solids contact unit. This allows overflow through the solids contact unit only. The differential head for filtration is greater for this type of filter (see Figures 4 and 5). The choice between these two types of filters depends on a number of factors, including: regulatory agencies, height limitations on the building, amount of operator attention available, and availability of backwash water.

If the automatic valveless gravity type filter is chosen, the solids contact effluent is directed to a pump suction tank. A low-head centrifugal pump is used to transfer the water to the inlet head tank of the filter. The head tank ensures a constant head to the filter inlet. The water then passes through the filter media and into the plenum chamber. The filtered water is directed upward through the effluent piping manifold and is discharged 14 ft (4.3 m) above the base of the filter. This allows gravity flow to an above-ground clearwell. If the conventional style gravity filter were chosen, the filter would have to be located above the clearwell.

MISCELLANEOUS

Appropriate chemical feeds (alum, lime, chlorine) must be selected. Package chemical feed equipment (mixing tank, agitator, metering pumps) is readily available from a number of manufacturers. Transfer, backwash, and service pumps must be provided where necessary. Instrumentation, such as level switches and flow metering (needed for operation and the water quality monitoring devices required to ensure good operation) must be added.

BUDGETARY INFORMATION

Budget estimates (base April 1977) for capital equipment only (fob: point of shipment) are given in Table II.

Table II. Approximate Capital Cost of Package Plants Includes Contact Unit, Gravity Filters, Chemical Feeders and Controls, fob. Excludes Freight and Installation.

Flow Rate		Estimated Budget Price
gpm	(l/sec)	April 1977
125	7.9	$ 70,000
250	15.8	95,000
350	22.0	110,000
500	31.5	130,000

CHAPTER 20

INSTRUMENTATION FOR AUTOMATIC OPERATION

Russell H. Babcock, P.E.

Consulting Engineer
165 Clapboardtree Street
Westwood, Massachusetts 02090

INTRODUCTION

Instrumentation for automatic operation has as its purpose the following:

- to control critical tasks
- to minimize tedious repetitive tasks
- to provide a tool for process supervision.

In order to accomplish these objectives, the variable that is to be controlled must first be measured. There are many variables that are measured and never directly controlled, *e.g.*, water consumption and reservoir level. These are the variables that establish the continuing demands on our systems. Others, such as chlorine feed rate and filtration rate, must be carefully and continuously controlled. The difference is that those variables which are monitored and not directly controlled are in fact often controlled but over a long time base, *e.g.*, years or decades for reservoir level. Those that are controlled in the usual sense differ in that the time base is short, such as in seconds or minutes.

It is apparent that in order to control any variable it must be capable of being continuously measured. With this fundamental fact in mind, the discussion of the physical variables of flow, level and pressure provides a basis for further understanding of automation.

FLOW MEASUREMENT

The measurement of flow can be accomplished in numerous ways. This discussion is limited to the so-called inferential methods that are most common in water works practice. These depend on the following relationship:

$$H = h_v + h_s$$

where: H = total available head
h_v = velocity head $= V^2/2g$
h_s = static head

This expression tells us that the total head at any point in a pressure conduit (pipe) is made up of the sum of the velocity head and the static head. If the velocity is increased, the velocity head is increased, and the static head must then decrease so that the total head remains constant. This is the principle of the venturi tube, the orifice plate and a number of proprietary flow tubes.

While the principle is simple, its implementation can be complex. The variables with which this problem is concerned are as follows:

D = pipe or inlet diameter in.
Q = flow gpm
h = differential across the primary device (in. of water)
G_l = specific gravity of liquid at reference temperature (usually $60°F$)
G_f = specific gravity of liquid at flowing temperature
d = throat diameter of primary device
k = discharge coefficient
S = $k/(d/D)^2$

The sizing factor, S, is used as a means of providing an equation with the flexibility required for solutions of actual flows, differentials and primary device sizes. These variables fit into the following equation:

$$Q = \frac{5.667 \ SD^2 \ \sqrt{G_f h}}{G_l} \tag{1}$$

The usual problem for the design engineer is to select a flowmeter, knowing the maximum flow, pipe size and fluid characteristics. Rearranging Equation 1 and solving for S,

$$S = \frac{QG_l}{5.667 \ D^2 \ \sqrt{G_f h}} \tag{2}$$

For the usual range of conditions encountered in water works practice, G_l and G_f are both equal to 1, which thereby simplifies the equations.

The values of S, as reported by Spink,[1] can be plotted against the Beta ratio (d/D). The engineer can thus make approximate calculations for sizing primary devices by solving a specific set of conditions for S and then selecting d/D.

The primary device in the pipeline produces a differential proportional to flow. To complete the measuring system, it is necessary to have a secondary instrument capable of detecting the differential and translating this into a chart record, an indicated flow, or a pneumatic or electronic signal for transmission. A complete measuring system, therefore, consists of a primary device (orifice plate, flow tube), the connecting piping, and the secondary instrument (recorder, transmitter). The usual problem facing a designer in this field is to select a primary device that balances initial vs operating cost. The exception to this is a gravity flow situation where the available head is fixed and economics do not usually become a factor. It is necessary to use a maximum Beta ratio to measure flow with minimum head loss. A Beta ratio of 0.75 is generally considered the maximum to use if pipe roughness is not to become a problem.

A second problem is to determine the maximum flow to be measured. This flow should not be that expected at some distant date but rather a realistic flow for the future because it is unrealistic to expect the desired accuracy to be achieved below 20% of scale. The higher the maximum flow selected, the higher the minimum accurate flow that can be measured. It is possible for actual flow at a new installation to be too low to be accurately recorded and totalized because the maximum flow selected is far too high. Selecting a lower maximum flow is not a problem with modern instrumentation, most of which can be recalibrated to a higher range when necessary.

The limitations described above are the result of the fundamental square root relationship between differential and flow. No computing relay or square root extracting devices that provide a linear flow record can alter this fundamental relationship. It is this relationship that states that if flow drops to 20%, differential drops to 4%. As this occurs the errors of zero shift and linearity become a sizable fraction of the total measurement.

The sequence of decisions and calculations to be made in selecting a primary device and associated instrumentation are as follows:

1. Assume a Beta ratio (maximum = 0.75).
2. Assume a reasonable maximum future flow so that the expected minimum is $\geq$ 20% of maximum. (This is not always possible.)
3. Make trial calculations for differentials generated at maximum flow for the primary device (orifice plate, venturi, flow tube).

4. From published data, determine permanent head loss expected from the primary device.

5. Compare initial costs of primary devices and associated instrumentation with power costs and other such viable considerations and make the equipment selection.

The secondary instruments are many and varied. While the mercury manometer is still used, it is rapidly being replaced by differential measuring devices of the mechanical type or force balance pneumatic and electronic transmitters. These devices are an improvement over the mercury flowmeter because they are smaller, lighter and capable of having their range changed to match future flows. Babcock[2] discusses the details of these devices.

The measurement of flow in many installations does not involve control but totalization for record or bookkeeping purposes. Totalizers commonly in use are pulse duration and pulse rate. Most totalizers used are of the pulse-duration type, which are often used to operate chemical feeders in proportion to flow[2].

LEVEL MEASUREMENT

The measurement of level in open tanks is fundamental to water treatment and distribution. All facilities have unique properties that impose limitations on the methods that can be used for level measurement. These are:

* The range of levels is apt to be relatively large.
. Freezing conditions can often interfere with or damage instrumentation.
. Usually it is necessary to transmit the information over relatively long distances.

These constraints make the following methods the most common:

* direct connected pressure element
* bubble tube or diaphragm box
* float and cable.

The direct connected pressure element is commonly used to measure the level in elevated reservoirs, in standpipes, and sometimes clearwells. In order to use this method, it is necessary that the point of measurement be *below* the *minimum level* to be measured, *e.g.*, the gate chamber of a standpipe. The pressure on the measuring element is indicated or recorded as feet of water. Once the measurement is available, it can be transmitted to a point where operating personnel are present, such as telemetering to the water treatment plant. It is common practice to add electric contacts to such instrumentation for starting and stopping pumps or sounding alarms.

The bubble tube and diaphragm box are similar in that the point of measurement is usually *above* the *maximum level*. This technique is commonly used to measure clearwell level or levels in buried tanks.

The bubble tube requires a small amount of air (1 ft^3/hr; 28 liter/hr) to be bubbled under pressure through a system of regulators to the end of a bubble tube or pipe located at the point of reference (or zero point) in the tank. The back pressure is equal to the head of liquid above the end of the bubble tube and hence indicates the level. This back pressure is measured with the same type of pressure element used in the direct connected instrument.

A variation of this arrangement is the diaphragm box, which consists of an open-ended box with a slack neoprene diaphragm closing the open end of the box. The box is connected to the pressure element of the measuring instrument by a capillary tube. The trapped air column in the box and tubing is at the same pressure as that produced by the column of liquid above the diaphragm box. This method is successfully used on those installations where no compressed air is available.

The float and cable is the oldest method and consists of a float whose vertical position moves a recording drum via a cable. The method, while common, has the inherent drawback of a cable that is subject to damage from numerous sources and the requirement that the instrument be mounted directly over the point of measurement. This method is usually limited to direct local measurements. Babcock[2] provides examples of all these methods.

ANALYTICAL MEASUREMENT AND CONTROL

The measurement of such variables as pH, conductivity, turbidity, chlorine residual, and selective ions is becoming commonplace in water treatment facilities. The methods used are essentially extensions of laboratory methods and devices. Because of the complexity of equipment, this entire class of measurements does not have the high degree of reliability experienced with the physical measurements of flow and level. This can be expected to change in time. Analytical measurements are commonly applied as follows:

- pH measurement and control of chemical coagulation
- pH measurement for corrosion control
- turbidity measurement of raw, backwash and finished waters
- chlorine residual measurement and control of finished water
- selective ion measurement of fluoride, hardness and other significant ions.

The possibilities for analytical measurement and control are almost limitless. The variety of physical-chemical principles incorporated in these devices are beyond the scope of this presentation. The reader is referred to any acceptable text on physical chemistry for background and to Babcock[3-5] for the specifics of their use in water works practice.

CONTACTING SYSTEMS, TWO-POSITION CONTROL OR ALARM

The contacting system, whether it be used for control or alarm, is quite often a component added to a measuring system. Such a control or alarm system can be added to a flowmeter, level recorder and pH recorder. There are a number of proprietary contacting systems available, most of which use either a mercury switch or relay to carry enough power to actuate auxiliary devices. A carrying capacity of 5 amps at 115 V is common for relays. The auxiliary devices can be alarms and motor starters.

Alarm devices are normally designed to be actuated at a given point, such as high or low alarm. Each alarm point usually requires a separate set of contacts. For two-position controls, the actuated device (water, valves) usually starts at a given point and continues to run or remain at the actuated position until reaching a second predetermined point. There must be a substantial change in measurement before the control action opposite to that initially produced is reversed. An example is level control of a standpipe by starting and stopping a high-service pump. The pump starts at a low point, such as 10 ft (3.05 m) from overflow, and continues until the level is nearly to the overflow point. If it were desired to sound an alarm at high level, then the alarm initiation point and alarm stop point would be set only a few inches apart.

The principal difference between alarm or two-position control is the *differential* between the make and break points. This differential is very small for alarms. In the case of two-position control, the differential can be any value that suits the problem. The principal distinction between alarm and two-position control is the size of the *differential.*

Another form of alarm and two-position control is that provided by so-called blind devices such as conductance probes. These devices operate by virtue of their location, such as probes where the circuit is completed by a rising water level. No indication of level is provided. The actuation of an alarm by these devices tells the operator that an alarm point has been reached, but it does not provide a continuing indication of the actual measurement.

The same devices that can be incorporated as a component in a level or flow instrument can also be added to a receiving instrument (telemeter). The device to be controlled is often far from the point of measurement. The control of reservoir levels is often dependent on the operation of a pump that may be miles away. Such requirements probably caused water works engineers to build the earliest telemeters.

MODULATING CONTROL SYSTEMS

The control of flow and level by holding the variable within limits on a continuous basis is called modulating control. Common examples are filter rate control or pressure control. Such control systems require the following:

- a measuring system (*e.g.*, flow or level)
- a control system
- a final operator (*e.g.*, valve or variable-speed pump).

The measuring system can be any type of flowmeter or level measuring system if it is continuous and not intermittent. The control system is basically a detector amplifier system that compares the desired value of a variable with the actual value and transmits a signal to take the necessary corrective action to bring the variable within limits. Depending on the type of control action desired, the variable can be made to equal the set point or merely to remain within an established displacement. Point control is usual for flow but control within a narrow band is common for level.

An often overlooked requirement for control is that the action of the final operator (*e.g.*, valve) must produce a measurement change. There cannot be anything between measurement and control to isolate one from the other. An example of this is the common error of attempting to control the level over a weir by locating a flow control device (valve) downstream from the weir. The weir hydraulically isolates the measurement point from the control element, thereby violating a fundamental requirement of control.

The water supply field is concerned with all types of modulating control, but the two basic types are open loop and closed loop. These terms are best defined by familiar examples. An example of closed loop control would be the rate of flow controller in common use on rapid-sand filters. It makes no difference whether this device is the nearly extinct mechanical type or a modern pneumatic or electronic control system; the following components are present:

- a flow primary device, such as a venturi tube
- a flow measuring instrument, such as a differential pressure transmitter
- a controller, such as a pneumatic or electronic controller
- a control valve and operator, such as a butterfly valve with pneumatic or electric operator

The flow through the rate control system is held at the desired value by being continuously measured and fed *back* to the controller. This device evaluates whether the flow is at the desired value and in turn manipulates the control valve through the valve operator to hold the desired value if measurement has wandered from the control point. In this example the actual flow is being measured. Such systems are said to be of the *closed loop* type.

Another common example is the feeding of chemicals, such as alum, in proportion to flow. It is common to vary the speed of the metering pump to maintain a constant dosage (concentration) as the plant flow rate varies. Normally no attempt is made to measure the flow rate of liquid alum. This type of system is often referred to as *pacing* and is a typical *open loop* system. There are many examples of such systems in water works practice: chlorine, fluoride, lime. The accuracy of such systems depends in large part on the accuracy of the type of feeder selected. The accuracy available from presently available feeding methods is normally adequate for waterworks installations. Maier[6] provides extensive detail on feeder types and their control.

The theory of control can become quite complex and the reader should consult Perlmutter,[7] Liptak[8] and Shinsky[9] for detailed theoretical material.

TYPICAL DESIGN EXAMPLES

Example 1. Booster Station

A common design problem is for a booster station to provide proper pressure over a wide range of flows for a residential area located at a higher elevation than the remainder of the distribution system. Let us consider the basic criteria involved in such a design:

- The station is to be equipped with two pump and motor installations.
- Normal operation requires only one pump and motor to meet the domestic requirements of the area served.
- The control criterion is to be pressure alone.

- Pressure is to be controlled by varying pump speed. Whether this is done by varying motor speed or by hydraulic or magnetic couplings between motor or pump is not germane to the control concept.

An important consideration is the mode of control to be used. There is no reason to require an absolutely fixed pressure control point. Practically, it is acceptable for pressure to fluctuate within narrow limits such as 55 to 65 psig (380 to 448 kn/m^2). By establishing such a criterion the following benefits result:

- The control system is simpler but adequate.
- Surges on both the suction and discharge mains are minimized.
- The pumping machinery can be expected to provide improved services because of gradual changes in speed.
- The control system is independent of time, which eliminates the phenomenon of "wind-up" as discussed by Babcock.[2]

Table I lists the principal items of instrumentation and their approximate costs, which do not include installation or control panels. For purposes of this discussion, it is assumed that the variable-speed pump is driven by a constant-speed electric motor and magnetic coupling. The instrumentation is essentially the same for all means of variable-speed pump control.

Table I.   Booster Station Controls

Item	Approx. Cost in 1977
Electronic pressure transmitter. Output 4-20 ma, range 0-100 psig.	$ 800
Electronic indicator receiver controller, proportional action. Output 4-20 ma.	1000
Electronic receiver recorder. Range 0-100 psig.	700
Total	$ 2500

In a typical booster station the control and recording instruments could be monitored in a section of the motor control center. The transmitter would normally be mounted adjacent to the point of measurement. A schematic diagram, using ISA (Instrument Society of America) symbols, is shown in Figure 1.

Example 2. Filtration Rate Control

A typical filter plant uses pneumatic filtration rate control. The filtration rate should be set in proportion to the clearwell level. A low level (clearwell nearly empty) causes a maximum filtration rate. Conversely, a high level (clearwell nearly full) causes a minimum filtration rate. The filters shut down completely at the highest allowable level. Such a control system is shown in Figure 2. The discussion in this example is limited to filtration rate control and neither backwash control equipment nor loss of head is considered. The reader is referred to Babcock[2] for an extended discussion on this aspect of filter plant instrumentation. The assumption is made that the filter effluent line size is 8 in. (203 mm) and that there are four filters (Table II).

In addition to the tabulated equipment necessary panels and accessories are required. Estimating these devices is difficult because it is a function of the plant design. A budget figure can be established by making the

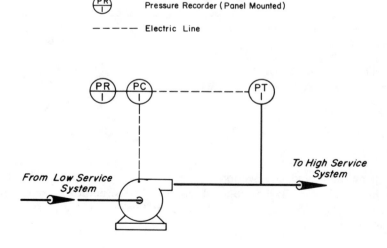

Figure 1. Schematic diagram of controls for a booster pumping station.

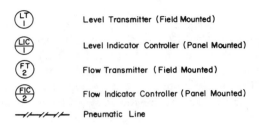

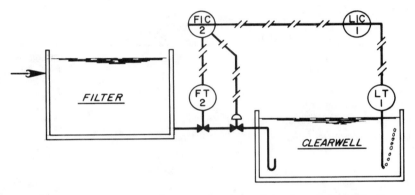

Figure 2. Schematic diagram of controls for clearwell level control of filtration rate.

Table II. Filter Rate Controls

Item	Approx. Cost in 1977
Four primary devices for 8-in.- (203-mm) diameter flow tubes. Four at $500.	$ 2,000
Four pneumatic flow transmitters. Range 0-50 in. (1270 mm) of water. Four at $575.	2,300
Four pneumatic indicator receiver controllers with remote set index, proportional integral action. Four at $690.	2,760
Four 8-in. (203-mm) butterfly control valves with pneumatic operators. Four at $1,000.	4,000
One pneumatic level transmitter.	500
One pneumatic indicator receiver controller, proportional action.	670
Total	$12,230

panels and accessories equal to the cost of the instruments and control valves, or approximately $12,000. The cost of backwash controls is not included because they are almost limitless in variety.

An important item required for pneumatic control is a source of compressed air. The flow requirement can be approximated by allowing 1 cfm of free air for each pneumatic relay. In the absence of detailed information on a specific device, it can be assumed that each transmitter, controller and positioner requires a pneumatic relay. In this example, Table II provides a basis for estimating air consumption.

Table III.   Estimated Air Requirement

Item		Air Required (free cfm)
Four flow transmitters; 4 at 1 cfm	=	4
Four flow controllers; 4 at 1 cfm	=	4
Four pneumatic operators; 4 at 1 cfm	=	4
One level transmitter	=	1
One level controller	=	1
		14 cfm

Since the life of a compressor is shortened by continuous operation, the air supply should consist of two compressors, each of which has a capacity equal to twice the expected air consumption. This would require two compressors, each with a capacity of 28 free cfm. In addition, an air storage tank of adequate capacity is required. Such an arrangement provides the necessary air supply with one compressor operating 50% of the time. The second compressor is in reserve should a mechanical failure of the operating compressor occur. The two compressors should be alternated weekly.

CONCLUSION

The automatic control of water treatment and distribution facilities is rapidly becoming a major factor in the design of a modern installation. This necessarily abridged presentation cannot provide the mass of detailed information the designer needs. But it can, perhaps, arouse the curiosity of the engineer to examine further this complex subject.

REFERENCES

1. Spink, L. K. *Principle and Practice of Flow Meter Engineering,* 9th ed. (Foxboro, Massachusetts: The Foxboro Company, 1967).
2. Babcock, R. H. *Instrumentation and Control in Water Supply and Wastewater Disposal* (New York: Dunn-Donnelley Publishing Corp., 1968).
3. Babcock, R. H. "Automation of Water Analysis and Analytical Process Instrumentation," *J. Am. Water Works Assoc.* 62(3):145-148 (1970).
4. Babcock, R. H. "Analytical Process Instrumentation Present and Future," *J. New England Water Works Assoc.* 82:151-165 (1968).
5. Babcock, R. H. "Ion Selective Electrodes for Quality Measurement and Control," *J. Am. Water Works Assoc.* 67(1):26-28 (1975).
6. Maier, F. J. *Manual of Water Fluoridation Practice* (New York: McGraw-Hill Book Co., 1963).
7. Perlmutter, D. D. *Introduction to Chemical Process Control* (New York: John Wiley & Sons, Inc., 1965).
8. Liptak, B. G. *Instrument Engineers Handbook* (Radnor, Pa.: Chilton Book Co., 1969).
9. Shinsky, F. G. *Process Control Systems* (New York: McGraw-Hill, Book Co., 1967).

CHAPTER 21

POTASSIUM PERMANGANATE FOR IRON AND MANGANESE REMOVAL AND TASTE AND ODOR CONTROL

Kenneth J. Ficek

Manager—Technical Services
Carus Chemical Company
La Salle, Illinois 61301

INTRODUCTION

The production of quality drinking water was emphasized as early as prehistoric times. Egyptian inscriptions and Sanskrit medical lore provide us with the earliest record of the quest for quality water.[1] The "Susruta Samhita," a body of medical lore dating to 2000 BC (but not recorded on manuscripts until 400 AD) states[1]: "Impure water should be purified by being boiled over a fire, or being heated in the sun, or by dipping a heated iron into it, or it may be purified by filtration through sand and coarse gravel and then allowed to cool."

Besides this prescription, other methods involving vegetable substances, bruised coral, pounded barley, macerated laurel, polenta, lime and aluminous earth were suggested as means of providing a good quality of drinking water.[1] In addition to these chemical treatments, elaborate aeration systems involving agitation, bronze water containers, stone vessels, wick syphons and triple filtration were some of the methods attempted to correct putrefaction.[1]

Although some of these methods do not sound like very scientific ways to approach a water quality problem, nonetheless they led to modern treatment methods presently employed in water treatment plants. Since raw water quality can be affected by a multitude of factors, it seems also intuitively obvious that no single treatment method nor chemical process will solve all problems on all occasions. Nevertheless, it is possible to design a treatment plant with the built-in flexibility required for all of the various treatment methods available. It may be more difficult, however, to convert the older plants to include some of the design features required to take advantage of modern technological advances.

The introduction of potassium permanganate as a water treatment chemical to remove taste and odors and oxidize iron and manganese can be incorporated into the plants that are presently being designed and, with minimum expense, be added to present treatment systems. To understand the advantages potassium permanganate offers in solving the above-mentioned problems, a basic understanding of the chemical, its reactions, the testing procedures, and the application will be discussed.

THE CHEMICAL, POTASSIUM PERMANGANATE

Potassium permanganate, represented by the chemical formula $KMnO_4$, is described as an oxidizing agent. In aqueous solutions, $KMnO_4$ provides solution oxygen to effect a chemical change in many organic and inorganic compounds. The chemical and physical properties[2] are described in Table I. When dissolved in water, potassium permanganate imparts a pink to purple color depending on the relative concentration of the chemical.

Table I.   Chemical and Physical Properties of Potassium Permanganate[2]

Formula	$KMnO_4$
Formula Weight	158
Bulk Density	90-100 lb/ft^3 (1440-1600 kg/m^3)
Physical Form	crystalline (about 80 mesh)
Stability	indefinite when stored in closed containers
Solubility	about 5% at room temperature

The first published mention of the "mysterious purple mass" was by Glauber[3] (sometimes called the father of German chemistry) in 1659. He noted that, when melting "Braunstein" (manganese dioxide) with saltpeter (potassium nitrate), and dissolving the product in water, the solution turned purple, then blue, red, and finally green. Other experimental work confirmed Glauber's findings and found some additional color changes, which eventually led to the chemical name, "Chameleon." The color changes are due to the various valence changes the manganese undergoes from its elemental metallic state (0) to its highest oxidation state (+7) in the permanganate ion.

In water treatment, the manganese in permanganate is reduced from this +7 state to the +4 state, which is the insoluble manganese dioxide. The color change is from pink or purple to yellow or brown, depending on the relative concentrations. In the pH ranges used in water treatment plants, the simple general reactions are written:

$$\overset{+7}{K}MnO_4 + H_2O + \begin{bmatrix} \text{organic compounds} & \rightarrow & \text{carbon dioxide} + \text{water} \\ \text{soluble iron} & \rightarrow & \text{ferric precipitate} \\ \text{soluble manganese} & \rightarrow & \text{manganese dioxide} \\ \text{hydrogen sulfide} & \rightarrow & \text{soluble sulfate} \end{bmatrix} + \overset{+4}{M}nO_2 \quad \begin{matrix}(1)\\(2)\\(3)\\(4)\end{matrix}$$

The manganese dioxide by-product is insoluble and is usually removed through coagulation, settling and filtration. This is discussed in more detail later.

HISTORY IN WATER TREATMENT

The first known reports discussing the application of potassium permanganate for taste and odor control were entered by Houston[4] in his reports to the London Water Board, 1920-1930. Houston emphasized the value of permanganate both as a "taste preventer and taste remover" in doses usually from 0.2 lb/MG to 0.8 lb/MG. Houston reported that, "It was first used in 1913 in connection with a 'geranium' (according to the scientist)—'castor oil' (as judged by the consumer) taste due to the excessive growth of Tabellaria at the West Middlesex Works." Later it was found to be most valuable in preventing (or removing) the iodoform taste of chlorinated water.

Since Houston's work over 50 years ago, the use of $KMnO_4$ for water treatment (except for iron and manganese removal) was not really promoted until about 1960. It was not until then that application techniques to facilitate its use were developed. Potassium permanganate has probably been tried on almost every type of taste and odor problem.[5] It is not a cure-all, but it can be used effectively to control some problems either alone or in conjunction with other treatment chemicals.

The prime requisite for the use of potassium permanganate is that the treatment process include filtration. The manganese dioxide produced by the reduction of permanganate is insoluble and should be removed. Otherwise the finished water may be colored and cause staining on laundry, bathroom and kitchen fixtures.

Often, manganese dioxide acts as a coagulant aid, reducing the amount of primary coagulant required to help clarify the water. Jar tests should be run to optimize the dosages of all chemicals, including potassium permanganate, required for the full treatment.

TASTE AND ODOR CONTROL

All surface supplies are plagued at least sometimes by taste and odor problems. The intensity and frequency of the problem is probably as varied as the chemicals causing the problems.[6] The causes of tastes and odors have been well documented in the literature,[7] with the most common cause being algae.[8] In addition to algae the following causes have also been cited:

- industrial wastes
- municipal sewage
- decomposing marine life
- decomposing aquatic weed
- decomposed animal life
- transient waterfowl
- falling tree leaves

With such a wide variety of chemicals causing taste and odor problems, it is impossible to suggest one type of treatment that would economically and efficiently solve all of the problems on all of the occasions. Although chlorine and activated carbon have been reported to be the most widely accepted treatment methods for taste and odor control, the use of potassium permanganate has increased in the last 15 years. Ozone and chlorine dioxide have also become more popular in recent years for specific applications.[9]

To determine if potassium permanganate alone or in combination with other chemicals will provide an answer to a specific problem, the standard jar test (Figure 1) to determine dosage requirements should be run. The

Figure 1. Jar test apparatus. **Potassium permanganate** added to five jars at left.

jar test procedure is well documented[10] and provides an indication of the approximate amount of chemicals required to solve the problem. Since the jar test procedure should be a standard practice in water laboratories, the details of running the test are not presented here.

It may be necessary to adjust the pH of the water in order to obtain the optimum results, because the $KMnO_4$ reactions can be affected by pH. Figures 2 and 3 show the oxidation rates of manganese and phenol

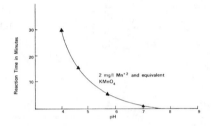

Figure 2. Manganese oxidation rate vs pH.

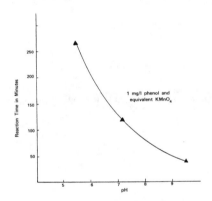

Figure 3. Phenol oxidation rate vs pH.

as functions of pH. Generally, as the pH increases, the $KMnO_4$ reactions go to completion in a shorter time.[11] Therefore, when conducting jar tests, if good results are not obtained in the normal operating pH range, an increase in pH or a change in the order of chemical addition should be tried. After the approximate amount of $KMnO_4$ is determined through the jar test, a plant trial is the next step.

POINTS OF APPLICATION

Potassium permanganate should be fed as early as possible in the system as shown in Figure 4, and before the addition of other chemicals. Some plants are so designed that potassium permanganate can be fed at the intake of the plant as shown in Figure 5. This permits the maximum time for oxidation of the more resistant taste- and odor-causing compounds.

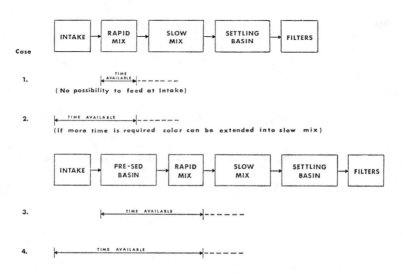

Figure 4. Time available for oxidation by potassium permanganate.

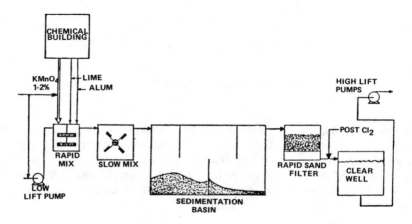

Figure 5. Gravity system water treatment plant.

Potassium permanganate should be fed ahead of chlorine where possible. Oxidation of these organics prior to chlorination may reduce the amount of chlorine being used and minimize the chances of forming chloro derivatives, which may be more offensive than the original compounds. Carbon should also be applied after the $KMnO_4$ is consumed as the $KMnO_4$ consumption increases when carbon and $KMnO_4$ are fed together as demonstrated in Figure 6.

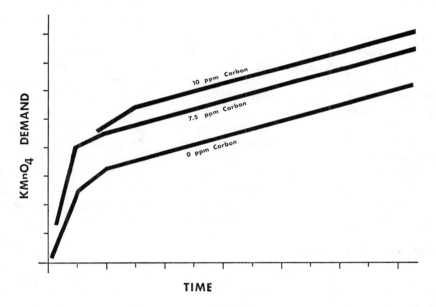

Figure 6. $KMnO_4$ demand vs time.

If it is not possible to feed at the intake, the potassium permanganate should be introduced either ahead of or into the rapid mix. The feed of the permanganate should be controlled so that the pink color does not go beyond the rapid mix. If more time is needed, the coagulation zone or slow mix can be used, as indicated by Figure 4, but the possibility of colloidal manganese dioxide passing into the filters becomes greater. This manganese dioxide could pass through the filters and impart a yellow color to the finished water.

In some plants, a color zone can be established. Since the color of the water changes from pink to brown, this color change zone can be seen, and the potassium permanganate feed rate changed as the potassium permanganate demand of the raw water fluctuates. Dosages of from 0.5 to 2.5 mg/l are usually sufficient to control most oxidizable taste- and odor-producing chemicals.[10] Some plants have used higher doses but only for short periods, when bad "sieges" of organics were present in the raw water.[12]

Potassium permanganate application can reduce or eliminate the need for activated carbon treatment. The manager[13] of a water plant using Lake Michigan water reported a substantial reduction in chemical cost and an improvement of the average finished threshold odor number (TON) by replacing the activated carbon treatment with potassium permanganate (see Figure 7). At this plant, dosages of approximately 0.3 to 0.5 mg/l consistently produced the desired quality in the finished water. On one occasion during the reported period, activated carbon was fed with the potassium permanganate. Even when the two chemicals were fed in combination, the total chemical cost was less than the cost of using activated carbon alone to control the odors. For many other water supplies, the combination treatment has been found to be much more effective and economical than either chemical alone.

A few water plants have reported that, because of trouble with feeding equipment or rapid reductions in $KMnO_4$ demand, they have overdosed to the point where pink water passed through the filter. Although permanganate is not toxic at the levels used in water plants, care should be taken to feed only the proper amount to avoid any residual in the finished water.[3] Activated carbon can be applied to the filters to remove any residual $KMnO_4$ that may be present in the settled water.

Feeding and Controlling

If more than about 25 lb of potassium permanganate are required per day, a dry feeder is recommended. There is a free-flowing grade of potassium permanganate available that has been developed specifically for dry feeding equipment. Although $KMnO_4$ itself is not hygroscopic, the minor amounts of impurities remaining on the $KMnO_4$ crystal can absorb some moisture and cause caking. An additive is used to coat the crystals and minimize dry feeding problems.

Potassium permanganate is also available in bulk hopper trucks for larger plants. The City of Pittsburgh has been using $KMnO_4$ delivered in bulk hopper trucks since 1964.[14] The bulk trucks are equipped with their own blowers to convey the material into the bulk storage bin. Such storage bins, shown schematically in Figure 8, should be equipped with appropriate dust collecting systems.

Proportioning pumps are used in situations where the daily demand of potassium permanganate is not more than 25 lb/day or where dry feeding is not practical. Most pump manufacturers are familiar with potassium permanganate and, when told that potassium permanganate solutions are to be fed, will supply the proper construction materials.[15]

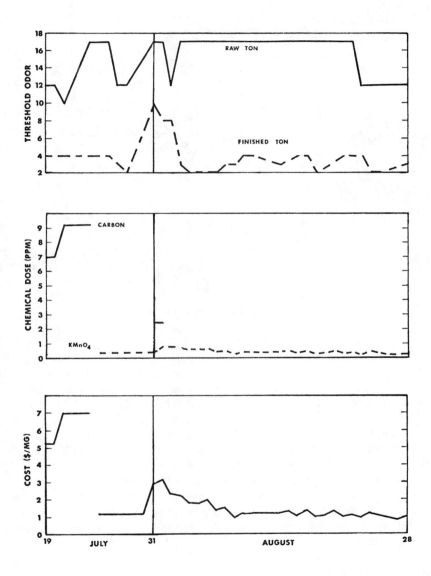

Figure 7. Odor control KMnO₄ vs carbon.

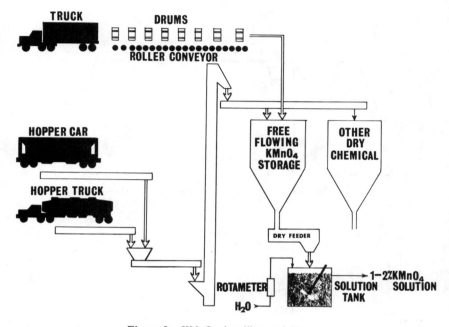

Figure 8. KMnO₄ handling and storage.

The maximum solubility of potassium permanganate at room temperature is about 5%. Normally, the solutions are made up to a convenient concentration such as 1/3 lb/gal (4%) for ease in computing feed rates. After the addition of the dry KMnO₄, the solution should be stirred for a minimum of 15 min with a mechanical agitator.

An analyzer to measure residual KMnO₄ has been developed to aid in the control of the chemical dosage.[16] Similar to residual chlorine analyzers, these instruments are capable of detecting KMnO₄ residuals as low as 0.01 mg/l. They can also be equipped to control a KMnO₄ feeder. The cost of the permanganate analyzer varies from $3220 for the analyzer and indicator to $6000 for the analyzer system, which includes a strip chart recorder and an electronic controller for the feeder. Figure 9 shows an analyzer-controller.

IRON AND MANGANESE REMOVAL

Iron and manganese exist in nature in both the insoluble (Fe^{+3} and Mn^{+4}) and soluble (Fe^{+2} and Mn^{+2}) oxidation states. The insoluble iron

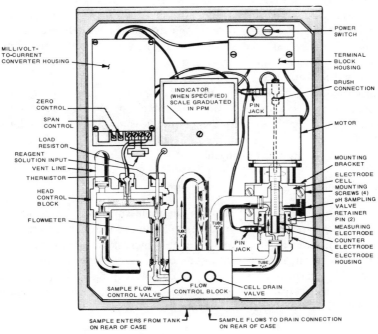

Figure 9. New Cairox® potassium permanganate analyzer and monitor.
(A) Cabinet. (B) Schematics of analyzer-controller.

and manganese are readily removed by coagulation, sedimentation and filtration. However, the removal of soluble iron and manganese poses a more serious problem.[11]

In ground and surface water supplies, soluble iron generally exists as divalent ferrous salt. Soluble manganese, with few exceptions, exists in the divalent manganous state. Iron and manganese also appear in water as organic complexes. These complexes are formed by the combination of iron and manganese ions surrounded by negatively charged organic units called ligands. These complexes, which are more difficult to remove than the inorganic compounds because of the organic "protective shell," are commonly referred to as "organically bound" iron and manganese.[17]

If the soluble iron and manganese are not removed from the water supply before distribution to the customer, several different complaints almost certainly result. The customers may complain of red or black water in the distribution system or of dark brown, red or gray stains on plumbing and fixtures. Housewives may complain of stains on clothes when bleach is added to the family wash. Textile plants may complain of graying of fabrics during the bleaching process and odd color shades during the dyeing process. Further, *Clonothrix* and *Crenothrix,* which are commonly called iron bacteria, use iron and manganese in their metabolism. If small amounts (0.1 mg/l) of iron and manganese are allowed to enter the distribution system, these forms of life may thrive and cause slimes. These slimes can take up chlorine and cause taste and odor problems in the distribution system.[17]

An oxidation process is the most effective means of removing soluble iron and manganese. During oxidation, soluble divalent iron and soluble divalent manganese are converted to insoluble states, which generally exist as the oxide hydrates. A number of techniques may be employed in order to oxidize iron and manganese.[11]

Air Oxidation

Air oxidation (also called aeration) in some situations is satisfactory although it is a relatively slow process. Simple aeration is generally ineffective in oxidizing the last undesirable increments of soluble iron and manganese. When high concentrations of manganese are present, and when the iron or manganese exist as organic complexes or chelates, air oxidation is generally ineffective. The cost of the capital equipment for aeration also precludes its use in some systems.

Catalysis

Catalysis using a variety of catalysts, particularly the copper ion, are known to enhance air oxidation.

Hypochlorites

Hypochlorites applied as the sodium or calcium salt hydrolyze to hypochlorous acid, which is a stronger oxidant than dissolved molecular oxygen. Therefore, soluble iron and manganese are more rapidly and completely oxidized by hypochlorite addition than by aeration. This process, too, offers certain limitations. More than the stoichiometric amount of chlorine is required, which leads to the possibility of forming chloro derivatives of organic contaminants, which cause taste and odor problems and impart a chlorinous taste to the water. Hypochlorous acid does not efficiently oxidize manganese and organically bound iron and manganese.

Chlorine

Chlorine, like hypochlorites, must first react with water to form hypochlorous acid, the active oxidizing agent. It is thus subject to the same limitations as hypochlorites. Chlorine gas is usually delivered in cylinders that require special handling and storage. It is metered into the water through chlorinators and, because of its toxic properties, special safety precautions must be followed.

pH

Adjustment of pH is an effective means of iron and manganese removal in lime or lime-soda softening plants. Generally, this gives satisfactory results since the pH range is above 9.5. This process is even more efficient when preceded by aeration. If lime or lime-soda softening is not practiced, pH adjustment for iron and manganese removal is seldom economically feasible.

Ion Exchange

Ion exchange resins are effective in removing small quantities of soluble iron and manganese. The use of ion exchange resins is extremely expensive, requires skilled personnel and can be further complicated by fouling because of high concentrations of oxide hydrates. Although the resins can be cleaned of the oxides, this is also expensive and the risk of losing the resins is great.

Chlorine Dioxide

Chlorine dioxide is a strong oxidizing agent that rapidly oxidizes soluble iron and manganese. It is seldom used for oxidizing soluble iron but is used to some extent for the oxidation of soluble manganese. The cost of obtaining chlorine dioxide by reacting sodium chlorite with chlorine in aqueous media has been generally prohibitive in the past for smaller water plants. Recently it has been reported that chlorine dioxide may cause blood disorders.[18] This is currently being studied and results will be reported in the near future.

Manganese Dioxide

Manganese dioxide affixed to zeolite or greensand is effective in removing soluble iron and manganese in either regenerative-batch or continuous processes. This process is treated in greater detail later in this chapter.

Potassium Permanganate

Potassium permanganate may be used in either of two conventional systems. It is utilized in pressure systems as well as the more conventional gravity systems used by large municipalities. The chemistry involved in both processes is virtually the same and the following discussion of the reactions may be applied to either system. The use of $KMnO_4$ for the removal of iron and manganese is very attractive because the reactions are complete, rapid, and require only a minimum quantity of chemicals. Theoretically, 1 mg/l of potassium permanganate oxidizes 1.06 mg/l of ferrous iron, as shown by Equation 5.

$$3 \ Fe^{++} + MnO_4^- + 4 \ H^+ \rightarrow \ MnO_2 + 3 \ Fe^{+++} + 2 \ H_2O \qquad (5)$$

Potassium permanganate is very effective in oxidizing soluble manganese. In theory, it requires 1 mg/l of $KMnO_4$ to oxidize 0.52 mg/l of soluble manganese.

$$3 \ Mn^{++} + 2 \ MnO_4^- + 2 \ H_2O \rightarrow \ 5 \ MnO_2 + 4 \ H^+ \qquad (6)$$

Table II shows a comparison of the different chemical oxidants required to oxidize 1 mg of either iron or manganese. The cost per pound of the various oxidants fluctuates not only with supply and demand,

Table II. Comparison of Oxidants for Oxidation of Iron and Manganese

	Theoretical weight to oxidize 1 mg of:	
Oxidant	Iron (Fe^{+2})	Manganese (Mn^{+2})
Oxygen, O_2	0.14 mg	0.29 mg
Chlorine, Cl_2	0.62	1.30
Calcium Hypochlorite, $Ca(OCl)_2$	0.64	1.30
Sodium Hypochlorite, NaOCl	0.67	1.36
Potassium Permanganate, $KMnO_4$	0.91	1.92
Chlorine Dioxide, ClO_2	1.21	2.45

but also with the quantity shipped and ultimate destination or use point. Therefore, a comparative cost analysis would be valid only for a given day at a given location. The relative effectiveness of the oxidation reactions and the equipment needed for a particular raw water supply must also be taken into account. Each situation therefore dictates a separate cost analysis.

Gravity Systems

The oxidation of both of these contaminants using $KMnO_4$ in gravity systems presents no real problem because the oxidative rates are fast and jar tests usually indicate the exact amount of $KMnO_4$ to be fed. If the demand is met, all of the ferrous iron and manganous manganese is oxidized and the resultant precipitates are coagulated and removed through settling and filtration.

Pressure Systems

The most popular type of pressure filter used in conjunction with potassium permanganate is the manganese zeolite or manganese-treated greensand filter shown in Figure 10. Although diatomaceous earth, regular sand, anthracite and more exotic filter media such as "Electromedia"® (Filtronics, Inc.) have been used, the greensand filter offers a number of advantages.

Manganese greensand is a mineral capable of exchanging electrons and thereby oxidizes iron and manganese to their insoluble and filterable states.[11] The greensand has the ability to do both the oxidation and the filtration. However, its oxidative capacity is limited and eventually the bed must be regenerated.

Since the bed obtains its original oxidative capacity from an oxidation step with potassium permanganate, it can be restored or regenerated in the

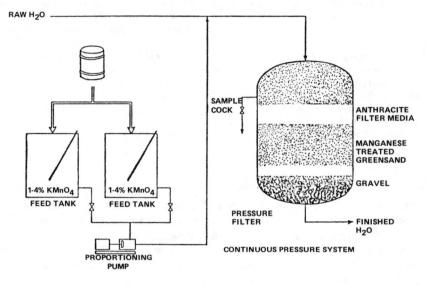

RAW H₂O

SAMPLE COCK

ANTHRACITE FILTER MEDIA

MANGANESE TREATED GREENSAND

GRAVEL

1-4% KMnO₄ FEED TANK

1-4% KMnO₄ FEED TANK

PRESSURE FILTER

FINISHED H₂O

PROPORTIONING PUMP

CONTINUOUS PRESSURE SYSTEM

Figure 10. Continuous KMnO₄ feed in manganese-greensand system.

same manner. After backwashing the insolubles from the filter, a solution of potassium permanganate is passed through the bed to oxidize the surface of the zeolite. Excess potassium permanganate is rinsed from the filter, and the unit is returned to service.

As the above process can be time-consuming, a more efficient process referred to as CR or continuous regeneration was developed and is described in a paper by Davis and Fackler.[19] A solution of potassium permanganate is continuously fed into the raw water line ahead of the filter to reduce the amount of soluble iron and manganese going to the filter. The greensand then acts as a buffer. If the feed of KMnO₄ does not oxidize all of the soluble iron and manganese, the greensand will oxidize and filter these minerals. If the feed of KMnO₄ is in excess of the demand, the excess pink color is used up in regenerating the greensand. If an excess of KMnO₄ is continuously fed, the pink color will eventually break through.

The manganese-treated greensand is usually capped with anthracite to remove the bulk of the insolubles so as not to blind the greensand. Most filters are equipped with supplemental air wash to ensure that the greensand is "scrubbed" clean during backwash.[20]

The first step in the design of any iron removal plant is to have a complete and reliable water analysis.   The second step is to determine the chlorine and permanganate requirements. The third is to treat this water through manganese greensand in a laboratory column. The above procedure enables the chemical requirements and the quality of the water to be accurately determined. Finally, for a complete evaluation, the additional step of conducting a pilot plant study to determine accurately the optimum flow rate and operating characteristics is desirable, especially for a large plant.

TOXICOLOGY

Solid potassium permanganate is considered a poison, and there is, apparently, some controversy about the degree of toxicity of the chemical. The majority of permanganate poisonings between 1924 and 1940 were attempted suicides. In nearly all instances, the product ingested was in the form of tablets or crystals and *none were fatal.*[3]

Toxicity tests on rats showed $KMnO_4$ to have an LD_{50} range of 700 to 1830 mg/kg.[3] *Extrapolating these numbers to humans, an average 175-lb man would have to swallow about 0.5 gal of a 5% solution, on an empty stomach, for death to occur.* A recommended safe concentration of permanganate solutions is 200 mg/l. Above this concentration, tissue damage can occur. However, concentrations as high as 10,000 mg/l have been recommended as an antidote for certain poisons. This is applied in low volumes so as not to exceed the LD_{50}.

SUMMARY

1. Potassium permanganate can be used effectively and economically to reduce taste and odors and to oxidize iron and manganese.

2. It is an oxidizing agent, and if it is applied prior to chlorination the amount of chlorine can be reduced.

3. Potassium permanganate can be used in conjunction with other water treatment chemicals, such as carbon, as another tool in producing a high quality water.

4. It is used in dosages normally ranging from 0.5 to 2.5 mg/l, but higher dosages may be required for more highly contaminated waters.

5. It can be fed with almost any dry feeder or proportioning pump that is constructed of chemically compatible materials.

6. It can be controlled visually or by the use of a continuous $KMnO_4$ analyzer.

7. The reaction produces manganese dioxide, which can act as a coagulant aid.

8. It can be used in conjunction with manganese-treated greensand pressure filters to produce iron- and manganese-free water.

9. Finally, potassium permanganate is safe and easy to handle. Bulk systems can be installed in water plants with high daily consumption.

REFERENCES

1. Baker, M. N. *The Quest for Pure Water* (Lancaster, Pennsylvania: Lancaster Press Inc., American Water Works Association, Inc., 1948), pp. 1-8.
2. "KMnO$_4$ Product Data." Form M-700, Carus Chemical Company, La Salle, Illinois (1974).
3. Emanuel, A. G. "The Chemistry and Applications of Potassium Permanganate in Water Treatment," Carus Chemical Company, La Salle, Illinois. Paper presented at Southwest American Water Works Association Meeting (1965).
4. Houston, Sir Alexander. 15th, 19th, 20th and 25th Annual Reports to the Metropolitan Water Board. London, England (1920-1930).
5. Atkinson, J. W., and A. T. Palin. "Chemical Oxidation in Water Treatment," International Water Supply Association Congress, 9th (1972), p. E9.
6. The American Water Works Association, Inc. *Water Quality Treatment* (New York: McGraw-Hill Book Company, 1940), pp. 228-229.
7. American Water Works Association Research Foundation. *Public Water Supply Treatment Technology* (New York: American Water Works Association, 1973), p. 74.
8. Research and Education Association Staff. *Pollution Control Technology* (New York: Research and Education Association, 1973), pp. 292-293.
9. Cheremisinoff, P. N., J. Valent, D. Wright, R. Fortier, and J. Magliaro. "Potable Water Treatment: Technical and Economic Analysis," *Water Sew. Works* 123:66 (1976).
10. "The CAIROX Method for Water Treatment." Form M-1050, Carus Chemical Company, La Salle, Illinois (1972).
11. Shrode, L. D. "Potassium Permanganate: Use in Potable Water Treatment," *Water Sew. Works* 119:R10-R-19 (Reference Edition, April 1972).
12. Popalisky, J. R., and F. Pogge. "Studies on Three-Phase Water Treatment at Kansas City, Missouri," Paper presented at Missouri Section American Water Works Conference (1967).
13. Private communication.
14. Beck, J. D. "Curious-Yellow," *The American City* 85(3):73 (1970).
15. "Handling and Storage." Form M-707, Carus Chemical Company, La Salle, Illinois (1974).
16. "Potassium Permanganate Analyzer." Fischer & Porter Company, Warminster, Pennsylvania (June 1973).

17. Shair, S. "Iron Bacteria and Red Water," *Ind. Water Eng.* (March/April 1975).
18. EPA. "Interim Primary Drinking Water Regulations. Control of Organic Chemical Contaminants in Drinking Water," *Federal Register* 43(28):5756 (1978).
19. Davis, R., and R. B. Fackler. "Removal of Iron and Manganese by Ferrosand CR and IR," Hungerford & Terry, Inc., Clayton, New Jersey (1974).
20. Wilmarth, W. A. "Two Methods for Removal of Iron, Manganese, and Sulfides," *Water Wastes Eng.* 5(8):52 (1968).

CHAPTER 22

ACTIVATED CARBON FOR
TASTE AND ODOR CONTROL

Billy H. Kornegay, Ph.D., P.E.
Technical Director
Carbon Department
Westvaco Corporation
Covington, Virginia 24426

INTRODUCTION

When Gorfield[1] of the School of Military Engineering in England pointed out in 1873 that drinking water must be free from tastes and odors, he was establishing a goal for the water treatment industry. Treatment methods devised to accomplish this goal through taste and odor removal include: (a) aeration, (b) oxidation and (c) adsorption. Adsorption is the only method to be covered in this chapter, and discussion of it is limited to systems employing activated carbons.

Although 22 water plants employed charcoal filters[2] in the United States as early as 1883, the material was not activated and the process was abandoned due to low adsorptive capacity. It was not until an "activation" process was developed that carbon became a viable remedy for taste and odor control. This "activated carbon" was first manufactured in the United States in 1913 by the Industrial Chemical Company, now the Chemical Division, Westvaco Corporation.

The first application of activated carbon for taste and odor control was in 1928. When chlorophenolic tastes in the Chicago water damaged half a million dollars in food products,[3] two Chicago meat packers successfully used activated carbon to remove these tastes, but the practice was deemed too expensive for municipal water systems. Granular carbon was considered to be more economical, and research by Sierp in Germany and

481

Harrison in the United States resulted in the construction of full-scale granular carbon filters at Hamm, Germany, in 1929[4] and Bay City, Michigan, in 1930.[5] The experience at these installations would probably have resulted in the large-scale application of granular carbon filters if the economics of powdered carbon treatment had not been demonstrated effectively. The first municipal use of powdered activated carbon was instituted by Spalding[6] in 1930 at the Hackensack Water Company at New Milford, New Jersey. The success of this application was followed by more than a thousand plants within the next decade[7] and established activated carbon as the most efficient means of ensuring palatability.

SOURCES AND DETERMINATION OF TASTES AND ODORS

There are numerous sources of taste and odor in water supplies, and most occur naturally or are of an industrial origin. Naturally occurring sources of taste and odor include vegetation, decomposed organic matter, mineral substances, hydrogen sulfide and algae. Algae constitute the most frequent source of naturally occurring taste and odor, and their occurrence may normally be related to conditions in the raw water source. Organics constitute the major source of industrial taste and odors with phenolic compounds particularly prevalent. However, the possible sources are so numerous and varied that it would be virtually impossible to list the compounds, or combinations of compounds, that produce tastes and odors. In addition, such a list would find limited utility in ensuring palatability.

Since substances causing taste and odor are so varied and objectionable at such low concentrations, chemical analytical techniques are seldom employed. The organoleptic senses, taste and smell, are normally used for both detection and measurement. The Threshold Odor Test has become the standard control test for several reasons. First, it is more convenient and does not involve the potential health hazard of tasting untreated water. Second, tastes and odors are normally closely related, and the removal of odors normally results in the removal of tastes. The "threshold odor number" is the extent to which an odor-bearing water must be diluted with an odor-free water to reduce the odor to a concentration that is just perceptible. The techniques employed are the control method, the short parallel test and the continuous monitor.

The details of these test methods may be found elsewhere[1] and are not included here. However, it should be noted that the threshold odor test is more objective than normally perceived and correlates reasonably well with organic content. This is demonstrated in Figure 1, where the threshold odor number of a paranitrophenol solution is plotted as a function of

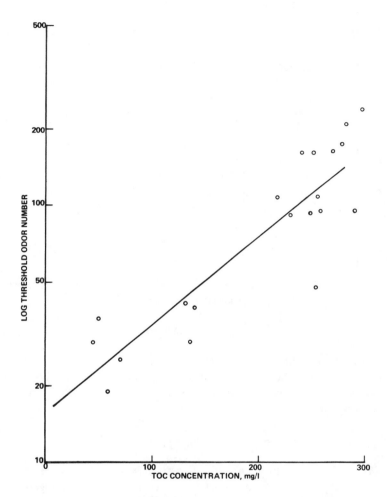

Figure 1. A correlation of the threshold odor number and total organic carbon concentration of paranitrophenol.

the total organic carbon content. The correlation coefficient between the log of the threshold odor number and TOC is 0.89 for the 20 samples shown.

Threshold odor data may also be indicative of organics removal from natural water as indicated by gross parameters as fluorescence. This is shown in Figure 2, which compares the reduction in fluorescence with the threshold odor reduction as the result of powdered activated carbon

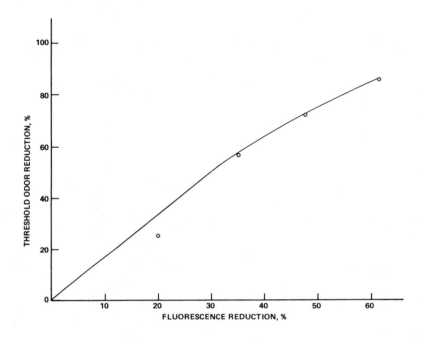

Figure 2. Comparison of the reduction in threshold odor and fluorescence.

additions. In this figure, the slope of the line is 1.63, indicating a fair agreement between the reduction of odor and organics as measured by fluorescence. This relation may vary with both time and location and is not always linear.

ACTIVATED CARBONS AND ADSORPTION

Activated carbons are carbonaceous materials subjected to selective oxidation to produce a highly porous structure, as shown in Figure 3. Some of the major raw materials are coal, wood, nut shells, peat, lignite and the residue from pulp and petroleum processes. These materials may be thermally or chemically "activated" to produce surface areas in excess of 2000 m^2/g. However, in most potable water applications economics dictate the use of carbons with surface areas from 500 to 1500 m^2/g. It is this high porosity and surface area that gives activated carbon its unique adsorptive properties. By changing the activating conditions, both the size and number of the pores may be controlled to provide the macro- and micropore distribution that is most suited for a particular use. The macropores are in excess of 100 Å and provide a ready access to the micropores (< 100 Å), which are of molecular dimensions.

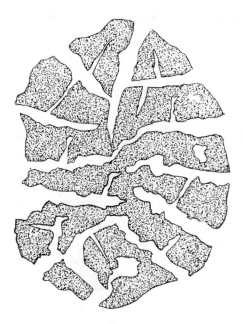

Figure 3. Activated carbon pore structure.

These small micropores provide the major surface area for adsorption, as shown in Figure 4. In fact, in water-grade carbons more than 70% of the available surface area occurs in pores having a radius of less than 50 Å. In good quality activated carbons the external surface area is negligible compared to that in the internal pores, and a particle size reduction would provide no appreciable increase in surface area. Therefore, the adsorption capacity of a given carbon is independent of particle size, except for very large or insoluble molecules, whereas the rate of adsorption is particle-size-dependent as demonstrated in Figure 5.

In order to understand the role of activated carbon in controlling tastes and odors, it is informative to consider the adsorption system. The total adsorption system involves three major components as shown in Figure 6. These are (1) the adsorbent, (2) the adsorbate or solute and (3) the solvent. The solvent and solute form the solution phase of the system, while the adsorbent and attached adsorbate make up the solid phase. Adsorption, then, is the process whereby an adsorbate moves from the solution phase to the surface of an adsorbent where it is held by attractive forces. The forces of attraction may be physical, chemical, electrical, or a combination of the three. Of the dilute solutions of nonelectrolytes, such as those found in water treatment systems, physical adsorption is the most

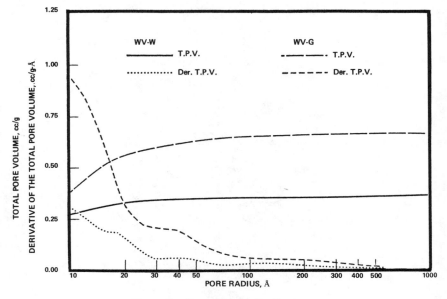

Figure 4. Pore volume distribution.

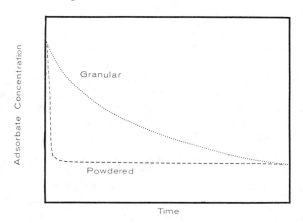

Figure 5. The effect of particle size on rate and capacity.

prevalent form and accounts for the vast majority of the observed adsorption. It is often helpful to view adsorption as a fight against solution. Therefore, any action that tends to increase the affinity of the organic impurities for the polar solvent (water) decreases adsorption. Adamson[8] related this in a more general form of Traube's rule, which states that a polar (nonpolar) adsorbent preferentially adsorbs the more polar (nonpolar) component of a nonpolar (polar) solution.

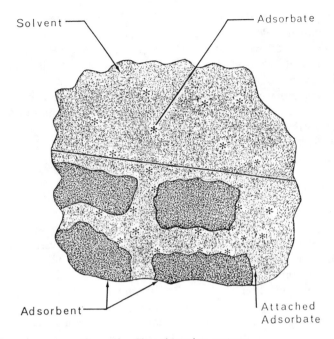

Figure 6. The adsorption system.

THE SELECTION AND DESIGN OF ACTIVATED CARBON SYSTEMS

Three of the major considerations in the application of activated carbon for potable water treatment are: (1) the form of activated carbon, (2) the carbon capacity and (3) the rate of adsorption. From the brief discussion of the adsorption system, it is apparent that the capacity and rate of adsorption are influenced by: (a) the nature and concentration of the activated carbon, (b) the nature and concentration of the impurities present, (c) the nature of the solvent (water), and (d) the environmental and operating conditions.

In treating large quantities of potable water, little control can be exercised over the nature of the solvent or the concentration of the impurities. In addition, environmental control has severe limitations due to costs and the requirements of the treated water. Therefore, the desired performance of the adsorption process is normally achieved by controlling the nature and concentration of the activated carbon, the method of operation, the allowable concentration of impurities in the final product, and, to a minor extent, the nature of the impurities.

The first decision that must be made is the form of activated carbon to be employed. Powdered activated carbons are normally considered to be smaller than 50 mesh, while granular carbons are larger. After selecting the form of activated carbons, the capacity, rate of adsorption and method of application may be determined. Powdered activated carbons are normally more economical for minor or normal taste and odor problems, and granular carbon is desirable for more severe conditions. When in doubt, both powdered and granular systems should be evaluated to ascertain which system is more cost-effective in achieving the desired results.

Powdered Activated Carbon Systems

When powdered activated carbon is used for taste and odor removal, the type of carbon, dosage, and point of application determine the quality of the finished water. Since the adsorptive capacity of various carbons are vastly different, the dosage required to achieve a palatable level varies widely. Adsorption parameters such as iodine number and phenol value provide some indication of the utility of carbons for potable water treatment.

Studies conducted with both experimental and commercial activated carbons at various locations revealed that the iodine number is the more significant parameter. Figure 7 shows the carbon dosage required to reach a palatable level as a function of iodine number at a Canadian city. It should be noted that the carbon dosage required to reach a palatable level is almost inversely proportional to the iodine number. The phenol value was not nearly as significant.

The best method of selecting a carbon is through comparative laboratory tests. In these comparative tests, various dosages of the carbons under consideration are added to the odor-bearing water and the residual threshold odor level determined. The residual threshold may be plotted vs the carbon dosage as shown in Figure 8 or as a Freundlich adsorption isotherm as shown in Figure 9. The dosage required to achieve a palatable level may be read directly from Figure 8 or calculated from the carbon loading, x/m at the palatable level from Figure 9. In this case, the bark char carbon required a dosage 2.4 times greater than Aqua Nuchar to reach a palatable level. However, in order to evaluate the relative cost-effectiveness, the price of the carbons must also be considered.

The results of more than 30 comparative studies and economic evaluations are shown in Figure 10. It is apparent from Figure 10 that Aqua Nuchar was more cost-effective than bark char carbons in every test and, on the average, cost almost 30-40% less to use. Another interesting point

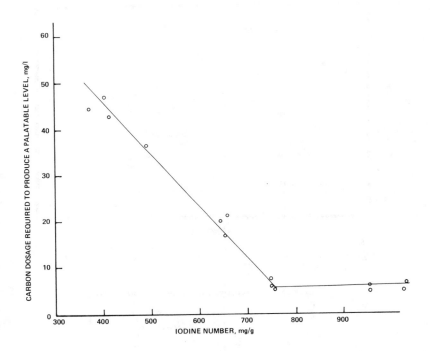

Figure 7. The influence of the iodine number on the required carbon dosage.

from Figures 8-10 is the high carbon dosage required to reach a palatable level in laboratory evaluations. These laboratory dosages are higher than required in plant operations and reflect the value of other unit operations and processes in taste and odor reduction. The Aqua Nuchar dosage required to achieve a threshold odor number of 3.0 in full-scale plant operations was 8.0 mg/l for the example cited above compared to a laboratory dosage of 44.5 mg/l. However, it is erroneous to assume that the taste and odor removal achieved without carbon additions is one minus the ratio of the plant dosage to the laboratory dosage. It should also be emphasized that these data are based on laboratory adsorption tests conducted on the raw water without the addition of other chemicals. When other chemicals are added in the same sequence and amounts as those used in full-scale plant operations, the required carbon dosages for laboratory and full-scale plant runs are very similar. However, carbon treatment of the raw water, without other chemicals added, is easier, provides a valid comparison of activated carbons, and

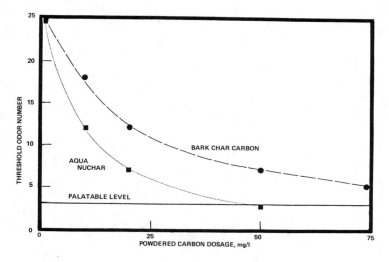

Figure 8. A comparison of carbons for taste and odor control.

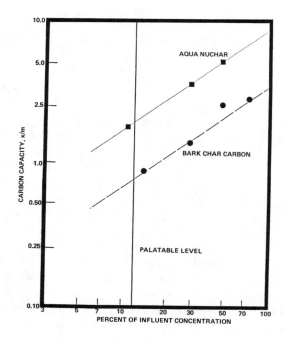

Figure 9. Freundlich adsorption isotherm for taste and odor evaluations.

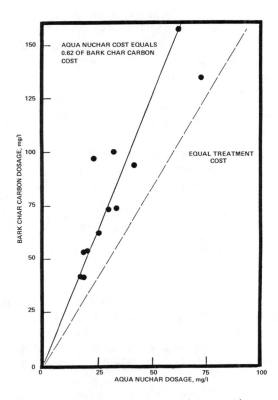

Figure 10. Comparison of carbon dosages and cost.

serves as a background for establishing the value of various points of application. By evaluating the cumulative comparisons of laboratory and plant
tests, it should be possible to quantify the advantages of various points of
application.

The point of application should be selected to ensure that: (a) adequate
mixing is provided between a small carbon addition and large volume of
water, (b) adequate contact times are provided (5-15 min) and (c) the
impurities present are most amenable to adsorption. It is this latter point
that provides the only control over the nature of the adsorbate. The
points of application that have been employed are: raw water, rapid mix
basin, flocculators, sedimentation basin, conduits to the filters, and top of
the filters.

Carbon applications to the raw water normally have been quite effective. The influent conduits provide long contact times and good mixing,
and adsorption can occur before the impurities have reacted with chlorine.

The rapid mix basin may be equally effective if chlorine is added at a later point, or if chlorine has a negligible influence on the adsorption characteristics of the taste- and odor-producing compounds. Application to the flocculators or sedimentation basins does not appear to be as effective except where color or high turbidity levels are present. Carbon additions to the top of the filters give a good final polish when used in conjunction with other applications, but they may not be adequate for total treatment due to the limited contact time. In the design of powdered carbon feed systems, several application points should be provided, and feed to the raw water is definitely recommended.

In a survey of 137 municipalities using activated carbon, Westvaco found the average dosage was 2.8 mg/l, the high 5.1, and the low, 0.5 mg/l. These figures may serve as a useful guide in designing powdered carbon systems. At a dosage of 3 mg/l and a delivered carbon cost of $0.35/lb, this would amount to a carbon cost of $0.87/1000 gal. This is a small fraction of the total treatment cost.

Granular Carbon Systems

Granular activated carbon may be installed in conventional sand filters or postfiltration columns. Carbon capacity and rate of adsorption must be determined for both installations, but the use of carbon as a filter medium must also be considered when carbon is installed in the filter. Although the filtration properties of granular activated carbons are virtually the same as a comparable-size sand or anthracite, many states still require the retention of some sand. This reduces the operating life and adsorption efficiency of the carbon due to reduced bed depths, but it does provide the advantages of a reverse-graded filter. Since there are over 23,000 existing filter installations representing more than $50 billion in investment, this will probably be the most widely used method of application in the near future. Information pertaining to this method of application is, therefore, stressed in this chapter.

Multimedia, or reverse-graded filters, employ a bed that decreases in particle size from top to bottom and in operation is somewhat analogous to a series of sieves. The larger particles of floc and debris are removed in the upper portion of the filter, and the smaller fraction passes through. As the material passes downward through the filter, it eventually encounters a restriction or is removed by gravity, inertia, or attractive forces. The mixed-media bed, therefore, provides removal in depth as opposed to surface removal of normal filters. As a result, mixed beds are usually more efficient and subject to longer runs between backwashing. To provide a reverse-graded filter, the various layers of media must be both larger and lighter than the material immediately below. However, these two criteria alone do not assure a well-designed filter. If a carbon particle (Sg = 1.5) is too much larger than the sand (Sg = 2.65), it may still have a faster settling velocity than the sand, and excessive intermixing may occur.

The major factor for consideration in designing a multimedia filter is backwash characteristics. If the carbon is too small, it may be washed out before appreciable sand expansion occurs, and "mud balling" could result. On the other extreme, if the carbon is too much larger than the sand, bed expansion would be retarded and poor cleaning of the carbon would occur. For best results, the sand and carbon should have approximately the same expansion at a given flow rate. From a comparison of the backwash curves for various sands and Nuchar granular carbons, an effective size ratio of approximately 2.5:1 is desirable for mixed-media filters. This is in general agreement with the findings of Camp,[9] who concluded that an anthracite (Sg = 1.5):sand ratio of between 2:1 and 3:1 is needed to provide expansion at the same wash rate. Since standard grades of Nuchar carbons are available in 20 x 50, 14 x 40, 12 x 40 and 8 x 30 mesh sizes, this provides a broad range of possibilities. However, most mixed-media filters employ either the 12 x 40 or 8 x 30 mesh sizes. This provides a method for selecting the particle size, and the carbon depth is normally governed by adsorption requirements.

The type of carbon is selected on the basis of carbon capacity from a Freundlich adsorption isotherm as previously described for powdered carbon. However, the carbon capacity, x/m, is determined at the influent concentration C_0, rather than the palatable level, C_e, because all or a major portion of the carbon will be in equilibrium with the influent before replacement.

Because of the inherent variability of waters, the determination of carbon depth required to provide a given length of taste- and odor-free service must be determined for each individual installation. This is accomplished by using a granular carbon pilot filter. By measuring the influent and effluent concentration at various depths with time, breakthrough curves such as those shown in Figure 11 may be constructed. The time of operation achieved before reaching an allowable breakthrough concentration, C_b, through each column and those through a known carbon depth may be determined from the various breakthrough curves. The operating time attained before reaching an unacceptable taste or odor breakthrough concentration, C_b, may then be plotted vs carbon bed depth as shown in Figure 12. It should be noted that at some critical bed depth, D_c, zero operating time is attained before reaching an unacceptable taste or odor, C_b. This dictates the minimum depth of carbon. The actual depth to be employed may be determined by summarizing the test

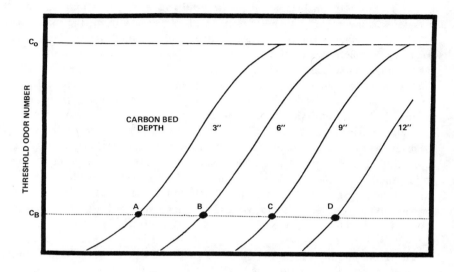

Figure 11. Granular column breakthrough curves.

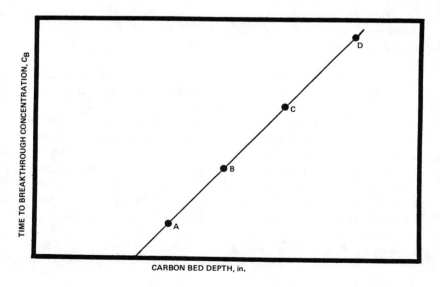

Figure 12. Carbon service time as a function of bed depth.

data in a plot similar to Figure 12. In water treatment plants the critical bed depth is measured in inches for taste and odor removal, but a minimum of 12 in. is recommended. As the depth increases, a greater percentage of the bed is totally exhausted before breakthrough, and better carbon utilization is obtained. Carbon filters have provided effective taste and odor control for over six years, but two to three years would be typical. At the end of this period, the carbon must be removed for regeneration or replacement. Therefore, a slurry transport system should be included in the plant design. If the exhaustion rate is less than 1500-2000 lb/day, on-site regeneration is not normally feasible.

The cost of a granular carbon filter 30 in. deep with a 350 ft^2 surface area would require 11-13 tons of granular carbon at a cost of $10,000-12,000. If this filter treated one mgd for two years, the carbon costs would be 1.4-1.6 ¢/1000 on a use and discard basis. Once again, this is a small percentage of the total cost of treatment, which normally ranges from 30 to 50¢/1000 gal.

SUMMARY

Although charcoal has been used since biblical times, the use of "activated carbon" is relatively new. The first activated carbon produced in the United States was manufactured in 1913 by the Industrial Chemical Company, now the Chemical Division of Westvaco Corporation. This activated carbon was used for industrial water purification in 1928, and the first municipal application was in 1930. Since that time powdered activated carbon has proven to be the most widely used method of taste and odor control. Comparative laboratory tests are the most effective method for evaluating the cost-effectiveness of various carbons, but, in the absence of these data, the iodine number should be used as the basis of selection.

Granular activated carbon systems are increasing in number and may be more cost-effective for severe taste and odor problems. The carbon also serves as a filter medium when applied in conventional systems. Therefore, adsorption, filtration, and backwash properties must be considered in design.

REFERENCES

1. Hassler, W. W. *Taste and Odor Control in Water Purification* (New York: Industrial Chemical Sales, West Virginia Pulp and Paper Co., 1947).
2. Croes, J. J. R. "The Filtration of Public Water Supplies in America," *Eng. News Am. Cont. J.* 10:277 (1883).

3. Baylis, J. R. "The Activated Carbons and Their Use in Removing Objectionable Tastes and Odors from Water," *J. Am. Water Works Assoc.* 21:787 (1929).
4. Sierp, F. "Improvement of the Smell and Taste of Drinking Water," *Tech. Gemeindebl* (Ger) 32:153 (1929).
5. Harrison, L. B. "Activated Carbon at Bay City's Filtration Plant," *J. Am. Water Works Assoc.* 23:1388 (1931).
6. Spalding, G. "Activated Char as a Deodorant in Water Purification," *J. Am. Water Works Assoc.* 22:646 (1930).
7. Baylis, J. R. "Development of Activated Carbon in Water Treatment," *Taste Odor Cont. J.* 5:2 (1938).
8. Adamson, A. W. *Physical Chemistry of Surfaces,* 2nd ed. (New York: Interscience Publishers, 1967).
9. Camp, T. R. "Theory of Water Filtration," *J. San. Eng. Div.,* ASCE 90(SA4) (1964).

CHAPTER 23

CHEMICAL CONDITIONING FOR WATER
SOFTENING AND CORROSION CONTROL

Douglas T. Merrill, Ph.D., P. E.
   Senior Engineer
   Brown and Caldwell, Consulting Engineers
   1501 North Broadway
   Walnut Creek, California 94596

INTRODUCTION

This chapter describes the use of Caldwell-Lawrence diagrams in the analysis and solving of corrosion control and water softening problems. Both subjects are of great importance to the water supply industry and to the general public. The diagrams provide a means to assess rapidly and with reasonable accuracy the condition of the raw water, the type and amount of chemicals needed to produce the finished water as well as the amount of chemical precipitate produced by the treatment—in short, all of the chemical information the operator needs. The calculations may of course be carried out analytically, but analytical calculations are complex and tedious. The diagrammatic approach is simpler, just as accurate, and far quicker.

Corrosion Control

The economic losses caused by corrosion of water systems are very large. Hudson and Gilcreas[1] cited the striking example of a 55 mgd (2.41 m^3/sec) water distribution system under attack from a corrosive water. They estimated that the distribution system (composed of reinforced concrete, asbestos-cement and cement-lined cast iron pipe) was losing 500 short tons (453 metric tons) of piping annually as $CaCO_3$.

Over a period of five years, the carrying capacity of the system had declined severely, as indicated by a decline in the Hazen-Williams coefficient from 130 to 80. In addition, large concrete storage tanks had corroded to the point that for safety reasons they could only be half filled. They estimated that the nation's economic loss to deterioration of distribution systems is $375 million annually, but that this could be avoided if corrosive waters were stabilized with only $27 million worth of lime.

Distribution system deterioration is, of course, only part of the problem. The products of corrosion are transmitted to the water, diminishing its quality, perhaps in some cases to the point where corrosion product concentrations are injurious to health. For example, a study of Boston's water distribution system conducted several years ago[2] indicated that 19% of the standing water samples exceeded the current drinking water standards for copper, 9% exceeded the limit for iron, 65% exceeded the limit for lead and that most of the contaminants were corrosion products from the distribution system. Similar results were obtained for Seattle.

It is not clear that such levels adversely affect health. For example, less than 5% of trace metals consumed by humans are ingested from water; much more comes from food. It is possible, however, that the ionized form in which the metal is obtained makes water-borne species more dangerous than those in food, which tend to be removed from the ionized state by chelating agents. For example, the Boston study[2] indicated that as the level of lead in tap water increased, the amount of lead in the bloodstream increased also. What *is* clear is that as long as the state of knowledge concerning the effects of waterborne corrosion products remains imperfect, the water industry's responsibility is to provide public protection through a program of corrosion control.

Corrosion occurs as the result of dissolution of a solid in a fluid, the most common fluid being water. It may be a strictly chemical phenomenon or a more complex electrochemical reaction. Corrosion can be arrested in many ways. The majority of treatments involve the placement of inert films at the solid-water interface. (An exception is cathodic protection—the use of galvanically or directly impressed current to arrest electrochemical corrosion.) In simplest terms the films prevent contact between the water and the solid surface, and the corrosive process cannot proceed. The film may be mechanically applied (for example, coal tar enamel or cement linings), derived from the deposition of chemicals (for example, polyphosphates, silicates or calcium carbonate), or formed from the products of the corrosive reaction itself. Combinations of methods may be used—for example chemical deposition upon a mechanically applied liner to cover holes or "holidays" in the liner. This chapter describes protection by deposition of $CaCO_3$ films. Chemical conditioning in this respect means the conditioning of waters so that $CaCO_3$ can be precipitated from them.

Water Softening

Water hardness presents many problems to water users. Hardness is defined as the sum of all multivalent cations in solution. From a practical standpoint, only Ca^{32} and Mg^{+2} need to be considered since in most waters they are present in an overwhelming majority. For domestic use the principal reasons for advocating softening have been soap-saving, better washing conditions, and the prevention of hot water heater scaling. With the advent of synthetic detergents the first two reasons have lost much of their importance. Industrial users soften waters to meet process requirements, which are many and varied. For example, low-pressure boilers require hardness levels below 20 mg/l, and hardness must be virtually absent in electric boiler feeds.[3] What constitutes "soft" water must therefore be defined by user requirements. "Soft" to some is "hard" to others. Table I shows one classification of hardness. AWWA drinking water goals are hardness < 80 mg/l as $CaCO_3$. No hardness standards are stated by the U.S. Public Health Service or the World Health Organization.

Table I. U.S. Geological Survey Classification of Hardness

Classification	Hardness, mg/l as $CaCO_3$
Soft	0-55
Slightly Hard	55-100
Moderately Hard	100-200
Very Hard	>200

$CaCO_3$. No hardness standards are stated by the U.S. Public Health Service or the World Health Organization.

SYSTEM CHEMISTRY

The basic components of the reaction system are water (H_2O), hydrogen (H^+), hydroxide (OH^-), carbon dioxide (CO_2), carbonic acid (H_2CO_3), bicarbonate (HCO_3^-), carbonate (CO_3^{-2}), calcium (Ca^{+2}), and magnesium (Mg^{+2}). Water dissociates to form hydrogen and hydroxyl ions, as in Equation 1.

$$H_2O \rightleftharpoons H^+ + OH^- \tag{1}$$

The laws of mass action dictate that the product of the hydrogen and hydroxyl ion concentrations be equal to a constant (K_W) according to Equation 2.

$$[H^+] \cdot [OH^-] = K_W \tag{2}$$

where $[H^+]$ and $[OH^-]$ are the molar concentrations of hydrogen and hydroxyl ions and K_W is a constant whose value depends on system temperature and salinity.

When CO_2 dissolves in water it forms a weak acid (H_2CO_{3*}), which is equal to the sum of dissolved CO_2 and H_2CO_3. This weak acid then dissociates according to Equations 3 and 4.

$$H_2CO_{3*} \rightleftharpoons H^+ + HCO_3^- \tag{3}$$

$$HCO_3^- \rightleftharpoons H^+ + CO_3^{-2} \tag{4}$$

The dissociation products distribute themselves according to Equations 5 and 6.

$$\frac{[H^+][HCO_3^-]}{[H_2CO_{3*}]} = K_1 \tag{5}$$

$$\frac{[H^+][CO_3^{-2}]}{[HCO_3^-]} = K_2 \tag{6}$$

where K_1 and K_2 are constants whose values depend on system temperature and salinity. At 25°C, $K_1 \cong 10^{-6.3}$ and $K_2 \cong 10^{-10.3}$. Inspection of Equations 5 and 6 indicates that where $[H^+]$ is $\geqslant 10^{-6.3}$ (i.e., pH is less than 6.3) H_2CO_{3*} is the predominant species, HCO_3^- dominates between pH 6.3 and 10.3, and above pH 10.3, the overwhelming majority of the carbonic species are in the CO_3^{-2} form. HCO_3^- and H_2CO_{3*} are present in equal quantities at pH = pK_1, that is, where $[H^+] = K_1$. Similarly, HCO_3^- and CO_3^{-2} are present where $[H^+] = K_2$, that is, where pH = pK_2. The pK points (pK_1, pK_2) have great significance in carbonic acid chemistry.

The calcium ion and carbonate ions react to form $CaCO_3$ according to Equation 7.

$$Ca^{+2} + CO_3^{-2} = CaCO_3 \downarrow \tag{7}$$

Precipitation can occur *if and only if* the product of the concentrations of Ca^{+2} and CO_3^{-2} in solution exceeds a given value K_{CA}, known as the solubility product constant. In equation form, the necessary condition for $CaCO_3$ precipitation is:

$$[Ca^{+2}] \cdot [CO_3^{-2}] > K_{CA} \qquad (8)$$

where $[Ca^{+2}]$ and $[CO_3^{-2}]$ are concentrations of dissolved calcium and dissolved carbonate and K_{CA} is a known and predictable value. For example, K_{CA} at 15°C for a water whose total dissolved solids concentration is 400 mg/l is 1.407 x 10^{-8}; at 25°C and TDS 40 mg/l, K_{CA} = 6.253 x 10^{-9}. All equilibrium constants (K_1, K_2, K_w, K_{CA}, etc.) can be rather reliably predicted.

A water whose Ca^{+2} and CO_3^{-2} concentration product exceeds K_{CA} is known as an *oversaturated* water. An oversaturated water is an unstable water; that is, it is in a state of stress and must change in some way so as to relieve that stress. In this situation the stress exists because the water contains more Ca^{+2} and CO_3^{-2} in solution than it can comfortably hold. It can and does spontaneously relieve this stress by expelling the excess Ca^{+2} and CO_3^{-2} ions from solution in the form of the precipitate, $CaCO_3$, as in Equation 7. Precipitation continues until a *saturated,* or equilibrium condition is reached, *i.e.,* $[Ca^{+2}] \cdot [CO_3^{-2}] = K_{CA}$. At this point the system is comfortable with itself and $CaCO_3$ precipitation ceases. Precipitation of $CaCO_3$ is favored where calcium and total carbonate concentrations are high and at high pH, where the carbonate distribution shifts to favor formation of CO_3^{-2}.

At the other extreme, $CaCO_3$ does not precipitate if the product $[Ca^{+2}] \cdot [CO_3^{-2}]$ is less than K_{CA}. Water in this condition does in fact have a tendency to dissolve $CaCO_3$, and is said to be *undersaturated* with respect to $CaCO_3$. If $CaCO_3$ is added to an undersaturated system, the $CaCO_3$ dissolves until a saturated condition is achieved, *i.e.,*

$$[Ca^{+2}] \cdot [CO_3^{-2}] = K_{CA}$$

Similarly, $Mg(OH)_2$ tends to deposit from systems where the concentration product $[Mg] \cdot [OH^-]^2$ exceeds the $Mg(OH)_2$ solubility product K_{MG}. The $Mg(OH)_2$ reaction is described by Equation 9.

$$Mg^{+2} + 2OH^- \rightleftharpoons Mg(OH)_2 \downarrow \qquad (9)$$

Systems in which $[Mg^{+2}] \cdot [OH^-]^2 = K_{MG}$ are saturated with respect to $Mg(OH)_2$. They are oversaturated if $[Mg^{+2}] \cdot [OH^-]^2$ exceeds K_{MG} and undersaturated if $[Mg^{+2}] \cdot [OH^-]^2$ is less than K_{MG}. Both $CaCO_3$ and $Mg(OH)_2$ become less soluble with increasing temperature and more soluble with increasing salinity.

CALDWELL-LAWRENCE DIAGRAMS

Caldwell-Lawrence (C-L) diagrams are diagrammatic solutions to problems involving $CaCO_3$ equilibria. All concentrations shown on the diagram refer to *soluble* components. Each point on the diagram represents a saturated condition, *i.e.,* each point describes a situation where

$[Ca^{+2}] \cdot [CO_3^{-2}] = K_{CA}$. Departures from the saturated condition can be accommodated as subsequently described. Each diagram is unique for a given temperature and salinity. Strictly speaking, results obtained from a diagram constructed for a temperature $25°C$ and a total dissolved salt (TDS) concentration of 40 mg/l are not valid for other conditions. But in fact, reasonable values can be obtained from diagrams depicting other conditions provided that departures from the baseline condition are not too large. Such compromises must be made since it is obviously impractical to construct diagrams for every possible situation.

The body of a C-L diagram (Figure 1) consists of a series of iso-calcium, iso-alkalinity and iso-pH lines. A saturated condition occurs at the point where three of the lines intersect, as at "A" and "B." Non-saturated waters can usually be represented by envelopes (instead of points), for example the envelope DEF. Before going further, we need to define some of the terms appearing on the diagram.

Concentration

Unless otherwise noted (and except for TDS and pH) *all quantities from this point on are expressed in mg/l as CaCO₃*. The advantage gained is convenience. One mg/l of a given substance reacts completely with one mg/l of any other substance, provided the concentrations of both are expressed in mg/l as $CaCO_3$. Under this convention, one need not write out and balance a chemical equation each time a reaction must be considered. For example, 25 mg of $Ca(OH)_2$ as $CaCO_3$ reacts with 25 mg of CO_2 or 25 mg of CO_3^{-2} when the concentrations of these chemicals are expressed as $CaCO_3$.

pH

pH is an indicator of the concentration of hydrogen ions in solution. More formally:

$$pH \cong -\log_{10} [H^+] \tag{10}$$

pH should be read to the nearest 0.01 unit with a pH meter having temperature compensation. The meter should be calibrated with a fresh buffer solution whose pH is in the desired operating range.

TDS (Total Dissolved Salts)

TDS is a measurement of system salinity and affects the shape of the various curves. Use the diagram developed for TDS values nearest to that of the water being processed. Generally TDS values are *not* expressed as mg/l $CaCO_3$ but simply in mass units, mg/l. This convention is followed in this chapter.

Temperature

Temperature strongly affects the shape of the curves. Again, use the diagram developed for temperatures nearest to that of the water being processed. Because temperature so strongly affects results, interpolation of results from two diagrams is suggested if the difference between the water temperature and temperature for which the diagram was derived is more than a few degrees.

Alkalinity

Alkalinity is defined as a water's acid neutralizing capacity and is equal to the amount of strong acid required to titrate a water from its initial condition to the carbonic acid end point. An equivalent expression for alkalinity is:

$$Alk = HCO_3^- + CO_3^{-2} + OH^- - H^+ \tag{11}$$

where the terms represent quantities expressed as mg/l $CaCO_3$. Alkalinity is a frequently used water quality parameter having the following valuable attributes:

1. Alkalinity is conservative. Thus changes in alkalinity occurring as a result of chemical addition or precipitation can be calculated in a simple and straightforward manner:

$$Alk_{final} = Alk_{init} + Alk_{added} - Alk_{pptd} \tag{12}$$

where the subscripts "init" and "final" refer to the initial and final water condition and the subscripts "added" and "pptd" refer to alkalinity added with the conditioning chemicals and removed by precipitation.

Example 1. A water with an initial alkalinity of 15 mg/l is treated with 30 mg/l Na_2CO_3 with the resultant precipitation of 5 mg/l $CaCO_3$. All quantities are expressed as $CaCO_3$. What is the alkalinity of the settled effluent, assuming a perfect solids separation?

    a. Alk_{init} = 15 mg/l.
    b. Alk_{added} = 30 mg/l, that is, we have added 30 mg/l of CO_3^{-2} ion.
    c. Alk_{pptd} = 5 mg/l.
    d. From Equation 12, Alk_{final} = 15 + 30 − 5 = 40 mg/l.

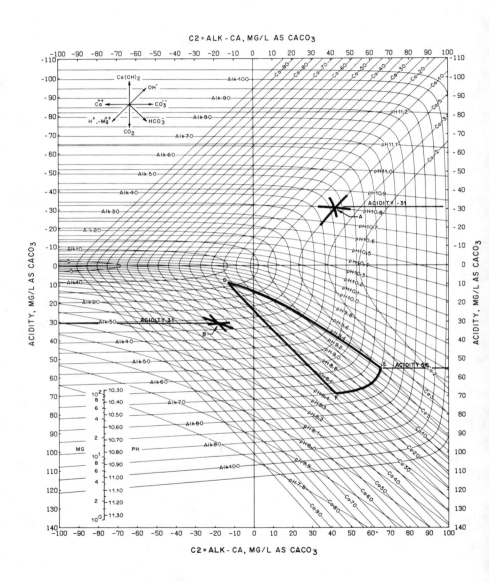

Figure 1. Caldwell-Lawrence water conditioning diagram; 25°C, 40 mg/l TDS.

Example 2. 35 mg CO_2/l (as $CaCO_3$) is dissolved into distilled water. What is the final alkalinity?

    a. Alk_{init} = 0. Distilled water contains only hydrogen and hydroxyl ions and they are present in equal concentrations. Thus, from Equation 11, Alk_{init} = 0.

    b. Alk_{added} = 0. No CO_2 term appears in the definition of alkalinity (Equation 11); thus no alkalinity has been added.

    c. No precipitation occurs; therefore Alk_{pptd} = 0.

    d. Therefore Alk_{final} = 0 + 0 − 0 = 0.

2. Alkalinity is independent of temperature; that is, the alkalinity value measured for a water at one temperature does not change as the water temperature is raised or lowered, providing that $CaCO_3$ precipitation does not occur. This has its practical effect in that alkalinity values calculated from titrations carried out at laboratory temperatures are valid even though the stream from which the sample was taken was at a different temperature.

Example 3. A sample is withdrawn from a stream whose temperature is $2°C$, then brought back to the laboratory where an alkalinity analysis is carried out at $25°C$. No $CaCO_3$ precipitation occurs during the warming process. The analysis shows an alkalinity of 25 mg/l. What is the alkalinity of the water at $2°C$? Answer: 25 mg/l.

Acidity

Alkalinity has its counterpart in acidity, the system's base neutralizing capacity. Acidity is expressed in terms of its various components by Equation 13.

$$\text{Acidity} = H_2CO_{3*} + HCO_3^- + H^+ - OH^- \tag{13}$$

where all quantities are expressed as mg/l $CaCO_3$. H_2CO_{3*} is the sum of H_2CO_3 and dissolved CO_2, as previously noted.

The properties that make alkalinity such a useful tool are also vested in acidity; that is, acidity is independent of temperature (providing CO_2 exchange with the atmosphere does not occur during heating or cooling). Furthermore, a water's final acidity is easily calculated knowing the acidity of the initial state and acidity of the chemicals added and precipitated.

$$\text{Acidity}_{final} = \text{Acidity}_{init} + \text{Acidity}_{added} - \text{Acidity}_{pptd} \tag{14}$$

In addition, acidity does not change during $CaCO_3$ precipitation or dissolution. This property is valuable in that $CaCO_3$ precipitation and dissolution processes can easily be traced on Caldwell-Lawrence diagrams since they move along lines of constant acidity. Note that acidity is frequently a negative value.

Example 4. The acidity of a lime-treated effluent is -200 mg/l. Addition of 60 mg/l of CO_2 is sufficient to precipitate 45 mg $CaCO_3$/l. What is the acidity of the carbonated effluent following separation of the precipitated $CaCO_3$?

    a. $Acidity_{init}$ = -200 (given).

    b. $Acidity_{added}$ = 60 mg/l CO_2.

    c. $Acidity_{pptd}$ = 0. No CO_3^{-2} term appears in the definition of acidity (Equation 13). Therefore no acidity is precipitated.

    d. $Acidity_{final}$ = -200 + 60 - 0 = 140.

Example 5. 35 mg CO_2/l is dissolved in distilled water. Calculate the acidity of the treated water.

    a. $Acidity_{init}$ = 0. Distilled water contains only H^+ and OH^- ions and they are present in equal concentrations. From Equation 13, acidity = 0.

    b. $Acidity_{added}$ = 35 mg/l.

    c. $Acidity_{pptd}$ = 0.

    d. $Acidity_{final}$ = 0 + 35 - 0 = 35 (mg/l as $CaCO_3$).

In practice, acidity is seldom measured because of a difficult end point. Acidity can be calculated, however, if alkalinity and pH are known. On the Caldwell-Lawrence diagrams, acidity is the vertical axis. The acidity of a water is found by extending a horizontal line from the *intersection of the system pH and alkalinity lines* to the vertical axis.

Example 6. Analysis of a water is: pH 9.33, alkalinity 70, Ca 28, temperature = $25°C$ and TDS = 40 mg/l. What is its acidity?· From Figure 1 (envelope DEF), acidity = 56 mg/l.

C2

C2 is the horizontal axis and is defined:

$$C2 = Alk - Ca = HCO_3^- + CO_3^{-2} + OH^- - H^+ - Ca \qquad (15)$$

Because CO_3^{-2} and Ca^{+2} must change by the same amount during $CaCO_3$ precipitation or dissolution, C2 remains constant while these processes occur. Thus $CaCO_3$ precipitation and dissolution processes are easily followed on Caldwell-Lawrence diagrams since they move vertically along constant C2 lines. The C2 value for any water can be found at the intersection of the measured calcium and alkalinity values.

pH-Mg Nomograph

The pH-Mg nomograph (lower left hand corner of Figure 1) is not included in the main part of the diagram, but is shown separately as a straight line scale. Listed are corresponding pH and Mg concentration values for which the system is saturated with respect to $Mg(OH)_2$. $Mg(OH)_2$ does not precipitate from systems where the actual Mg concentration present is less than the listed value.

Example 7. A water has the following analysis: Mg 20, pH 10.53, temperature 25°C, TDS 400 mg/l. Will $Mg(OH)_2$ precipitate under the given conditions? From the nomograph, the saturation value for Mg at pH 10.53 is 40 mg/l, so $Mg(OH)_2$ cannot be present in this system until Mg is $\geqslant$ 40 mg/l. Therefore, no $Mg(OH)_2$ will precipitate.

Direction Vectors

The direction vectors (upper left hand corner of Figure 1) indicate the direction that points describing the system condition must follow upon addition of a given chemical *provided* the system remains saturated with respect to $CaCO_3$ during the process. For example, the addition of $NaHCO_3$ to a saturated system moves the point describing the system condition downward and to the right at a 45° angle. Note that all angles shown are 45° or 90°.

Caldwell-Lawrence diagrams can be used for both corrosion control and water softening calculations. Understanding of the techniques used to solve corrosion control problems is a prerequisite for the working of water softening calculations. Accordingly, corrosion control calculations are discussed first.

CHEMICAL CONDITIONING FOR CORROSION CONTROL

Corrosion Phenomena

Metallic corrosion is an electrochemical phenomena. The two most common types are uniform corrosion, which produces an even thinning of the metal, and nonuniform corrosion, which produces both pin-hole leaks and tubercular accumulations of the corrosion products. In iron systems, nonuniform or pitting corrosion is the more common and more serious. Not only is early failure of the pipe wall more likely, but the tubercular protuberances formed compromise the pipe's water carrying capacity.

In the nonuniform corrosion of iron pipe, localized iron dissolution produces pitting, as in Equation 16. The pitted areas become anodic, *i.e.,* positively charged.

$$Fe^\circ - 2e^- \rightarrow Fe^{+2} \qquad (16)$$

The ferrous ion produced then reacts with water, as in Equation 17. Hydrogen ion is generated and the solution in the immediate vicinity of the anodes becomes acidic.

$$Fe^{+2} + 2H_2O \rightarrow Fe(OH)_2 \downarrow + 2H^+ \qquad (17)$$

This reaction is subsequently followed by the stepwise oxygenation of the bivalent corrosion product to the trivalent state.

$$\frac{1}{2}O_2 + 2Fe(OH)_2 + H_2O \rightarrow 2Fe(OH)_3 \qquad (18)$$

and by further slow transformation to iron oxide.

$$2Fe(OH)_3 \rightarrow Fe_2O_3 + 3H_2O \qquad (19)$$

Electrons released as a result of the reaction described in Equation 16 transfer through the pipe wall to surfaces (the cathodes) surrounding the anodic pit, and are taken up by oxidizing agents. A typical cathodic reaction is:

$$\frac{1}{2}O_2 + H_2O + 2e^- \rightarrow 2OH^- \qquad (20)$$

Cathodic reactions produce hydroxyl ion; thus the regions in the immediate vicinity of the cathodes become basic. It is likely that the potential difference between electrodes (the corrosive driving force) increases as the pH difference between anodic and cathodic areas increases. Because bicarbonate neutralizes both OH^- and H^+, it reduces the corrosive driving force. Therefore, waters containing significant quantities of bicarbonate are likely to be less corrosive.

Depositions on cathodic areas mitigate corrosion by hindering electron transfer between the metallic surface and the electron acceptor (oxygen). Depositions on anodic areas retard corrosive reactions by hindering migration of the anodic products into solution and by creating conditions at the interface that decelerate electron transfer or inhibit chemical reactions.[4] Fe_2O_3, formed by the subsequent slow oxidation of the intermediate corrosion products $Fe(OH)_2$ and $Fe(OH)_3$, is known to provide effective protection; the intermediate products themselves are porous and do not provide as effective protection. The protection provided by Fe_2O_3 films is reduced by the peptizing action of chlorides and sulfates.

Oxygen exerts a dual influence on iron pipe corrosion. It accelerates corrosion by driving the cathodic reaction (Equation 20) but helps suppress corrosion by contributing to the formation of anodic films. Its overall effect is determined by which set of reactions (anodic or cathodic) is controlling at the particular time. Stumm[4] suggests that corrosion is cathodically controlled during initial stages, anodically controlled later on.

Depositions purposely laid down for corrosion protection can be controlled to produce the desired effect. Stumm[4] indicates that total coverage is not required to reduce corrosion. Corrosion rates are reduced in proportion to the coverage of the rate-limiting electrode. However, Larson[5] suggests that while partial coverage may be helpful in minimizing "red water," unless coverage is complete some tuberculation will appear; partial protection can be worse than no protection at all from the standpoint of water-carrying capacity.

The description just completed is keyed to pitting corrosion of iron pipe. The principles outlined are generally applicable, however. Similar electrochemical reactions occur during corrosion of other metals. The differences between uniform and nonuniform corrosion are more related to the number and size of the anodic and cathodic areas than to any difference in underlying principles.

The corrosion of concrete and cement structures appears to be by a less complex mechanism, *i.e.,* simple dissolution. Waters undersaturated with respect to the constituents of concrete and cements will tend to corrode structures containing such materials.

Characteristics of a Well-Conditioned Water

What are the characteristics of a well-conditioned water? Conditioning requirements must be viewed in terms of the equipment to be protected and the fluid in contact with it. Metallic corrosion is an electrochemical phenomenon requiring a complete electrical circuit in order to proceed. Neutral waters devoid of oxygen (*e.g.,* some ground waters) are not likely to be corrosive because they lack the cathodic electron acceptor (oxygen) required to complete the circuit. Little or no conditioning is generally required for such waters *provided* oxygen can be totally excluded during the water treatment and distribution processes. On the other hand, oxygenated waters (*e.g.,* surface waters), which are by far more common, contain this electron acceptor so that corrosion must be prevented by another technique. This technique consists of placing a film between the fluid and the metal. This barrier interferes with the migration of corrosion products from anodic areas and the flow of oxygen to cathodic areas, which thus reduces the rate of the corrosive reaction. To be effective, the film should be dense, tenacious and provide uniform coverage. The literature suggests that such $CaCO_3$ films can be produced under the following conditions:

1. The water should be oversaturated with respect to $CaCO_3$. The oversaturation should place the theoretical precipitation potential in the range 4-10 mg/l.
2. Calcium and alkalinity values should each be at least 40 mg/l as $CaCO_3$, and more if economically feasible. They should be present in approximately equal concentrations.
3. The ratio alkalinity/(chlorides + sulfates) should be at least 5/1 (where all concentrations are expressed as $CaCO_3$).
4. pH values should be in the range 6.8-7.3.

Note that some of these conditions may be mutually exclusive, *i.e.,* one may be achievable only at the expense of another. Thus the well-conditioned water (as represented by items 1, 2, 3 and 4 above) may sometimes be impossible to prepare, or, if possible, be prepared only at great expense. If this be so, which conditions should be preserved, and which compromised? Items 1, 2 and 3 appear to be the most important and should be preserved. As a practical matter item 4 can be preserved only if the untreated water is initially high in calcium and alkalinity. Otherwise, economics demand that conditioning of low calcium-low alkalinity waters be carried out at pH 8.5 and above, even though films produced at such pH values are not so protective. Excessive pH values should be avoided, however, because the more destructive nonuniform corrosion prevails at higher pH values and because subsequent troublesome $Mg(OH)_2$ deposition in domestic water heaters must be prevented. As noted previously, $Mg(OH)_2$ becomes less soluble with increasing temperature and could form scale in hot water heaters unless the plant effluent pH is kept below the limits described by Figure 2.

Example 8. A water contains 40 mg/l Mg and has a temperature of 13°C. What pH limit should be applied to the finished water? From Figure 2, if conditioned water pH exceeds approximately 9.2, scaling problems might occur in the consumers' hot water heaters. Therefore the pH of the finished water should be limited to less than 9.2.

The pH range 8.0-8.5 should also be avoided if at all possible. In this range water has little ability to absorb anodic acids and cathodic bases produced by the corrosive reaction, which, if not neutralized, continue to accelerate and perpetuate the destructive process. Figure 3 shows that:

1. Water's neutralizing ability (also known as buffering capacity) is a function of the pH and total carbonate concentration (C_T) of the system. Buffering capacity, formally defined as the quantity

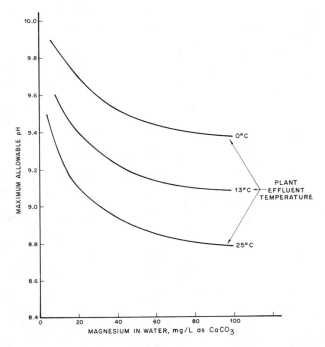

Figure 2. Approximate pH limits to avoid $Mg(OH)_2$ fouling in domestic water heaters. (After Reference 6.)

of strong base or strong acid required to change system pH by one pH unit, is a quantitative measure of a water's ability to resist pH change. A large buffering capacity denotes high resistance to change, thus good neutralizing ability.

2. Buffering capacity is greatest at the pK points and least at the equivalence points.

3. The pH region 8.0-8.5 alluded to above brackets the HCO_3^- equivalence point.

4. Although C_T (essentially bicarbonate ion at pH 6-10) increases buffering capacity enormously on either side of the pH range 8.0-8.5, within this range it has very little effect, and buffering capacity is low no matter what the C_T value. We have previously stated that waters containing significant quantities of bicarbonate are less likely to be corrosive. Figure 3 allows the limitations of this statement to be clearly seen. Within the range pH 8.0-8.5 it has little validity.

Guidelines for the protection of cement are less well documented. Waters low in calcium, aluminum, silica and iron are undersaturated with

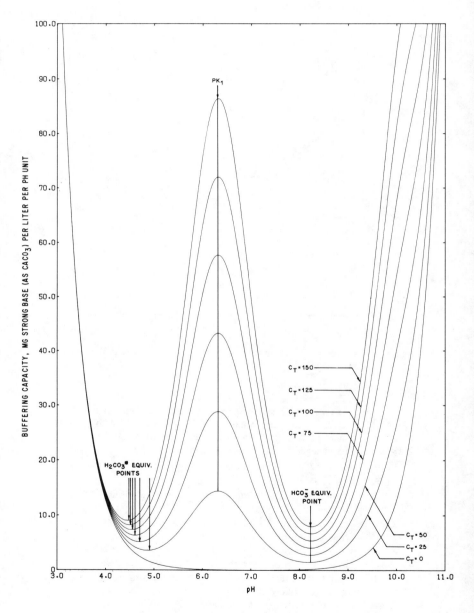

TEMP. = 25.0 DEG. CENTIGRADE, IONIC STRENGTH =.01000

Figure 3. Buffering capacity vs pH for the carbonic acid system.

respect to the constituents of cement (*e.g.,* $3CaO \cdot SiO_2$, $2CaO \cdot SiO_2$, $3CaO \cdot Al_2O_3$, $4CaO \cdot Al_2O_3 \cdot Fe_2O_3$) and thus corrosive. The absence of solubility data for these compounds makes it difficult to determine the concentrations of Ca, Al, Si, and Fe that must be present to prevent the undersaturated (corrosive) condition. However, protection *can* be provided by the deposition of a $CaCO_3$ film. Therefore water conditioned to conform with these guidelines presented for the protection of metallic surfaces is effective in protecting cement as well. For cement, less emphasis need be placed on maintaining a minimum alkalinity/($Cl^- + SO_4^{-2}$) ratio. This rule pertains to maintaining the passivating properties of protective metallic oxide films. Such films are of no consequence in cement systems.

Before plunging further into the intricacies of $CaCO_3$ protection it seems prudent to emphasize that simply because a water is undersaturated with respect to $CaCO_3$ does not automatically mean it is aggressive. It may contain other protective materials that obviate the need for $CaCO_3$ protection. For example, silica and color are natural components of water that appear to possess corrosion inhibiting properties. Waters containing sufficient quantities of such components (or waters having high alkalinity/($Cl^- + SO_4^{-2}$) ratios may be effectively protective, even without the assistance of $CaCO_3$ films. Unfortunately such inhibiting components may be partially removed in the coagulation-filtration process, thus turning a naturally protective water into an agressive one.

The protective state of the unconditioned water should, of course, be evaluated prior to embarking on any conditioning program. The most rigorous evaluation is inspection of the water distribution system itself (including household plumbing, where corrosion is often most severe). An alternative approach is by measurement of the rate of corrosion of test specimens in the laboratory, or better yet, specimens mounted in the distribution system. Techniques for the measurement of corrosion rates are discussed by Fontana and Greene[7] and in the ASTM standards.[8]

If the water is not naturally protective, and a program to **provide** protection by $CaCO_3$ films is decided upon, the engineer must evaluate the various routes by which the protective condition can be achieved. He should examine all approaches to determine which will best satisfy his requirements as to cost and other factors. Caldwell-Lawrence diagrams can be used as a tool in this analysis.

Using Caldwell-Lawrence Diagrams to Solve Corrosion Problems

The methods for solving water conditioning problems are illustrated by several examples, using a step-by-step procedure. Note that before calculations can proceed, the untreated water's temperature, calcium, magnesium, alkalinity, TDS and pH values must be measured.   Techniques for the measurement of these parameters are described in *Standard Methods*.[9]   A better procedure for alkalinity is given by Sawyer and McCarty.[10]

Example 9.   A water has the following characteristics: alkalinity 60 mg/l, calcium 40 mg/l, magnesium 10 mg/l, temperature 25°C, pH 8.1, TDS 40 mg/l.   Chlorides and sulfates are negligible.   You think the water is corrosive because the customers are complaining about "red water" and your distribution system is iron.   Determine if this is so and see what kind of protection would be afforded by the addition of, say, 5 mg/l $Ca(OH)_2$.

Step 1.   *Locate the untreated water on the C-L diagram.*   Note that three parameters must be defined in order to make this location.   These are calcium, alkalinity and pH.   The appropriate diagram is that for 25°C, TDS = 40 mg/l.   (Unmarked diagrams for other temperature and TDS conditions are shown on pages 555 to 564.)   The water is represented by the three-sided envelope ABC (solid lines) in Figure 4, bounded by the concentration line Alk 60, Ca 40 and pH 8.1.

Step. 2.   *Determine the saturation state of the untreated water.*   If the water is described by a point, then it is saturated.   If an envelope is required to describe the water, it is either undersaturated or oversaturated. If the measured calcium is greater than the calcium value found at the intersection of the pH and alkalinity lines, the solution is oversaturated, and if it is less, the solution is undersaturated.

Because the water of this example cannot be described by a single point it is not saturated, *i.e.,* it is either oversaturated or undersaturated. This water is undersaturated because the measured calcium value is less than the value of the calcium line (80 mg/l) passing through the intersection of the alkalinity and pH lines. This water requires treatment to promote an oversaturated (protective) condition.

Step 3.   *Determine the acidity of the untreated water.*   This is necessary for subsequent calculations.   Acidity is found at the intersection of the alkalinity and pH lines.   In this example, acidity = 62 mg/l.

Step 4.   *Determine the values of calcium, alkalinity and acidity that would be achieved as a result of adding a given amount of conditioning chemical.*   Impose the artificial (and temporary) restriction that $CaCO_3$ precipitation will *not* be allowed to occur during the addition.   Under this presumption all the chemicals added go into solution and stay in solution until precipitation begins.   (Note that this restriction is simply

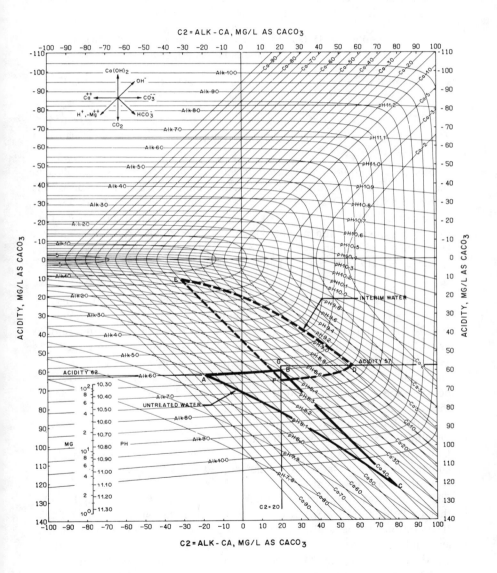

Figure 4. C-L diagram for Example 9; 25°C, 40 mg/l TDS.

a ploy to facilitate calculations. In reality, precipitation could begin as soon as the chemicals are added.)

As a result of adding 5 mg/l $Ca(OH)_2$ the following temporary (interim) values of calcium, alkalinity and acidity are achieved.

- Ca = 40 + 5 = 45 (adding 5 mg/l Ca increases calcium by 5 mg/l)
- Alk = 60 + 5 = 65 (by Equations 11 and 12, adding 5 mg/l OH^- increases alkalinity by 5 mg/l)
- Acidity = 62 – 5 = 57 (by Equations 13 and 14, adding 5 mg/l OH^- decreases acidity by 5 mg/l)

The pH of the interim state is found at the point (acidity 57, Alk 65). The pH is 9.08. The interim state (Envelope DEF, dashed lines) can now be located on the C-L diagram.

Step 5. *Determine the saturation state of the interim system.* Because the calcium value (45 mg/l) of the interim water exceeds the value of the calcium line (8 mg/l) at the intersection of the alkalinity and pH lines, the interim water is oversaturated and precipitation can occur. Note that the $CaCO_3$ precipitation potential is not 45 – 8 = 37 mg/l, but something less, as discussed in Step 6. The CZ value (65 – 45 = 20 mg/l) is found at the intersection of the calcium and alkalinity lines.

Step 6. *Allow the precipitation to proceed* once oversaturation has been achieved. (Remove the "no precipitation" restriction.) During $CaCO_3$ precipitation the envelope representing the water condition shrinks, eventually converging to a single point (Point G) that represents the saturated condition. This point is located at the intersection of the C2 and acidity lines. The potential for $CaCO_3$ precipitation is the difference between the calcium value of the interim state described in Step 4 and the calcium value of the saturated state.

If the $CaCO_3$ potential calculated is a desirable value, and other factors (alkalinity, Ca and pH) are also acceptable, the calculation is complete. If not, then return to Step 4, add more or less conditioning chemical, and repeat the calculations until the desired potential is achieved.

In the example, Point G represents the saturated condition. The difference between the calcium value of the interim state (45 mg/l) and that of Point G (38 mg/l) is equal to the precipitation potential (7 mg/l). If conditions at the wall and conditions in the bulk solution were equivalent and the liquid residence time in the distribution system were sufficiently long for the $CaCO_3$ precipitation to go to full completion, 7 mg/l $CaCO_3$ would be precipitated. Check to see that the alkalinity, calcium, alkalinity/(Cl^- + SO_4^{-2}) ratio and pH of the interim water are conducive to formation of a protective film.

1. The alkalinity (65) and calcium (45) values are above the 40 mg/l minimum and approximately equal to one another.
2. The alkalinity/(Cl^- + SO_4^{-2}) ratio is sufficient. (Chlorides + sulfates as given are negligible.)
3. From Figure 2, a water at $25°C$ containing 10 mg/l Mg will not cause scaling problems in hot water heaters until a pH of approximately 9.3 is exceeded. The pH of the interim water (9.08) is below that limit, so $Mg(OH)_2$ fouling should be no problem. The pH is also above the range of minimum buffering capacity (pH 8.0-8.5).

All the conditions for a protective water have thus been fulfilled, and therefore, the water is satisfactory. If it were not satisfactory, Steps 4 through 6 would be repeated with either more or less conditioning chemical (or different types of chemical) until a satisfactory condition is achieved.

Step 7. At the end of each exercise, *transform chemical requirements from mg/l $CaCO_3$ to real mass units* (those that can be measured on a balance or a scale) by using Equation 21.

$$\frac{\text{lb chemical required}}{\text{million gal } H_2O \text{ treated}} = \frac{16.7 \text{ (mg/l chemical as } CaCO_3\text{)(equiv wt chemical)}}{\text{(chemical purity,\%)}} \quad (21)$$

Chemical usage may be obtained in metric units (kg/m^3) by multiplying the mass in English units by 1.2×10^{-4}. Equivalent weights and approximate bulk chemical costs are found in Table VIII (page 553). The cost data can be used for making economic comparisons between the various processing alternatives.

There are two means for obtaining lime: $Ca(OH)_2$ may be purchased directly or it can be created by slaking quicklime (CaO). Quicklime is less expensive, but must be put through the sometimes aggravating slaking process. Each operator must determine for himself if the chemical cost saved by using quicklime is worth the cost of the slaking equipment and difficulties encountered in slaking. Determine the chemical cost of both options, assuming the purity is 90% for both kinds of lime. From Equation 21 and Table VIII:

1. If quicklime is used, CaO required = 16.7 (5 mg/l)(28)/90% = 26.0 lb/mil gal; chemical cost is 26.0 lb/mil gal x 2.25¢/lb = 59¢/mil gal. In metric units, CaO required = $(1.2 \times 10^{-4})(26) = 3.12 \times 10^{-3}$ kg/m^3; chemical cost = $(3.12 \times 10^{-3})(4.96) = 0.015$¢/$m^3$.
2. If hydrated lime is used, $Ca(OH)_2$ required = 16.7(5)(37)/90 = 34.3 lb/mil gal; chemical cost = 34.3 x 2.80 = 96¢/mil gal. In metric units, $Ca(OH)_2^2$ required = $(1.2 \times 10^{-4})(34.3) = 4.12 \times 10^{-3}$ kg/m^3; chemical cost = $(4.12 \times 10^{-3})(6.17) = 0.025$¢/$m^3$.

Table II.   Operational Summary for Example 9

Parameter	Untreated Water	Effect of Chemicals Added	Interim State	Saturated State
Chemicals Added, as $CaCO_3$	–	5 mg/l $Ca(OH)_2$	–	–
pH	8.10	–	9.08	8.45
Ca, mg/l as $CaCO_3$	40	+5	45	38
Alkalinity, mg/l as $CaCO_3$	60	+5	65	58
Acidity, mg/l as $CaCO_3$	62	–5	57	57
C2, mg/l as $CaCO_3$	20	–	20	20
pH_s	8.42	–	8.32	8.45
Saturation State	under-saturated	–	over-saturated	saturated
$CaCO_3$ Precipitation Potential	0	–	7	0
Langelier's Index	–0.32	–	0.76	0.00
Ryznar's Index	8.74	–	7.56	8.45

If plant water quality varies, then the lime required to maintain the desired $CaCO_3$ potential changes also and must be recalculated. The conditioning process for this example is summarized in Table II.

Water Conditioning Indices

Some of the indices commonly used in water conditioning can be calculated from the C-L diagrams.

Langelier's Saturation Index

Langelier's Index is defined by Equation 22.

$$LI = pH - pH_s \tag{22}$$

where   pH $\equiv$ pH of the system (as measured by a pH meter)

pH_s $\equiv$ pH the system would have at the measured values of calcium and alkalinity.

pH_s is equal to the value of the pH at the intersection of the calcium and alkalinity lines. The pH_s for the untreated water of Example 9 is 8.42, and for the interim water is 8.32 (Figure 4). Thus, the index for the untreated water = 8.10 - 8.42 = -0.32, and for the interim water the index = 9.08 - 8.32 = 0.76. A positive Langelier Index usually connotes an oversaturated state, and a negative index, an undersaturated state.

Ryznar's Stability Index

The Ryznar Index is given by Equation 23.

$$RI = 2 \, pH_s - pH \tag{23}$$

where pH and pH_s are as defined above. Dye and Tuepker[11] have suggested that waters with an index greater than about 7 are corrosive and less than 7 are scale-forming. The index for the interim water (2 x 8.32 - 9.08 = 7.56) suggests it is not scale-forming, a notion in conflict with the conclusions drawn from use of the C-L diagram, and the Langelier Index. The conditioning process for Example 9 is summarized in Table II.

Example 10. Could the water described in Example 9 have been conditioned with sodium hydroxide? If so, would it be more or less cost effective than lime in producing an approximately equivalently conditioned water?

Caustic soda *can* be used as a conditioning chemical. In order to see whether it will be effective, the amount needed to provide a $CaCO_3$ precipitation potential of about 7 mg/l must be determined, as must the effect of the substitution on the other parameters (alkalinity, calcium and pH) and the chemical cost.

Step 1. Start with the step-by-step procedure outlined in the previous examples and locate the untreated water on the C-L diagram. It is the same water as in example a, that is Ca = 40, Alk 60, pH 8.1, acidity 62 ppm, and it is undersaturated. This water is represented by the solid envelope ABC shown in Figure 5.

Step 2. As a first trial, assume an addition of 5 mg/l NaOH. If the "no precipitation" restriction is applied, the values of calcium, alkalinity and acidity resulting are:

- Ca = 40 + 0 = 40 (no change, as NaOH contains no calcium)
- Alk = 60 + 5 = 65
- acidity = 62 - 5 = 57

The pH of this interim water can be located at the intersection of the acidity and alkalinity lines; pH is 9.08. The interim state (envelope DEF, dashed lines) can now be plotted on the diagram. The interim water is oversaturated because its calcium concentration (40) exceeds the calcium

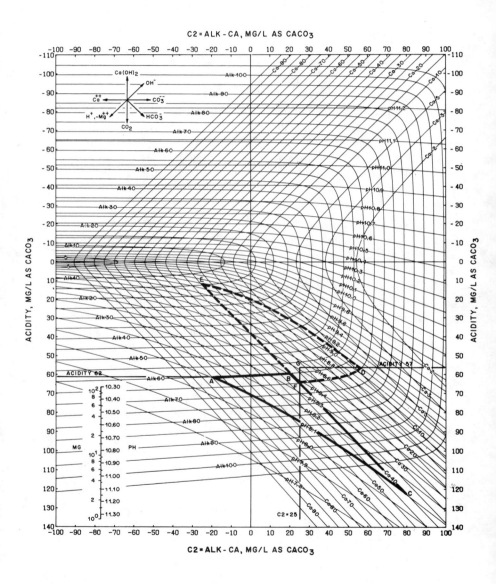

Figure 5. C-L diagram for Example 10; 25°C, 40 mg/l TDS.

value (9) at the intersection of the alkalinity 65 and pH 9.08 lines
(Point D). C2 is found at the intersection of the interim water's calcium
and alkalinity lines and equals 65 – 40 = 25.

Step 3. Remove the "no precipitation" restriction and allow the precipitation to proceed. If all the $CaCO_3$ that could precipitate did in
fact do so, the envelope would shrink to Point G, which represents the
saturated state. The precipitation potential is equal to the difference in
calcium values between the interim state (Ca 40) and the saturated state
(Ca 33) so the precipitation potential = 40 – 33 = 7. This is equal to
the target precipitation potential of 7. The interim water's alkalinity
and pH values are the same as those of the interim water of Example 9.
Its calcium value (40 mg/l) is slightly less, however, and thus its protective
ability is also slightly less.

Step 4. From Equation 21 and Table VIII, the amount of 98% pure
NaOH required = 16.7(5)(40)/98 = 34.1 lb/million gal (0.00409 kg/m^3)
and the cost =(34.1 x 7.50)/100 = $2.56/million gal (0.068¢/m^3). This
is about 4.3 times the cost of CaO and 2.7 times the cost of $Ca(OH)_2$.

The conditioning process for this example is summarized in Table III.
Fortunately, the first trial (5 ppm NaOH) gave the approximate $CaCO_3$
potential desired. The reader may verify that other additions provide
more or less potential. For example, an addition of 10 ppm NaOH
provides a precipitation potential of 17 mg $CaCO_3$/l, which is too much.
Note that Na_2CO_3 was not considered as a conditioning agent. Would
this have been more cost effective than NaOH? Than lime? What about
lime plus Na_2CO_3? The feasibility of all these alternatives can be
examined via Caldwell-Lawrence diagrams.

The Effect of Temperature

Temperature drastically alters the shape of the Caldwell-Lawrence
diagram. Therefore, calculations for a 2°C water should not be done on,
say, a 25°C diagram. A 2°C diagram (or one reasonably close to it)
must be used. This raises the following question. Must the analysis of
the untreated water be carried out at stream temperature in order for
the parameters measured to be valid? This has practical implications in
that water analyses are usually carried out at laboratory temperature,
whereas the temperature of the stream is quite different.

The answer is that parameters determined at laboratory temperatures
are useful, because with one exception (pH) they are under most conditions
exactly the same as at any other temperature. More specifically, alkalinity
and calcium do not change with temperature (provided $CaCO_3$ precipitation
has not occurred) nor does acidity change (provided there has been no

Table III. Operational Summary for Example 10

Parameter	Untreated Water	Effect of Chemicals Added	Interim State	Saturated State
Chemicals Added, as $CaCO_3$	–	5 mg/l NaOH	–	–
pH	8.10	–	9.08	8.52
Ca, mg/l as $CaCO_3$	40	0	40	33
Alkalinity, mg/l as $CaCO_3$	60	+5	65	58
Acidity, mg/l as $CaCO_3$	62	–5	57	57
C2, mg/l as $CaCO_3$	20	–	25	25
pH_s	8.42	–	8.38	8.52
Saturation State	under-saturated	–	over-saturated	saturated
$CaCO_3$ Precipitation Potential	0	–	7	0
Langlier's Index	–0.32	–	0.72	0.0
Ryznar's Index	8.74	–	7.66	8.52

CO_2 transfer from the atmosphere). Thus alkalinity, calcium and acidity values determined at 25°C are the same as those at 2°C or any temperature under the conditions noted.

Example 11. A sample is brought from a slush ice-laden stream having a temperature of 2°C. The sample bottle is sealed to prevent the transfer of CO_2 and left overnight in the laboratory, where it comes to room temperature. Inspection shows that no precipitation has occurred. The analysis is: temperature 25°C, TDS 40, pH 8.8, alkalinity 35, and Ca 10. The treatment plant intake is on this stream. Determine if this is an aggressive water at stream temperature.

Step 1. First, find the acidity of the 25°C sample. Acidity (33 mg/l) is located at the intersection of the lines representing pH 8.8 and Alk 35 on the 25°C, TDS 40 C-L diagram.

Step 2. Because no $CaCO_3$ precipitation or CO_2 transfer has occurred, the calcium, alkalinity and acidity values at 2°C are exactly the same as at 25°C. Draw in the calcium 10 and alkalinity 35 lines on the 2°C, TDS 40 diagram. The intersection of the lines acidity 33, alkalinity 35 locates the pH (9.22). The envelope locating the untreated 2°C water can now be drawn and any conditioning calculations can proceed from there in the usual manner.

Table IV.   Operational Summary for Example 11.

Parameter	Warm Water	Cold Water
Temperature, °C	25	2
pH	8.80	9.22
Ca, mg/l as $CaCO_3$	10	10
Alkalinity, mg/l as $CaCO_3$	35	35
Acidity, mg/l as $CaCO_3$	33	33
C2, mg/l as $CaCO_3$	25	25
pH_s	9.34	9.97
Saturation State	undersaturated	undersaturated
$CaCO_3$ Precipitation Potential	0	0
Langelier's Index	−0.54	−0.75
Ryznar's Index	9.88	10.72

The calculations are summarized in Table IV.   Note that the main effect of bringing the water to room temperature was to decrease pH by 0.42 units.

Low Calcium-Low Alkalinity Waters

Low calcium-low alkalinity waters, such as those derived from snow melt, pose special problems for the operator because only part of the problem can be worked on C-L diagrams.   In addition the treatment process is more complex, since two conditioning chemicals are often required.   Example 12 shows how such problems can be handled.

Example 12.   The analysis for a water is alkalinity 10, calcium 2, magnesium 3 (all as $CaCO_3$), pH 8.3, TDS 40, temperature 25°C. Chlorides and sulfates are negligible.   The distribution system (cast iron) is corroding.   Provide protection by $CaCO_3$ deposition.

Step 1.   The diagram for 25°C and TDS 40 should be used.   But the untreated water condition cannot be located on the Caldwell-Lawrence diagram because:

- the intersection of the alkalinity and pH lines occurs at some point to the left of the chart, and
- no intersection of the calcium-alkalinity line occurs.

This is typical of low calcium-low alkalinity systems.

However this causes no great problems if one remembers that the purpose for locating the untreated water is to determine the initial acidity and estimate the saturation state. Both can be established in other ways. Acidity can be calculated directly from Equation 24.

$$\text{Acidity} = \frac{\left(\text{Alk} - \frac{K_W}{H^+} + H^+\right)\left(1 + \frac{H^+}{K_1}\right)}{\left(1 + \frac{K_2}{H^+}\right)} + H^+ - \frac{K_W}{H^+} \qquad (24)$$

where K_W, K_1 and K_2 are mass action constants as shown in Table IX (page 553), and alkalinity, acidity and H^+ are expressed in terms of mg/l $CaCO_3$. H^+ (as mg/l $CaCO_3$) can be found as a function of pH from Figure 11 (page 554). From the latter, $H^+ = 2.5 \times 10^{-4}$ mg/l $CaCO_3$ at pH 8.3, TDS 40 mg/l. From Table IX, $K_W = 2.68 \times 10^{-5}$, $K_1 = 1.15 \times 10^{-2}$, $K_2 = 5.37 \times 10^{-6}$, and therefore

$$\text{Acidity} = \frac{\left(10 - \frac{2.68 \times 10^{-5}}{2.5 \times 10^{-4}} + 2.5 \times 10^{-4}\right)\left(1 + \frac{2.5 \times 10^{-4}}{1.15 \times 10^{-2}}\right)}{\left(1 + \frac{5.37 \times 10^{-6}}{2.5 \times 10^{-4}}\right)}$$
$$+ 2.5 \times 10^{-4} - \frac{2.68 \times 10^{-5}}{2.5 \times 10^{-4}} = 10 \text{ mg/l as } CaCO_3$$

The water can be characterized as being undersaturated or oversaturated by noting that any water whose pH-alkalinity intersection occurs to the left of the chart has a calcium value exceeding 95 mg/l at saturation. The water (Ca 2) is clearly undersaturated.

The alkalinity and calcium concentrations of the untreated water are extremely low. They must be increased to at least 40 mg/l, and at the same time $CaCO_3$ precipitation potential must be held within limits (4-10 mg/l). From Figure 2, pH values of about 9.5 must not be exceeded. All objectives can be met, but more than one chemical is needed in order to do so. For example, if lime only is used, the water can be conditioned to the proper range of $CaCO_3$ potential, but only at pH values exceeding 9.5. Furthermore, minimum alkalinity and calcium values cannot be achieved. (The reader may verify this for himself.) If lime and carbon dioxide are used, however, all goals can be achieved.

Step 2. First, add sufficient lime to satisfy alkalinity and calcium requirements. As usual, the "no precipitation" restriction is in force during chemical addition. Adding 38 ppm Ca (OH)$_2$ to the initial water satisfies the minimum calcium requirements and produces interim water I (Figure 6, envelope ABC) which is oversaturated.

- Ca = 2 + 38 = 40
- Alk = 10 + 38 = 48
- acidity = 10 − 38 = −28

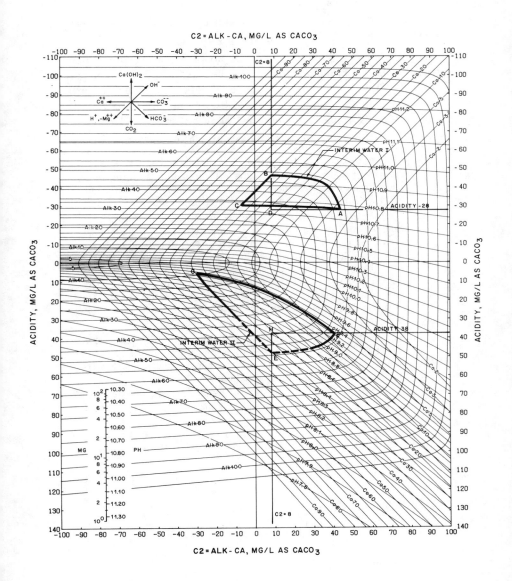

Figure 6. C-L diagram for Example 12; 25°C, 40 mg/l TDS.

- pH = 10.78
- C2 = 48 − 40 = 8
- $CaCO_3$ precipitation potential is 17 mg/l

Although lime addition satisfied minimum calcium and alkalinity requirements, the pH and $CaCO_3$ precipitation potentials are too high.

Step 3.   Next, still maintaining the "no precipitation" restriction, add sufficient CO_2 to bring simultaneously the $CaCO_3$ precipitation potential into the 4-10 mg/l range and pH to below 9.5. How much is sufficient? This may be done by the usual trial and error method. However, a more direct method is available if one uses following simple facts:

    a. Adding CO_2 does not change alkalinity (48) or calcium (40). The intersection of the alkalinity and calcium lines (Point E) and at least the first segments of these lines can therefore immediately be drawn in for the carbonated water. These segments are represented by the heavy dashed lines of the lower envelope of Figure 6. The line C2 = 48 − 40 = 8 can also be drawn.

    b. The precipitation potential is always the difference between the calcium value of the interim water and the calcium value at the intersection of the C2 and acidity lines. Therefore, simply draw in the acidity line whose intersection with the C2 line gives the desired precipitation potential. For example, if a potential of 8 mg/l $CaCO_3$ is desired, select an acidity line that intersects the C2 line at a calcium value (32) that differs from that of the interim water (40) by 8 mg/l. The acidity line satisfying this condition is acidity = 38 mg/l (line FH).

    c. Extend the acidity line and the alkalinity lines until they intersect (Point F). This locates the pH (9.35). The envelope FGE) representing interim water II can now be completed (although this is not necessary to finish the calculation).

    d. From Equations 13 and 14, the differences between the acidities of interim waters I and II is equal to the amount of CO_2 added, i.e., 38 − (−28) = 66 mg/l CO_2.

Assuming 90% pure CaO and 100% pure CO_2 are used, chemical requirements and costs are:

    a. **CaO required** $= \dfrac{16.7\ (38)(28)}{90} = \dfrac{197.4\ \text{lb}}{\text{million gal}} = \left(\dfrac{0.0237\ \text{kg}}{\text{m}^3}\right)$

    CaO cost $= \dfrac{197.4(2.25)}{100} = \$4.44/\text{million gal}\ (0.117¢/\text{m}^3)$

    b. CO_2 required $= \dfrac{(16.7)(66)(22)}{100} = \dfrac{242.5\ \text{lb}}{\text{million gal}} = \dfrac{0.0291\ \text{kg}}{\text{m}^3}$

However in a gas transfer operation all of the reagent added is generally not absorbed. Thus the CO_2 delivered must be boosted to compensate for the less-than-perfect transfer efficiency. Assuming a 90% transfer efficiency,

$$CO_2 \text{ delivered } = 242.5 \left(\frac{100}{90}\right) = \frac{269.4 \text{ lb}}{\text{million gal}} = \frac{0.0323 \text{ kg}}{m^3}$$

$$CO_2 \text{ cost } = \frac{(269.4)(4.10)}{100} = \frac{\$11.05}{\text{million gal}} = \frac{0.292\cent}{m^3}$$

(c)   Total chemical cost = 15.49/million gal=(0.41$\cent$/m^3).

This is 15-25 times the cost required to condition the water described in Example 9. This example vividly illustrates the technical and economic problems involved in treating low calcium-low alkalinity waters.

The operational summary for Example 12 is presented in Table V.

Although Interim Water I is clearly oversaturated, Langelier's Index is negative. At pH values above the carbonate equivalence point the Index reverses, i.e., Langelier's Index $\equiv$ pH$_s$ - pH. This has not always been appreciated and has caused some confusion among those wishing always to apply a fixed rule. The advantage of the Caldwell-Lawrence diagrams can be seen here. The technique used to determine a water's saturation state is always the same, and ambiguity is not a problem.

The fact that this calculation was carried out in separate steps (first calculations involving lime, then calculations involving CO_2) does not imply that chemical addition should also be staged. Better treatment is provided if chemicals are added simultaneously. In this way problems associated with extreme variation in pH (initial over-precipitation, for example) can be avoided.

Suppose that the magnesium concentration had been 30 mg/l instead of 3. From Figure 2, the maximum pH value suitable for Interim Water II would have been 9.0 instead of 9.35. At this reduced pH, can the same $CaCO_3$ precipitation potential (8 mg/l) be maintained, and can the other criteria for a well-conditioned water still be satisfied? The answer is yes. Simply add more lime and more CO_2. This will produce Interim Water II of higher calcium and alkalinity and, of course, lower pH. The reader can verify this for himself (and also that conditioning costs will increase).

The example suggests that treatment to provide an ideally conditioned water may sometimes be an economic strain. If that is so, limited conditioning should be carried out on the premise that some conditioning is better than none at all. Increasing the calcium and alkalinity concentrations of waters low in these components usually reduces corrosion to some extent, even when the interim conditions produced are less than ideal for precipitation of $CaCO_3$ films. For example, the City of San Francisco,

Table V. Operational Summary for Example 12

Parameter	Untreated Water	Effect of Chemicals Added	Interim Water I	Effects of Chemicals Added	Interim Water II	Saturated State
Chemicals Added, mg/l as $CaCO_3$	—	38 $Ca(OH)_2$	—	66 CO_2	—	—
pH	8.3	—	10.78	—	9.35	8.70
Ca, mg/l as $CaCO_3$	2	+38	40	0	40	32
Alkalinity, mg/l as $CaCO_3$	10	+38	48	0	48	40
Acidity, mg/l as $CaCO_3$	10	−38	−28	+66	38	38
C2, mg/l as $CaCO_3$	8	—	8	—	8	8
pH_s	—	—	10.95	—	8.50	8.70
Saturated State	undersaturated	—	oversaturated	—	oversaturated	saturated
$CaCO_3$ Precipitation Potential	0	—	17	—	8	0
Langelier's Index	—	—	−0.17	—	0.85	0.00
Ryznar's Index	—	—	11.12	—	7.65	8.70

California, controls a red water problem with modest lime additions and by blending with harder and more alkaline ground waters. This scheme is apparently successful, even though the water as distributed is undersaturated with respect to $CaCO_3$. In such situations the precipitation of dissolved pipe materials (for example, iron and zinc) and remarkably, calcium,* is enhanced as the calcium, alkalinity and pH are raised. Precipitates of these materials ($FeCO_3$, $ZnCO_3$, $CaCO_3$) deposit on and seal off corroding areas, reducing the rate of corrosion in comparison to previous rates. Some protection is afforded, but of course not so much as under ideal conditions. The degree of protection provided in such cases is dictated by the municipality's ability or willingness to pay.

It should now be apparent that almost any water condition can be produced, given the proper mix of conditioning chemicals. The routes traveled in arriving at a solution to the problem are limited only by the imagination of the designer or operator. Once the various routes have been identified, then such real world factors as chemical availability and cost determine which one is to be followed.

Operational Guidelines

Water should be brought to the oversaturated condition just prior to entering the system to be protected. If the oversaturated condition is achieved earlier, $CaCO_3$ deposition could begin prematurely and in equipment where protection is not really required. Premature deposition represents lost protection and sometimes causes operating problems. For example, if the oversaturated state were attained prior to filtration, deposition of $CaCO_3$ on the filter media could cause clogging, excessive pressure drop and difficult filter cleaning.

Chemicals should be added so that they are rapidly and uniformly distributed throughout the *entire* flow. In this way localized overdosing and underdosing is avoided. If initial mixing is poor, water near the chemical injection point is highly oversaturated while water remote from the feed point receives little chemical and may perhaps even be undersaturated. The overall result is rapid precipitation in the areas of gross oversaturation and little, if any, elsewhere. A loose, nonuniform coating

---

*Under certain circumstances, $CaCO_3$ precipitation may occur at the wall, even though the bulk or average water is undersaturated with respect to $CaCO_3$. This apparent contradiction can be explained by the fact that as a result of the corrosive reaction the pH in the immediate vicinity of certain areas (the cathodes) of a corroding surface is always high in comparison to the pH of the bulk solution. The addition of nominal amounts of calcium and alkalinity may be sufficient to produce a locally oversaturated condition at the cathode even though the bulk water remains undersaturated. Small but significant amounts of $CaCO_3$ can then be deposited.

results. Chemicals should therefore be injected at points of maximum turbulence—for example, in hydraulic jumps or mechanically mixed basins. Pulsating chemical feeders should be avoided or, if this is not possible, a desurger should be installed in the feed line.

Lime slaking and feeding equipment should be sealed as well as possible to prevent contact and reaction of the lime slurry with atmospheric CO_2. The resulting reaction product, $CaCO_3$, has a discouraging tendency to plug such equipment. This is also wasteful of lime, since $CaCO_3$ formed externally is of little or no value in protecting the distribution system.

The effect of velocity on corrosion depends upon the relative proportion of corrosion stimulating to corrosion inhibiting factors in the water. If the water is aggressive, velocity accelerates corrosion, since the corrosive elements (oxygen, chlorides) are more rapidly transferred to the pipe surface. If the water is well-conditioned, however, the protective ingredients (calcium, alkalinity) also are more rapidly transported to the surface and overwhelm the corrosive factors. McCauley and Abdullah[12] found good protective coatings formed from waters oversaturated with respect to $CaCO_3$ and moving at about 2 ft/sec. Good protection is generally not obtained from waters moving at low velocities, even when the ratio of inhibiting to corrosive factors is high.

When the water is chlorinated with chlorine gas, the pH is reduced, and the extent of the reduction depends upon the chemical composition of the water and the amount of chlorine used. Chlorination with hypochlorite effects little pH change, but, if anything, the pH increases slightly. If gaseous chlorination follows chemical conditioning for corrosion control, the $CaCO_3$ precipitation potential is reduced and might even be eliminated altogether. For this reason, chemical conditioning should follow chlorination after an interval sufficient to stabilize the pH. Of course, the calculations must reflect the altered condition of the feed water. The initial water condition is not that of the raw water but that of the chlorinated water just prior to conditioning. Remember that chlorination also unfavorably alters the alkalinity/(Cl^- + SO_4^{-2}) ratio, and such additions should be compensated by adding extra alkalinity.

Larson[13] conducted experiments indicating that free chlorine residuals also accelerate corrosion of steel. This appears to be additive to the adverse effects of chlorination on the $CaCO_3$ precipitation potential and the alkalinity/(Cl^- + SO_4^{-2}) ratios. In Larson's experiments, corrosion was accelerated when the free chlorine residual exceeded 0.4 mg/l. He theorized that the maximum free residual level might not always be 0.4 mg/l. This level could be affected by such factors as pH and the alkalinity/(Cl^- + SO_4^{-2}) ratio. The lesson is that no more chlorine should be added than that necessary to obtain disinfection in critical parts of the distribution system. Overdosing could increase the corrosion problem.

Marble Test

Occasionally the calculations should be checked with a simple experiment to assure that the chemical condition of the finished water has been correctly assessed. There are several reasons why variances between expected and actual results might occur.

* The diagrams available might not match the exact temperature and TDS of the water.
* Chemical species not considered when constructing the diagram might be present.
* Errors can be made in chemical analyses or in calculations, chemical feed equipment may not be operating as expected, or flow measurements could be wrong.

The saturation state of the finished water (in fact, of any water) with respect to $CaCO_3$ can be easily determined by the Marble test.

1. Divide a sample of the water into two parts.
2. Measure the calcium concentration of the first part.
3. Place the second portion in a stoppered bottle and add 0.1-0.2 g of finely divided $CaCO_3$ prior to sealing the bottle. The sample should fill the bottle completely. Mix by rapidly inverting the bottle several times. Mix several times over the next few hours. Allow the bottle to stand overnight and at the temperature of the process stream, if possible.
4. Next day, filter the supernatant from the second portion and measure the calcium concentration of the filtrate.
5. $CaCO_3$ precipitation has occurred if the calcium concentration of the second (filtered) portion is less than that of the first. When this occurs, the $CaCO_3$ precipitation potential is equal to the difference between calcium values. If the calcium concentration of the second portion exceeds that of the first, the water was undersaturated. If the concentrations are equal, the water was saturated. If the actual potential as measured by the Marble test is much different than the calculated potential, then a problem exists that must be resolved. As a start, check those items listed as potential sources of error. If the problem cannot be found, adjust the chemical dose to provide the desired precipitation potential as indicated by the Marble test.

COPING WITH LIMITATIONS OF THE CALCULATIONS

The calculations predict the $CaCO_3$ precipitation potential—that is, the amount of $CaCO_3$ which could precipitate if there were sufficient time for the reaction to go to full completion. In many water systems, liquid residence time is insufficient to allow full precipitation, so the amount of $CaCO_3$ available for protection may be somewhat different than predicted by the calculation.

Second, it must be emphasized that conditions at the precipitation site control the reactions that occur there. For example, if $CaCO_3$ is to precipitate on the pipe wall, the water in the immediate vicinity of the wall must be oversaturated with respect to $CaCO_3$. As Equations 17 through 20 suggest, however, the pH (and thus the saturation state) at the solid-water interface is often considerably different than in the bulk solution, particularly if the solid is metallic and corroding. Because conditions at the pipe wall are difficult to measure, the local saturation condition is seldom, if ever, evaluated. Bulk conditions are used instead, even though they are but approximations to the condition at the water-solid interface.

Finally, the degree of protection afforded by the scale that does form depends upon its quality (hardness, porosity, uniformity, tenacity), which is influenced by factors that for various reasons have not been considered in the calculations.

Thus, it appears that no one set of chemical conditions affords the same degree of protection in all systems. Furthermore, the calculations presented here do not describe the effects of all the variables. It appears, then, that the best conditions for each system must be found through trial and error.

The time and effort involved in finding these best conditions can be minimized if a logical, systematic approach to the problem is used. As a start, the bulk water should be conditioned to approximate closely those conditions described previously and summarized below:

1. A $CaCO_3$ precipitation potential of 4-10 mg/l.

2. Calcium and alkalinity concentrations of at least 40 mg/l as $CaCO_3$. Note that the greater the calcium and alkalinity the more protective the water is.

3. An alkalinity/(Cl^- + SO_4^{-2}) ratio of at least 5/1, where all concentrations are expressed as mg/l $CaCO_3$.

4. pH values less than those which could produce subsequent $Mg(OH)_2$ scaling in hot water heaters (see Figure 2). The pH range 8.0-8.5 should also be avoided, if possible.

5. Overdosing of chlorine must be avoided.

The above conditions apply to the most common situation—oxygenated waters carried in metallic distribution systems. They do *not* apply to neutral waters devoid of oxygen (*e.g.,* some ground waters) in metallic systems. Such waters may not need conditioning, *provided* that oxygen can be excluded during water treatment and distribution processes. Note, however, that elimination of oxygen transfer is a formidable task. It may well be less expensive to condition the water than to build an air-tight system. For cement systems, condition 3 is less important, and there appears to be little need to avoid the pH range 8.0-8.5.

After maintaining the desired condition for a period of time, the results must be evaluated. An evaluation might consist of measuring the weight loss of corrosion coupons, determining by inspection that surface conditions had ceased to deteriorate (or become worse), or observing that a pipe's hydraulic characteristics had stabilized or continued to decline. (Note that elimination of red water, while aesthetically pleasing, is not a sure sign that corrosion has been arrested. This is particularly true for waters conditioned at high pH. Corrosion may be continuing, but the corrosion products, in lieu of finding their way into the water, may instead be retained at the pipe surface as tubercules. Under these conditions the interface becomes rougher and the carrying capacity of the pipe diminishes.) Such evaluations could last from a few days to over a year, depending on the intent of the test. If feedback is satisfactory, the conditioning program would be maintained, but if not, a different set of conditions should be tried.

Although this discussion has been limited to protection by deposition of $CaCO_3$ films, objectivity requires that a point made earlier be reemphasized, namely that other protective techniques are also available. In certain situations these techniques may be more cost-effective than $CaCO_3$ deposition. Example 12 showed that bringing low calcium-low alkalinity waters to a protective condition was expensive where reagents that precipitate $CaCO_3$ were used. Alternative techniques such as silicate addition or treatment with zinc-polyphosphate[13] might in this particular situation provide equivalent protection at lower cost. An impartial analysis requires that such alternatives be considered.

WATER SOFTENING

Softening Chemistry

Hardness is removed in the lime-soda process by precipitation. The reactions that occur in the lime-soda process are described in Equations 25 through 30. These equations are written in molar quantities.

$$Ca^{+2} + 2HCO_3^- + Ca(OH)_2 \rightarrow 2CaCO_3\downarrow + 2H_2O \tag{25}$$

$$Mg^{+2} + 2HCO_3^- + 2Ca(OH)_2 \rightarrow 2CaCO_3\downarrow + Mg(OH)_2 + 2H_2O \tag{26}$$

$$Mg^{+2} + SO_4^{-2} + Ca(OH)_2 \rightarrow Mg(OH)_2\downarrow + Ca^{+2} + SO_4^{-2} \tag{27}$$

$$Ca^{+2} + SO_4^{-2} + Na_2CO_3 \rightarrow CaCO_3\downarrow + 2Na^+ + SO_4^{-2} \tag{28}$$
$$CO_2 + Ca(OH)_2 \rightarrow CaCO_3\downarrow + H_2O \tag{29}$$
$$2HCO_3^- + Ca(OH)_2 \rightarrow CaCO_3\downarrow + CO_3^{-2} + 2H_2O \tag{30}$$

It is important to distinguish between the reactions described by Equations 25 and 26, the reactions described by Equations 27 and 28, and those described by Equations 29 and 30.

Before making this distinction, however, one more parameter, total carbonate (C_T) must be defined. C_T is the sum of the carbonic acid derived species (CO_2, H_2CO_3, HCO_3^- and CO_3^{-2}) and may be determined by Equation 31.

$$C_T = \frac{\text{alkalinity} + \text{acidity}}{2} \tag{31}$$

If C_T in the untreated water exceeds total hardness (where both are expressed as mg $CaCO_3$/l) then Equations 25 and 26 apply. Sufficient total carbonate is usually present in the untreated water to precipitate the calcium which must be removed from the initial water plus the calcium of the added lime. Lime will generally (but not always) be the only reactant required.

Hardness present in quantities less than or equal to C_T is known as carbonate hardness. If initial hardness exceeds C_T, Equations 27 and 28 apply to that fraction of hardness greater than initial C_T. This fraction is known as noncarbonate hardness. Lime by itself cannot remove noncarbonate hardness. As Equation 27 indicates, the addition of lime to noncarbonate hardness simply results in the exchange of one form of hardness (Mg^{+2}) for another (Ca^{+2}). Subsequent addition of carbonate, in the form of Na_2CO_3 (soda ash) is required for removal of the exchanged calcium as shown by Equation 28. Both lime and soda ash are generally used if noncarbonate hardness is present in the untreated water. Equations 29 and 30 indicate that not all of the reactant used acts to remove hardness. Nonhardness constituents exert reactant demands that must be satisfied. Examples are free CO_2 (Equation 29) and C_T in excess of hardness (Equation 30).

Softening can also be obtained with caustic soda as shown by Equations 32, 33 and 34, which are also written in molar quantities.

$$Ca^{+2} + 2HCO_3^{-2} + 2NaOH \rightarrow CaCO_3\downarrow + 2Na^+ + CO_3^{-2} + 2H_2O \tag{32}$$
$$Mg + 2HCO_3^{-2} + 4NaOH \rightarrow Mg(OH)_2\downarrow + 4Na^+ + 2CO_3^{-2} + 2H_2O \tag{33}$$
$$Mg^{+2} + SO_4^{-2} + 2NaOH \rightarrow Mg(OH)_2\downarrow + 2Na^+ + SO_4^{-2} \tag{34}$$

The CO_3^{-2} ion created in Equations 32 and 33 can react to remove noncarbonate hardness present. Equations 32, 33 and 34 show that caustic soda can be used to remove both carbonate and noncarbonate hardness and may be used in place of soda ash and as a substitute for part or all of the lime requirements. This presents the obvious advantage of needing only one conditioning feed system as opposed to two for the lime-soda process. Other advantages of NaOH are ease of handling and feeding, lack of deterioration on storage, and the production of less $CaCO_3$ sludge.

The chemical choice must be determined by an analysis of all factors involved, including the raw water condition, effluent goals, reactant costs, and relative amounts of sludge produced. Such an analysis can be carried out rapidly and with reasonable accuracy by using Caldwell-Lawrence diagrams.

Caldwell-Lawrence Diagrams for Water Softening Problems

The methods for solving water softening problems are illustrated by several examples, using a step-by-step procedure. The Mg-pH nomograph and chemical addition vectors (ignored in the corrosion control problems) must be used in water softening problems. As with corrosion control calculations, there is no single correct way to do things. The designer should examine all the options to find the best suited for his particular situation.

Example 13. A raw water has the following analysis: pH 7.1, Ca 260, Mg 80, alkalinity 360, temperature 25°C, TDS 400 mg/l. An effluent magnesium value of < 15 mg/l is required with total (Ca + Mg) hardness not to exceed 50 mg/l. Calculate the chemical dose and cost required to produce the finished water and the amount of chemical precipitate formed.

Step 1. *Locate the untreated water on the Caldwell-Lawrence diagram* using the same technique employed for the corrosion control problems. The untreated water is represented by a point if saturated with respect to $CaCO_3$ and is represented by an envelope if it is not.

Locate the untreated water on the 25°C, TDS 400 mg/l diagram. The water condition is shown by the envelope ABC bounded by the lines Ca 260, pH 7.1, alkalinity 360 (Figure 7). The acidity of the untreated water is 480 mg/l. From Equation 29, C_T = (480 + 360)/2 = 420 mg/l. Because C_T (420) exceeds total hardness (260 + 80 = 340), all hardness is carbonate hardness and lime is normally the only reagent required. This will be confirmed by subsequent calculations.

Step 2. *Determine the saturation state of the untreated water* using the techniques described in the corrosion control discussion. At pH 7.1 and

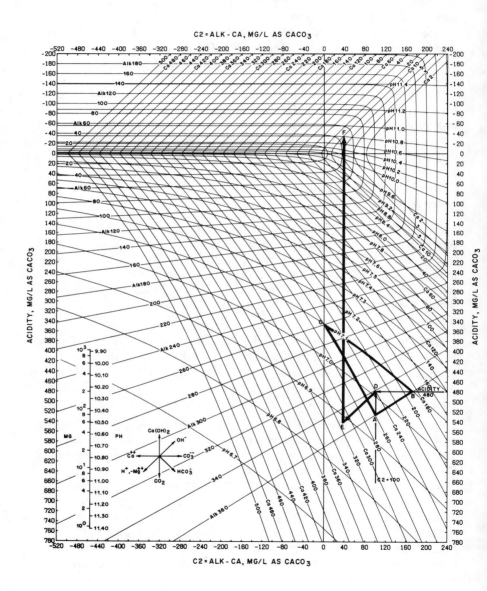

Figure 7. C-L diagram for Example 13; 25°C, 400 mg/l TDS.

alkalinity 360, a saturated water contains 186 mg Ca/l. The raw water contains 260 mg Ca/l. Therefore it is oversaturated.

Step 3. *Equilibrate waters which are not saturated.* For the purposes of calculation, unsaturated waters must be brought to a saturated condition (equilibrated) so that they can be described by a point. This point is the beginning point for one or a series of addition vectors. A statement made earlier must be repeated and emphasized: addition vectors can *only* be used when the water is in a *saturated* condition.

Equilibration in water softening plants is almost always by contact with solid $CaCO_3$. Other chemicals could be used, but they would represent an expense while $CaCO_3$, which is always present in the reaction zone, does not. $CaCO_3$ particles serve as catalysts to initiate precipitation from oversaturated waters; they dissolve to bring undersaturated waters to equilibrium. Either way, the equilibration process proceeds with acidity and C2 constant throughout. The equilibrated condition is always found at the intersection of the acidity and C2 lines. In this example, Point D describes the condition of the water following equilibration.

Step 4. From the Mg-pH nomograph, *determine the pH that will produce the required effluent magnesium concentration.* This is done by simply checking off the pH value opposite the desired effluent magnesium concentration. The desired Mg concentration of 15 mg/l can be obtained by $Mg(OH)_2$ precipitation if system pH is raised to 10.80.

Step 5. Starting at the point describing the equilibrated raw water condition, *draw the magnesium removal vector* downward and to the left at a 45° angle as described by the direction diagram. The distance to be moved is such that the vector's projections on the acidity and C2 axes are equal to the amount of magnesium to be removed. The head of the vector designates a second point, which has no particular physical significance (unlike all other points) but which must be established so that the calculation can proceed.

The amount of magnesium to be precipitated = 80 - 15 = 65 mg/l as $CaCO_3$. From Point D draw a vector DE downward and to the left at 45 degrees. Its projection on the acidity axis must be equal to 65 mg/l; its projection on the C2 axis must also be 65 mg/l.

Step 6. From the point just established extend a direction vector to the line representing the pH value required to produce the final Mg concentration. The direction of this vector depends on the nature of the chemical used and is indicated by the direction diagram. For example, if $Ca(OH)_2$ is used, the vector is vertical upward. If NaOH, it is upward and to the right. The tip of the vector locates a point describing the water condition following chemical addition. This water has the pH, Ca, and alkalinity values of the lines that intersect at the point, and it has the magnesium value determined from the Mg-pH nomograph. The amount of chemical required to achieve this condition is equal to the projection of the addition vector on the acidity or on the C2 axis. (The applicable axis is obvious.) The water may or may not be satisfactory in terms of effluent quality. If it is unsatisfactory, other chemicals (represented by more addition vectors) can be added until the desired condition is achieved. Here the value of the diagrammatic approach can be fully appreciated as the types and amounts of chemicals required to link the raw and finished water conditions quickly become evident.

Use lime to raise the pH to 10.80 (the pH that produces a soluble magnesium value of 15 mg/l). The lime addition vector EF extends vertically upward, intersecting the line representing pH 10.80 at Point F. The amount of lime required to reach Point F is equal to the projection of the lime addition vector on the acidity axis (570 mg/l, as $CaCO_3$). By Equation 21 and Table VIII (page 553) $Ca(OH)_2$ required is 16.7(570)(37)/90 = 3913 lb/ million gal (0.469 kg/m^3) assuming 90% pure $Ca(OH)_2$.

The calcium concentration at Point F is 10 mg/l. The water condition at Point F satisfies effluent requirements, *i.e.*, Mg 15 and total hardness (Ca 10 + Mg 15 = 25) is less than 50 mg/l. Note that lime was the only reagent required.

Step 7. Finally, use material balances to calculate the quantity of chemical sludge produced. The general material balance equation is:

$$\text{Chemical}_{pptd} = \text{Chemical}_{initial} + \text{Chemical}_{added} - \text{Chemical}_{final} \qquad (35)$$

For this example,

a. $CaCO_3{}_{pptd}$ = 260 + 570 – 10 = 820 mg/l as $CaCO_3$

This can be expressed in actual lb/million gal by the use of Table VIII and Equation 21 with the term "chemical precipitated" substituted for "chemical required" in the equation.

$$CaCO_3{}_{pptd} = \frac{(16.7)(820)(50)}{100} = 6847 \text{ lb/million gal } (0.821 \text{ kg/m}^3)$$

b. $Mg(OH)_2{}_{pptd}$ = 80 + 0 – 15 = 65 mg/l as $CaCO_3$
        or 316 lb/million gal (0.038 kg/m^3)

Comment

If no magnesium removal is required, Steps 4 and 5 are eliminated and the direction vector for the chosen chemicals is drawn directly from the point representing the equilibrated raw water to the vicinity of the finished water condition. As before, directions can be changed (*i.e.*, other chemicals used) if such a change is useful in "homing" in on the final water condition.

Example 14. A raw water analysis is: pH 7.2, Ca 340, Mg 100, alkalinity 100, TDS 400, temperature 25°C. A water containing no more than 40 mg/l Mg hardness and 30 mg/l Ca hardness is required. Calculate the chemical doses required to achieve the finished water condition as well as the chemical costs and the amount of chemical sludge precipitated.

Step 1. The 25°C, TDS 400 mg/l diagram should be used. The envelope ABC representing the raw water is shown on the diagram in Figure 8. A saturated water at alkalinity 100 and pH 7.2 would contain in excess 500 mg Ca/l. Since the raw water contains only 340 mg Ca/l, it is undersaturated. Acidity of the raw water is 128 mg/l; therefore C_T = (alkalinity + acidity)/2 = (100 + 128/2 = 114. Because initial hardness exceeds C_T, noncarbonate hardness is present and two conditioning chemicals are required.

Step 2. Equilibrate the water to the condition represented by Point D, the intersection of the acidity and C2 lines.

Step 3. The finished water must contain no more than 40 mg/l of Mg. From the nomograph, the pH required to produce a soluble magnesium residual of 40 mg/l is 10.59. The amount of magnesium to be precipitated = 100 - 40 = 60. From Point D, draw a vector DE (solid line) downward and to the left so that its projection on the acidity and C2 axis is equal to the magnesium to be removed (60 mg/l). For the time being ignore the dashed vectors as they are part of another problem.

Step 4. Use lime to raise the pH to 10.59 (the pH producing the desired Mg concentration). The lime addition vector extends upward vertically, intersecting the line representing pH 10.59 at Point F. The amount of lime required to reach Point F is equal to the projection of the lime addition vector on the acidity axis, 205 mg $Ca(OH)_2$/l as $CaCO_3$.

Step 5. The amount of soluble calcium present at Point F (324 mg/l) exceeds the effluent requirement. Because noncarbonate hardness is present, low calcium levels cannot be achieved by adding lime only. But calcium can be reduced to the desired value by adding Na_2CO_3. The Na_2CO_3 addition vector is extended from Point F in a horizontal direction until it intercepts the appropriate calcium value (30 mg/l). The amount of Na_2CO_3 required is equal to the projection of the Na_2CO_3 addition vector FG on the C2 axis (302 - 2) = 300 mg/l as $CaCO_3$.

Step 6. The total chemical requirement is 205 mg $Ca(OH)_2$/l and 300 mg Na_2CO_3/l, both as $CaCO_3$. The total chemical cost (based on

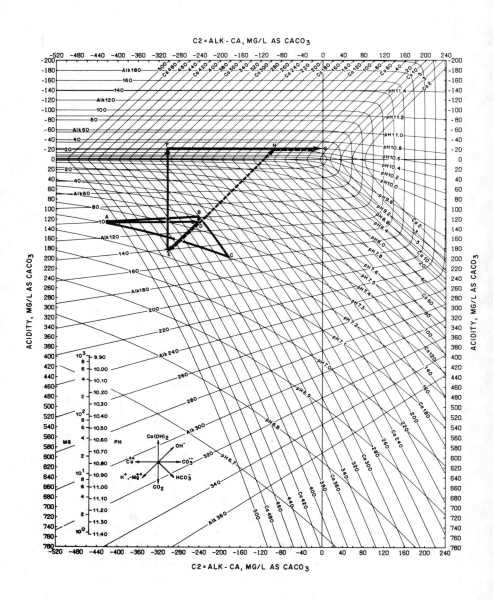

Figure 8. C-L diagram for Example 14; 25°C, 400 mg/l TDS.

Equation 21 and data from Table VIII, Page 553) is \$128/million gal
($3.39\cancel{c}/m^3$) if 90% pure $Ca(OH)_2$ and 90% pure Na_2CO_3 are assumed.
$Mg(OH)_2$ produced is 60 mg/l as $CaCO_3$. From Equation 35, $CaCO_3$
produced in the precipitation is equal to 340 + 205 – 30 = 515 mg/l as
$CaCO_3$.

Step 7. All chemical possibilities have by no means been exhausted.
Better ways of achieving the same end result may be available. Inspection
of the diagram indicates that a shorter path from Point E to Point G
would be obtained using caustic soda (NaOH) instead of lime. The
amount of soda ash required would also be reduced. The NaOH direction
vector EH (dashed) is extended from Point E upward and to the right at
a 45° angle until it intersects the pH line 10.59 at Point H. (Vector EH
is shown slightly displaced from its true path for the sake of clarity.)
Na_2CO_3 is then added to bring the water from Point H to Point G. The
amount of NaOH required (205 mg/l) is equal to the projection of the
NaOH addition vector on the acidity or C2 axis. Ninety mg/l Na_2CO_3
are required to bring the system to Point G. Total chemical costs are
\$143/million gal ($3.78\cancel{c}/m^3$), assuming 90% pure NaOH and Na_2CO_3 are
used.

The amount of $Mg(OH)_2$ sludge produced is the same as produced
by the lime-soda ash treatment. However, the $CaCO_3$ sludge production (340 - 30 = 310 mg/l) is substantially less because the treatment chemicals (NaOH, Na_2CO_3) contain no calcium to contribute to
$CaCO_3$ sludge· production. In choosing the chemicals to be used, the
advantages of less chemical cost for the lime-soda ash case would have to
be weighed against the disadvantage of greater sludge production.

Neutralization

Examples 13 and 14 are incomplete because the treated waters have pH
values so high as to be unsatisfactory for most domestic or industrial uses.
pH reduction to a more neutral state is required. Neutralization can be
carried out with any acid, for example, H_2SO_4 (strong) or CO_2 (weak).
The latter has advantages where high doses of lime are required for magnesium removal. Treated waters from magnesium removal processes may
contain large amounts of calcium noncarbonate hardness, which can be
precipitated during a two-stage recarbonation. Additional softening is thus
obtained. In the first stage, the high pH water is neutralized to the point
of minimum calcium solubility, then passed to a settling tank where the
precipitated $CaCO_3$ is removed. The softened overflow is then carbonated
once more to reduce the pH to an acceptable level. No further $CaCO_3$
precipitation takes place in the second stage.

Neutralization need not always be carried out by two-stage recarbonation. One-stage neutralization with strong acid or CO_2 can be used if an economic or operating analysis indicates no advantage is gained by two-stage processing. Consider Example 15.

Example 15. A softened water has the following characteristics: pH 11.24, Ca 112, Mg 3, alkalinity 100, TDS 400, temperature 25°C. The pH must be reduced to 7.8 before the water can be released. Neutralization can be carried out using CO_2 (one- or two-stage recarbonation) or by strong acid. Recommend the best treatment.

A. Two-stage recarbonation

Step 1. The water condition is located on Figure 9 by a point (A), and therefore the water is saturated. For the first stage, the CO_2 addition vector AB (solid line) is extended vertically downward cutting across lines of decreasing calcium concentration (which indicates that Ca is being precipitated) until the line "acidity = zero" is reached (Point B).* If the vector is extended beyond this line, calcium values begin to increase, indicating that the $CaCO_3$ just precipitated is beginning to redissolve. The "acidity = zero" line is thus a line of minimum $CaCO_3$ solubility, and the lowest possible calcium values are attained here. The first stage recarbonation should be stopped here and the water passed on to a sedimentation tank for removal of the precipitated $CaCO_3$ (112 – 28 = 84 mg $CaCO_3$/l). The condition of the settled water is Ca 28, Alk 16, pH 9.8. Note that the alkalinity, which is very difficult to read from the alkalinity lines in the vicinity of the C2 axis, can be calculated from the C2 value, which is easily read. Remember that C2 = Alk – Ca, therefore Alk = C2 + Ca = –12 + 28 = 16. The CO_2 required for the first stage is equal to the projection of the addition vector on the acidity axis (100 mg CO_2/l).

Step 2. The settled effluent is then passed to the second stage of recarbonation. Because all $CaCO_3$ formed during the first stage has been removed and because no further $CaCO_3$ precipitation is possible, further CO_2 additions produce a state that is undersaturated with respect to $CaCO_3$. This means that addition vectors can no longer be used to describe the reaction path during Stage 2. How then do we complete the calculation? Simply by using the techniques developed in the previous section on corrosion control. Any waters produced by further CO_2 addition will be undersaturated and can be represented by an envelope. What are the bounds of the envelope? Ca will remain at 28 mg/l because no $CaCO_3$ precipitation can occur. Because CO_2 added does not contribute to alkalinity and because no $CaCO_3$ is precipitated, alkalinity also remains constant. The pH value desired (7.8) represents the lower boundary. The CO_2 required is equal to the difference between the acidity at Point B

---
*For the time being, ignore the dashed lines. They are part of the one-stage recarbonation problem.

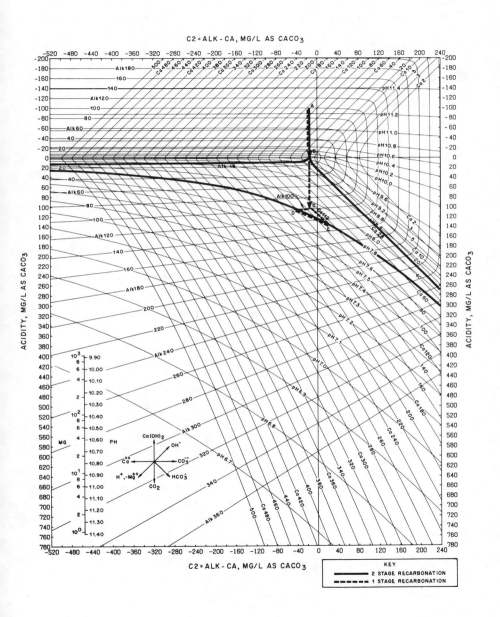

Figure 9. C-L diagram for Example 15; 25°C, 400 mg/l TDS.

(zero) and the acidity of the undersaturated final state. The acidity of the final state cannot be obtained directly since the intersection of the pH and alkalinity lines is off the diagram. Acidity may instead be calculated from Equation 24 and is equal to 17 mg/l. The CO_2 required to proceed from Point B to the final state = 17 - 0 = 17 mg CO_2/l. The overall CO_2 requirement (first plus second stage CO_2 usage) is, therefore, 100 + 17 = 117 mg/l.

B. One-stage recarbonation

Step 1. CO_2 is added continuously (*i.e.*, without interruption for solids separation at Point B) until the desired pH is achieved. $CaCO_3$ precipitated along the path AB begins to redissolve once the addition vector (dashed line)* progresses beyond the point of minimum $CaCO_3$ solubility. (Note that the vector begins to cut across lines of increasing Ca once past Point B.) At Point C all the $CaCO_3$ that has been precipitated has been redissolved (Ca at Point C = Ca at Point A) and the water passes from a saturated state to an undersaturated one. Further CO_2 addition produces an undersaturated water with the same calcium and alkalinity values as the water at Point C (and also Point A) and a pH 7.8. The final effluent is represented by the envelope CDE. The CO_2 required to each Point C from Point A is 100 - (-105) = 205 mg/l. The CO_2 required to move from Point C to the final condition is equal to the differences in acidity of those states, 1 mg/l. The total CO_2 requirement is 206 mg/l.

C. Neutralization with strong acid.

If no solid $CaCO_3$ is present at the beginning (which is true for the water at Point A) the first drop of acid produces an undersaturated water. In this particular example any condition between the first drop and the neutralized effluent is represented by an envelope. Every X mg of acid increases the acidity by X and decreases the alkalinity by X. The calcium value remains constant, as calcium is neither added nor precipitated.

The procedure to be followed is to add increments of acid until the alkalinity and acidity values are such that the intersection of the acidity and alkalinity lines is at pH 7.8. The acid required is equal to the acidity difference between the neutralized and initial water. Unfortunately in this example, the intersection of the alkalinity and acidity lines is to the left of the diagram, and therefore, the pH value corresponding to the calculated acidity and alkalinity values cannot be determined diagrammatically. Instead the calculation must be carried out analytically by making the following rearrangement of Equation 24:

---

*Construction lines for the one-stage carbonization example have been slightly displaced for the sake of clarity.

a. The acidity after addition of X mg/l strong acid is equal to the initial acidity plus X. The alkalinity is equal to alkalinity minus X. Thus for any addition X:

$$(Acidity_{init} + X) = \frac{\left[(Alk_{init} - X) - \frac{K_W}{H^+} + H^+ \right] \left[1 + \frac{H^+}{K_1} \right]}{\left[1 + \frac{K_2}{H^+} \right]} + H^+ - \frac{K_W}{H^+} \qquad (36)$$

b. As the values of initial acidity, initial alkalinity, final H^+, K_1, K_2 and K_W are all known, X may be calculated by a trial and error procedure. In this example, 100 mg/l H_2SO_4 is required to neutralize from the initial condition to pH 7.8.

The neutralization calculations are summarized in Table VI. From a chemical standpoint, the least cost solution is two-stage recarbonation. The water from the two-stage process is also softer, which is presumably consistent with process goals. On the negative side, more chemical sludge is produced (thus more must be handled) and an additional clarifier must be provided to settle the $CaCO_3$ produced by first-stage recarbonation. The final choice of process must be determined by the weighing of a multitude of factors, including chemical and capital costs, sludge production and effluent quality requirements.

Table VI. Comparison of Three Neutralizing Schemes

Parameter	Two-Stage Recarbonation	One-Stage Recarbonation	Strong Acid Addition
pH	7.8	7.8	7.8
Ca, mg/l as $CaCO_3$	28	112	112
Mg, mg/l as $CaCO_3$	3	3	3
Alkalinity, mg/l as $CaCO_3$	16	100	0
$CaCO_3$ Sludge Produced, mg/l as $CaCO_3$	84	0	0
Chemical Requirement (mg/l as $CaCO_3$)	114 mg/l CO_2	206 mg/l CO_2	100 mg/l H_2SO_4
Chemical Cost	$17/million gal	$31/million gal	$27/million gal

Split Treatment

Split treatment has been used effectively where magnesium removal is a prime processing objective. In split treatment the water is divided into two streams:

- **the** treated portion, which is raised to a high pH to effect substantial Mg removal (carbonate hardness is also removed).
- the bypassed fraction, which is not treated at all.

The treated and bypassed fractions are then recombined. The recombining process effects a neutralization and some additional calcium hardness removal. The degree to whcih the treated portion is processed and the "split" are so arranged that the combined effluent has the magnesium concentration desired.

Advantages claimed for this process are lower overall lime and soda ash requirements (if the latter is needed) and the elimination of the need for any neutralizing chemical. This process will be examined through the Caldwell-Lawrence diagrams to see if such claims are valid.

Example 16. Find the least cost method to produce a water having a magnesium concentration of 40 mg/l and a pH of 8.3 given a water of the following initial analysis: Mg 100, Ca 180, alkalinity 240, pH 7.1, TDS 400 mg/l, temperature 25°C.

The magnesium balance is given by Equation 37:

$$Mg_e = Mg_t (1 - X) + Mg_r X \tag{37}$$

where Mg_e ≡ magnesium concentration of the effluent

Mg_t ≡ magnesium concentration of the treated portion

Mg_r ≡ magnesium concentration of the raw water

X ≡ fraction of the water bypassed.

The magnesium concentration of the treated portion can be found by rearranging Equation 37 to:

$$Mg_t = \frac{Mg_e - Mg_r X}{1 - X} \tag{38}$$

The maximum fraction that can be bypassed can be found by setting $Mg_t = 0$. (Mg_t cannot be reduced further.)

$$X_{max} = \frac{Mg_e}{Mg_r} \tag{39}$$

The desired value of Mg_e may be obtained as long as $0 \leqslant X \leqslant X_{max}$. For the water given, $X_{max} = 40/100 = 0.4$. In the following example, the fraction bypassed is 0.3.

Step 1. First consider the portion of the water to be treated. (Ignore the bypassed fraction for the present.) X = 0.3, therefore $Mg_t = [40 - 100(0.3]/0.7 = 14.3$ The raw water condition (envelope ABC) is first located on the Caldwell-Lawrence diagram (25°C, TDS 400, Figure 10). This water is undersaturated and can be equilibrated to the condition described by Point D.

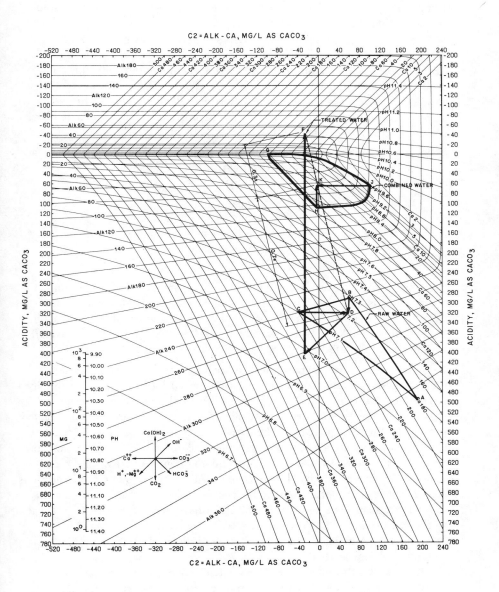

Figure 10. C-L diagram for Example 16; 25°C, 400 mg/l TDS.

Step 2. In order to bring the water to the required Mg concentration of 14.3 mg/l, 85.7 mg Mg/l must be removed. Lime is used to bring the system to the pH (10.83) where the desired magnesium solubility is obtained.

The magnesium removal vector DE is extended downward and to the left at 45° until its projection on the acidity and C2 axes is 85.7 mg/l. Then the lime addition vector EF is extended vertically upward until it intersects the line representing pH 10.83. The condition of the treated water is: Ca 70, Mg 14.3, pH 10.83, alkalinity 44, acidity –42. The lime dose required to raise the treated portion to pH 10.83 is equal to the length of the lime addition vector, 447 mg/l.

Step 3. Combine the treated and bypassed portions (first imposing a "no precipitation" restriction) to obtain:

1. Ca_{comb} = 70(0.7) + 180(0.3) = 103
2. Alk_{comb} = 44(0.7) + 240(0.3) = 104
3. $Acidity_{comb}$ = –42(0.7) + 320(0.3) = 67
4. Mg_{comb} = 14.3(0.7) + 100(0.3) = 40

The combined water condition is located by envelope GHJ. The combined water is oversaturated. When equilibrated, its condition is represented by Point K. The equilibrated combined water condition at K is: Ca 68, Mg 40, pH 8.3, alkalinity 69, acidity 67. Note that 103 – 68 = 35 mg/l of $CaCO_3$ precipitated when the waters were combined.

By examining the diagram one can see that the equilibrated waters (bypassed, treated and combined effluent) are all located on a straight line (line FKD, dashed). Furthermore, the ratio of the distance between points representing the equilibrated combined effluent and the equilibrated treated portion to the total length of the line is equal to the fraction bypassed (0.3). This suggests a quick and easy method for finding the composition of waters produced by blending. To find the composition of the blend, first draw a line between the points representing the equilibrated conditions of the waters to be blended (F,D). The resultant blend (K) must lie on this line at a distance such that the ratio of the distances $\overline{FK}/\overline{FD}$ is equal to the ratio (mass of D) (mass of D + F). This law is known as the Inverse Lever Arm Rule and is used extensively in chemical and metallurgical engineering

The lime dose prorated over the treated and bypassed portions = 447(0.7) + 0(0.3) = 313 mg/l. From Equation 35, overall $CaCO_3$ production = 180 + 313 – 68 = 425 mg/l. The overall $Mg(OH)_2$ production = 100 – 40 = 60 mg/l.

For comparison, a similar calculation was carried out where the same residual magnesium concentration and pH was obtained using conventional lime treatment and one-stage recarbonation. A comparison of important parameters for the equilibrated effluents is shown in Table VII. Ninety percent pure CaO and 100% pure CO_2 were assumed for treatment chemicals.

The value for the fraction bypassed used in this example (0.3) is not necessarily the optimum one. Other values may produce somewhat more effective treatment or the same treatment for less cost. The optimum value for fraction bypassed must be found by trial and error.

Table VII. Comparison of Split and Conventional Treatment

Parameter	Split Treatment	Conventional Treatment
Ca, mg/l as $CaCO_3$	68	25
Mg, mg/l as $CaCO_3$	40	40
pH	8.3	8.3
Lime Required, mg/l as $CaCO_3$	313	400
CO_2 Required, mg/l as $CaCO_3$	none	48
$Mg(OH)_2$ Produced, mg/l as $CaCO_3$	60	60
$CaCO_3$ Produced, mg/l as $CaCO_3$	425	555
Chemical Cost, $/million gal	$37	$54

Working "Off-the-Chart" Softening Problems

For very practical reasons (size, cost) the range of parameters covered by the diagrams is limited. For example, the maximum calcium and alkalinity values displayed on the charts in this book are 500 and 360 mg/l, respectively. Nature, however, respects no such boundaries. The investigator may be required to carry out softening calculations for waters of 1000 mg Ca/l or 750 mg alkalinity/l. If so, how does he proceed?

This is not as large a problem as might initially appear. Even though the raw water condition cannot be plotted, the equilibrated raw water's condition can almost always be located. Once this important point is located, calculations can be carried out in the usual manner.

The reason equilibrated water is easily located can be seen from the following argument.

1. The equilibrated raw water condition (a point) can be located if its C2 and acidity values are known.
2. A water's C2 and acidity values do not change during equilibration. Therefore the C2 and acidity values of the equilibrated and initial raw water are identical.
3. The C2 and acidity values of the initial raw water are easily calculated by Equations 15 and 24.
4. The parameters C2 and acidity are much less likely to be off the chart than the parameters calcium and alkalinity.

As an example, consider an initial water of Ca 750, alkalinity 400, pH 7.0, TDS 780, temperature 25°C. The calcium and alkalinity values are too high for the initial water condition to be plotted. The C2 (–350 mg/l) and acidity (566 mg/l) values can be calculated however, and they locate a point on the diagram that represents the equilibrated water condition. Magnesium removal calculations can begin from this point as usual.

Consider an even more extreme example. The initial water condition is Ca 800, alkalinity 400, pH 6.5, TDS 820, temperature 25°C. For this water C2 = –400 mg/l and acidity 915 mg/l. The equilibrated water condition plots below the chart, as the acidity value (915 mg/l) exceeds the acidity limit of the chart (780 mg/l). This causes no problems, however. Simply extend the acidity axis to 915 mg/l and plot the equilibrated water condition. Unlike the calcium and alkalinity lines, the acidity and C2 axes are linear and can be extended indefinitely with no loss of accuracy. Calculations initiated from equilibrated waters that are off the chart should be no less accurate than these begun within its boundaries.

Limitations of Softening Calculations

All calculations are approximations to reality and the softening calculations presented here are no exceptions. While the calculations predict that calcium and magnesium residuals of less than 2 mg/l can be obtained, practical lower limits appear to be around 20 to 30 mg/l for calcium and 10 mg/l for magnesium, at least for temperatures prevalent at domestic treatment plants. Several factors would tend to increase the disparity between calculations and field results:

1. Calculations presented here pertain to dissolved constituents. Because solids separations are imperfect, effluents from water softening plants have higher residuals than those predicted by the calculations.

2. Kinetic factors retard the approach to the equilibrated conditions predicted by the diagrams.

3. Complexing phenomena, not considered in the calculations which led to the construction of the diagrams, increase solubilities to levels above those predicted by the diagrams. This increase is probably not significant, however.

4. Each diagram is constructed at fixed ionic strength where in fact ionic strengths vary as ions are added and removed as the result of chemical addition and precipitation.

5. The mass action constants used in the calculations are not exactly known. For example, the apparent solubility product of $CaCO_3$ is known to vary as the ratio Mg/Ca in the water being treated changes.[14] In addition, there is considerable variation in the $Mg(OH)_2$ solubility product values reported in the literature.[6]

6. Water temperatures and ionic strengths are often different from those for which diagrams are available.

Despite the problems just mentioned, the diagrams remain powerful problem-solving tools. They are considerably more accurate than the stoichiometric method (bar diagrams) traditionally employed in water softening calculations. They are no less accurate than other analytical solutions operating under the same restrictions, yet they can be carried out in a fraction of the time. The few studies reported in the literature[15,16] have, in fact, found reasonable agreement between calculations and field results.

LARGE-SCALE DIAGRAMS

The Caldwell-Lawrence diagrams included here are necessarily small; there may be some difficulty in working at this scale. Large-scale diagrams (about 14 x 18 in.) can be obtained from Brown and Caldwell Consulting Engineers, 1501 North Broadway, Walnut Creek, CA 94596, by writing to the attention of Douglas Merrill. Diagrams at other temperatures and TDS concentrations can be produced upon request. The diagrams wear out quickly if written on directly. They will last longer if problems are worked on overlying tracing paper or mylar film.

SUMMARY

Caldwell-Lawrence diagrams can be used in the analysis and solving of corrosion control and water softening problems. The diagrams provide a means to assess rapidly and accurately the condition of the raw and finished waters, the type and amount of chemicals required to produce the finished water and the amount of chemical precipitates formed. Insights into the workings of a very complex chemical system are thus obtained. Such insights are not generally available simply through the inspection of gross analytical parameters. Perhaps the greatest attribute of the diagrams is the sharpening of the investigator's perception of the problem and the developing of his problem-solving intuition. When all aspects of the problem can be visually laid out and the development of the solutions is neither tedious nor time-consuming, the investigator is encouraged to examine many alternatives in arriving at the most suitable process selection. This chapter shows many ways that the diagrams can be used in the rational analysis of water processing alternatives.

ACKNOWLEDGMENTS

Appreciation is expressed to Dr. John F. Ferguson of the University of Washington, Seattle, Washington, for a helpful critical review of the original manuscript, to Dr. Robert L. Sanks of Montana State University, Bozeman, Montana, for useful suggestions, to Mr. Robert Niclas of Metropolitan Engineers, Seattle, for modifications necessary to prepare the diagrams for publication, and to Ms. Patricia Asper, Patricia Morrison, Peggy Shepherd and Linda Henry, also of Metropolitan Engineers, for typing the manuscript.

The diagrams used in this paper were generated by the CDC 6400 computer and Calcomp plotter at the University of Washington. The computer program for the diagrams was adapted from a program taken from a report by Loewenthal and Marais[15] of the University of Cape Town, South Africa. This report is now a book.[17]

Table VIII. Equivalent Weights of Some Conditioning Chemicals and Chemical Precipitates and Approximate Bulk Chemical Costs f.o.b. Seattle, January 1976[18]

	Equiv. Weight Chemical	Approx. Chemical Cost	
		(¢/lb)	¢ /kg
CaO (quicklime)	28	2.25	4.96
Ca(OH)$_2$ (slaked or hydrated lime)	37	2.80	6.17
Na$_2$CO$_3$ (soda ash)	53	3.00	6.62
NaOH (caustic soda)	40	7.50	16.54
CO$_2$ (carbon dioxide, commercial grade)[a]	22	4.10	9.04
CaCO$_3$ (calcium carbonate)	50	–	–
Mg(OH)$_2$ (magnesium hydroxide)	29.1	–	–
H$_2$SO$_4$ (sulfuric acid)	49.0	3.30	7.28

[a]Less expensive CO$_2$ might be obtained where CO$_2$ is generated from underwater burners using natural gas.

Table IX.   Values for K_w, K_1 and K_2 at Various Temperatures and TDS Concentrations[a]

	Temperature			
TDS	2°C	5°C	15°C	25°C
	K_w			
40	3.73×10^{-6}	4.96×10^{-6}	1.20×10^{-5}	2.68×10^{-5}
400	3.86×10^{-6}	5.67×10^{-6}	1.38×10^{-5}	3.07×10^{-5}
1200	4.82×10^{-6}	6.40×10^{-6}	1.55×10^{-5}	3.46×10^{-5}
	K_1			
40	7.39×10^{-3}	7.74×10^{-3}	9.96×10^{-3}	1.15×10^{-2}
400	8.45×10^{-3}	9.17×10^{-3}	1.14×10^{-2}	1.32×10^{-2}
1200	9.54×10^{-3}	1.03×10^{-2}	1.29×10^{-2}	1.49×10^{-2}
	K_2			
40	2.90×10^{-6}	3.20×10^{-6}	4.26×10^{-6}	5.37×10^{-6}
400	3.80×10^{-6}	4.19×10^{-6}	5.58×10^{-6}	7.03×10^{-6}
1200	4.84×10^{-6}	5.34×10^{-6}	7.11×10^{-6}	8.96×10^{-6}

[a]To be used in equations where alkalinity, acidity and hydrogen ion concentrations are expressed as mg CaCO$_3$/l.

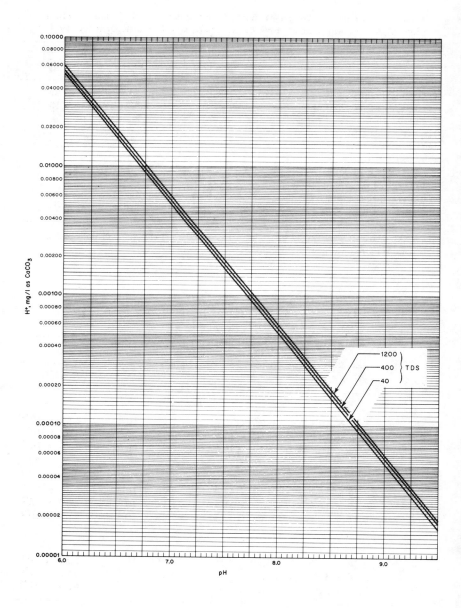

Figure 11. pH vs H$^+$ (mg/l as CaCO$_3$) at different TDS values.

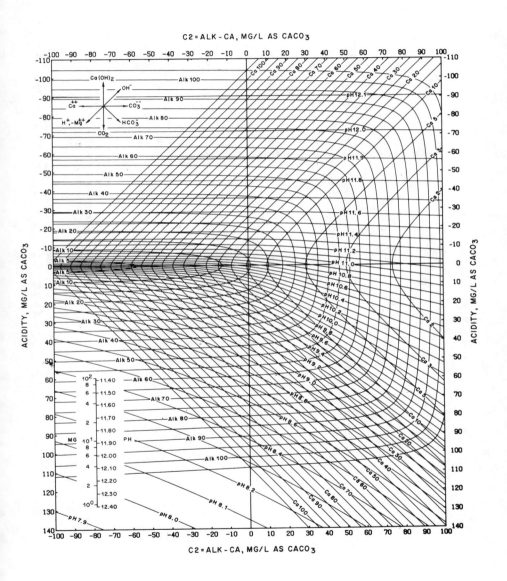

Figure 12. Water conditioning diagram for $2°C$ and 40 mg/l total dissolved solids.

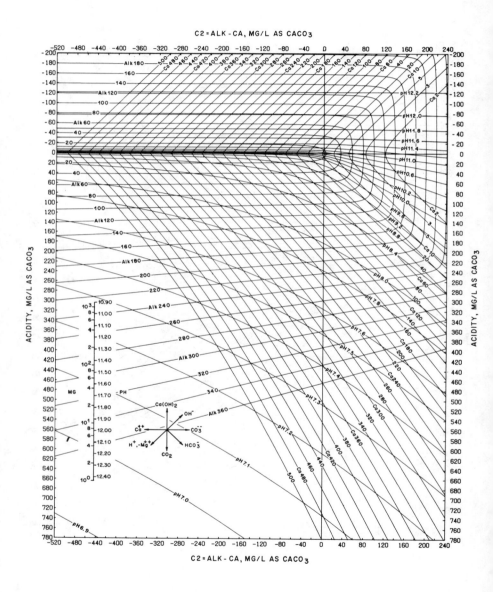

Figure 13. Water conditioning diagram for 2°C and 400 mg/l total dissolved solids.

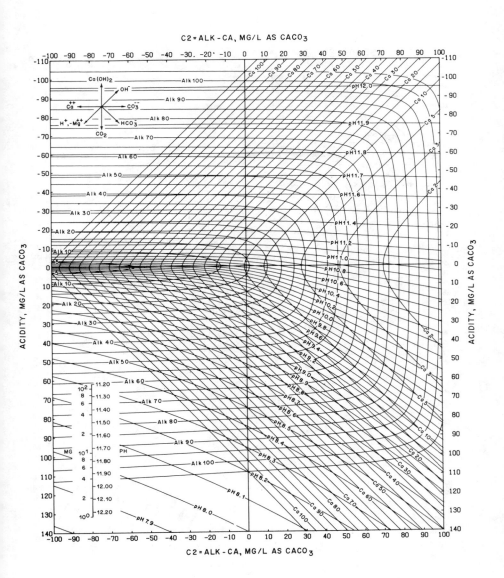

Figure 14.   Water conditioning diagram for $5°C$ and 40 mg/l   total dissolved solids.

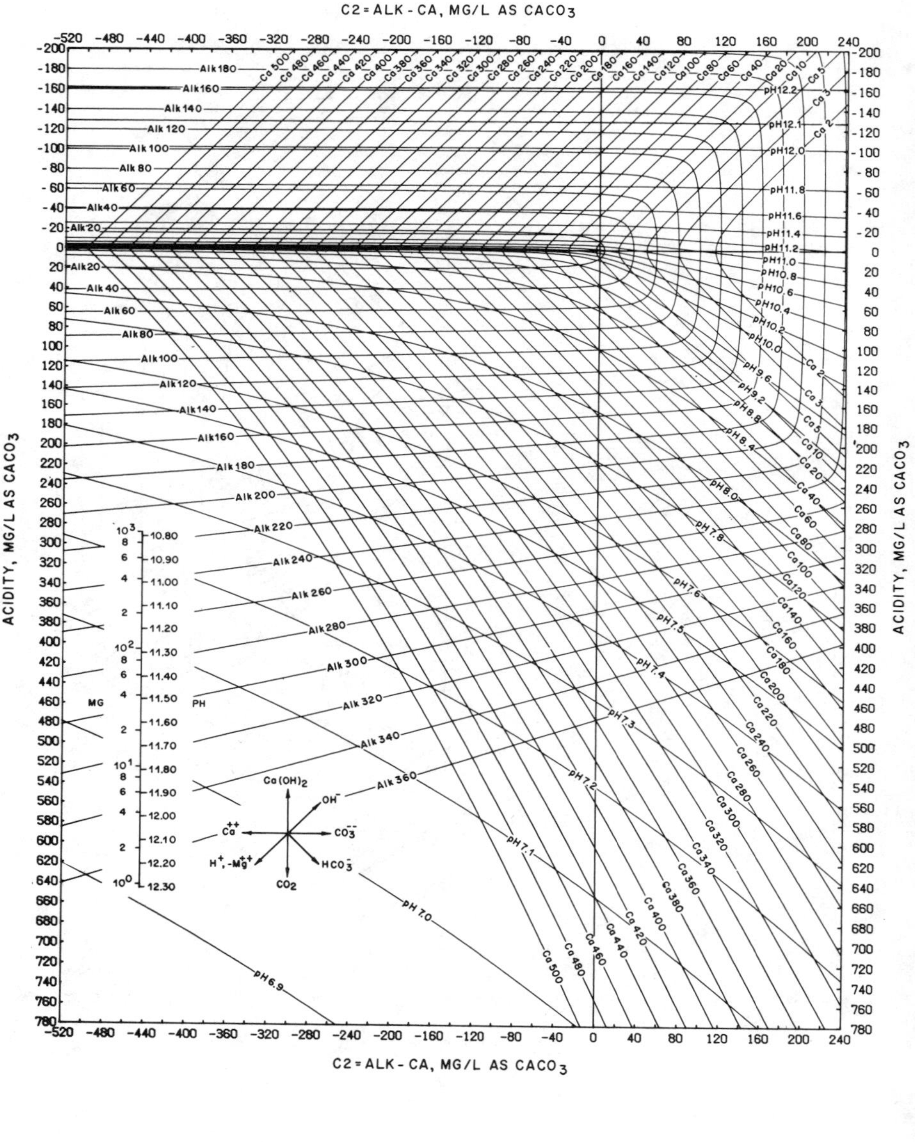

Figure 15. Water conditioning diagram for 5°C and 400 mg/l total dissolved solids.

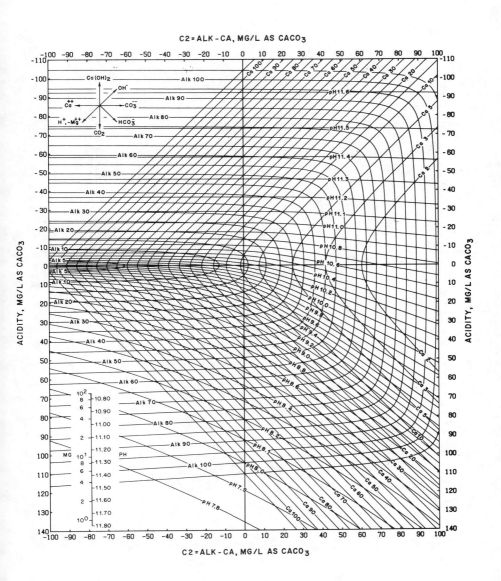

Figure 16. Water conditioning diagram for $15°C$ and 40 mg/l total dissolved solids.

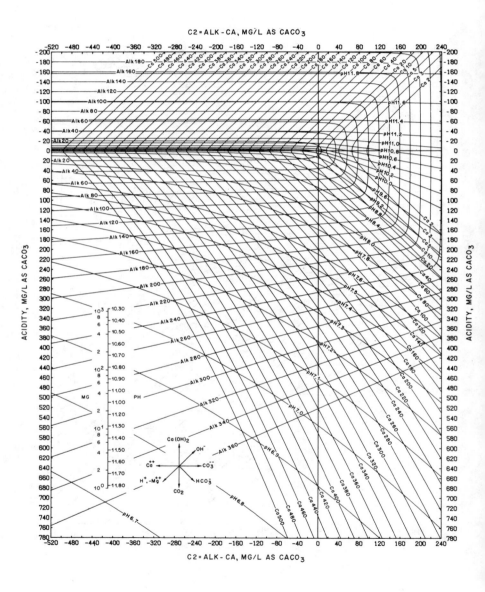

Figure 17. Water conditioning diagram for 15°C and 400 mg/l total dissolved solids.

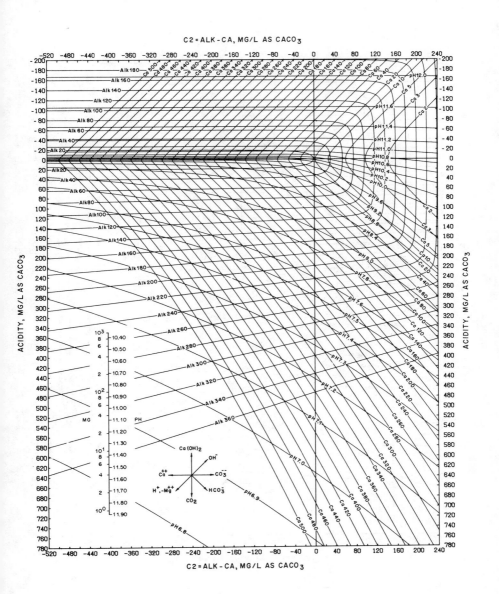

Figure 18.   Water conditioning diagram for 15°C and 1200 mg/l total dissolved solids.

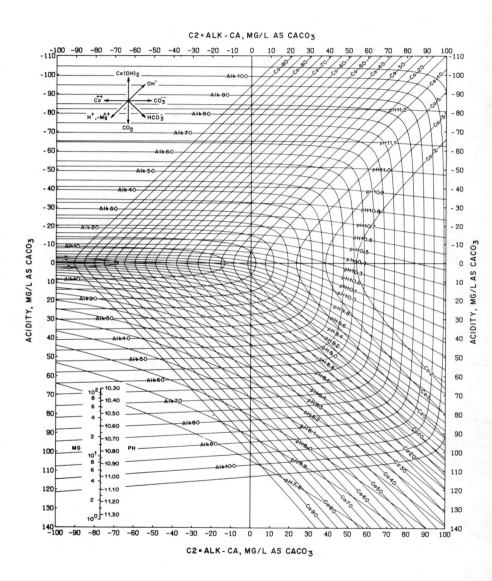

Figure 19. Water conditioning diagram for 25°C and 40 mg/l total dissolved solids.

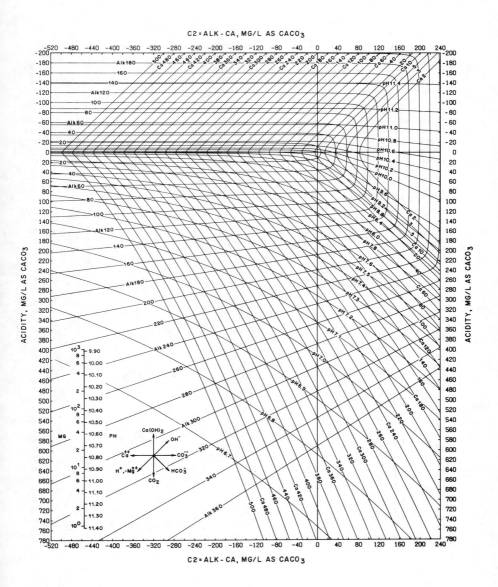

C2 = ALK - CA, MG/L AS CACO₃

Figure 20. Water conditioning diagram for 25°C and 400 mg/l total dissolved solids.

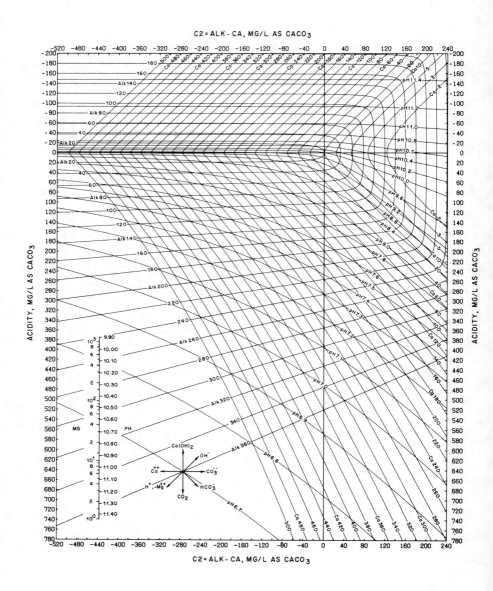

Figure 21. Water conditioning diagram for 25°C and 1200 mg/l total dissolved solids.

REFERENCES

1. Hudson, H. E., and F. W. Gilcreas. "Health and Economic Aspects of Water Hardness and Corrosiveness," *J. Am. Water Works Assoc.* 68(4):201 (1976).
2. Craun, G. F., and L. J. McCabe. "Problems Associated with Metals in Drinking Water," *J. Am. Water Works Assoc.* 67(11):593 (1975).
3. Linstedt, K. D., E. R. Bennett, and S. W. Work. "Quality Considerations in Successive Water Use," *J. Water Poll. Cont. Fed.* 43(8): 1681 (1971).
4. Stumm, W. "Investigation of the Corrosive Behavior of Waters," *JSED, Proc. Am. Soc. Civil Eng.* 86(11):27 (1960).
5. Larson, T. E. In *Water Quality Treatment,* prepared by the American Water Works Association (New York: McGraw-Hill Book Company, 1971), p. 295.
6. Larson, T. E. "The Ideal Lime-Softened Water," *J. Am. Water Works Assoc.* 43(8):649 (1951).
7. Fontana, M. G., and N. D. Greene. *Corrosion Engineering* (New York: McGraw-Hill Book Company, 1967).
8. *1976 Annual Book of ASTM Standards,* part 10 (Philadelphia: American Society for Testing and Materials, 1976).
9. *Standard Methods for the Examination of Water and Wastewater,* 13th ed. (New York: American Public Health Association, 1971).
10. Sawyer, C. N., and P. L. McCarty. *Chemistry for Sanitary Engineers,* 2nd ed. (New York: McGraw Hill Book Company, 1967), pp. 329-330.
11. Dye, J. F., and J. L. Tuepker. In *Water Quality Treatment,* prepared by the American Water Works Association (New York: McGraw-Hill Book Company, 1971), p. 313.
12. McCauley, R. F., and M. O. Abdullah. "Carbonate Deposits for Pipe Protection," *J. Am. Water Works Assoc.* 50(11):1421 (1958).
13. Larson, T. E. "Corrosion by Domestic Waters," Bulletin 59, Illinois State Water Survey, Urbana, Illinois (1975).
14. Ferguson, J. F., and P. L. McCarty. "The Precipitation of Phosphate from Fresh Waters and Waste Waters," Technical Report No. 120, Dept. of Civil Engineering, Stanford University, Stanford, California (1969).
15. Loewenthal, R. E., and G. v. R. Marais. "The Carbonic System in Water Treatment," Research Report No. W4, The University of Cape Town, Department of Civil Engineering, Rondebosch, Cape Town, South Africa (May 1973).
16. Caldwell, D. H., and W. B. Lawrence. "Water Softening and Conditioning Problems," *Ind. Eng. Chem.* 45(3):535 (1953).
17. Loewenthal, R. E., and G. v. R. Marais. *Carbonate Chemistry of Aquatic Systems: Theory and Applications* (Ann Arbor, Michigan: Ann Arbor Science Publishers, Inc., 1976).
18. Internal memorandum. Brown and Caldwell, Seattle, Washington (1976).

CHAPTER 24

LIME-SODA SOFTENING PROCESSES

Carl W. Reh, P.E.
Partner
Greeley and Hansen, Engineers
222 South Riverside Plaza
Chicago, Illinois 60606

INTRODUCTION

The design of lime softening facilities for municipal plants involves the application of the basic water chemistry to continuous, uninterrupted flow regimens to produce a finished water with desirable, consistent, and predictable characteristics and properties. Lime-soda ash-caustic softening includes a number of flow patterns, a wide variety of equipment components, and much accessory equipment for process control and sludge disposal. This chapter discusses the following:

- alternate processes and bases for selection
- process elements, specialized equipment and tankage for the softening operations
- process control.

Separate coagulation may be desirable for surface supplies when the raw water is highly turbid, when the water quality is highly variable, or when recalcining of sludge is practiced. Softening may be provided for well waters and for surface supplies. Clarification may be accomplished in the same basins as softening reactions, or separate clarification and softening trains may be provided. The AWWA[1] suggests that presedimentation be provided if high turbidities (up to 3000 TU or more) continuing for several days are expected several times a year. Rapid agglomeration of calcium carbonate floc appears to require that the $CaCO_3$ crystals remain relatively free of organic colloids. The presence of organic colloids may impede crystallization.[2]

567

PROCESS DESCRIPTION

The lime-soda ash-caustic processes for the removal of hardness are summarized as follows:

- Carbonate hardness may be precipitated as calcium carbonate or magnesium hydroxide by the addition of lime as illustrated by the following:

$$CO_2 + Ca(OH)_2 \rightarrow CaCO_3 \downarrow + H_2O \tag{1}$$
$$Ca(HCO_3)_2 + Ca(OH)_2 \rightarrow 2CaCO_3 \downarrow + 2H_2O \tag{2}$$
$$Mg(HCO_3)_2 + 2Ca(OH)_2 \rightarrow 2CaCO_3 \downarrow + Mg(OH)_2 \downarrow + 2H_2O \tag{3}$$

- Noncarbonate calcium hardness may be precipitated by the addition of soda-ash, as illustrated by the following:

$$CaSO_4 + Na_2CO_3 \rightarrow CaCO_3 \downarrow + Na_2SO_4 \tag{4}$$

- Noncarbonate magnesium hardness may be precipitated as magnesium hydroxide by the addition of lime, with the concurrent production of calcium sulfate, which, in turn, can be precipitated as calcium carbonate by the addition of soda-ash. These reactions are illustrated as follows:

$$MgSO_4 + Ca(OH)_2 \rightarrow Mg(OH)_2 \downarrow + CaSO_4 \tag{5}$$
$$CaSO_4 + Na_2CO_3 \rightarrow CaCO_3 \downarrow + Na_2SO_4$$

- Carbonate and noncarbonate hardness may be removed by the addition of caustic soda, as illustrated by the following:

$$CO_2 + 2NaOH \rightarrow Na_2CO_3 + H_2O \tag{6}$$
$$Ca(HCO_3)_2 + 2NaOH \rightarrow CaCO_3 \downarrow + Na_2CO_3 + 2H_2O \tag{7}$$
$$Mg(HCO_3)_2 + 4NaOH \rightarrow Mg(OH)_2 \downarrow + 2Na_2CO_3 + 2H_2O \tag{8}$$
$$MgSO_4 + 2NaOH \rightarrow Mg(OH)_2 \downarrow + Na_2SO_4 \tag{9}$$

Note that calcium sulfate would react with the soda ash formed and be precipitated as calcium carbonate. See Equation 4.

Softened water is usually saturated with calcium carbonate at the high pH values. Therefore, the water must be stabilized prior to filtration. This may be accomplished by the addition of carbon dioxide or acids to the water, as follows:

$$CaCO_3 + CO_2 + H_2O \rightarrow Ca(HCO_3)_2 \tag{10}$$
$$2CaCO_3 + H_2SO_4 \rightarrow Ca(HCO_3)_2 + CaSO_4 \tag{11}$$
$$2CaCO_3 + 2HCl \rightarrow Ca(HCO_3)_2 + CaCl_2 \tag{12}$$

The use of carbon dioxide for stabilization is usual, hence the term "recarbonation." The principal reasons for the preferential use of carbon dioxide instead of acids are that acids increase the dissolved solids in the finished water, and, if mineral acids are used would increase the hardness more than two-stage recarbonation.

PROCESS FLOW ARRANGEMENTS

Softening can be selective with reference to the removal of calcium and magnesium hardness. Water relatively low in magnesium hardness can be treated with lime to remove the calcium carbonate hardness. If magnesium is to be removed, excess lime is added. If noncarbonate hardness is to be reduced, soda-ash is added. Caustic soda (NaOH) may be substituted for lime [Ca(OH)$_2$] and soda ash (Na$_2$CO$_3$). The softened water needs to be stabilized before the water is applied to filters. The several processes may be accomplished in a number of flow patterns, as illustrated in Figures 1-3.

Figure 1 shows single stage conventional softening. Depending upon the kinds and amounts of chemical added, the calcium and magnesium carbonate and noncarbonate hardness can be removed.

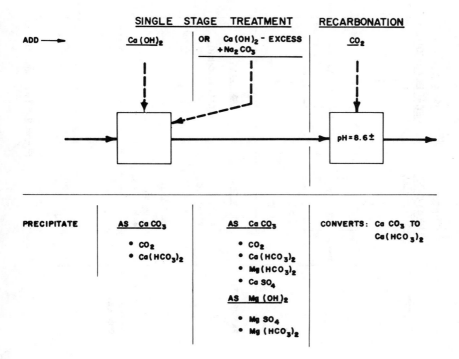

Figure 1. Single-stage softening and recarbonation.

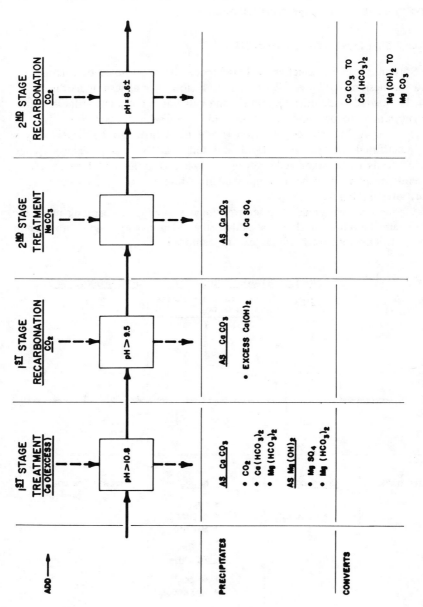

Figure 2. Two-stage softening and recarbonation.

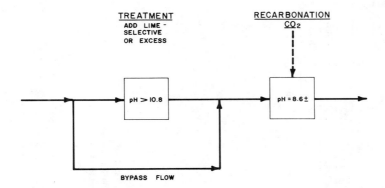

Figure 3. Split treatment and recarbonation.

Figure 2 depicts two-stage treatment and two-stage recarbonation. This arrangement permits greater removal of the magnesium hardness and promotes greater control of the effluent characteristics. It also would provide the lowest practical level of hardness in the effluent. Figure 3 shows a "split treatment" arrangement. This system also may be tailored to provide selective calcium carbonate hardness reduction or calcium and magnesium hardness reduction. Feeding excess lime to the main stream removes the magnesium hardness. The softened and unsoftened water is mixed prior to recarbonation, with CO_2 added in the minimum amount necessary to provide stable water. This process flow arrangement frequently obviates the need for CO_2 addition.

Other modifications have been used for specific problems. Where efficient lime recovery and maximum hardness reduction are necessary, three-stage processes have been proposed.[3] This process involves the precipitation of $CaCO_3$ in the first stage, followed by the addition of excess lime in the second stage to precipitate magnesium hydroxide, followed by two-stage recarbonation.

Although not shown on any of the diagrams, aeration may be considered to be a part of the softening process. The Virginia Water Works Regulations[4] states, "When concentrations [of CO_2] exceed 10 milligrams per liter, the economics of removal by aeration as opposed to removal of CO_2 with lime should be considered."

Choice of Reagents

The choice of reagents, lime-soda ash vs caustic, is affected by three parameters.

Cost. Lime and soda ash remove all elements of hardness, except magnesium noncarbonate hardness, at lower cost than can be achieved with caustic. See Table I.

Table I. Comparison of Lime Soda and Caustic Softening: Stoichiometric Dosages and Costs of Pure Chemicals for 1 mg/l (as $CaCO_3$) of Demand Source

| Source of Demand | Chemical Dosage Required lb/million gallons[a] | | | Chemical Costs $/million gallons[b] | | |
| | Lime-Soda Softening | | Caustic Softening | Lime-Soda | | Caustic |
	CaO, as CaO	Na_2CO_3, as Na_2CO_3	NaOH, as NaOH	CaO^c	$Na_2CO_3{}^d$	$NaOH^e$
$CO_2{}^f$	10.61	None	15.16	$0.212	None	1.21
Ca–CH	4.67	None	6.67	0.093	None	0.53
Mg-CH^g	9.34	None	13.34	0.187	None	1.57
Mg-NCH^g	4.67	8.84	6.67	0.093	0.486	0.53
Ca-NCH	None	8.84	None	None	0.486	None
Na-Alk	4.67	-8.84	None	0.093	-0.486	None

^{a}lb/million gallons = 1.2 x 10^{-4} kg/m^3

^{b}$/million gallons = 2.64 x 10^{-4} $/m^3

^{c}@ $40/T

^{d}@ $110/T

^{e}@ $160/T

^{f}As CO_2; all others expressed as $CaCO_3$.

^{g}In addition to the amounts stated in the table, and without regard to the hardness, 219.9 lb CaO or 417 lb NaOH at $5.84 and $33.36/million gallons, respectively, would be required to provide the equivalent of the 35 mg/l CaO excess necessary for Mg removal.

Ca-CH = calcium, carbonate hardness

Mg-CH = magnesium, carbonate hardness

Mg-NCH = magnesium, noncarbonate hardness

Ca-NCH = calcium, noncarbonate hardness

- Total Dissolved Solids. The total dissolved solids increase with the use of caustic because all reactions produce soluble soda ash (Na_2CO_3), sodium sulfate (Na_2SO_4) or sodium chloride (NaCl).
- Sludge Production. Reaction with caustic soda produces less sludge than reaction with lime-soda ash, as follows:

  − No sludge is produced for CO_2 or Na-alk.
  − About 30-40% as much sludge is produced for Mg removal.
  − Half as much sludge is produced for calcium carbonate hardness removal.
  − The same amount of sludge is produced for calcium noncarbonate hardness removal.

The comparative quantities of sludge production are summarized in Table II.

Table II. Sludge Production: Stoichiometric Amounts from 1 mg/l as $CaCO_3$[a] of Demand Source

Source of Demand	lb Dry Solids/million gallons[b]	
	Lime Soda	NaOH
CO_2	18.95	None
Ca-CH	16.68	8.34
Mg-CH	16.68	4.84
Mg-NCH	13.18	4.84
Ca-NCH	8.34	8.34
Na-Alk	8.34	None

[a]Except for CO_2, which is expressed as mg/l CO_2
[b]lb/million gallons = 1.2 x 10^{-4} kg/m^3

The economics of lime-soda ash vs caustic are illustrated in Table III. Softening with caustic is most competitive for either very low or very high alkalinity waters. For the lime-soda ash system, the amount of sludge increases with increasing alkalinity; for the caustic system, the quantity of sludge is independent of alkalinity.

A further consideration of the choice of reagents relates to the properties of the chemicals. Caustic soda is easier to store and feed than lime; caustic soda does not deteriorate in storage. Hydrated lime absorbs water and CO_2 from the air and reverts to $CaCO_3$. Quick lime is unstable and can slake in storage. Soda ash is usually sold in dry form. It is hygroscopic and is quite soluble.

Table III. Comparison of Lime-Soda and Caustic Softening

TH = 250 mg/l
Ca-H = 150 mg/l
Mg-H = 100 mg/l
CO_2 = 10 mg/l
Q = 1 mgd

| | Lime-Soda Softening | | | | | | NaOH Requirement | | | |
| | CaO Requirement | | Na$_2$CO$_3$ Requirement | | | | | | | |
	Dry (tons/day)[a]	1-Month Storage (ft³)[b]	Dry (tons/day)	1-Month Storage (ft³)	Cost (¢/1000 gal)[c]	Sludge Dry (tons/day)	Dry (tons/day)	1-Month Storage (ft³)	Cost (¢/1000 gal)	Sludge Dry (tons/day)
Case (A) Alk = 150 mg/l Stoichiometric	0.64	710	0.44	415	2.56 +4.84 =7.40	2.00	0.91	1,147	14.6	0.87
Excess hydroxide[d]	0.79	878	–	–	8.00	2.26	1.12	1,411	17.9	0.87
Case (B) Alk = 250 mg/l Stoichiometric	0.87	970	–	–	3.48	2.42	1.24	1,562	19.8	0.87
Excess hydroxide[d]	1.02	1,133	–	–	4.08	2.68	1.45	1,827	23.2	0.87
Case (C) Alk = 350 mg/l Stoichiometric	1.10	1,230	–	–	4.40	2.84	1.24	1,562	19.8	0.87
Excess hydroxide[d]	1.25	1,389	–	–	5.00	3.10	1.45	1,827	23.2	0.87

[a] tons = 0.907 T
[b] ft³ = 0.0283 m³
[c] ¢/1000 gal = 3.785 ¢/m³
[d] 35 mg/l CaO or 50 mg/l NaOH. Values are totals for stoichiometric and excess hydroxide requirements.

Note: 1. Hardness and alkalinity expressed as mg/l of $CaCO_3$.
2. Chemical requirements as 100% pure dry tons per day.
3. 1-month storage assumes
   A. 60 lb/ft³ for 90% CaO
   B. 65 lb/ft³ for 99.4% Na$_2$CO$_3$
   C. 1 dry ton of NaOH = 42 ft³ of 50% NaOH solution
4. Costs assumes
   A. $40/dry ton CaO
   B. $110/dry ton Na$_2CO_3$
   C. $160/dry ton NaOH

Factors for Process Design

The softening process is influenced by the following factors:

1. the rate of dissolution of chemical reagents
2. the rate of precipitate formation
3. the growth rate of precipitated particles to a size that can be removed
4. the settling rates of the floc.

CaO and $Ca(OH)_2$ dissolve relatively slowly. Although 30 sec may be adequate to dissolve coagulant in the process stream, 5-10 min may be required for lime.[1] Similarly, while high "G" levels are required to disperse coagulants, softening reactions generally require mixing only to keep the lime particles in suspension and distributed uniformly during the time of mixing.

The rate of precipitate formation of $CaCO_3$ is temperature dependent. Cole[5] reports that water at a temperature of 1°C took about 90 min to reach its minimum hardness, whereas at 22°C minimum hardness was reached in about 30 min. The data are summarized in Figure 4.

The rates of formation and growth of precipitate are usually addressed only in qualitative terms. The recirculation of sludge enhances the growth of the precipitate. Sludge recirculation is an integral part of solids contact basin system design. The reaction produced by lime in raw water is very slow in the absence of crystallization nuclei. However, if the lime and the water are mixed in the presence of precipitated $CaCO_3$, the reaction time is reduced and the stability of the softened water is enhanced. Hartung,[6] in his studies on Meramec River water, concluded that a more stable effluent water was obtained with a slurry concentration of 5% or more by weight.

Settling rates are a function of particle size and particle density. As either increases, the rate of settling increases. Inasmuch as the water density increases with lowering temperatures and the rate of formation of $CaCO_3$ decreases with lowering temperatures, low temperature operations usually require larger settling tanks and the use of coagulant aids to increase the size and density of the floc.

In comparing lime-soda and caustic systems, the equilibrium concentration and rate of $CaCO_3$ formation would be expected to be similar. However, inasmuch as in the caustic system there is no slow dissolving lime involved, a conventional, short-term, high (coagulant type) "G" rapid mix may be used.

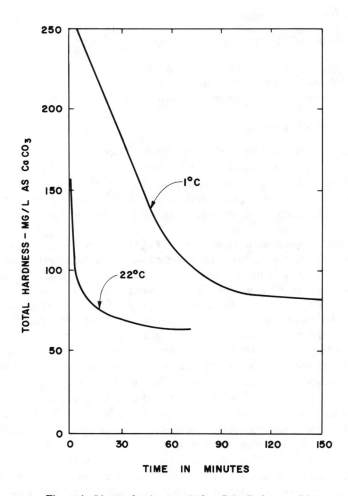

Figure 4. Lime-softening rate (After Cole. Reference 5.)

Coagulants

Coagulants are necessary to agglomerate the fine calcium and magnesium precipitates into particles that will settle more readily than non-coagulated water. The two categories of coagulants are metal coagulants and polymeric coagulants. Coagulants lower the suspended solids of water applied to the filters and foster softening efficiency by coagulating and removing fine $CaCO_3$ and $Mg(OH)_2$ particles so that these particles are not dissolved during subsequent recarbonation and pH adjustment.

Metal coagulants form inorganic polymers after addition to water. Those commonly used are aluminum and iron salts. Generally, 10 mg/l or less of an iron or aluminum salt (except sodium aluminate) is used. About 5 mg/l sodium aluminate appears to be adequate.

Metal coagulants commonly used include the following:

- alum $Al_2(SO_4)_3 \cdot 14.3\ H_2O$
- sodium aluminate $Na_2OAl_2O_3$
- ferric sulfate $Fe_2(SO_4)_3 \cdot 9\ H_2O$
- ferrous sulfate $FeSO_4 \cdot 7\ H_2O$
- ferric chloride $FeCl_3$
- magnesium carbonate $MgCO_3 \cdot 3\ H_2O$

With the exception of sodium aluminate, all of these lower alkalinity and release CO_2. The change in alkalinity and CO_2 release is shown in Table IV.

Table IV. Effect of Coagulant on Alkalinity and Carbon Dioxide

Coagulant	Alkalinity Reduction (mg/l $CaCO_3$ per mg/l of coagulant)	CO_2 Release (mg/l CO_2 per mg/l of coagulant)
Alum	0.55	0.44
Sodium Aluminate	-0.61	-0.54
Ferrous Sulfate	0.66	0.58
Ferric Sulfate	0.77	0.66
Ferric Chloride	0.93	0.81
Magnesium Carbonate	0.72	0.64

Ferric salts are useful in a wide pH range (4 to 11). If ferrous salts are used, the water must have a pH of 8.5 or more or the salts must be used with chlorine to allow oxidation to the ferric state. Alum is used near a neutral pH, 6.0 to 7.8. At both higher and lower pH, alum tends to form relatively soluble compounds. Sodium aluminate is used in softening reactions because it does not consume alkalinity.

Magnesium carbonate represents the newest technology. It is used primarily to reduce or eliminate sludges.[7] The process was developed by Black,[8] starting in Dayton, Ohio, in the 1950s. Magnesium reacts with lime to form magnesium hydroxide and calcium carbonate precipitates.

The subsequent carbonation of the sludge dissolves the magnesium hydroxide. The reactions are as follows:

$$MgCO_3 + Ca(OH)_2 \rightarrow Mg(OH)_2 \downarrow + CaCO_3 \downarrow \qquad (13)$$
$$Mg(OH)_2 + CO_2 \rightarrow MgCO_3 + H_2O \qquad (14)$$

Thus the coagulant is recovered and if the lime sludge is recalcined, essentially all components of the sludge are reclaimed. By the use of froth flotation, clay, silt, and other suspended contaminants can be separated from the calcium carbonate sludge, thus facilitating recalcining. See Chapters 5 and 7.

Polymeric coagulants are in their polymer form before addition to the water to be treated. Organic polymers are not in general use in softening plants. However, there are several reasons for considering the use of polymers:

- The volume of sludge is less than obtained with metal coagulants.
- Polymers generally have a minor effect on pH; the need for further pH adjustment in the finished water may be reduced.
- Polymer-coagulated sludge tends to be more readily dewatered than metal-flocculated sludge.

A recent report by the AWWA Research Foundation[9] stated that of 17 American Water Works Company's plants in the midwestern and western regions, 12 used a liquid cationic polymer. The polymers include Betz 1190, Catfloc B, Catfloc T, Magnifloc 575C, Nalco 607 and Nalco 8101. Nalco 8113 proved effective at the St. Louis County Water Company plant in clarifying lime-softened water, although this polymer was not effective for raw water clarification.

Black and Christman[10] investigated a number of coagulants and coagulant aids. They observed that activated silica was effective when added prior to lime additions.

Although it appears that the use of polymers for coagulation of softening sludges will become increasingly attractive, the final selection of a polymer must be made on the basis of plant scale testing.

Stabilization

Stabilization of softened water can be accomplished by recarbonation, by acid addition, and by phosphates to avoid precipitation of mono-carbonates. The most common method is recarbonation with CO_2. Mineral acids (sulfuric, hydrochloric, phosphoric and others) have been used but not intensively. The use of mineral acids for stabilization increases the total dissolved solids of the finished water and may increase the hardness. Cost and safety of handling may also tend to reduce the popularity of acids for stabilization.

Plant Data

Data on water characteristics, reagents, and types of illustrative water softening plants are summarized in Table V.

PROCESS ELEMENTS

The principal process elements include aeration, rapid mix, flocculation, sedimentation and stabilization. Ancillary process elements include chemical storage, feeding and sludge disposal. Further, the individual process elements and the process flow diagrams mutually affect each other. Although the process elements are discussed individually, the effects of the position and function of the separate elements in the flow diagram need to be kept in mind. Figures 5-8 are illustrative of the variations in flow diagrams.

The finished water requirements and the raw water characteristics affect the selection of both the process flow diagram and the process elements. A surface water supply subject to wide variations in hardness and turbidity may require clarification prior to any softening. Alternatively, if the softening sludge is not to be recalcined, close monitoring and control may obviate the need for stage treatment. Where design factors are suggested, these should be regarded as average or illustrative. Rules of regulatory agencies may control the sizing of plant elements.

Conventional Process Elements

Separate elements for rapid mix, flocculation and sedimentation are termed "conventional." Generally the elements are similar to those required for clarification. The discussion herein emphasizes the special requirements of softening.

Rapid Mix

The function of rapid mix is to disperse the chemicals and maintain this dispersal for a sufficient period of time to dissolve the reagents and start the reaction. The displacement time and the kind of mixing equipment are influenced by temperature and reagent. The rate of dissolution of lime is quite low. The minimum displacement time, therefore, should be considered to be 5 min, and longer displacement times may be desirable. The addition of metallic coagulants to a rapid mix basin ahead of softening reagents is often found to be advantageous. Mixing equipment similar to that used for clarification would be satisfactory for this purpose. In-line blenders are not used for flash mixing of softening reagents.

Table V. Data for Illustrative Municipal Water Softening Plants

Place	Source of Supply	Hardness (mg/l) $CaCO_3$ — Raw Water			Softened Water			Softening Process — Type[b]	Re-agents[c]	Coagulants[d]	Stabilization[e]	Plant Capacity mgd
		T-H[a]	Ca-H[a]	Mg-H[a]	T-H	Ca-H	Mg-H					
Cocoa, FL	Wells	330	274	56	85	35	50	A,SN	L,SA	PS	CO_2-1	40
Daytona B., FL	Wells	320	284	36	106	60	46	SN	L	PM	ND	12
Gainesville, FL	Wells	207	156	51	96	52	44	SN	L	PS	CO_2-1	40
Miami, FL[f]	Wells	244	212	32	75	55	20					
St. Petersburg, FL	Wells	203	187	16	102	86	16	A,SP	L	PM	None	55
Champaign, IL	Wells	250	125	125	81	37	45	SP	L	AS	H_2SO_4	10.8
Olympia Field, IL	Wells	480	325	150	150			SN	L,SA	AL	ND	0.7
Park Forest, IL	Wells	565	340	225	225[g]	90	135	A,LZ	L	NA	CO_2-2	3.7
Salina, KS	River & wells								L,SA	AL,PM	ND	10
Topeka, KS	Kansas R.	272	201	71	104	66	38	MS	L,SA	AL or NA	CO_2	40
Jackson, MI	Wells	437	292	145	85	35	50	SP	L,CS	ND	ND	24
Independence, MO	Wells	254			105	50	55	SN	L	None	ND	6
St. Louis Co., MO	Meramec R.	191	94	93	106	37	69	MS	L	AS,F	ND	24
Columbus, OH	Scioto R.	256	162	94	100	73	27	MS	L,SA,CS	AL	CO_2-1	65
Dayton, OH	Wells	358	227	131	107	74	33	SR	L,SA	AS	CO_2-1	96
Sioux Falls, SD	Wells	530	450	80	285	215	60	SN	L	AS,PM	CO_2-2	40.2

[a]T-H = total hardness; Ca-H = calcium hardness; Mg-H = magnesium hardness.

[b]A = aeration; SN= single stage; SP= split treatment; SR = split recarb; MS = multistage; LZ = lime zeolite.

[c]L = lime; SA = soda ash; CS = caustic.

[d]AS = activated silica; AL = alum; F = ferric salts; PM = polymer; NA = sodium aluminate; PS = potato starch.

[e]ND = not determined; CO_2 = source not determined; CO_2-1 = submerged combustion burners; CO_2-2 = separate CO_2 generator.

[f]CO_2 in by-pass stream.

[g]Lime-softened water only.

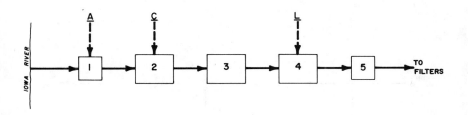

1. RAPID MIX
2. FLOCCULATION
3. SEDIMENTATION
4. SOFTENING
5. RECARBONATION

A = ALUM
C = CARBON
L = LIME

PLANT CAP. = 4.0 MGD

Figure 5. Flow diagram. Water treatment plant, Iowa State University. (After Kross, Fisher, and Paulson. Reference 11.)

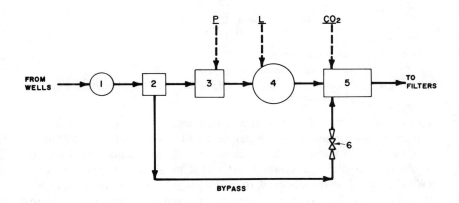

1. AERATOR
2. FLOW SPLITTER
3. RAPID MIX
4. SOLIDS CONTACT BASIN
5. RECARBONATION BASIN
6. FLOW CONTROLLER

L = LIME
P = POLYMER

Figure 6. Flow diagram. Water treatment plant, St. Petersburg, Florida.

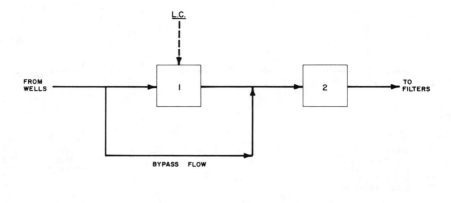

Figure 7. Flow diagram. Water treatment plant, Jackson, Michigan. (After Bhatia and Cooperwasser. Reference 12.)

For displacement times of 1 min and longer, a "G" of 700 is suggested.[1] G, the mean temporal velocity gradient, is defined as follows:

$$G = \sqrt{\frac{550\,P}{V\mu}}$$

where: P = water horsepower
V = volume of rapid mix basin in ft^3
μ = viscosity (at 50°F, the value is 0.273 x 10^{-4} lb sec/ft^2 (1.307 centipoise)

To maintain a "G" of 700 with a displacement period of 5 min would require a power input to the water of about 11 hp/mgd (187 kW/m^3 sec) of capacity. At a "G" of 900, for which a 30-sec displacement period would be adequate, the power input would be about 1.9 hp/mgd (32 kW/m^3 sec). For the dispersal of metal coagulants, the use of a high "G" rapid mix for coagulant dispersal, followed by a longer period rapid mix basin for softening, may be cost effective.

The power required for "blending" is not needed to maintain the dispersal of lime. For general application Foster[13] suggests a power input of 0.25-1.0 hp/mgd (17 kW/m^3 sec) for rapid mix. The lower level would appear adequate for mixing if the coagulant and the softening reagents were added separately.

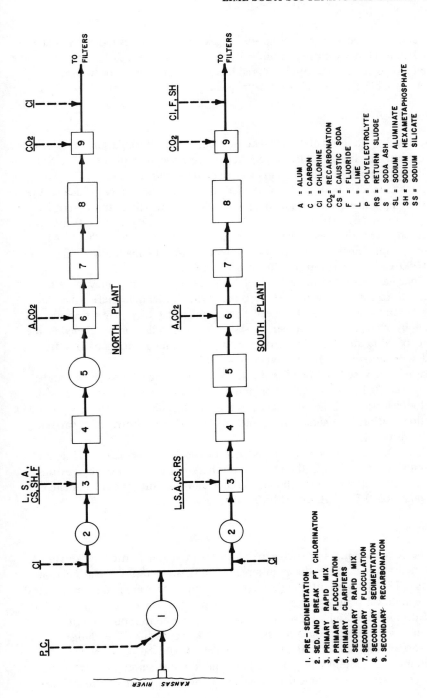

Figure 8. Flow diagram. Water treatment plant, Topeka, Kansas.

1. PRE-SEDIMENTATION
2. SED. AND BREAK PT CHLORINATION
3. PRIMARY RAPID MIX
4. PRIMARY FLOCCULATION
5. PRIMARY CLARIFIERS
6. SECONDARY RAPID MIX
7. SECONDARY FLOCCULATION
8. SECONDARY SEDIMENTATION
9. SECONDARY RECARBONATION

A   = ALUM
C   = CARBON
Cl = CHLORINE
CO_2 = RECARBONATION
CS = CAUSTIC SODA
F   = FLUORIDE
L   = LIME
P   = POLYELECTROLYTE
RS = RETURN SLUDGE
S   = SODA ASH
SL = SODIUM ALUMINATE
SH = SODIUM HEXAMETAPHOSPHATE
SS = SODIUM SILICATE

Variable speed drives with speed ranges of 1:3 are desirable. Variable speed motors, eddy-current couplings, fluid drives and V-belt variable speed drives have been used. Variable speed motors may be wound rotor, variable frequency or dc with package ac/dc static type rectifiers.

Flocculation

Generally, state regulatory agencies require that flocculation times and equipment be similar to those required for clarification plants. The State of Virginia specifies that flocculation shall be based upon a "GT" value of 20,000 to 200,000. (GT is defined as the product of G, which is the mean temporal velocity gradient in sec^{-1}, and T, which is the detention time in seconds.) Minimum flocculation and reaction time of 40 min is specified. The time is probably of greater importance than "G." At 40-min displacement time, the power requirement to maintain a "GT" of 200,000 would be about 1.0 hp/mgd (17 kW/m^3 sec).

Flocculators of the paddle-wheel type are usual. They may be either of the horizontal- or vertical-shaft type. With horizontal shafts, the flow may be transverse or axial. Axial flow lends itself to tapered flocculation either by changing the number of slats on the paddles or by changing the speed of subsequent shafts. Both fixed baffles and rotating baffles are used with horizontal units to avoid "streaking."

Peripheral speeds of paddles are usually maintained between 0.5 and 2.0 ft/sec (0.15-0.6 m/sec). Variable speed drives are needed to provide for the selection of the best speed for good flocculation. The hp required for flocculation is generally low, and the use of some form of a variable V-belt drive, therefore, is usual.

Once floc has formed, the avoidance of local high velocities minimizes breaking up of the floc, thereby promoting good clarification by settling. The velocity of flow through ports, pipes and conduits should remain in the range of 0.5-1.5 ft/sec (0.15-0.46 m/sec).

Sedimentation

The requirements and design parameters for sedimentation tanks for softening are similar to the requirements and parameters for clarification plants. The quantity of sludge from softening, however, makes mechanical sludge removal a necessity even for small plants. Depending upon the final disposal method, sludge withdrawal may be continuous or intermittent. Provisions should be included for back flushing the sludge draw-off piping, the minimum size of which should be 6 inches.

Sedimentation tanks may be round, square with center influent, square with cross feed, or rectangular. Sludge removal in round or square tanks may be the center scraper type as shown by Figure 9. This type may be arranged in multiple units within a single rectangular tank. Rectangular tanks may also be arranged for drag chain type scrapers, as shown by Figure 10, or monorakes similar to the equipment used for clarification plants.

Velocities through inlet ports and effluent ports should not exceed 1.5 ft/sec (0.46 m/sec) at maximum flow rates. Outlets may be arranged with weirs and troughs. Velocity control, however, is facilitated by the use of submerged ports. The water level can be controlled at the filters, thus obviating the need for additional level control facilities at the sedimentation tanks.

A displacement period of 2-4 hr is usually adequate. If a water depth of 12 ft is used, the surface loading rates with 2 and 4 hours of displacement time would be 1077 and 538 gpd/ft^2 (43.9 and 21.9 m^3/m^2 day), respectively. Both displacement time and surface loading rates need to be considered in establishing the geometry of the tanks.

Solids Contact Basins

Solids contact basins are well suited for the precipitation of hardness. The presence of previously precipitated solids provides nuclei for the development of additional precipitates of calcium carbonate and magnesium hydroxide. The improvement of softening by maintaining a slurry of precipitated material has been demonstrated by Tuepker and Hartung[14] and others.

Many types of solids contact basins are available. Each of the several manufacturers claims advantages of his design over designs of other manufacturers. The several designs, however, may be classified into two general types:

1. Sludge blanket type, wherein the raw water is mixed with lime or other reagents and then rises through a "sludge blanket" of previously precipitated material.
2. Slurry recirculation type, wherein the raw water and reagents are added to a relatively large recirculating slurry of precipitated material.

Both types have been used successfully for softening. The early work by Spaulding[15] was done on a sludge blanket type of basin. Subsequent developments have added rakes for positive removal of sludge, turbine mixers, and differing designs of hoods to separate the reaction from the settling zones.

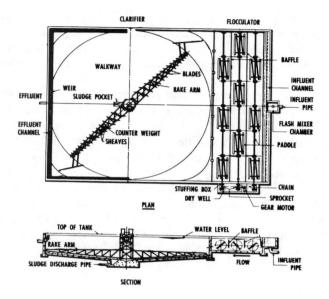

CROSS-FLOW ARRANGEMENT

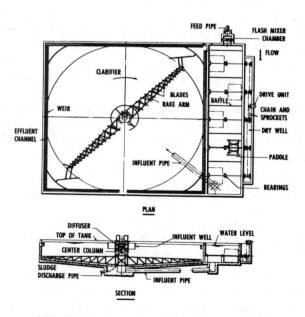

CENTER FEED ARRANGEMENT

Figure 9. Center sludge draw-off. Reproduced by permission from Dorr-Oliver, Inc. Bulletin No. 6971.

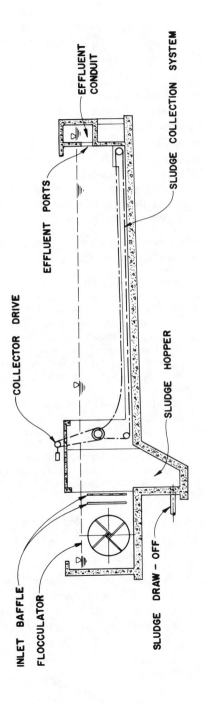

Figure 10.   Chain-flight sludge collector.

State regulatory agencies have adopted certain standards for the design, of solids contact basins. The "10-State Standards"[16] stipulate the following for softeners:

- Detention period: surface water, 4 hr
  ground water, 1 hr
- Coagulation: separate chamber or baffled zone
- Flocculation: minimum of 30 min
- Slurry concentration: minimum of 1%
- Sludge concentration for draw off: minimum of 5%
- Total water losses: maximum of 3%
- Weir loading: maximum of 20 gpm/ft (358 m^3/m day)
- Up-flow rates: maximum of 1.75 gpm/ft^2 (103 m^3/m^2 day) at slurry separation line.

Certain additional features need to be considered in the design of solids contact basins for softening, among which are the following:

1. Chemical application points: The addition of softening chemicals directly to the reaction zone generally is satisfactory. Addition of the coagulant ahead of the solids contact basin is preferable.
2. Rake torque: The minimum torque should be specified. The rake and drive should be adequate to start the rake after a 24-hr shutdown with the expected slurry concentration. Variable speed may not be necessary for the rake drive. Provisions should be made for reversing the motor to assist in freeing the rake after shutdown.
3. Desludging: The sludge should be drawn off frequently and for short durations. The minimum time between sludge drawoff intervals should be about 15 min. The maximum interval should be about 4 hr. The time for desludging should be as short as possible to maintain the desired slurry concentration. Not over 10% of the total cycle time should be used for each desludging interval. Provisions should be made for back-flushing with water at each desludging.

Illustrative data are summarized in Table VI.

Stabilization

The almost universal method of stabilizing softened water is to use CO_2, which may be obtained from the combustion of a variety of fuels and from a number of sources. The approximate yield of CO_2 from various fuels is given in Table VII. The combustion gasses may contain between 10 and 12% CO_2.

Table VI. Illustrative Data on Solids Contact Basins

Place	Manufacturer	Capacity (mgd)	Size - ft		Total Detention Time (min)	HP		
			Plan	SWD		Turbine	Rake	Total
Ft. Lauderdale, FL	Dorr	8.0	50 ft^2	15.0	73			3.0
Ft. Lauderdale, FL		25.0	80 ft. diam.	22.0	50			7.5
Cocoa, FL	Infilco	20.0	100 ft. diam.	21.0	90			40.0
Daytona Beach, FL	Eimco	6.0	65 ft. diam.	15.0				
Gainesville, FL	Eimco	20.0	85 ft. diam.	17.5	53	30	5	
Park Forest, IL	Infilco	2.0	31.5 ft. diam.	14.5	60			3.0
Park Forest, IL	Infilco	1.14	24.5 ft. diam.	14.5	64			3.0
St. Louis County, MO Meramec	Eimco	12.0	90 ft. diam.	17.0	98	30	3	
Champaign-Urbana, IL	Eimco	4.0	64 ft. diam.			15	¾	
		8.0	82 ft. diam.			20	3	

Table VII. Sources of Carbon Dioxide[a]

Fuel	Quantity	CO_2 Yield (lb)
Natural Gas	1000 cf	115
Coke	1 lb	3
Fuel Oil (No. 2)	1 gal	23
Propane	1000 cf	141
Butane	1000 cf	143

[a]After Reference 17.

The source of the CO_2 is important. Stack gases produced from a variety of fuels (wherein the choice of fuels may not be determined by the requirements of water treatment) should be avoided. The presence of sulfur and phenolic materials may impart objectionable tastes and odors to the water. CO_2 may be obtained as a by-product from the recalcining of lime sludge. The CO_2 in the exhaust of recalcination may approach 30%.

The types of CO_2 generators developed for water works use are submerged burners and forced-draft generators. Gas is the fuel choice for both types and is virtually mandatory for submerged combustion. Both the submerged burners and the forced-draft generators use compressors for the combustion air. The forced-draft generator has a pressure-tight combustion chamber and therefore gas compressors for the combustion gasses are not required. The control of output capacity is through the control of the fuel supply, and secondarily by the control of the output of the air compressor. The maximum turn-down ratio is about 5:1.

The use of liquid CO_2 appears to be increasing. It may be purchased in 20- and 50-lb cylinders and in bulk. Insulated tank trucks have nominal capacities of 10 tons and 20 tons. Rail shipment is also feasible.

Liquid CO_2 requires refrigeration for storage. Above 87.8°F (31°C), CO_2 does not exist in any state other than gaseous. At that temperature the vapor pressure is 1072 psia (7391 kN/m^2). Bulk storage, therefore, should be insulated, should be outdoors in a weather-protected area, and should have an automatic refrigeration system to maintain the pressure within design limits. Storage vessels designed for a working pressure of 300 psig (2069 kN/m^2 gage), which is considered usual, requires that the temperature of the CO_2 be maintained at approximately 0°F (-18°C).

Liquid CO_2 can be fed to the recarbonation basin as a gas, or the CO_2 may be dissolved in water and fed as a solution. The dissolution of 1 lb of CO_2 (0.45 kg) requires about 60 gal (227 l) of water.

The basins required for recarbonation are related to the purpose. Recarbonation for the conversion of excess lime to $CaCO_3$ (in split recarbonation processes) may include a small basin for the introduction of CO_2 and a large basin for reaction and settling. At Columbus, Ohio, the displacement time in the recarbonation basin is about 4 min and the reaction time is about 2 hr.[18]

For conversion of monocarbonates to bicarbonates, the time required for absorption and reaction is quite short. The 10-States Standards[16] suggests a total detention time of 20 min and a mixing compartment with a minimum detention time of 3 min.

CO_2 from submerged burners may be introduced into the water directly from the burner or through a pipe grid. Forced-draft generators usually have a pipe grid for the diffusion of CO_2. Hoover[19] suggests a pipe grid of 3/4 in. laterals with 3/32 in. holes spaced on 1-ft (0.3 m) centers. The total area of the holes should be about 80% of the header area. Header size is related to the total quantity of gas handled. Other systems for diffusing the gas have been developed. If the gas is dissolved prior to introduction into the basin (as described for liquid CO_2) the diffusion is probably not critical in the full utilization of CO_2.

The efficiency of solution or utilization of CO_2 is largely a function of the diffusion efficiency. If gas is introduced into the basin, utilization efficiency may be between 75 and 50%. The bulk of the loss of CO_2 is by escape to the atmosphere. If the CO_2 is dissolved, the reaction efficiency, with time, probably will be high.

The escape of CO_2 to the atmosphere is important in the design of ventilation systems. Prolonged exposure to atmospheres in which the concentration of CO_2 is 5% may cause unconsciousness and death. Exposure to a concentration of 10% can be endured for only a few minutes. If the concentration increases to 12-15%, unconsciousness occurs rapidly and death may result. These higher concentrations are obviously more important if liquid CO_2 is used. CO_2 is heavier than air so ventilation systems are necessary to avoid the build-up of potentially hazardous levels of CO_2 even when combustion methods of CO_2 generation are used.

Chemical Storage and Feeding

The storage and feeding of chemicals used specifically for softening are discussed in this section. Metallic coagulants and polymers used for clarification are discussed elsewhere. Magnesium carbonate, although a coagulant, is discussed because of its use in connection with recalcining of lime sludge.

Lime is the principal reagent used in softening. Because of the large quantities used, it is usually purchased as "quick-lime," 90% available CaO, and slaked to $Ca(OH)_2$ in conjunction with the feeding. Quick-lime is usually purchased as pebble lime, to avoid excessive dusting. Lime from recalcining calcium carbonate sludge is usually in powdered form. Lime is generally fed by gravimetric feeders to slakers, which in turn may be paste type or retention type. Paste-type slakers use a water-to-lime ratio of 2 to 2.5:1 and slake near boiling temperatures. Detention-type slakers use a water-to-lime ratio of 4 to 5:1. Detention-type slakers should be provided with a source of supplementary heat (for example, hot water) to maintain a slaking temperature of 160°F (71°C) or higher. After slaking, the lime slurry should be diluted to approximately 10% $Ca(OH)_2$ for transportation and feeding. Lower concentrations are practical, inasmuch as the solubility of $Ca(OH)_2$ is 0.17% at 50°F (10°C). Lime slurries may be pumped by the use of all-iron centrifugal pumps. The slurry tends to cake on all surfaces. Flat hose, open troughs and easily dismantled piping systems are needed for maintenance at the systems.

For very small systems, hydrated lime may prove economical. Hydrated lime is a light powder and is available in 100-lb bags. It must be stored in tight containers inasmuch as it absorbs CO_2 from the moisture in the atmosphere and is converted to $CaCO_3$. The slurry from hydrated lime is similar to the slurry from quick-lime.

Soda Ash chemically is sodium carbonate, Na_2CO_3. It is available as dense granules, medium granules and powder, and light powder. It can be purchased as 99.2% Na_2CO_3, and is available in dry form in 100-lb bags or drums and in bulk form.

The dry powder may be fed by gravimetric feeders. Rotolock bin gates are desirable to prevent flooding. Bin agitators are desirable.

Soda-ash is very soluble [12.5% at 50°F (10°C)] and is not corrosive. The solution can be pumped. Relatively large dissolvers appear desirable for the dense granules. The solution rate of 0.25 lb/gal (0.43 kg/l) and a 10-min displacement time appear adequate. Bulk handling equipment should include dust collection equipment.

Caustic Soda chemically is sodium hydroxide, NaOH. It is available in dry form in flakes, lumps and powder in 25-, 50-, 350-, 400- and 700-lb drums. All dry forms are available in 98.9% NaOH. NaOH is deliquescent and dangerous to handle. It is very soluble (51.5% at 50°F or 10°C), and its solubility increases rapidly with higher temperatures. On dissolving, large amounts of heat are generated. Methods of dissolving must include careful control and cooling to avoid boiling and spattering.

The solution form is preferable. It is available in liquid form as 50% NaOH. At this concentration, it begins to crystallize at 54°F (12°C). The solution is shipped in insulated tanks and may be diluted upon delivery. At Fremont, Ohio,[20] provisions are made for dilution to 2, 2.5, 3 or 4 lb/gal (0.24, 0.30, 0.36 or 0.48 kg/l). At Jackson, Michigan,[12] the caustic solution is diluted to 25% and then fed to day tanks. Caustic soda is not corrosive to steel or iron; concrete tanks may be lined with steel plate. Complete protective clothing is necessary to avoid severe burns to skin and eyes. Protective clothing should include respirator with goggles, rubber gloves and sleeves, rubber aprons, rubber work shoes, and cotton clothing including head covering.

Magnesium carbonate (MgCO₃) as a coagulant has been proposed as "a recycled coagulant."[8] Magnesium carbonate trihydrate ($MgCO_3 \cdot 3H_2O$), which was used for experimental and pilot plant work, is apparently not commercially available. Inasmuch as it is proposed to be reclaimed from the precipitation of $Mg(OH)_2$, it is likely that $MgCO_3$ will be handled in liquid form. The softening sludge that contains the $Mg(OH)_2$ would be treated with CO_2, which would convert the $Mg(OH)_2$ to $Mg(HCO_3)_2$. The use of magnesium carbonate, therefore, would be particularly beneficial in connection with recalcining, which would produce the necessary CO_2 for the treatment of the $Mg(OH)_2$.

Acids are sometimes used for stabilization. Acids are hazardous chemicals, and when used they should be handled with care. Protective equipment should include goggles, rubber gloves and sleeves, aprons, rubber work shoes and complete protection by appropriate clothing.

Hydrochloric acid (HCl) is shipped in glass bottles, carboys, rubber-lined steel drums, trucks, and tank cars. Its concentration is about 30%, and it is usually diluted to 20% (by adding acid to water) for feeding. It is highly corrosive, and pumps, valves and piping should be rubber-lined.

Sulfuric acid (H₂SO₄) is available in concentrations of about 77-98%. In concentrations of 93%, it is noncorrosive to steel. Once diluted, it is highly corrosive and should be housed and transported in rubber, plastic or glass-lined containers, pumps, valves and piping. It is shipped in glass bottles, carboys, drums, trucks and tank cars. It may be diluted to any strength for feeding.

PROCESS CONTROL

The control of lime-soda ash-caustic softening is dependent upon several variables under the control of the plant operators, as follows:

- flow, whether total flow or (in split treatment) the by-pass flow
- chemical feed, including chemicals for recarbonation
- desludging to maintain the optimum level of solids in the tanks for removal of hardness

Control of the softening process requires a combination of laboratory procedures and determinations and monitoring systems to ensure that the finished water is maintained at desired hardness levels.

If raw water characteristics change frequently, close observation and frequent adjustments in chemical feed rates are necessary. If raw water tends to remain uniform throughout a day, adjustments are necessary solely for changes in rate of flow and to accommodate malfunctions.

Monitoring instruments should include the following:

- rate-of-flow meters for total flow and by-pass flow
- pH meters for raw water and for the effluent from each step of the process
- turbidity meters for each step of the process including raw water
- feed rates for all chemicals

The flow can be split automatically by proportionate controls on rate-of-flow controllers. Chemical feed rates can be arranged to respond automatically to changes in flow rates. Regardless of the degree of automation, however, monitoring devices are necessary to ensure that malfunctions of mechanical equipment do not go uncorrected. Turbidity and pH monitoring are essential for all plants, and for those plants with a high order of automatic devices, alarms for "off-limits" are a necessity. Gilman[21] suggests that in-plant monitoring and control computers offer the following:

- automatic generation of periodic logs, trends and summary reports
- tool for computed variables
- application of complex mathematical models for process control
- centralized alarm monitoring and display of diagnostic data and corrective recommendations
- elimination of panel hardware such as indicators, recorders, controllers, timers and alarm lights, except as back-up

REFERENCES

1. *Water Treatment Plant Design* (Denver: American Water Works Association, Inc., 1969), p. 210.
2. *Water Treatment Plant Handbook* (Degrement, 1973), p. 284.
3. Ferguson, J. F., and D. N. Giva. "Chemical Precipitation in Water Softening and Iron and Manganese Removal," *Proceedings, 18th Annual Public Water Supply Engineers' Conference* (University of Illinois at Urbana-Champaign: Dept. of Civil Engineering, 1976).
4. *Water Works Regulations.* Commonwealth of Virginia, Section 9.06.01(b), p. 134 (June 1977).
5. Cole, L. D. "Surface Water Treatment in Cold Weather," *J. Am. Water Works Assoc.* 68:22 (1976).
6. Hartung, H. O. "Calcium Carbonate Stabilization of Lime Softened Water," *J. Am. Water Works Assoc.* 48:1523 (1956).
7. Thompson, C. G., J. E. Singley and A. P. Black. "Magnesium Carbonate: Recycled Coagulant—II," *J. Am. Water Works Assoc.* 64:93 (1972).
8. Black, A. P., and C. G. Thompson. "Plant Scale Studies of the Magnesium Carbonate Water Treatment Process," *Environmental Protection Technology Series EPA—660/2-75-006* (May 1975).
9. "Jar Test Evaluation of Polymers." An Am. Water Works Assoc. Research Foundation Report, *J. Am. Water Works Assoc.* 68:46 (1976).
10. Black, A. P., and R. F. Christman. "Electrophoretic Studies of Sludge Particles Produced in Lime-Soda Softening," *J. Am. Water Works Assoc.* 53:737 (1961).
11. Kross, R. E., N. B. Fisher and W. L. Paulson. "Solids Concentration and Performance of Solids Contact Softeners," *J. Am. Water Works Assoc.* 60:597 (1968).
12. Bhatia, I. S., and N. I. Cooperwasser. "Michigan City Picks Lime-Caustic Soda," *Water Wastes Eng.* 12:25 (1975).
13. Foster, J. H., Jr. "Conventional Flocculation and Sedimentation in Water Treatment," *Proceedings, 18th Annual Public Water Supply Engineers' Conference* (University of Illinois at Urbana-Champaign: Dept. of Civil Engineering, 1976), p. 63.
14. Tuepker, J. L., and H. A. Hartung. "Effect of Accumulated Lime Softening Slurry on Magnesium Reduction," *J. Am. Water Works Assoc.* 52:106 (1960).
15. Spaulding, C. H. "Conditioning of Water Softening Precipitates," *J. Am. Water Works Assoc.* 29:1697 (1937).
16. *Recommended Standards for Water Works,* 1976 ed. Great Lakes-Upper Mississippi River Board of State Sanitary Engineers (1976).
17. Haney, P. D., and C. L. Harmann. "Recarbonation and Liquid Carbon Dioxide," *J. Am. Water Works Assoc.* 61:512 (1969).
18. Cosens, K. W., and F. Farr, Jr. "Columbus Replaces Historic Water Treatment Plant," *Public Works* 105:60 (1974).
19. Hoover, C. P. *Water Supply and Treatment* (National Lime Association, 1936).

20. Hess, J. S. "Lime and Caustic Soda Softening at Fremont, Ohio," *J. Am. Water Works Assoc.* 60:980 (1968).
21. Gilman, H. D. "Application of Process Control and Automation in Water Treatment," *Proceedings 18th Annual Public Water Supply Engineers' Conference* (University of Illinois at Urbana-Champaign: Dept. of Civil Engineering, 1976), p. 151.

CHAPTER 25

ION EXCHANGE

Robert L. Sanks, Ph.D., P.E.

Professor of Engineering and Engineering Mechanics
Montana State University
Bozeman, Montana 59715

Senior Engineer
Christian, Apring, Sielbach & Associates
2020 Grand Ave.
Billings, Montana 59102

INTRODUCTION

Although the principles of ion exchange were discovered and investigated more than 100 years ago and the first commercial use of ion exchange for water conditioning was made 70 years ago, extensive use of the process has been limited to the last three decades. The development of polystyrene-divinyl benzene resins by D'Alelio in 1944 ushered in the modern era of ion exchange. The process is now used extensively for water softening, demineralization of water for high-pressure boiler feed, the manufacture of pharmaceuticals, the purification of chemicals, as organic "traps" for such purposes as whitening sugar, and in the laboratory for chromatography or other analytical procedures. This chapter is concerned with practical use of ion exchange in municipal water softening and for the removal of specific undesirable ions in municipal water supplies and/or individual homes.

As the name implies, ion exchange involves the transfer of one ion for another. For example, a cation in solution attaches to the solid exchanger (resin) which in turn releases a different cation into the solution. All exchangers exhibit selectivity; they may prefer one ion over another by a factor of 15 or more. This preference is not, however, a fixed number but varies with ionic strength, relative amounts of ions, the kind of exchanger, and to a lesser extent with other factors such as temperature. Some of these factors influence preference to such an extent that the selectivity for one ion at a given solute concentration may even reverse if the solution is greatly diluted. The ratios of one ion to others in solution also has a significant effect on selectivity. Some exchangers have a high affinity for a particular ion, while others do not. For example, clinoptilolite (a natural zeolite) has a very strong preference for ammonium, whereas synthetic resin exchangers do not. This preference for one ion over another is the property that makes ion exchange so valuable.

Many common materials are exchangers including: natural zeolites such as greensand, clay (especially montmorillonite), sulfonated coal (crushed anthracite treated with concentrated sulfuric acid), and peat. But the synthetic resins with their high capacity and controllable properties make large-scale ion exchange practical and feasible. Were the benefits of ion exchange more fully understood, the municipal use of ion exchangers would more widespread and popular.

Ion exchange has several advantages:

- It is selective. It can often remove unwanted ions preferentially. Examples of preferential removal include nitrate, iron, manganese, ammonium and heavy metals.
- There are many manufacturers of ion exchange equipment, which keeps costs competitive.
- The process and the equipment have been tested over many years. Designs are well developed into "off-the-shelf" units that are rugged and reliable.
- Manual and completely automatic units are available.
- Temperature effects from $0°C$ to $35°C$ are negligible.
- The process is excellent for softening in small and large installations, for example, the Culligan home water softeners and the 400-mgd Diemer plant that once softened Colorado River water.

On the other hand, ion exchange has a number of disadvantages that must be weighed carefully:

- Chemicals for regeneration may be expensive, corrosive or even dangerous, and the waste regenerant may present a difficult disposal problem.
- Automatic plants require expert maintenance and nonautomatic plants require knowledgeable operation and frequent attention.
- Capital cost is high (although it can be reduced by optimizing design).
- The expensive ion exchange resins can be easily ruined by ignorant operation. Examples are fouling of strong base resins by humic acids and oxidation of iron and manganese in cation exchangers.

The optimization of ion exchange plants for municipal operation by many of the equipment suppliers is often unsatisfactory. Most ion exchange plants are used for industrial purposes, and designers are conditioned to furnish plants that produce very high-quality effluent. They find it difficult to adjust to the concept that low-quality effluents produced by starvation regeneration, high flow rates and extended break-throughs represent satisfactory operation for municipal plants. Hence their designs are likely to be overly conservative and costly. Consequently, city engineers and consultants to municipalities must be prepared to make preliminary designs to ensure wise decisions on equipment sizing.

Objectives

The objective of this chapter is to provide sufficient background so that environmental engineers can make preliminary designs, evaluate their costs, optimize capital and operating costs, and guide the manufacturer (or equipment supplier) toward a final design that approaches the optimum.

A second objective is to evaluate the desirability of ion exchange for problem waters and for centralized municipal plants vs individual home units.

Limitations

The rigorous theory of ion exchange is so complex that it is of little value for practical problems where waters contain many competing ions. Fortunately, calculations for softening have been simplified to a series of charts or tables by resin manufacturers and these are easy to use. Practical solutions for more complex problems require pilot plant tests, but these are not difficult to make and the apparatus need not be expensive. The presentation in this chapter is limited to these simplified calculations and to explanations of pilot plant tests using simple equipment and simple analytical methods. The discussion is limited to fixed-bed ion exchange plants in which the resin in the reactor is operated in the same manner as a rapid sand filter. The resin bed is alternately backwashed, regenerated and rinsed, and placed into service for exhaustion. This process is repeated as often as necessary. There are many alternates not discussed herein, such as continuous ion exchange, in which the resin is pulsed around a closed loop within which exhaustion, regeneration, and rinse take place simultaneously. There are hybrids in which a "bank" or "train" of reactors are used alternately in partial exhaustion, regeneration, or rinse or in which resin is pumped from one reactor to the next.

Most ion exchangers occur as small beads or granules usually between 16 and 50 mesh in size, but some exchangers are liquids. One of the most common schemes for simple reactors charged with beads or granules is the duplex system, which provides an uninterrupted flow. One reactor is backwashed, regenerated and rinsed while the other reactor is in service. For municipal use a single reactor is adequate because continuity of flow can be ensured by storage. Hence, the less costly single reactor is often a logical choice.

PROPERTIES OF EXCHANGERS

There are many kinds of exchangers but for the purpose of this discussion the most important ones are shown in Table I. Exchangers are generally divided into four classifications depending upon the kind of functional group, which determines whether cations or anions are exchanged and whether the resin is a strong or weak electrolyte (strong acid, strong base or weak acid, weak base). Strong acid and strong base resins operate at all pH values, but their capacity is limited so they must be regenerated more frequently. As their regeneration is inefficient, the cost of chemicals is high. Weak electrolyte resins have much higher capacity and regenerate almost stoichiometrically (efficiency approaches 90%), but they operate only over a limited pH range.

Strong Acid (Cation) Exchangers

The strong acid exchangers operate at any pH, split strong or weak salts, require excess strong acid regenerant (typical regeneration efficiency varies from 25-45% in concurrent regeneration), and they permit low leakage. In addition, they have rapid exchange rates, are stable, and may last 20 years or more with little loss of capacity, exhibit swelling less than 7% going from Na^+ to H^+ - form, and are useful for softening and demineralization (removal of all cations with little leakage).

Weak Acid Exchangers

The weak acid exchangers do not remove cations satisfactorily below pH 7, and hence do not remove cations of strong electrolyte salts unless the solution is buffered, can be regenerated with strong or weak acids with high efficiency (usually more than 90%), and have a high affinity for Ca^{+2} (which must be removed with acid). In addition, they permit high leakage of sodium but low leakage of calcium, have capacities about twice that of strong acid resin, exhibit high volume change (swell 90% going from Na^+-form to H^+-form), are more resistant to oxidants (*e.g.,* chlorine) than strong acid types, and in the presence of sufficient alkalinity (more than 20%) are useful in the first of two cation exchangers in demineralization. (They can be regenerated with the waste acid from the second, strong acid exchanger.)

Table I. Common Types of Synthetic Exchangers

Type Resin	Functional Group	Drained Density lb/ft³	Drained Density kg/m³	Operating pH Range	Maximum Exchange Capacity me/g	Maximum Exchange Capacity me/ml	Maximum Exchange Capacity kgr/ft³	Regeneration	Trade Name Example
Strong Acid	-SO₃⁻H⁺ Sulfonic acid	49-53	790-850	0-14	4.8	2.0	43.7	Excess strong acid	Duolite C-20 Amberlite 120 Dowex 50
Weak Acid	-COO⁻H⁺ Carboxylic acid	45	720	7-14	11	4.5	98.3	Weak or strong acid	Duolite C-433 Amberlite IRC-50 Zeo Carb 226
Strong Base Type I	-CH₂N(CH₃)₃ + OH⁻ Quaternary ammonium	45*	720	0-14	4.3	1.3	28.4	Excess strong base	Amberlite IRA-410 Duolite A-101 D
Strong Base Type II	-CH₂N(CH₃)₂CH₂CH₂OH⁺ OH⁻ Modified quaternary ammonium	45	720	0-14	3.4	1.4	30.6	Excess strong base	Amberlite IRA 140 Duolite A-102 D
Weak Base	-N(CH₃)₂H⁺OH⁻ Tertiary amine	32	510	0-6	9	2.5	54.6	Weak or strong base	Duolite A-7 Amberlite IRA-93
Intermediate Base	Mix of above two	43	690	0-14	8.8	2.7	59.0	Strong base	Duolite A-30B

Strong Base Exchangers

The strong base exchangers operate at any pH, can split strong or weak salts, require excess high-grade NaOH for regeneration (with the typical efficiency varying from 18-33%), and irreversibly sorb (fouled by) humic acids from decay of vegetation and hence lose capacity. A cure is to remove humic acids first by an organic "trap" such as a weak base exchanger or activated carbon. In addition, they are less stable than cation resins, with life probably not exceeding three years in severe service.

Type I exchangers are for maximum silica removal. They are more difficult to regenerate and swell more (from Cl^- to OH^- form) than Type II. The principal use of Type I is to make the highest quality water. When they are loaded with silica, they must be regenerated with warm NaOH.

Type II exchangers remove silica (but less completely than Type I) and other weak anions, regenerate more easily, are less subject to fouling, are freer from the odor of amine, and are cheaper to operate than Type I, but they have greater thermal lability. They are particularly useful for food products.

Weak Base Exchangers

The weak base exchangers: do not remove anions satisfactorily above pH 6, are often based on phenol-formaldehyde or epoxy matrices instead of polystyrene-divinyl benzene, regenerate with a nearly stoichiometric amount of base (with the regeneration efficiency possibly exceeding 90%) and are resistant to organic fouling. In addition, they swell about 12% going from OH^- to salt form, they do not remove CO_2 or silica, and they have capacities about twice as great as for strong base exchangers. They are useful for following strong acid exchangers to save cost of regenerant chemicals, as organic "traps" to protect strong base exchangers, and to remove color.

Intermediate (Difunctional) Base Exchangers

Intermediate base exchangers contain both weak base and strong base groups, remove anions only below pH 9 after their strong base capacity is exceeded, have high capacities, and are about twice as efficient in regeneration as strong base exchangers. They also can sorb silica, CO_2, and phenol up to their strong base capacities, are dense (enabling use of higher backwash rates) and are useful as substitutes for weak base resin in a multiple-bed series.

Consideration of the foregoing discussion leads to the conclusion that a weak acid resin would sorb Na^+ from $NaHCO_3$ but not from a $NaCl$ solution, and a weak base resin sorbs Cl^- from HCl but not from $NaCl$. This occurs because stripping either Na^+ or Cl^- from brine drastically alters the pH and reduces the capacity of weak electrolyte resins to function. Obviously, then, resins must be selected for the conditions of operation. For most applications, this is readily accomplished after a study of the properties of the resins and the water to be treated. Resin manufacturers are always glad to help.

THEORY

For water softening and the removal of iron and manganese, a single reactor containing a strong acid resin is needed. The chemical reaction is

$$2 \; Na \cdot R + Ca^{+2} \quad = \quad Ca \cdot R_2 + 2 \; Na^+ \tag{1}$$
$$Ca \cdot R_2 + Fe^{+2} \quad = \quad Fe \cdot R_2 + Ca^{+2} \tag{2}$$
$$Ca \cdot R_2 + Mn^{+2} \quad = \quad Mn \cdot R_2 + Ca^{+2} \tag{3}$$

If the reactions tend to drive toward the right, the resin prefers calcium over sodium and iron or manganese over calcium. The preference for cations increases with valence and with atomic weight in the following manner:

$$H \; < \; Na \; < \; K < \; Mg \; < \; Ca \; < \; Mn \; < \; Fe$$

A complete "history" for a cation resin initially in H^+-form is shown in Figure 1. After sodium has broken through, the resin can continue to soften the water. But for water softening, the resin would normally be regenerated with brine rather than with acid, and therefore the cycle would begin with resin in Na^+-form instead of H^+-form.

The exchange of anions by a strong base resin for a water containing approximately 500 mg/l TDS including 23.7 mg/l NO_3^--N is shown in Figure 2. The resin, initially in chloride form, passed nearly 250 bed volumes (BV) before the nitrogen broke through. At higher ionic concentrations, the selectivity of the resin for nitrate and sulfate reversed, and sulfate broke through before nitrate. At still higher concentrations sulfate and bicarbonate broke through almost simultaneously, and the breakthrough of nitrate occurred long afterward.

Data of this kind are rather scarce in the literature, and mathematical analyses are either too simplified or excessively complex. Consequently, for such specialized ion exchange applications, it is more practical to base the design upon laboratory investigations. But calculations for softening

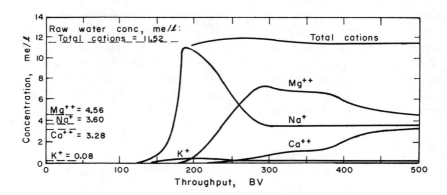

Figure 1. Cation concentrations in effluent of Duolite C-20 resin in virgin H⁺-form. (After Reference 2).

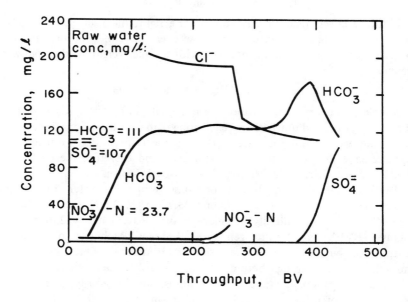

Figure 2. Anion concentration in effluent of Dowex-21K resin in Cl⁻ form. (After Reference 3).

are simple and the data have been well documented, so it is relatively easy to demonstrate calculations for softening.

SOFTENING

The approach used herein is taken from Duolite literature.[1]   However, appropriate data are also published by other companies: Rohm & Haas Company, Philadelphia, PA 19105; Dow Chemical U.S.A., Midland, MI 48640; The Permutit Company, E. 49 Midland Ave., Paramus, NJ 07652.
Important definitions and conversion factors are given in Tables II and III, respectively.   Before any calculations can be made, chemical analyses of the water must be obtained, and preferably there should be several over a period of time.   Resin manufacturers report their data in terms of kilograins calcium carbonate per cubic feet of resin (kgr/ft^3 or kgrpcf) and in terms of grains calcium carbonate per gallon (grpg) of water. Ionic concentrations of the hypothetical water in Table IV are given in terms of me/l, mg/l and gr/gal $CaCO_3$.   The requirement of electroneutrality means that total concentration of cations in me/l or gr/gal $CaCO_3$ should equal the anions, and a check should always be made.   If the error is greater than 3%, the analysis is insufficiently accurate.

Table II.   Abbreviations

Symbol	Definition
BV	total volume (including voids) of resin bed
BV/hr	bed volumes per hour
eq/l	equivalents per liter = normality = 1 mole H^+ per liter, 0.5 moles Ca^{+2} per liter, or the equivalent
g	gram
$gal/min/ft^3$	gallons per minute per cubic foot of resin
$gal/min/ft^2$	gallons per minute per square foot
gr	grains
gr $CaCO_3$	grains as calcium carbonate
gr/gal	grains as calcium carbonate (except as noted) per gallon
kgr/ft^3	kilograins as calcium carbonate (except as noted) per cubic foot of resin
meq/l	milliequivalents per liter
mg	milligrams
mg/l $CaCO_3$	milligrams per liter as calcium carbonate
ml	milliliter = 1/1000 liter
ppm	parts per million $\cong$ mg/l
TDS	total dissolved solids

Table III.   Conversion Factors from Customary Units to Metric

To Convert	To	Multiply By
BV/hr	gpm/ft^3	0.1247
	g/l	17.12
kgr/ft^3	meq/ml (of resin)	0.04577
gr/gal	meq/l	0.342
gr/gal	ppm $\cong$ mg/l	17.12
g	gr	15.43
$gal/min/ft^2$	m^3/m^2	0.04074
$gal/min/ft^3$	m^3/m^3	8.02

Table IV.   Raw Water Analysis

Cation	Concentration (as the ion)			Anion	Concentration (as the ion)		
	me/l	mg/l	gr/gal $CaCO_3$		me/l	mg/l	gr/gal $CaCO_3$
Na^+	3.42	78.7	10.0	HCO_3^-	1.68	102.5	4.91
Ca^{+2}	3.49	69.8	10.2	Cl^-	4.78	169.5	13.98
Mg^{+2}	3.32	40.4	9.7	SO_4^{-2}	3.42	164.2	10.00
Fe^{+2}	0.07	2.0	0.2	NO_3^-	0.61	37.8	1.78
Mn^{+2}	0.03	0.8	0.1	F^-	–	Tr	-
Σ	10.33	191.7	30.2	Σ	10.49	474.0	30.67

Note: hardness = 10.2 + 9.7 + 0.2 + 0.1 = 20.2 grpg

Design Procedure

1. Select the resin to be used.   If cations are to be exchanged, the exchanger must be cation resin.   Strong acid resins in sodium form are recommended by manufacturers for softening, but there is a choice of gel or porous types.   Either would do, but in this example, Duolite C-20 resin is selected.   It is the same kind of resin as Dowex HCR, Permutit Q100, and Amberlite 120.

2. Determine pretreatment requirements.   Resins are good solid coagulants and dirt attaches strongly to them, so the suspended solids must be virtually zero and the turbidity should be less than 1 or 2 TU. There should be no oxidants such as chlorine in the water.   Iron and manganese in the water can prove troublesome unless special provision is made in the operation for them.   Of special difficulty is Fe(II), which oxidizes to Fe(III) within the resin bed.   If this happens within the resin bead, the resin is likely to be irreversibly fouled.   If the iron is in the trivalent state before it enters the resin bed, it coats the particles with a sheath of insoluble hydroxides.   These can be removed but

only with some difficulty. However, if the water contains no dissolved oxygen and is clear and colorless as it comes from the well, and if iron or manganese is in the bivalent state, then ion exchange is an excellent way to remove iron and manganese effectively. If the iron and manganese are in the trivalent state, or if the water contains dissolved oxygen or oxidants, or if oxygen cannot be completely excluded from the system, both iron and manganese must be removed prior to softening.

3. Select the regenerant level. Figure 3 and Table V indicate the effects of regeneration level. Above 5 lb $NaCl/ft^3$ (80 kg/m^3), the efficiency drops. So for operating economy choose 5 lb $NaCl/ft^3$ at 11.2% (1 lb of salt/gal, which allows for expansion due to addition of salt), to give 20.3 kgr/ft^3 (348 g/l) exchange capacity.

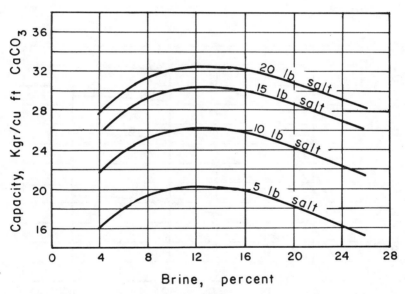

Figure 3. Exchange capacity vs brine concentration for Duolite C-20 resin. (After Reference 1).

4. Select the regeneration contact time with the aid of Figure 4. The total contact time is equal to the time to pump the regenerant brine through the reactor plus an equal volume of rinse water at the same flow rate. At 1 lb/gal, the total volume is 5 gal/ft^3 of regenerant plus 5 gal/ft^3 of rinse for a total of 10 gal/ft^3. If 45 min contact time is selected from Figure 4, the flow rate is 0.22 $gal/min/ft^3$. Rates much less than 0.2 $gal/min/ft^3$ are not justified. Experience indicates that 5 gal of rinse/ft^3 of

Table V. Efficiency of NaCl Regeneration[a]

Regeneration Level NaCl		Capacity		Average Salt Requirement	Salt/Hardness Removed
(lb/ft³)	(kg/m³)	(kgr/ft³)	(kg/m³)	(lb/kg removed)	(kg/kg)
4	64	16-18	274-308	0.24	0.22
5	80	19-21	325-360	0.25	0.23
6	96	21-23	360-394	0.27	0.25
8	128	24-26	411-445	0.32	0.30
10	160	26-28	445-479	0.37	0.35
12	192	28-30	479-514	0.41	0.39
15	240	30-32	514-548	0.48	0.45
20	320	32-34	548-582	0.60	0.56
50	801	36-38	616-651	1.35	1.26

[a]After Reference 1.

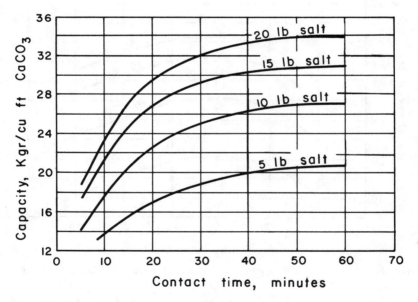

Figure 4. Exchange capacity vs regeneration time for Duolite C-20 resin. (After Reference 1.)

resin is too low, so another 8 gal/ft³ is arbitrarily added but at a higher rate. Even this may not be enough and it might be desirable to add a "sweetening on" stage, which is essentially more rinse to produce an effluent with a suitably low TDS content. To determine whether this arbitrary additional volume of rinse is adequate, a check can be made in the laboratory with all of the regeneration and rinse effluent collected in a series of fractions. Chemical analyses of each fraction would show a rapidly declining TDS curve, and the designer simply selects a suitable point on the curve.

5. Recheck the capacity of the resin as a function of the raw water analysis. Competition for exchange sites makes the sodium/hardness ratio important. The capacity decreases with an increase in sodium/hardness ratio and, as shown by Figure 5, with an increase in TDS (or hardness). At 5 lb salt/ft^3 and a hardness of 20 gr/gal, the capacity is 19.8 kgr/ft^3 at a sodium/hardness ratio of 1/3 and (from a similar plot not shown here) 18.8 kgr/ft^3 at a sodium/hardness ratio of 1/1. Interpolation gives 19.3 kgr/ft^3. This capacity is lower than the previously calculated capacity and now becomes the controlling value.

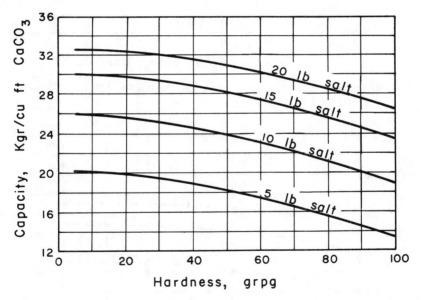

Figure 5. Exchange capacity vs influent hardness for sodium/hardness ratio 1/3 and Duolite C-20 resin. (After Reference 1.)

6. Recheck the capacity vs the service flow rate as shown in Figure 6. Flow rates up to approximately 8 gal/min/ft^3 are acceptable, but for the minimum depth of 24 in. for the resin bed, this becomes 16 gal/min/ft^2, and such high rates tend to promote channeling. Above approximately 10 gal/min/ft^2, a long run might require intermediate backwash.

7. The relation between bed depth, regeneration level and capacity is given in Figure 7. In order to realize the capacity of

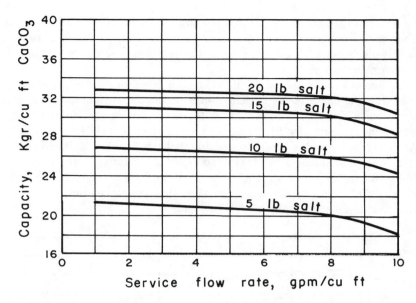

Figure 6. Exchange capacity vs service flow rate for Duolite C-20 resin. (After Reference 1.)

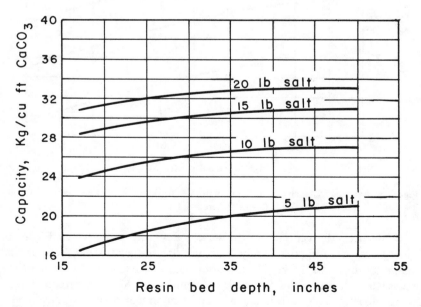

Figure 7. Exchange capacity vs bed depth for flow rate of 2.5 gal/min/ft^3 in Duolite C-20 resin. (After Reference 1.)

19.3 kgr/ft^3 from Step 5, select a bed depth of 28 in. or 2.33 ft (71 cm).

8. Check the leakage of the hardness cations. Ion exchange is so effective that, for municipal use, high leakages are not only acceptable but desirable. (Minimum calcium should be 40 mg/l as CaCO$_3$ as explained in Chapter 23, and if this is not achieved by leakage, it must be achieved in other ways—by blending, for example.) Expert ion exchange designers, who design mostly for high-quality industrial requirements, must be prodded into designing for high leakage because in most of their experience they want the lowest possible leakage. Leakage increases with a reduction of regenerant level, an increase in flow rate, an increase of sodium/hardness ratio in the feed water, and a decrease of resin bed height. From Figure 8, the leakage is only 1 mg/l as CaCO$_3$. This is less than desirable. No iron or manganese can leak until long after the total capacity of the resin is utilized. One remedy would be to blend raw and finished waters (limited by the iron and manganese content), and another would be to condition the finished water with lime. Extending the service flow beyond breakthrough is unacceptable because magnesium would emerge before calcium (see Figure 1).

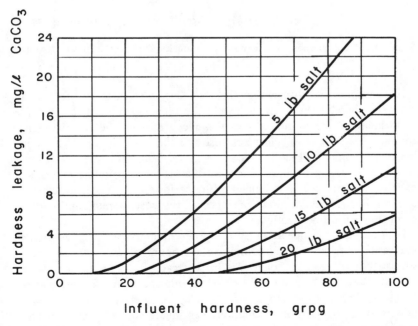

Figure 8. Leakage vs influent hardness for sodium/hardness ratio 1/3 and Duolite C-20 resin. (After Reference 1.)

9. The backwash flow rate is obtained from Figure 9. At 50°F (10°C), 50% bed expansion is achieved at 6 gal/min/ft^3 (48 m^3/m^2 hr).

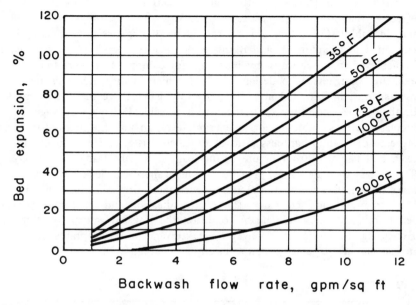

Figure 9. Bed expansion vs backwash flow rate for Duolite C-20 resin. (After Reference 1.)

10. The pressure drop within the resin bed is given in Figure 10. In addition, there is the pressure loss in the piping that must be computed or estimated. The optimum design is reached when the amortization of capital cost, labor cost, chemical cost and the cost for power for pumping is a minimum. Hence, one should investigate different sizes of reactors and operating conditions to find the least cost. In the author's experience, least cost is obtained with the lowest level of regeneration and the highest service flow rate because the cost of pumping is low compared with amortization.

11. Compute the operating cycle by determining the time required for the various operations and compute the amount of water treated per square foot of cross-sectional area in the reactor. From the numerous resulting choices, the designer can select the optimum. To summarize the foregoing steps: resin capacity is 19.3 kgr/ft^3, depth is 2.33 ft, and hardness is 20.2 gr/gal. The calculations per square foot of resin bed are as follows:

Capacity = 19.3 kgr/ft^3 x 2.33 ft^3 = 4969 gr

Volume throughput = $\dfrac{4969 \text{ gr}}{20.2 \text{ gr/gal}}$ = 2226 gal

Flow rate at 8 gal/min/ft^3 = 8 x 2.33 ft^3 =

18.6 gal/min

Time of service flow (throughput) = $\dfrac{2226}{18.6}$ =   119 min

Backwash at 6 gal/min/ft^2 (arbitrarily)   10 min

        Volume backwash = 6 gal/min x 10 min =   60   gal

Regeneration

        Volume = (5 gal 11.2% brine + 5 gal water)

                x 2.33 =   23.3 gal

        Time (from Figure 4)

        Regeneration flow rate = 23.3 gal ÷ 45 min =

                0.52 gal/min   45 min

Added rinse

        Volume = 8 gal/ft^3 x 2.33 ft^3 =   18.6 gal

        Time = 18.6 gal ÷ 18.6 gal/min   1 min

Totals   101.9 gal   175 min

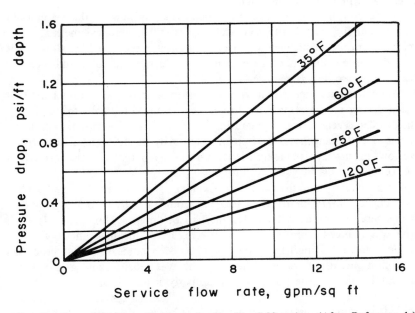

Figure 10. Pressure drop vs flow rate for Duolite C-20 resin.   (After Reference 1.)

A total cycle time of 175 min, nearly 3 hr, can provide two cycles per 8-hr operating day with plenty of time for servicing equipment. If fully automatic operation is used with fail-safe alarms and shut-down mechanisms, the ion exchange plant can operate 24 hr/day. However, at least 10% of the time should be allocated for servicing, automatic shut-downs and other considerations. If the plant is operated on the basis of two cycles per day, it produces 2226 x 2 = 4452 gal/ft^2 of resin bed with a total volume of waste regenerant, rinse and backwash of 101.9 x 2 = 204 gal. If it is operated for seven cycles over a period of 20 hr/day, it can produce 15,580 gal/day. This water would contain only 1 mg/l hardness as $CaCO_3$ and the iron and manganese would be almost zero. As municipalities do not need such soft water, it might be well to continue to operate the ion exchanger until the iron or manganese begins to break through or until the level of magnesium becomes intolerable. Computations break down at this point and only a pilot plant test can predict performance with accuracy. Any such test should be run at the site using the actual water and precisely the same conditions that would be realized in the prototype plant. With readily oxidizable ions such as iron and manganese in the water, it is quite impractical to transport the water to a laboratory. Fortunately, the pilot plant tests are not difficult, and a locally available scientist (for example, a high school science teacher) might be employed to run them.

If the consumer is unconcerned about the hardness of the water (which is, incidentally, very hard), ion exchange is still a practical method of removing iron and manganese. Depending upon the variables already discussed it is possible that this exchanger could operate for a full week before regeneration is required.

When ion exchangers remove iron and manganese, oxidation may take place during regeneration with brine because there is likely to be some dissolved oxygen in the regenerating solutions. To prevent oxidation of the iron and manganese, 0.01 lb of sodium hydrosulfite/gal is effective. Various proprietary reducing agents are marketed that are also effective for keeping resins in good condition.

As an example of iron removal, Camden, New Jersey, installed ion exchangers. In 1953 iron was reduced from 37 mg/l to 0.5 mg/l and manganese was reduced from 1.2 mg/l to 0.1 mg/l. Three years later iron was reduced from 52 mg/l to 0.1 mg/l and the manganese was reduced from 1.3 mg/l to zero. By 1960 each cubic foot had processed 2 million gallons of water. Then a pump was repaired and a gasket on the intake side of the pump was improperly installed, allowing air to enter the feed water stream. Within three months the resin bed was fouled by iron. This lesson illustrates the reason resin manufacturers are reluctant to recommend ion exchange for the removal of iron and manganese. But with an adequate operations manual, responsible operators, and the use of reducing agents as a regular practice in regeneration, ion exchange is feasible, economical and altogether appropriate for the removal of Fe(II) and Mn(II).

Other methods of iron and manganese removal include: oxidation with dissolved oxygen, chlorine, or potassium permanganate (although these methods may not work if the iron is in organic form); lime-soda softening; and the use of manganese zeolite. Manganese zeolite is made by treating greensand with manganous sulfate then potassium permanganate, which forms manganic oxide films that in turn yield oxygen to oxidize Fe(II) and Mn(II) to Fe(III) and Mn(III). An added advantage accrues because the greensand filters the hydroxides so formed. The greensand is regenerated with potassium permanganate. This process is satisfactory for the removal of manganese, but it is less so for the removal of iron. Furthermore, it is a preferred process only when iron and manganese are less than 2 or 3 mg/l.

NITRATE REMOVAL

The ion exchange process has not been well developed for the removal of nitrate because the selectivity of commercial resins for nitrate at usual ionic strength is only fair. Walitt and Jones[4] used a 1-naphthylmethylamino derivative of polystyrene, which is highly selective for nitrate. Unfortunately, regeneration with 1 N hydrochloric acid was poor. Evans and Korngold[5] developed the Ducol process in which a strong acid cation-weak base anion pair of reactors is used in two stages—first for demineralization and subsequently for softening. Working with Beersheba water spiked with 70 mg/l of nitrate, they were able to produce partially demineralized and softened water with nitrate reduced to 0.1 mg/l in the demineralization cycle and to 0.7 mg/l in the subsequent softening cycle. But Buelow et al.[3] who studied both low and high TDS water spiked with high nitrate content, found resins were less selective for nitrate at moderate TDS levels. For water containing high TDS (more than 40,000 mg/l) a strong base resin (Dowex 21-K) had a high preference for nitrate over sulfate, bicarbonate or chloride. At a lower TDS (approximately 20,000 mg/l) the preference was less marked, and at a TDS of approximately 500 mg/l, the preference was

$$HCO_3^- < Cl^- < NO_3^- < SO_4^=$$

Some preliminary and crude experiments were run in the Montana State University laboratories. The feed for Run 2 was made with distilled water containing 4.7 me/l $NaHCO_3$, 1.0 me/l $MgSO_4$, 0.5 me/l $MgCl_2$ and 1.4 me/l $Ca(NO_3)_2$. The feed water for Run 3 was made of distilled water and 1.4 me/l of nitric acid, 1.0 me/l of sulfuric acid and 0.5 me/l of hydrochloric acid. In Run 2 with Duolite ES-308 weak base resin in chloride form, the throughput was 400 BV to a breakthrough of 7 mg/l NO_3-N. In Run 3 the throughput (for the same resin but in hydroxide form)

Chemical costs for a two-bed system exceeded those for single beds, but the two beds yielded softened, nitrate-free water and a waste regenerant valuable for fertilizer.[6]

PILOT PLANT TESTING

For simple applications such as softening, calculations alone may constitute a sufficient basis for the design of prototype ion exchange plants. But for determining the leakage of unwanted ions such as iron and manganese in complex waters, for establishing the exact amount of rinse or "sweetening on" (the first part of a service cycle in which effluent must be discarded), or for operations not completely encompassed by the literature, pilot plant tests are necessary. These tests can be carried out with small diameter reactors, but the height of the resin bed in both model and prototype must be the same. The smallest column diameter used should be 1 in. (2.5 cm). Larger diameters, up to 4 in. (10 cm), are easier to operate because there is less trouble with bubbles of air that disrupt the bed. But the larger columns require more water, chemicals and resin. However, if the larger quantities are not troublesome, it is better to use the larger reactors.

For pilot plants, it is desirable to use transparent reactors so the condition of the bed can be observed. Corning Conical Pyrex Pipe* is excellent, as are tubes of plexiglass. Fittings for pyrex pipe can be made with rubber stoppers tightly driven into the pipe. End fittings for plexiglass can be made of plexiglass plates glued into place with ethylene dichloride. The simplest bed support is a wad of glass wool at the bottom of the column. The height of the reactor should be about twice as great as the depth of the resin bed to allow for some freeboard above the 50% expansion required of the backwash.

The feed or regenerant solutions can be either pumped or applied by gravity flow. Pumping (particularly with a positive displacement pump) is a more accurate means of control, but gravity feed is satisfactory if it

---

*Corning Glass Works, Process Equipment, Industrial Products Department, Corning, New York 14830.

is controlled during the run and adjustments are made to compensate for compaction of the bed caused by swelling of the resin. Weak acid and weak base resins may swell enough to break the tubing unless the bed is loosened occasionally by backwashing. The schematic diagram of Figure 11 shows the feed regulated by a Mariotte jar, which has the advantage of providing a constant head as the feed container empties. Fine adjustments of flow can be made by raising or lowering the effluent hose or by adjusting the screw-type tubing clamps.

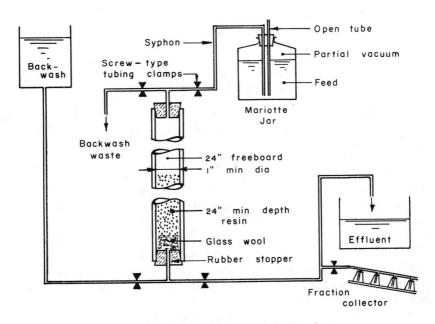

Figure 11. Schematic diagram of pilot plant.

During both the service run and the regeneration and rinse, it is usually necessary or desirable to collect fractions of the effluent. This can be done by manually directing the effluent stream into a series of containers at regular intervals, but such a procedure requires constant attention, and a fraction collector operating on a part of the waste stream is better. Commercial fraction collectors are best, but they are also expensive. Figure 12 shows how to make an inexpensive, reasonably satisfactory fraction collector. The volume collected in each flask is regulated by the position of the open glass tube. When the container fills to the bottom of the glass tube, the air trapped in the plenum forces the liquid up the glass tube, and when it reaches the level of the water flowing through the

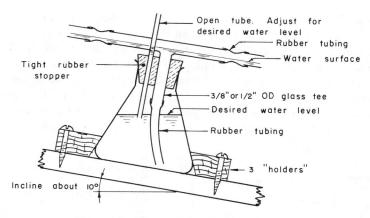

Figure 12. One unit of a fraction collector.

tee, all subsequent effluent will pass to the next flask. Usually a dozen flasks connected end to end and fastened to a board inclined at about 10° is sufficient.

Several relatively inexpensive kits, including the Hach Portable Engineer's Laboratory,* are capable of all of the chemical analyses necessary with sufficient accuracy except for sodium and potassium. Potassium is not usually important, but sodium is, and the most satisfactory way to analyze sodium is by flame photometry. Unfortunately, instruments for flame photometry are expensive and the use of such devices requires skilled laboratory technicians. However, samples for sodium can be put into small polyethylene (not glass) bottles and sent to a commercial laboratory for analyses. If the pH is reduced to less than 2 with nitric acid, delays of several weeks are acceptable.

The results of the chemical analyses should be plotted as shown in Figure 1 wherein the abscissa is given in BV and the ordinates in either meq/l or mg/l as $CaCO_3$. The first run with a virgin resin gives a long service run that cannot be repeated in subsequent runs unless the regeneration level is excessively high. Consequently, it is necessary to condition the resin bed with two or three successive runs until the capacity associated with the conditions of operation have reached "cyclical stability." As demonstrated by Table V and Figures 3 to 8, many factors influence effluent quality and, of course, economics. In summary, regeneration level and concentration, flow rate, sodium/hardness ratio, bed depth and extent of breakthrough are all important factors in operation.

---

*Hach Chemical Company, P.O. Box 907, Ames, Iowa 50010.

There are many excellent treatises on operating laboratory or bench-scale ion exchange units. Among them are: Duolite Ion Exchange Manual[7] and "Amberlite Ion Exchange Resins Laboratory Guide."[8] These provide helpful hints on the actual operation of the equipment.

COSTS

Reliable, up-to-date cost data are scarce in the literature. Sanks[9] showed 1957 data gathered by Peak and David,[10] and noted costs may be adjusted to present-day prices by means of the Engineering News-Record Construction Cost Index (ENRCCI), usually with surprising accuracy. Data on the ENRCCI are given in Table I and Figure 1 in Chapter 30 by Dickson. The most economical and satisfactory way to purchase ion exchange reactors is to buy skid-mounted units that require only a foundation and water and electrical hook-ups to be ready to operate. The owner is thus assured of a properly functioning unit. Costs for skid-mounted units in Figure 13 are based on data from Illinois Water Treatment Company* for preengineered, standard packages. Custom engineering for specific applications would increase costs by 15% in smaller sizes and 7 to 8% in larger sizes. Costs from other manufacturers** are likely to be different for a number of reasons, including differences in specifications and quality. For final cost estimates, bids should be obtained from several manufacturers, and prices should be compared on the basis of similar specifications and performance requirements.

Several considerations increase the cost of an ion exchange installation which include:

- Add 15% for installation and hook-up.
- Add the cost of power, raw water pipe, wastewater pipe, the cost of a raw water pump, and the cost of pretreatment facilities if required.
- Either add a second unit to make a duplex system or (better for municipalities) add enough treated water storage to supply the

---

*840 Cedar Street, Rockford, Illinois 61105.
**The Permutit Co., E. 49 Midland Ave., Paramus, New Jersey 07652.
  Cochrane Division, Crane Co., P. O. Box 191, King of Prussia, Pennsylvania 19406.
  Graver Water Conditioning Co., U.S. Highway 22, Union, New Jersey 07083.
  Infilco Division Degremont Corp., Box K-1, Koger Executive Center, Richmond, Virginia, 23288.
  US Filter Corp., Fluid Systems Division, 12442 E. Putnam St., Whittier, CA 90608.
  L*A-Water Treatment Div., Chromalloy, 17400 E. Chestnut St., City of Industry, California 91749
  Others are found in the literature.

demand when the ion exchange reactor is being regenerated or is not producing for other reasons.

- Add controls and equipment for blending raw water with treated water or for adding conditioning chemicals to treated water for corrosion control.
- Add concrete foundations. Consider curbs and drains to confine spills or leaks.
- For moderate climates, add insulation and heating elements for out-of-doors installations, and for severe climates add a heated and lighted building to protect the equipment from freezing.
- Add the costs for final disposal of waste streams. See the section on disposal of waste regenerant.
- Add the cost of land, if appropriate.
- To above first three items add about:
   1% for freight
   10% for contingencies
   10% for engineering
   30% for contractor's profit and overhead
   1% for interest during construction.

Finally, the engineer should compare the costs and intangibles of a central plant that must treat all of the municipal water (including that used for lawn watering) to the cost of individual home water softeners that treat only the water used in the house. One intangible factor is: softened water (with its high sodium/hardness ratio) tends to make clayey soils more impervious and hence is less desirable for irrigation. The large plant is, however, more cost-effective per gallon treated than is a system of small plants. But, on the other hand, it must treat a larger quantity.

DISPOSAL OF WASTE REGENERANT

In times past, disposal of waste regenerant was usually a simple matter. But now, disposal into water courses (even with neutralization) is forbidden, particularly from large plants. The cost of disposal is important and for some installations it may be critical. It may be most practical to dry the salts by evaporation in lagoons. One means of "ultimate" disposal would be burial of the dried salt at a suitable site or use of the salt for deicing roads. At some plants, the wastes are pumped into deep, saline aquifers. Ocean disposal should be satisfactory for plants near a coast. Clifford and Weber[6] proposed using certain waste regenerants for fertilizer.

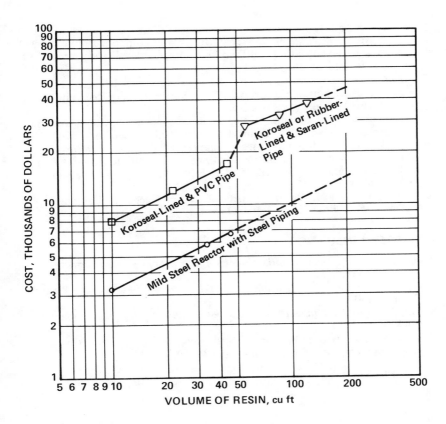

Figure 13. Cost of SKID-MOUNTED REACTORS (fob). Includes brine regenerant tank, regenerant pump and mixer, fully automatic controls based on time sequence, strong acid gel-type resin, and all process pipe and valves, but no raw water pump. 1976 prices.

REFERENCES

1. Diamond Alkali Company (now Diamond Shamrock Chemical Co.) *Duolite C-20 for Softening Water*, Duolite Data Bulletin No. 24, Diamond Shamrock Chemical Company, Resinous Products Division, 19701 Spring Street, Redwood City, California 94063 (1966).
2. Sanks, R. L. "Partial Demineralization of Brackish Water by Ion Exchange," Ph.D. Thesis, University of California-Berkeley (1965).
3. Buelow, R. W. *et al.* "Nitrate Removal by Anion-Exchange Resins," *J. Am. Water Works Assoc.* 67:528 (1975).
4. Walitt, A. L., and H. L. Jones. *Basic Salinogen Ion Exchange Resins for Selective Nitrate Removal from Potable and Effluent Waters*, Water Pollution Control Research Series ORD-17010 FKF 12/69 USDI Federal Water Quality Administration (December 1969).

5.   Evans, S., and E. Korngold.   U.S. Patent 228,304.
6.   Clifford, D. A., and W. J. Weber, Jr.   "Multicomponent Ion Exchange: Nitrate Removal with Land-Disposable Regenerant." *Ind. Water Eng.* 15:18-26 (March 1978).
7.   *Duolite Ion-Exchange Manual.*   Resinous Products Division, Diamond Shamrock Chemical Company, 19701 Spring Street, Redwood City, California   94063 (1969).
8.   Amberlite Ion Exchange Resins Laboratory Guide.   Rohm and Haas Company, Philadelphia, Pennsylvania   19105 (December 1974).
9.   Sanks, R. L.   "Ion Exchange," Chapter 14 in *Process Design in Water Quality Engineering New Concepts and Developments*, E. L. Thackston and W. W. Eckenfelder, Eds. (Austin, Texas: Jenkins Publishing Co., 1972).
10. Peak, R., and M. David.   "Costs of Cation Exchange Equipment," *Chem. Eng. Prog.* 53:37J-40J (1957).

SUGGESTED READING

Applebaum, S. B.   *Demineralization by Ion Exchange* (New York: Academic Press, 1968).
Arden, T. V.   *Water Purification by Ion Exchange* (New York: Plenum Press, 1968).
Kunin, R.   *Ion Exchange Resins,* 2nd ed.   (New York: John Wiley and Sons, Inc., 1958).
Page, J. S.   *Estimator's Manual of Equipment and Installation Costs.* (Houston: Gulf Publishing Company, 1963).

CHAPTER 26

WATER QUALITY IMPROVEMENT
BY REVERSE OSMOSIS

Isadore Nusbaum, P.E.

Consulting Engineer
4105 Collwood Lane
San Diego, California 92115

Alan B. Riedinger

Technical Staff Consultant
Fluid Systems Div., UOP
2980 North Harbor Drive
San Diego, California 92101

There has been a marked interest in water quality improvement and in processes that are capable of water quality enhancement above and beyond that attainable by conventional water treatment. Among the methods that have recently reached commercial status are the pressure driven membrane systems. These are known as "reverse osmosis" and "ultrafiltration (q.v. page 649). This discussion is primarily concerned with the application of reverse osmosis (or hyperfiltration as it is also called) to water treatment for municipal and industrial water supplies. The purpose of the chapter is to acquaint the consulting engineer with information that will aid him in applying and selecting equipment to meet the water quality requirements. Some information is provided on the factors used in designing the membrane system itself. However, it should be understood that the design of those elements of the reverse osmosis specifically involved with membrane operation are the function of the manufacturer. Nevertheless the consulting engineer and the client have an equally responsible role in establishing the water quality requirements, in the application and use of the processed water, in the design and utilization of pre- and posttreatment methods and other factors that contribute to the successful use of the process.

As a commercially available system for water treatment, reverse osmosis (RO) is less than 10 years old. Investigations into the application of RO to desalting sea water and demineralizing brackish water were sponsored by the U.S. Office of Saline Water (now the Office of Water Research and Technology) as a part of the program for developing new methods of producing municipal and industrial water supplies from saline water. However, it was the development by Loeb[1,2] and his associates at UCLA of the asymmetric cellulose acetate membrane in the early 1960s which demonstrated reasonable water permeation rates and semipermeability to solutes that made reverse osmosis a feasible and economic process.

Osmosis and reverse osmosis depend on the presence of a barrier or membrane that is selective so that the solvent of a solution can pass through the membrane while other components of the solution or the solutes cannot. Such a membrane is described as semipermeable. The osmotic pressure is the pressure required to stop the flow of solvent through a semipermeable membrane separating two solutions of different concentrations. To separate water from dissolved solids by reverse osmosis the applied pressure must be greater than the osmotic pressure.

Two basic simplified equations are used to describe the flow of water and solutes across the membrane in reverse osmosis. The product water or permeate flow through a semipermeable membrane may be expressed as:

$$Fw = (A) \, (\Delta P - \Delta \pi) \tag{1}$$

where Fw = water flux $(cm^3/cm^2 \; sec)$
  A = water transport coefficient $(cm^3/cm^2 - sec\text{-}atm)$
  ΔP = pressure differential applied across the membrane (atm)
  $\Delta \pi$ = osmotic pressure differential across the membrane (atm)

The flow of water is thus primarily pressure-dependent. There is some transport of solute across the semipermeable membrane and this is expressed as:

$$Fs = B \, \Delta C \tag{2}$$

where Fs = salt (solute) flux $(g/cm^2 \; sec)$
  B = salt (solute) transport coefficient (cm/sec)
  ΔC = $C_1 - C_2$ = concentration gradient across the membrane (g/cm^3)

The transport of solute is primarily dependent on the concentration differential.

It should be obvious from these equations that the greater the net driving pressure above the solution osmotic pressure, the greater the flow of product water. Since the transport of salt is only slightly affected by the pressure, the higher the applied pressure that can be tolerated the better the water quality will be. Sea water has an osmotic pressure of 398 psi (2744 kN/m²); brackish waters for the most part have osmotic pressures of less than 50 psi (345 kN/m²). To get water of good quality from sea water, operating pressures of 800-1500 psi (5500-10,300 kN/m²) are required. Brackish water reverse osmosis systems are normally designed to operate in the range of 300-600 psi (2100-4100 kN/m²). For extensive discussions on the fundamentals of reverse osmosis reference is made to several publications.[3-6]

At present, commercial reverse osmosis membranes are made from two types of polymers. The first membranes developed were cellulose acetate with an acetyl content of about 40%. Subsequently, cellulose triacetate membranes have been cast and used, and these have an acetyl content of 43.2%. Both membranes can be prepared in sheet form with water fluxes of 10-20 gpd/ft² at 400 psi (0.41-0.81 m³/m² day at 2800 kN/m²). The cellulose triacetate membrane is also made in a hollow fine fiber. The cellulose acetate membranes are organic esters and subject to hydrolysis at very low pH levels and at pH in the alkaline range. The membranes are resistant to some oxidizing agents and can tolerate up to one mg/l of free or available chlorine for extended periods. Periodic exposure to 10-20 mg/l of chlorine for short periods can also be tolerated.

The other major commercial membrane in use at this time is made of a polyamide polymer in a hollow fine fiber. This membrane can tolerate a pH range of 4.0-11.0. The membrane is sensitive to degradation by oxidizing agents, and chlorine residuals in excess of 0.1 mg/l are not recommended. Polyamide membranes have recently been prepared in sheet form.

The commercial membranes in general use have been limited to operating temperatures of 85°-95°F (29°-35°C). The membranes suffer from compaction during service over extended periods at the high operating pressures. In compaction the membranes undergo a densification, causing a reduction in the water flux. With the possible exception of the cellulose triacetate membrane, the commercial membranes have not met the standards required for processing very high salinity waters such as sea water. However, the cellulose triacetate membranes suffer excessive compaction at the driving pressures required for sea water.

A new generation of membranes is presently semicommercial and should be available for general use in the near future. These are the thin-film composite membranes in which a very thin film (about 200 Å) of a polymer is formed on a porous substrate of a relatively incompressible substance. The new membranes, which have been made of several new polyamide–type materials as well as cellulose triacetate, have produced potable water in a single pass through the membrane from sea water at pressures of 800-1000 psi (5500-6900 kN/m²). They have also shown very good membrane properties at pressures of 300-400 psi (2100-2800 kN/m²). The thin film polymers have shown excellent pH resistance (from 2.0 to 12.0), improved stability to oxidizing agents (except chlorine), improved stability to elevated temperatures up to 140°F (60°C), improved chemical resistance to organic substances such as solvents, greater resistance to compaction, and improved rejection of organic solutes.

Four distinctly different modular designs for incorporating the membrane into a useful package have been developed. These are the plate-and-frame, tubular, spiral-wound and the hollow fine fiber. The plate-and-frame, which resembles a filter press, had a number of disadvantages that weighed against commercial development. It has been used mostly for laboratory and study purposes.

The tubular concept consists generally of small diameter (about 1/2 in. or 1.3 cm) porous or perforated tubes with the membrane placed inside the tube which is also the membrane support and the pressure vessel. It has been most widely used for special applications in pharmaceutical and food processing and in waste treatment. It has not been widely used for municipal and industrial water treatment applications primarily because of the relatively high capital and operating costs and the space requirements for large plants.

The spiral-wound and hollow fine fiber equipment are presently in extensive use throughout the world for municipal and industrial water treatment. There are over 30,000,000 gpd (1.3 m³/sec) capacity of each type in operation. Several plants of 1 mgd (44 l/sec) capacity are operating in Florida with a 3 mgd (131 l/sec) plant under construction. A 5-mgd (220 l/sec) plant, which will operate on tertiary effluent for water reclamation by ground water recharge, is under construction in southern California. About 5 mgd (220 l/sec) is operating at a large industrial complex in Japan at a steel mill and a large power station. Recent news releases stated that a contract had been let for a 30-mgd (1.3 m³/sec) reverse osmosis municipal water demineralization plant in Saudi Arabia.

Probably the most significant project now under consideration is the 100-mgd (4.4 m³/sec) Yuma Desalting Project under the direction of the U.S. Bureau of Reclamation. This project, for which proposals were submitted in 1976, will consist of 100 mgd of membrane desalting capacity divided between spiral-module RO, hollow fine fiber RO and electrodialysis. Although the capacities for each method have not yet been established, the plant may ultimately have 35 mgd (1.53 m³/sec) each of spiral-wound and hollow fine fiber RO for a total of 70 mgd (3 m³/sec). The plant is being designed to demineralize the Wellton-Mohawk return irrigation flow to provide Mexico with a water of improved quality from the Colorado River.

The spiral-wound membrane element (Figure 1) was developed by Fluid Systems Division, UOP Inc. (formerly General Atomic Co.), under the sponsorship of the Office of Saline Water, U.S. Department of the Interior. It has now become a generic type due to its general acceptance and its manufacture by several companies. The spiral-wound unit consists of one or more membrane envelopes each formed by enclosing a channelized product water-carrying material between two large flat membrane sheets. The membrane envelope is sealed on three edges with a special adhesive and attached with the adhesive to a small diameter pipe, which has openings to collect the permeate. The envelopes are wound around the pipe to form a cylinder 2, 4, 8 or 12 in. (5, 10, 20 or 30 cm) in diameter and up to 40 in. (102 cm) in length. A polypropylene screen is used to form the feed water channel between the membrane envelopes. A wrap is applied to the membrane element to maintain the cylindrical configuration. The center tube or pipe is also the permeate collecting channel and several elements can be connected in series within a single-pressure vessel as shown in Figure 2.

The hollow fine fiber membrane element was developed by the Du Pont Company, using a polyamide polymer. Subsequently the Dow Chemical Company introduced a hollow fine fiber system using cellulose triacetate as the membrane material. The polyamide hollow fibers have been made with diameters of 50-85 μ with the i.d. about one-half the o.d. Triacetate fibers are estimated to be about 200-300 μ o.d. The membrane and pressure vessel are an integrated unit (Figure 3). The fibers are formed into a U-shaped bundle, with the open ends of the fibers potted in an epoxy tube sheet. The bundle attached to the tube sheet is arranged in a cylindrical pressure vessel. Feed water enters the center of the vessel through a porous or perforated pipe and is distributed radially through the fiber bundle. Under pressure the water flows into the hollow fine fibers and out through the capillaries.

One of the major differences in spiral wound and hollow fine fiber membrane elements is the character of the flow within the element. Spiral wound elements operate (or attempt to operate) under turbulent flow conditions. Hollow fine fiber units are essentially laminar flow.

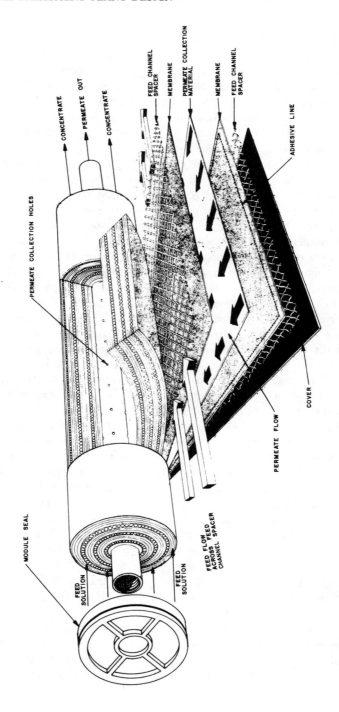

Figure 1. Spiral wound membrane module.

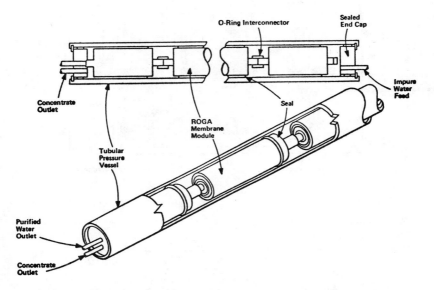

Figure 2. Permeate collecting channel of the spiral wound membrane module, showing several elements connected in series within a-single pressure vessel.

Although RO (hyperfiltration) and ultrafiltration belong in a general way to the field of filtration methods, they have another characteristic that makes them different from conventional filtration methods. In conventional filtration there are two streams—the feed and the product. All of the flow passes through the filter media. RO and UF belong to cross flow filtration methods in which there are three streams—the feed, the product and the concentrate. The concentrate stream remains to be disposed of after processing and must be considered in the total project.

Figure 4 is a schematic illustration of a reverse osmosis system. The basic components are a high pressure pump to deliver the feed water to the membrane elements at a pressure which assures an adequate driving force and water permeation rate in the system, a pressure regulating valve on the concentrate stream to maintain the driving force, and, of course, the pressure vessel assembly or module containing the membrane elements. In actual practice, depending on the size and other performance requirements of the plant, the RO system consists of a parallel or parallel-series arrangement of pressure vessel assemblies to produce the permeate quantity and quality required. Figure 5 shows a schematic representation of a larger system. Scale-up of RO systems is essentially linear, and an arrangement of parallel units or blocks is easily made.

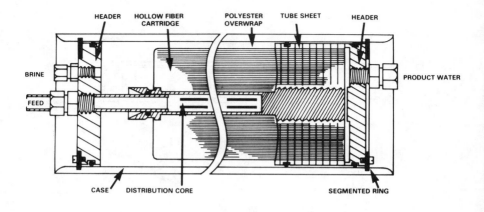

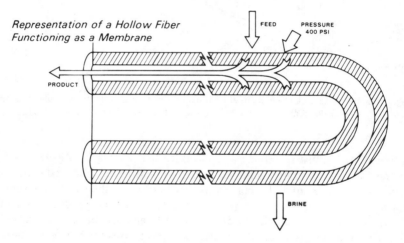

Figure 3. Dowex RO-20K, reverse osmosis permeator.

There are certain characteristics of reverse osmosis membrane elements and systems with which the engineer should have some familiarity in order to understand the operational requirements and performance.

CONCENTRATION POLARIZATION

As in any hydrodynamic system a boundary layer exists next to the membrane surface, and it does not mix completely with the bulk fluid stream. The thickness of this layer decreases with increasing velocity and turbulence. In a reverse osmosis element the permeate is removed from

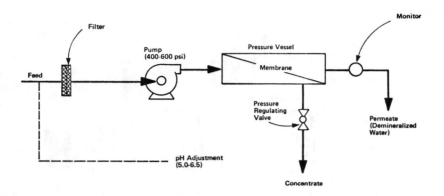

Figure 4. Schematic of reverse osmosis system.

this layer and then through the membrane. The dissolved salts accumulate in the boundary layer and can only escape back to the bulk stream by diffusion. The effect is known as concentration polarization and is defined by the ratio of the salt concentration at the membrane surface to the average salt concentration in the bulk stream. This surface solution concentration controls membrane performance. In addition, excessive concentration polarization can result in the deposition of slightly soluble compounds on the membrane, although the concentration in the bulk stream is below saturation.

SPECIFIC ION AND COMPOUND REJECTION

Studies have been made and results published showing individual ion rejections (or transport) that can be expected in reverse osmosis with cellulose acetate and other membranes. The rejection of individual ions is not meaningful since they are present in the water as a result of their solution as a compound (salt, acid, base). The ions do not pass through the membrane as individual ions but as charge balanced combinations. Although the exact mechanism is not known, the resultant effect is that a charge balance is maintained.

Table I shows the rejection of a number of the more common inorganic compounds using a commercial spiral-wound membrane element with cellulose acetate membrane at 400 psi (2800 kN/m²) net. Similar results would be obtained with other membranes and configurations. The membrane element was operated at a low recovery and the data represent a feed-brine average across the element. In general these follow the trend

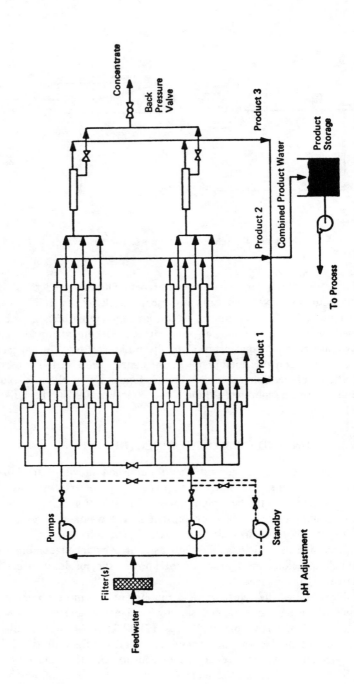

Figure 5. Large reverse osmosis system.

Table I. Rejection of Various Compounds at 400 psi (2800 kN/m^2) Net,
pH 5 to 6, ROGA® #4101

Compound	Rejection %	Compound	Rejection %
LiCl	96.4	SiO_2	92.3
NaF	97.1	H_3BO_3	44
NaCl	96.0	NH_4Cl	93.3
NaBr	92.3	$CuCl_2$	98.9
NaI	88.9	$CuSO_4$	99.6
$NaNO_3$	93.3	$NiCl_2$	99.4
$NaHCO_3$	97.4	$NiSO_4$	99.9
KCl	95.5	$FeCl_2$	99.7
$NaSO_4$	99.6	$AlCl_3$	99.2
$MgCl_2$	98.8	CO_2	0
$CaCl_2$	98.8	H_2S	0
$MgSO_4$	99.8	HCN	0
NaH_2PO_4 *	99.8		

that, within a chemical family, the rejection decreases with increasing molecular weight. The rejection of an ionized species increases as the charge on the ions increases. The salts containing divalent or trivalent anions or cations are better rejected than those containing only monovalent ions. Compounds that may be only slightly ionized, such as weak acids and bases (*i.e.*, H_2CO_3, H_2S, NH_4OH), are only slightly rejected, if at all. Dissolved gases are poorly rejected or, rather, have high transport rates.

A considerable variation in the rejection of individual ions is shown when processing natural waters that contain a large number of dissolved substances. Table II illustrates the results obtained by taking a sample of a city tap water and preparing it for RO processing by acidifying separate samples with sulfuric acid and hydrochloric acid and softening another portion with cation resin and again acidifying with the two acids. This has the effect of showing the differences in apparent ion rejection that can occur with a varying relationship in composition.

TEMPERATURE

The performance of the membrane element or a reverse osmosis system is usually normalized to a temperature of 77°F (25°C). The water transport and salt transport coefficients increase with an increase in temperature and decrease with a decrease in temperature for the commercial membranes in use. Figure 6 is a plot showing the deviation

Table II.   Variation in Ion Rejection; RO Product Results[a]

	San Diego Tap Water	Acidified with H_2SO_4		Acidified with HCl		Softened and Acidified with H_2SO_4		Softened and Acidified with HCl	
	(mg/l)	(mg/l)	% Rej.[b]	(mg/l)	% Rej.	(mg/l)	% Rej.	(mg/l)	% Rej.
Ca	72	0.6	99.2	0.7	99.1	–	–	–	–
Mg	34	0.3	99.2	0.3	99.2	–	–	–	–
Na	93	4	95.9	5.1	94.8	7.4	97.1	9.3	96.3
K	6.9	0.4	94.5	0.5	93.1	0.3	95.9	0.3	95.9
HCO_3	150	1.3	96.9	1.1	97.3	1.9	95.4	1.5	96.4
SO_4	245	2.2	99.4	1.5	99.4	2.3	99.4	1.6	99.4
Cl	85	6.1	93.2	8.7	94.7	8.8	90.2	12.6	92.3
SiO_2	10	1	90.5	1	90.5	1	90.5	1	90.5
TDS	686	15.3	97.8	18.4	97.3	20.8	97.2	25.6	96.4

[a]Tests were run using a spiral wound membrane element (ROGA® 4160 S) at 405 psi (2800 kN/m²).
[b]Rejection based on feed/brine average.

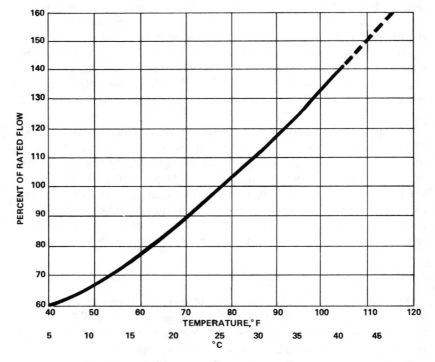

Figure 6.   Effect of temperature on ROGA® elements flux at temperature/flux at 77°F.

in flow that occurs for a cellulose acetate spiral wound element with temperature. Similar curves have been plotted for hollow fine fiber elements. An approximation is 2.5% for each degree centigrade. Increases in the feed water operating temperature may also accelerate the rate of membrane degradation and compaction. Membrane element specifications usually show limits for the operating temperatures of 40°-90°F (4°-32°C). Recently there has been some relaxation of the upper limit for special cases to 100°-105°F (38°-41°C). It has found that materials of construction other than the membrane may be limiting under high pressure at elevated temperatures.

COMPACTION

Some densification of the membrane structure may take place while operating at elevated pressures. The change is known as compaction and is accompanied by a reduction in the water permeation rate. The commercial membranes in use are all subject to compaction. Thin film composite membranes are an effort to minimize compaction.

EFFECT OF pH

The pH of the solution may have a significant effect upon the durability and the performance of the membrane whether made of cellulosic esters or polyamides. The importance of pH in pretreatment of the feed water is discussed in another section. For cellulose acetate, the pH limits normally set are 3 to 8, a range in which the rate of membrane hydrolysis can be tolerated. The membrane has a minimum hydrolysis rate at a pH of about 5. For polyamide hollow fine fiber membranes the recommended pH limits have been 4 to 11.

The pH of the feed has very little effect on the rejection of salts of strong acids and strong bases. It may, however, have a pronounced effect on rejection of relatively weak acids or bases. In general, the nonionized species are poorly rejected. The ionized salts of the weak acids and bases are in general well-rejected.

EFFECT OF TOTAL DISSOLVED SOLIDS ON
SALT REJECTION AND PERMEATE PRODUCTION

Figures 7 and 8 show the effect of feed water dissolved solids concentration on system performance at four water recovery levels. The water permeation rate decreases with increasing feed salinity and with increasing recovery. Product water quality decreases with increasing feed salinity and increasing recovery.

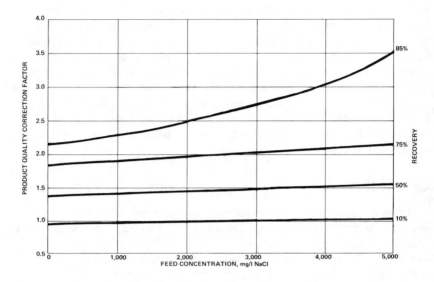

Figure 7. Effect of feed salt concentration and recovery on RO system performance. Permeate quality correction factor 400 psi (2800 kN/m²) feed pressure at 77°F (25°C).

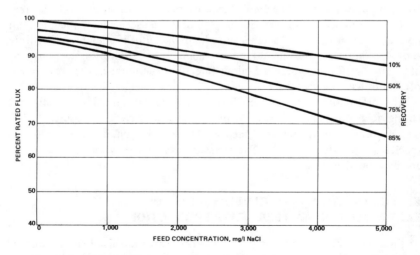

Figure 8. Effect of feed salt concentration and recovery on RO system performance. Percent of rated product flow per element.

The decrease in performance with feed water salinity and permeate recovery is due to the hydraulic pressure losses and the increase in the average salinity and osmotic pressure through the membrane elements in the system. Although these figures were derived with spiral wound elements, similar data have been published for hollow fine fiber systems.

ORGANIC COMPOUNDS

In general, for organic compounds, the low molecular weight, nonpolar, water-soluble species tend to pass through the membrane. Although the statement is frequently made that rejection below a molecular weight of 200 is poor and that above is good, this should not be regarded as absolute. Rejection of low molecular weight compounds such as organic acids and amines appears to follow the same pattern as that of weak inorganic acids and bases. The undissociated species are poorly rejected and the salts are well rejected.

There is a growing literature on membrane performance and References 3-6 as well as the current literature should be reviewed. Studies have been under way by the Water Quality Division of the EPA using reverse osmosis for concentration of organics. Much of this information has not yet been published.

Many of the compounds that would be of major interest in water treatment such as phenols, low and high molecular weight chlorinated hydrocarbons and other halogenated species, including many pesticides and low molecular weight alcohols, are poorly rejected or appear in the permeate without diminution with the current commercial membranes. However, data from new membrane experiments indicate that it may be possible to tailor membranes to permit rejection of some of the above substances.

Generally, large complex organic substances such as those that cause color in water supplies or interfere with the performance of ion exchange equipment or interfere with coagulation and filtration are almost completely removed. Large nitrogen-containing molecules that interfere with chlorination are removed. Lignins, humic and fulvic acids, MBAS and many organics deriving from waste discharges are rejected by reverse osmosis membranes.

Membrane damage or degradation may be caused by some organics. The polymers used may be soluble in many organic compounds. Other organics cause "swelling," which may result in a loss of rejection. These effects do not appear to be significant for some of the materials where concentrations are low. However, pilot studies should be considered or carried out where there is any doubt.

BACTERIA, VIRUS AND PYROGENS

One can expect complete rejection of bacteria, viruses and fungi; however, complete rejection would depend on having a membrane free of all imperfections. The probability of having a perfect membrane in a large reverse osmosis unit is very low. For example, a 1-mgd RO plant (based on permeate flow) using spiral wound elements would have a membrane area between 50,000 and 100,000 ft^2 (4650 and 9300 m^2). A hollow fine fiber plant may have an area approximately 10 times as great. As a result the small numbers of organisms that get through the membrane may colonize in the large surface area of the backing material or the hollow fibers and proliferate. Continuous or intermittent disinfection would relieve this problem. Since many of the compounds or physical conditions that interfere with disinfection are removed by the membrane, disinfection of the permeate can be very effective.

The reverse osmosis membrane elements provide a large surface area for the attachment and growth of bacterial slimes and molds, which can cause fouling of the membrane. There is also evidence that some organisms or their enzyme systems attack cellulose acetate membranes.

Pyrogens are substances that, when they occur in fluids being administered, such as plasma, vaccines and antisera, cause reactions that produce high temperatures and chills. They are thought to be of microbial origin. A pyrogen isolated from a typhoid preparation had a molecular weight of 62,000. A recent report indicated that, following disinfection of the reverse osmosis system, complete pyrogen removal was experienced. These results would be expected in a leak-free system.

RADIOACTIVE SUBSTANCES

There is no reason to believe that radioactive species of compounds or ions should behave any differently than nonradioactive species. In one series of experiments on dilute radioactive wastes, the activity in a waste stream contaminated primarily with uranium and thorium and their daughter products was reduced by a factor of more than 10,000.

GASES IN REVERSE OSMOSIS

Questions are frequently raised concerning the rejection of dissolved gases such as chlorine, oxygen, carbon dioxide, hydrogen sulfide and sulfur dioxide. As gases, they are poorly rejected. Some of the gases react with water, and the neutralized salts of the acids formed are rejected, *e.g.,* sodium bicarbonate and sodium sulfite. A study made of chlorine rejection revealed that free chlorine is poorly rejected below a pH of 7 as the equilibrium is shifted toward undissociated HOCl and Cl_2. Data obtained on free carbon dioxide show virtually the same CO_2 concentration in the product as in the feed. Experiments on sulfite pulp wash water at a pH of about 3 showed that most of the SO_2 permeated the membrane.

CONSULTING ENGINEER'S ROLE IN APPLICATION OF REVERSE OSMOSIS

In the application of reverse osmosis to water quality improvement, the major role of the engineer is: (a) defining the problem and the requirement, (b) determining the quality and the quantity of the new water source, (c) designing the pretreatment and posttreatment systems, (d) overseeing the site planning and preparation, and (e) providing for the storage and distribution of the product and for the disposal of the concentrate.

A report issued by the U.S. Office of Water Research and Technology[7] reviewed the operation of 11 commercial reverse osmosis and electrodialysis plants. Among the many findings and recommendations were the following:

> "The quality and quantity of the raw water source must be firmly established before the specification of a treatment plant. This is the most important single technical consideration in the planning of a treatment facility. Extreme care should be exercised in the design of the raw water pretreatment systems, especially with respect to the equipment selected to control and monitor feedwater chemistry. Raw water pretreatment systems have not been satisfactory as installed in the majority of these plants."

Pretreatment

It should be recognized that the pretreatment requirements and problems of reverse osmosis are essentially conventional. Water chemistry considerations and physical conditions that must be considered in processing the water are those that the engineer considers in the design of water treatment systems for municipal and industrial uses. Table III lists the data that should be obtained on a water supply for any purpose; much of this data is frequently not obtained. It should be required for reverse osmosis plant design. However, many plants have been designed and operated with far less data than listed. As the water recovery demand becomes higher and the water quality requirements become greater, more complete data become more important.

Table III. Water Analysis Requirements for Reverse Osmosis

Data Should Include:
   Sample Identification
   Raw Water Source, such as wells and surface
   Temperature
   Color
   Turbidity and Suspended Solids (if any)
   Specific Conductance
   pH
   Plugging Factor for Hollow Fine Fiber Designs
   Bacterial Analysis

The Following Constituents in mg/l Should be Included:

Calcium	Strontium	Chloride
Magnesium	Barium	Fluoride
Sodium	Ammonia	Nitrate
Potassium	Carbonate	Silica
Iron	Bicarbonate	Hydrogen Sulfide
Manganese	Sulfate	O-Phosphate

Include also any other data that may affect performance or the presence of which may be important to the water supply, such as other trace heavy metals and oil.

Water Quality

Using Table III as a basis for discussion, a basic requirement is identification of the water source to be used. Well waters and surface water sources have special problems and advantages of their own. Well waters may require little or no filtration, if properly handled, even when they contain dissolved iron and manganese that readily precipitates upon exposure to air.

Methods were developed in pilot studies with RO on a ground water containing up to 8 mg/l Fe and 5 mg/l Mn as shown in Table IV. A submersible well pump was used to pump the water directly to the suction of the high-pressure pump without exposure to air. The pH of the feed water was reduced in the line by acid injection, and sodium hexametaphosphate (a threshold precipitation inhibitor for calcium carbonate and calcium sulfate) was also added. Both acid and inhibitor also act to retard ferrous and manganous ion oxidation. This same technique has been used successfully for both spiral wound and hollow fiber equipment. If it is not possible to use this method, the iron and manganese must be precipitated and removed before demineralization. Almost without exception surface waters have a greater tendency to foul RO membranes due to colloidal and suspended matter and they frequently require

Table IV. Water Analyses—Brackish Well, San Diego
(75% Recovery @ 600 psi)

Concentrations (mg/l)	Well	Feed[a]	Product	Concentrate	Feed-Conc. Average	Percent Rejection
Cations:						
Calcium	360	360	7.6	1,400	880	99.1
Magnesium	230	240	0.5	940	590	99.9
Sodium	900	900	110	3,400	2,150	94.9
Potassium	26	26	3.8	91	59	93.6
Iron (Fe^{+2})	6.5	6.5	0.12	24	15	99.2
Manganese (Mn^{+2})	3.8	3.8	0.0	16	10	100
Anions:						
Bicarbonate	460	340	12	1,150	745	98.4
Sulfate	620	630	0.0	2,580	1,605	100
Chloride	1,960	2,020	170	7,850	4,935	96.6
Nitrate	0.53	0.53	0.26	3.3	1.9	86.3
Fluoride	0.70	0.70	0.12	1.3	1.0	88.0
Boron (B)	0.3	0.3	0.14	0.4	0.35	60.0
Silica (SiO_2)	33	34	7.6	120	77	90.1
Phosphate (Total)	0.36	10.0	0.18	40	25	99.3
Total Alkalinity—$CaCO_3$	380	280	10	940	610	98.4
Total Hardness—$CaCO_3$	1,830	1,880	20	7,350	4,615	99.6
Dissolved Solids (180°C)	4,460	4,580	310	17,720	11,650	97.3
Conductivity @ 25°C (umhos)	7,400	7,560	610	24,000	15,780	96.1
pH	7.0	6.7	5.3[b]	6.8	—	—

[a]Reflects pretreatment-acid addition and sodium hexametaphosphate.
[b]pH affected by CO_2 concentration.

filtration before processing. Surface waters are frequently contaminated by wastewaters. A typical problem is the acid mine waters of the Eastern coal mining states.

The temperature of the water source affects the operation of the RO system and may also affect pretreatment performance. In some areas of the U.S. and in some parts of the world such as Saudi Arabia, temperatures greater than 90° to 100°F (32° to 38°C) are found. Therefore, cooling may be required before demineralization. Surface waters in the temperate zone and some well waters may have temperatures approaching 40°F (4°C). These can be treated by RO but the water flux rate may be very low. Sometimes it may be economical to use heat to raise the temperature and the water permeation rate. The effect of low temperature on conventional water treatment processes for pretreatment must also be kept in mind.

Color

Color may indicate the presence of humic and fulvic acids, which may coat the membrane and require periodic removal. Hollow fiber elements are particularly susceptible to this problem. Turbidity and suspended solids must be reduced to low levels since their presence in excessive amounts causes plugging of the elements and fouling of the membranes. Turbidity specifications of 1 TU or less are frequently used but are not a guarantee of freedom from fouling.

Hollow fiber system equipment sources have developed a parameter based on the ASTM Standard Method of Test for Silting Index[8,9] called the plugging factor. Spiral wound elements appear to be relatively insensitive to the plugging factor. Turbidity and zeta potential, among others, have been evaluated for use as fouling parameters, but they have limited value. Jar tests have also been used to evaluate pretreatment, but as in any use for water treatment must be used with caution. Pilot scale studies with reverse osmosis systems with permeate capacities of 5000 gpd or greater have been useful not only in predicting operational problems but also in determining the recovery rate and the water quality at various recovery rates. In general, however, unless special problems are encountered, computer programs are now used to predict water quality and water recovery rates.

Calcium

Since calcium and bicarbonate and even carbonate ions are to be found in virtually all waters that are to be considered for reverse osmosis, the need for pH control and threshold inhibition becomes essential. Ion exchange softening to remove calcium to prevent both carbonate and sulfate scaling has been used. However, it is not always desirable since the calcium and other ions removed are replaced with sodium with different permeate quality results. Chemical softening reduces the calcium and alkalinity but does not completely remove them. Thus the potential for depositing scale on the membrane and in the system still persists although it is somewhat reduced.

There has been a great deal of reliance on the Langelier Saturation Index (and others) to predict the need for pH control and inhibition without understanding that these indices are qualitative estimates of the tendency of a water to cause corrosion or deposit scale. In addition, there is a continuing change at the membrane surface, in which calcium is almost totally concentrated in the brine stream, bicarbonate is well rejected and carbon dioxide passes through the membrane almost without deterrent. The net effect is to create conditions in the boundary layer amenable to precipitation. Adjustment of the pH into the 5 to 6 range minimizes scaling and the addition of a small amount (less than 10 mg/l and frequently less than 5 mg/l) of sodium hexametaphosphate effectively extends the range limiting calcium carbonate precipitation. Some manufacturers have published tables recommending sodium hexametaphosphate concentration up to 20 mg/l based on solubility product considerations, but these should be used with caution. Other polyphosphates can be used if hexametaphosphate is not available but they should be evaluated since they are generally not as effective.

Some of the new synthetic threshold inhibitors now on the market for industrial purposes are very effective but have not been approved for potable water applications. All the inhibitors appear to be completely rejected by the membranes. Control of pH down to the 5 to 6 range also has the salutary effect on cellulose acetate membranes of minimizing hydrolysis. The threshold inhibitor is also effective on calcium sulfate and on barium and strontium sulfates.

Calcium carbonate scaling causes plugging of hollow fiber elements and may cause plugging of spiral wound elements. Deposition of calcium carbonate scale on cellulose acetate membranes may accelerate hydrolysis since the equilibrium pH that can occur at the membrane surface is about 10. The scale is easily removed by using an acid flush. (Manufacturer's recommendations should be followed.) Organic acids such as citric and hydroxyacetic are effective, and phosphoric acid, sulfamic acid, acid salts and even sulfuric and hydrochloric acid have been used with caution.

Sulfuric acid is by far the most common acid used for pH control since it is readily available and low in cost. However, it may increase the sulfate concentration to an undesirable extent. Under these circumstances hydrochloric acid may be used. It is more costly and the increase in the chloride content of the feed water leads to an increase of chloride in the permeate. Carbon dioxide has been suggested for pH control but is normally not practical because of the large amount required. In lime-softened waters it can be used for part of the control (carbonate to bicarbonate) with final pH adjustment with sulfuric acid.

Among the sparingly soluble salts, calcium sulfate is relatively soluble at about 2000 mg/l in distilled water. However, it is frequently present in natural waters at relatively high levels and, on concentration in RO systems (as calcium sulfate is almost totally rejected) exceeds the solubility limit. More often than any other compound, it may limit the water recovery of the system. Once the calcium sulfate has precipitated, it is not readily removed. The calcium sulfate solubility does increase with the increasing ionic strength of the concentrate as it proceeds through the RO system. In fact the solubility increases by as much as a factor of 3 ($\sim$ 6000 mg/l) with an increase in salt concentration or ionic strength. Many studies have been made on the solubility of calcium sulfate in various concentrations of sodium chloride and other electrolytes. The data of Marshall et al.[10] has been found to be applicable over the range of interest to reverse osmosis. The data have been found to fit the following:

$$K_{sp} = 1.8 \times 10^{-3} \ S^{0.75} \tag{3}$$

where K_{sp} is the solubility product of $CaSO_4$ in $(mol/l)^2$ and S is the ionic strength in mol/l.

The extent to which a water can be concentrated until saturated with calcium sulfate can be calculated from:

$$R_{CaSO_4} = \left[\frac{(1.8 \times 10^{-3} \ S^{0.75})}{[Ca^{++}] [SO_4^{=}]} \right]^{0.8} \tag{4}$$

where R_{CaSO_4} is the ratio of concentration of a saturated solution to the feed solutions, $[Ca^{++}]$ is the molar concentration of calcium in the feed solution, and $[SO_4^{=}]$ is the molar concentration of sulfate in the feed solution. The recovery to reach saturation is:

$$\text{Recovery} = 1 - \frac{1}{R_{CaSO_4}} \tag{5}$$

Fortunately calcium sulfate has a tendency to supersaturate and is readily inhibited from precipitation by sodium hexametaphosphate. In fact, with some well waters that are pumped from the ground almost saturated with $CaSO_4$ it has been possible to run spiral wound element systems up to twice saturation. The short residence time within the RO system aids in minimizing calcium sulfate scaling. Cleaning solutions that contain chelating compounds for descaling have been used, but they have only been effective if the scaling has not proceeded too far.

Strontium and Barium

Due to improved analytical methods strontium and barium have been found to be present in ground waters far more extensively than had been thought. Since the sulfates of these two elements are far less soluble than calcium sulfate, they may cause scaling. Methods similar to that used for calcium sulfate have been used to determine whether barium and strontium will cause a problem at the water recovery desired. There is some evidence that barium and strontium sulfate precipitation is inhibited by sodium hexametaphosphate.

Iron and Manganese

Iron and manganese were discussed previously with regard to handling ground waters containing ferrous and manganous ion. However, their presence in a water source may limit the use of chlorine or other oxidizing agents for pretreatment before RO. Manganous ion at the pH in the normal operating range for RO is oxidized by air at a very low rate. However, some caution must be exercised since it appears that in the presence of Fe^{+3} the manganous ion oxidation may be catalyzed. In any event, Mn^{+2} is frequently present when Fe^{+2} is present and if ferric hydroxide is precipitated it may be accompanied by precipitated manganese oxide. Precipitated ferric hydroxide is usually amenable to filtration by granular media filters since it is also frequently used as a coagulant. Cleaning with solutions with a pH of less than 3 has been effective in removing iron and even manganese, particularly where a complexing acid such as citric or hydroxyacetic has been used. Also, reducing agents have been used to solubilize precipitated iron and manganese for removal.

Fluoride and Nitrate

The presence of ions such as fluoride, nitrate and trace heavy metals for which water quality standards have been established should be noted in a water analysis. RO can be useful in reducing fluoride to acceptable limits but its rejection is variable depending on water composition. Pilot studies may be necessary to determine that fluoride limits can be met. Nitrate rejection is also variable depending on the water, and pilot studies may be necessary.

Silica

Silica as SiO_2 is a common constituent of ground and surface waters. Silica solubility in water is estimated to be about 125 mg/l but this is dependent upon a number of factors not easily evaluated. Silica has also become a limiting factor in RO water recovery. Very few specific data are available and there is much reluctance to operate beyond apparent silica saturation since it would be difficult to remove once precipitated. Spiral wound element systems have been operated at apparent silica concentrations in excess of saturation without serious problems. Silica has been credited with causing serious fouling problems in hollow fiber units on some occasions. Silica removal by pretreatment could add complexity and costs to the plant. Ammonium bifluoride has been suggested as a silica cleaning agent, but as yet there is no data on its effectiveness.

Hydrogen Sulfide

Hydrogen sulfide is frequently present in ground waters and should be handled, if possible, in a manner similar to that for iron and manganese by pumping the water directly to the high-pressure pump without exposure to air. Once exposed to air (particularly at the pH frequently used for RO) very finely divided sulfur may precipitate. If this is permitted to enter any RO membrane element, it can result in severe fouling problems; it is almost impossible to remove without damaging the elements. The H_2S will be present in the permeate and, depending on concentration, is best removed by a degasification tower and/or chlorination.

Microorganisms

The presence of significant numbers of microorganisms in the feed water should be established since disinfection as pretreatment may be required. In a ground water containing reduced iron and manganese and hydrogen sulfide along with numbers of bacteria it is difficult to use oxidizing disinfectants without creating other problems.

Organics

Sometimes organics detrimental to RO membrane performance may be present in the feed water and activated carbon has been used to remove the interfering substances. RO has often proven itself an effective method of reducing or eliminating organics.

Aluminum

Aluminum as the ion is not frequently found in significant concentrations in either surface or ground waters. It is frequently used as a coagulant for municipal and industrial water treatment. Usually, the finished water has a pH between 7 and 8. If the water is then to be treated by RO and the pH is reduced to between 5 and 6, aluminum hydroxide may precipitate because it is much less soluble in that range than between 7 and 8. Aluminum hydroxide has a minimum solubility at about pH 5. This was first observed in pilot plants where a municipal water that had been treated with aluminum sulfate at the community water treatment plant was being used for some studies. The aluminum hydroxide can be removed by filtration or it can be removed from the membrane elements by acidification in the same way that iron hydroxide is removed.

Pretreatment Summary

Reverse osmosis systems are normally supplied with the instrumentation and equipment for addition of acid for pH control. They also include equipment for the use of a threshold inhibitor and may include cartridge filtration of 5 to 25 microns for protection of the elements as a last resort. The engineer must play an important role in the design and selection of pretreatment equipment for membrane plants. Site preparation and location of equipment may be essential to the success of the project. The engineer should understand the membrane process, its operation and application in order to use it to its best advantage.

Storage and Distribution

The permeate or product water from an RO system is characteristically low in solids and pH and high in carbon dioxide. It may also contain trace amounts of hydrogen sulfide and it should be almost free of turbidity. By any criteria it can be classified as a corrosive water and requires stabilization before distribution. Again the Langelier and other indices have limited value and must be used with caution for waters low in solids, alkalinity and hardness. Methods used to stabilize the permeate are: (a) decarbonation or degasification for reduction of CO_2 and increase in pH, (b) addition of calcium carbonate, lime or caustic to adjust pH, and (c) addition of silica and other corrosion inhibitors. Blending of the product water with the raw water to produce a suitable water can sometimes be utilized. When this is done the water quality may be reduced somewhat but a water meeting potable or usable water requirements can be produced at lower cost and may be easier to stabilize with less corrosion potential.

The water produced for potable use, particularly from surface or suspect ground water sources, should be treated with chlorine or other acceptable disinfectant before distribution. RO-treated water is almost free of turbidity and other interferences to good water disinfection practice.

Disposal of Concentrate

Another problem that must be faced by the engineer for every demineralization process and often for conventional water treatment plants is the disposal of wastes. For reverse osmosis systems this is the brine or concentrate that is usually low in pH. One factor that may influence concentrate disposal in an area is that RO is a process in which there is little change in the total salt budget as compared to ion exchange demineralization where chemicals are used for regeneration. Thus far, no serious problems have been encountered in disposing of the concentrate into municipal wastewater systems, into the ocean or estuaries, into surface waters and into evaporation ponds. However, this problem must be faced in the design of any demineralization plant. The waste stream from an RO plant is comparable in volume to that from the other demineralization processes.

COSTS

Water costs for reverse osmosis are usually expressed in cents per 1000 gallons (3.8 m^3) of permeate produced. For brackish waters up to 8000 to 10,000 mg/l TDS the following are the principal cost factors:

- membrane replacement
- power
- pre- and posttreatment including chemicals
- fixed charges for capital recovery of the initial investment
- operation and maintenance including labor and materials
- cleaning when required.

Power costs depend on water recovery limits and at 75% recovery it can be expected that plant electrical energy requirements will be about 6 to 8 kWh per 1000 gal (3.8 m^3) for large units. For very large RO systems and moderate-to-low water recovery, some studies have been made of energy recovery systems. They appear feasible.

The most fruitful area for significant cost decrease is in membranes.[11] Membranes producing higher water flux rates at lower pressures and with longer membrane life markedly improve system economics. There has been a steady introduction of such improvements into available equipment and some are in the semicommercial stage at this time. Usually these improvements are introduced in a configuration that can make use of existing equipment.

Unrealistic performance specifications and equipment warranties should be avoided. Both spiral wound element and hollow fine fiber element systems have demonstrated that high quality water can be produced from marginal and poor water sources at moderate cost. Suitable warranties on performance, costs and durability are available from the equipment suppliers and they should be consulted. Where unusual problems may be anticipated, pilot scale studies for both water quality and recovery should be run. These can usually be performed at low cost and may result in substantial savings or design improvements for the client.

ULTRAFILTRATION (UF)

Some confusion exists about the distinctions between ultrafiltration and reverse osmosis (hyperfiltration). At one time all permeation phenomena "of a solution under a pressure gradient through thin semipermeable membranes, resulting in at least a partial separation of solute molecules from the solvent or from the solvent and other solute molecules," was called ultrafiltration.[12] Ultrafiltration was also designated as a "process of separation whereby a solution containing a solute of molecular dimensions significantly greater than those of the solvent is depleted of solute by being forced under a hydraulic-pressure gradient to flow through a suitable membrane."[6] Reverse osmosis, ultrafiltration and ordinary filtration were said to differ superficially only in the size scale of the particles that are separated and that differentiation was in a large measure arbitrary.

It should be recognized that there are considerable differences between ordinary filtration and the membrane processes and although the line of demarcation between ultrafiltration and reverse osmosis may appear (and is) vague, there are important differences.

In ultrafiltration and reverse osmosis, a velocity vector must exist parallel to the plane of filtration. Since this is generally provided by a flow of fluid across the membrane and the separation of the fluid into a permeate and concentrate, the method has been termed cross flow separation. It is essential that the cross flow be maintained to minimize the effects of concentration polarization in the boundary layer on permeate quality and permeation rate.

Ultrafiltration is usually applied to the membrane separation of solutes above a molecular weight of 500. The osmotic pressures of the solutions are negligible and operating pressures for ultrafiltration systems are usually about 100 psi or less.

Reverse osmosis is the term usually applied to the separation of ionic or low molecular weight species from a solvent, usually water. It has been suggested that reverse osmosis is the separation of solutes whose molecular dimensions are within one order of magnitude of that of the solvent.[13] There is usually a significant osmotic pressure to overcome in reverse osmosis and this pressure must be exceeded. Operating pressures for reverse osmosis as applied to dissolved solids concentrations up to 10,000 mg/l are usually in the range of 300 to 600 psi (2100 to 4100 kN/m^2). For sea water applications and other relatively high TDS solutions, the pressures may range from 800 to 1200 psi (5500 to 8300 kN/m^2).

Ultrafiltration membranes are porous in structure and the mechanism itself is considered a sieving one. Since the apparent pore size of the membrane may be as small as 10 Å, it can be readily visualized that ultrafiltration can play an important role in water and wastewater purification. Colloidal and high molecular weight dissolved organic substances are rejected. Bacteria and viruses can be almost totally excluded from the permeate. However, there have been few if any large-scale applications of ultrafiltration in water and wastewater treatment thus far. Ultrafiltration has been applied to some industrial wastes and to the concentration of pharmaceuticals, enzymes and foods.[14-16]

At several plants, ultrafiltration was used to remove humic and fulvic acids from a water prior to demineralization by reverse osmosis. The humic acids were causing fouling in hollow fine fiber RO systems. Ultrafiltration has been suggested for reverse osmosis pretreatment to limit fouling in a number of applications and development of charged membrane concepts for this purpose is under way.

UF membrane materials are sometimes the same as those used for RO but they receive a modified treatment to give them a different structure. A large number of other polymeric film-forming substances have also been developed for UF. Membrane configurations are similar to those used in RO. Spiral, tubular and hollow fiber UF systems have been developed and marketed. A very extensive literature already exists on ultrafiltration and the small number of references attached should only be considered a brief sampling.

Reverse osmosis and ultrafiltration as membrane processes represent a new generation of water treatment techniques. In the last few years great emphasis has been placed on the need for an upgrading in potable water quality, culminating in the passage in 1975 of the U.S. "Clean Water Act." Conventional water treatment methods are limited in their ability to meet future demands for improved potable water quality. Membrane processes will become an important tool for water quality enhancement. Reverse osmosis, in a single process:

- reduces excess TDS
- reduces hardness
- reduces excess fluorides
- reduces or removes many organics
- removes or reduces nuisance and toxic heavy metals
- produces almost turbidity-free water
- removes or reduces microorganisms including virus
- improves disinfection practice.

REFERENCES

1. Loeb, S., and S. Sourirajan. "Sea Water Demineralization by Means of an Osmotic Membrane," *Adv. Chem. Series* 38:117 (1962).
2. Loeb, S., and S. Manjikian. "Brackish Water Desalination by an Osmotic Membrane," UCLA Dept. of Engr. Report 63-22, University of California at Los Angeles (1963).
3. Merten, U. *Desalination by Reverse Osmosis* (Cambridge, Mass: The M.I.T. Press, 1966).
4. Cruver, J. E. *Physicochemical Processes for Water Quality Control*, W. J. Weber, Jr., Ed. (New York: Wiley-Interscience, 1972), p. 307.
5. Lonsdale, H. K., and H. E. Podall, Eds. *Reverse Osmosis Membrane Research* (New York: Plenum Press, 1972).
6. Lacey, R. E., and S. Loeb, Eds. *Industrial Processing with Membranes* (New York: Wiley-Interscience, 1972).
7. Hornburg, C. D., O. J. Morin and G. K. Hart. "Commercial Membrane Desalting Plants—Data and Analysis," OWRT Contract No. CPFF/14-30-3276 (Washington, D.C., 1975).
8. Beach, W. A., and A. C. Epstein. "Summary of Pretreatment Technology for Membrane Processes," *Ind. Water Eng.* 12(4):13 (1975).
9. Luttinger, L. B., and G. Hoche. "Reverse Osmosis Treatment with Predictable Water Quality," *Environ. Sci. Technol.* 8(7):614 (1974).
10. Marshall, W. L., R. Slusher and E. V. Jones. "Solubility and Thermodynamic Relationships for $CaSO_4$ in $NaCl-H_2O$ Solutions from $40°-200°C$, 0 to 4 Molal $NaCl$," *J. Chem. Eng. Data* 9(2): 187 (1964).
11. Channabassappa, K. C. "Need for New and Better Membranes," *Water and Sewage Works—Reference Number—1976*, p. 153 (April 30, 1976).

12. Perry, E. S., and C. J. Van Oss, Eds. *Progress in Separation and Purification,* vol. 3 (New York: Wiley-Interscience, 1970), p. 97.
13. Michaels, A. S. *Progress in Separation and Purification,* vol. 1, E. S. Perry, Ed. (New York: Interscience, 1968), p. 297.
14. Porter, M. D., P. Schratter and P. N. Rigopulos. "Byproduct Recovery by Ultrafiltration," *Ind. Water Eng.* 8(3):18 (1971).
15. Okey, R. W. *Water Quality Improvement by Physical and Chemical Processes,* E. F. Gloyna and W. W. Eckenfelder, Jr., Eds. (Austin, Texas: University of Texas Press, 1970), p. 327.
16. Sachs, S. B., and E. Zisner, "Wastewater Renovation by Sewage Ultrafiltration," presented at the First Desalination Congress of the American Continent, Mexico City, Mexico (October 24-29, 1976), VVI-9.

CHAPTER 27

TECHNIQUES AND MATERIALS
FOR PREVENTING CORROSION

Abel Banov

Senior Editor
American Paint Journal Company
370 Lexington Avenue
New York, New York 10017

Herbert J. Schmidt, Jr.

President
MCP Facilities Corporation
25 Glen Head Road
Glen Head, New York 11545

INTRODUCTION

Chemicals involved in water treatment put a severe burden on the surface materials that protect metal or concrete tanks, conduits, basins and other equipment directly used in the various processes or located nearby. The use of any coatings except high-performance types to protect these structures lead to short and costly repainting cycles.

OBJECTIVES

This chapter is intended to provide sufficient guidance for selecting the proper coating for surfaces under mild service conditions. But the selection for severe service conditions must be based on so many considerations that a choice is too complex for other than experienced personnel. The safest policy is to consult an expert coatings technologist who is experienced in the protection of water treatment plants and distribution systems. To do his job properly, he must be informed of every detail and circumstance of use, and this chapter should aid the engineer in recognizing important use factors. Before accepting recommended coatings, insist on test results; if none are available, demand objective tests by an independent laboratory.

SELECTION OF MATERIALS

The selection of the most effective materials for the protection of water treatment plant structures is a multifaceted task. The task is made more difficult by: (a) the often severe conditions of exposure, (b) our primitive knowledge of the kinetics of corrosion and general inability to forecast corrosion rates, and (c) the effect of water quality (and soil quality for buried structures), which varies greatly. Since the severity of corrosion is so difficult to predict and since some water treatment chemicals (alum, for example) are so aggressive, it is wise to select high-quality coatings. At the very least, designers should weigh the extra cost of better materials against the frequency of repainting, with due consideration for the high cost of surface preparation in comparison to the cost of the coating.

The conditions of exposure are important in the selection of coatings. Some coatings resist alkali, others are preferred for protection against acids, and some resist both. Furthermore, some alkalis require more costly coatings than others.[1] Both temperature and chemical concentration are important, and a coating that is satisfactory for a chemical environment at a low temperature may fail miserably at a higher temperature. A coating may perform well under a splashing contact but fail under immersion. Or (as has occurred in covered steel reservoirs) a coating may provide a long service life below the usual water levels but permit corrosion in the damp atmosphere above the water.

The following considerations should be given to the selection of a protective coating.

- Resistance to the corrosives to which it will be exposed
- Resistance to abrasion or impact if likely to be exposed to carts and pallet trucks
- Estimated life expectancy of the object to be coated
- Estimated recoating time
- Total applied cost of a coating system (primers, intermediate and top coats, plus surface preparation and application)
- Comparative cost of coating off-site (in closed buildings) and on-site
- Advantages of off-site coating
- Film thickness required for desired performance

- Number of coats required to reach that film thickness
- Hundreds of square feet coverage per gallon/mil of film.
- Hiding pigment per gallon
- Binder content per gallon

Surface Preparation

Selecting the proper protective material is only part of the job. Surface preparation is equally important and sometimes more so. Without proper surface preparation, the coating may not adhere properly, and permeation by a corrosive liquid or fumes may cause damage that the coating was intended to prevent. Improper surface preparation often leads to early failure because of lifting, blistering and cracking of the coating.

Surface preparation often has to be more exacting with the more effective coatings. Usually the better coating demands more costly surface preparation (such as sandblasting) to a near white or white metal blast to roughen the surface so the harder coating has a better perch to grip. Improper surface preparation will, on the other hand, almost certainly nullify the use of a fine, high-performance coating. Hence, a prudent engineer demands preparation to match the product.

Application

A third aspect of surface protection (in addition to coating selection and surface preparation) is application. Most coatings in a large facility can be applied by spraying, but sometimes brushing or rolling may be necessary or advisable. In some instances, airless spray is preferred to the conventional air spray equipment. A trained applicator can control the build-up of the coating, whereas unskilled application can lead to cascading, where some areas may have too much material and others are virtually undefended.[2]

Specifications should include dry film build, which is usually expressed in mils (thousandths of an inch). Some surfaces may require only 3 or 4 mils. Surfaces immersed in mild alkali may need 7 to 12 mils, and surfaces exposed to other chemicals may need as many as 40 mils or more. In general, higher builds give more lasting protection. Balancing the cost of such high builds against the effectiveness of the protection is an exacting task for the specifier.

Some superior products perform as well or better in relatively thin films (or builds) than others in thick films, and, since fewer coats are needed, the cost of labor is lower. Some fine coatings can only be applied in very thin films which require several coats, whereas others can be applied heavily in multipass applications.

Inspection

Engineers should require proper inspection, particularly inspection of film thickness with both wet and dry film gauges. Other inspections that should be specified include: (a) adhesion, to ascertain that the material is really holding to the surface, (b) atmosphere, to assure that humidity is not excessive because it can reduce adhesion and film integrity and can cause undersurface rusting, and (c) excessive wind, which can lead to over-rapid surface drying, trapped solvent within the film, and eventual rupture.

The prospects of poor weather in the field and the difficulty of enforcing inspection procedures makes it highly desirable to apply paint off-site in closed buildings where contractors can carefully supervise personnel in surface preparation and material application under nearly perfect conditions. Complete systems consisting of primer, midcoats and topcoats can be applied in the shop even on large structures. After transportation to the job site, any breaks or scuff marks in the coating can be easily touched up. Because water treatment facilities usually require meticulous preparation and application, the use of off-site painting is an especially important consideration.

Cost Effectiveness

The task of the designer (or specifier) is to choose the most cost-effective surface protection with due regard for the costs of material, surface preparation, application, expected life and repair or replacement.

A general rule about surface protection is that the coating itself constitutes about 20% of the cost, and surface preparation and application make up about 80%. Expenditures for more durable coatings or linings are more than justified if they mean postponing the higher expenses of field surface preparation, application, and the paint required for recoating. Some areas are inaccessible and cannot be recoated.

Given a $100,000 coating operation (of which about 80% is for preparation and application), the use of a coating with an eight-year life instead of a four-year life saves one recoating job every eight years. The more durable coating might cost $25,000 whereas the less durable one might cost $20,000, which means an added initial expenditure of $5,000 would save $100,000 every eight years. Also to be considered, of course, is the time and perhaps cost saved by avoiding shutdowns.

Another example of cost-effectiveness is shown by Table I, in which an epoxy system was replaced by a urethane system for storage tanks at an industrial chemical plant. For example, (a) the time and labor required to drain an alum storage tank, (b) the waste in adherent alum

Table I. Comparative Painting Costs[a]

	Previous Epoxy System			Urethane System		
	Labor/ft²	Materials/ft²	Total	Labor/ft²	Materials/ft²	Total
Rigging	$0.09	$0.00	$0.09	$0.09	$0.00	$0.09
Surface Prep. SSPC SP6-63 commercial sandblast	0.25	0.00	0.25	0.25	0.00	0.25
Epoxy prime coat 2 mils	0.12	0.04	0.16	0.12	0.04	0.16
Epoxy intermediate coat 5 mils	0.12	0.09	0.21	0.12	0.09	0.21
Epoxy finish coat 2 mils / Urethane finish coat 2 mils	0.12	0.04	0.16	0.12	0.08	0.20
	$0.70	$0.17	$0.87/ft²	$0.70	$0.21	$0.91/ft²
Total cost (incl. $0.26/ft² for insurance, taxes, profit)			$1.13/ft²			Total cost (incl. $0.27/ft² for insurance, taxes, profit) $1.18/ft²
Estimated Annual Costs						Estimated Annual Costs
Epoxy system for 2- to 3-yr life span			$0.377/ft²			Urethane system calculated for 8-yr minimum life span $0.147/ft²

[a] After Mobay Corporation, Penn Lincoln Parkway, West Pittsburgh, Pennsylvania 15205.

that must be flushed from the tank and the pipes, (c) the personnel time required to supervise draining and refilling, (d) the management time required for coating contracts, (e) the possibilities of delays due to materials or labor, (f) the many large and small tasks attendant upon stopping and starting one stage of operation, and (g) the requirement and nuisance of a temporary means for continuing operation all combine to make the best coating the most cost-effective one. Certainly, the effort and care necessary to select the proper coating and the proper surface preparation and application techniques will pay off in lower maintenance costs and longer periods between shutdowns.

PROTECTIVE COATING FAMILIES

Types of Coatings

Protective coatings can be divided into two types: those materials useful for decoration and those useful primarily for protection against the environment. Those that decorate are usually known as architectural coatings. Those that protect may also be architectural coatings, but those that protect in moderately severe to harsh environments are called high-performance coatings. This chapter is confined to high-performance coatings since information about coatings for decoration is readily available.

Various families or systems of coatings and their comparative performances under a variety of exposures are rated in Table II. Exterior durability ratings are based on empirical observations of weathering and abrasion resistance. Solvent, base and acid resistance ratings are based on four-hour "spot tests" in which a pool of each of three test reagents is applied to a steel panel painted in accordance with the manufacturer's instructions. The maximum rating for each separate spot test is 5, which is equivalent to "no effect." A rating of 1 indicates complete failure. Thus, the maximum rating for the three test reagents is 15. The salt spray tests are made according to ASTM D 1654.[*]

Most high-performance materials are applied as liquids, but some are applied as powders with heat used to fuse the particles to themselves and to the surface. Others are used as protective linings. Liquids may be applied at room temperature; some dry or cure without heat whereas others do require heat. Some high-performance coatings may be applied in the field. Others, especially those needing heat, are usually applied off-site in a coating application plant.

---

[*]American Society for Testing Materials, 3 Parkway, Philadelphia, Pennsylvania 19102.

Table II. Comparative Test Data[a]

	Topcoat	Urethane High-Build	Epoxy Mastic High-Build	Alkyd	Alkyd 30% Silicone	Vinyl High-Build	Chlorinated Rubber	Epoxy Mastic	2-Component Aliphatic Urethane	Urethane High-Build	2-Component High-Build Epoxy	Maximum Rating
	Intermediate Coat:	2-Component Epoxy/Polyamide					Chlorinated Rubber	Urethane Moisture Cure	Moisture Cure	Moisture- Zinc-Rich Epoxy	2-Component Aromatic Urethane	
	Primer	Inorganic Zinc	Inorganic Zinc	Alkyd	Alkyd	Vinyl	Chlorinated Rubber	Cure	Cure	Cure	Epoxy	
Exterior Durability Basis: Weathering and abrasion resistance		13	7	3	7	7	7	7.	13	13	7	15
Solvent Resistance Basis: Total of tests with IPA MEK toluene		11	14	5	6	7	6	11	14	13	14	15
Base Resistance Basis: Total of tests with NaOH 50% NH$_4$OH 60% Aniline		13	13	7	7	10	11	12	14	11	13	15
Acid Resistance Basis: Total of tests with H$_2$SO$_4$ 50% HCl 37%, Acetic 100%		15	11	8	9	11	15	10	15	13	11	15
Salt Spray Basis: ASTM D 1654		10	10	3	1	5	6	7	7	7	10	10
Total Rating		62	55	26	30	40	45	47	63	57	55	70

[a]After Mobay Corp., Penn Lincoln Parkway West, Pittsburgh, Pennsylvania 15205.

Powder coating is almost always applied off-site because heat is required for fusion. Powders are either based on synthetic resins or frit-like metals or ceramics. Synthetic powders provide porcelain-like surfaces capable of withstanding extremely harsh environments for a long time. Facilities are now available for coating large structural members. Before long, large beams and sections of tanks to be erected in corrosive environments will be able to have this highly effective means of protection. Frit-like powders are more difficult to apply (and usually are more expensive) because they must be melted in high-temperature arcs and sprayed onto surfaces by specially trained operators. But these frit-like materials provide long-term protection against highly corrosive chemicals. They are especially effective on moving parts such as valves, gates and pump parts that are directly in contact with chemicals and also subject to physical wear.

Depending upon the synthetic material and the pigment used, liquid and powder coatings, can be opaque, translucent, transparent or clear. Frit-like coatings are the color of the metal or ceramic used. Some colorants and hiding agents also serve as anticorrosives by reacting with other ingredients in the formulation or by establishing a physical relationship with the surface. Lead oxide, for example, reacts with some binders to reduce the permeability of the coating. Zinc dust, a greyish primer pigment, sets up a tiny cathodic current with iron and steel, and when the surface is wet, the zinc is sacrificed instead of the metal surface.

Vehicle Families for Harsh Environments

Families of coatings and linings for use with corrosive chemicals are identified by the synthetic vehicle or category of material used.

Acrylics

Waterborne or solvent-borne acrylics offer some degree of protection in harsh environments. Solvent-reducible acrylics usually are applied in the factory where heat can be used. Waterborne acrylics for exterior protection against mild acids and for solvent-affected environments may not be as effective as epoxies or urethanes, but in circumstances where surfaces to be protected cannot be thoroughly dried, they are desirable because they cure over light moisture.

Epoxies

Tough and chemically-resistant, epoxies are available in several types: liquid epoxies, low-melting solid epoxies, epoxy ester, water-reducible epoxies, and epoxy powder coatings.

Liquid epoxies can be applied up to 100-mils (2.5-mm) thick in a single coat, or (with sufficient solvent) can be sprayed in thin coats. Epoxies are used effectively to prevent deterioration of platform surfaces, concrete and steel tanks, and surrounding equipment. In coal tar blends, they constitute high-build coatings for immersion or for underground service.

Low-melting solid epoxy resins, diluted in solvents, are used extensively for heavy-duty maintenance coatings and floor paint. They are often used for metal priming (aided by suitable reactive pigments). These epoxies are usually identified by the material that cures them. One is epoxy polyamide (cured by a polyamide resin), and the other is epoxy polyamine (cured by a polyamine resin). The polyamide is more flexible and water-resistant. The polyamine is harder and more chemical resistant. Cured with polyesters they provide hard, tile-like surfaces resistant to acids, alkalies and general abuse.

The reactive portion of the epoxy ester is precured by the fatty acids in one of the vegetable oils. That gives the esters some of the characteristics of alkyds, the "old war-horse" resins, except that they are considerably more chemical resistant. Epoxy esters are sometimes added to acrylics, polyesters or phenolics to improve their chemical resistance.

Other epoxies now growing in importance are the water-reducible versions, valuable where moisture is likely to be found on surfaces, and the epoxy powder coatings, which have an important role in protecting the steel reinforcing bars used in concrete tanks and structures. These "rebars" are often corroded by chemicals or salts that inevitably permeate through minute pores in the concrete. The expansion due to the rust forming on the rebars causes the concrete surfaces to spall and when the bars fail, the structure may fail.

Fluorocarbons

These rank among the highest in resistance to most of the chemicals likely to be faced, but they are costly. They are used where resistance to chlorine, bromine, brine, sulfuric acid, and other commonly used acids and solvents is required. Resistance to gamma and other radiation has led to their use in nuclear power plants. Mixing tanks, valves and piping protected with fluorocarbons have provided long-term, maintenance-free service. Fluorocarbons can safely withstand temperatures from -80°F to 300°F (-62°C to 149°C) in prolonged exposure. Versions currently available require factory application with touchup in the field.

Chlorinated Rubber

Where mild-to-medium strength alkalis in water are encountered, chlorinated rubber is useful. It is sometimes specified where microorganisms are likely to be a problem or where radioactive materials may have to be processed. It has also been used with zinc dust as a metal primer in systems for protecting metals against water immersion.

Phenolic Resins

Combinations of phenolic resins and formaldehyde, often modified with vegetable oils or a synthetic resin, can be tailored for a variety of uses where alkalis, chemicals and moisture are fairly troublesome but not extreme.

Polyesters

A versatile family, the polyesters impart various properties to coatings. They may be used in powder coatings, but in water treatment plants they are most likely to be useful in high-performance combination resins such as epoxy polyesters, polyester-cured urethanes, or silicone polyesters where toughness, and sometimes a glaze-coating, results, with distinctive characteristics derived from the co-monomers used.

Silicones

Improved chemical resistance and high baking resistance for single-package epoxies are imparted by silicones.

Vinyls

Where resistance to moisture, salt water, chemicals and oil is required in a high-performance coating, vinyl resin-based materials are often selected. One advantage of the vinyl family is that versions are available for either factory or field applications. A former disadvantage (that has now been overcome) was their limited "build," wherein three or four coats were needed to achieve the 7-mil (0.18-mm) dry film that is the minimum for immersion. Now, "high-build" vinyls permit single coats. For high performance in corrosive atmospheres or for immersion service, vinyls should be seriously considered.

Urethanes

Urethanes offer effective resistance to acids and alkalis, and they can be applied in the field. Two-package urethanes are those that need a polyester, an acrylic or an epoxy for curing. They can be applied in the field or in the factory without heat, and they are as effective as most oven-cured coatings. Of the two types of urethanes, aliphatic and aromatic, the aliphatics are the newest in this country, but extensive use abroad has proven they are superior to the aromatics.

Single-package urethanes are an improvement over alkyds, the "old reliables" of the industrial maintenance field, because they substitute an improved component for the one normally used in alkyds. For exteriors and interior areas not subjected to spills or severe fumes, these are effective although somewhat more expensive than the commonly-used alkyds.

Elastomeric Urethanes

These are urethane materials compounded to have the characteristics of tough rubber. They are applied as linings for tanks and other vessels used for acids, alkalis and abrasive materials.

Metallizing Sealers

Special spray guns fitted with high-heat sources use zinc or aluminum powder or wire to coat surfaces exposed to extremely corrosive atmospheres. Such coatings have been used successfully for fresh or salt water immersion in bilges and in tanks for strong acids and alkalis. Coating the intimately bonded metal with clear or aluminum-pigmented vinyl eliminates any porosity and adds another protective dimension. Applied with new plasma-spray devices, exotic anticorrosive metals such as tungsten carbide-cobalt, chromium carbide, nickel chrome, and magnesium zirconate are costly but effective ways to obtain protection of surfaces against extreme corrosion, radiation and temperatures beyond 1000°F (540°C), particularly where wear or abrasion is involved.

Neoprene

Where resistance to ozone and abrasion is needed as well as ability to resist a variety of strong chemicals, then neoprene (polychloroprene) is serviceable. Like natural rubber, neoprene is vulcanized but with better results. Neoprene's properties can be tailored to the needs of the user by altering the modifiers and production procedures. Neoprene is provided in solvent-base coatings, water-reduced dispersions, cured sheets and uncured sheets. Chlorinated rubber (both solvent-reduced and latex) frequently is used as a primer for neoprene coatings. Neoprene coatings that are to be exposed to sunlight and other atmospheric conditions are often overcoated with hypalon, an elastomer that is compatible with neoprene.

Chlorinated Polyether

A wide variety of harsh chemicals can be used with coatings and linings of this material. The application is more difficult than with most coating materials. Fusion of the resin solids occurs at 375-425°F (190-218°C).

Dip or knife-coating can be accomplished with a warm solution of the chlorinated polyether in cyclohexanone. Dispersions of chlorinated polyether in chlorinated hydrocarbons can be applied at 30-35% solids. Single coats of 20 or more mils can be applied. Each coat should be fused at the recommended temperature. A coat of better appearance can be obtained if the object is quenched immediately after fusion of the final coat. Since fusion heating limits the size of the object to the size of the available oven, dispersions can only be used on relatively small objects. Powder coating is also used to apply chlorinated polyethers. Here, too, the size is limited.

Linings of 40 mils and more are extruded from molding pellets of chlorinated polyether. These are applied to sandblasted metal coated with rubber-based adhesive primers. The adhesive layer should be heat-reactivated at 250°F (121°C) just prior to the application of the liner.

Inorganic Zinc

Inorganic resin-like materials are used with heavy concentrations of zinc dust to provide outstanding primers for steel. Ethyl silicate is the most frequently used binder. The use of zinc affords a type of cathodic protection. Near white or white metal sandblasting is mandatory for surface preparation. One feature of inorganic zinc as a factory primer is that welding can be done in the field.

Chlorosulfonated Polyethylene (Hypalon)

Chlorosulfonated polyethylene results from the reaction between polyethylene, chlorine and sulfur dioxide. With curing agents and suitable conditioning, it can be made into tough, chemical-resistant linings that are effective barriers against ozone and oxidizing chemicals. For the highest chemical resistance, litharge (lead oxide) is used for curing. Magnesium oxide instead of litharge is selected for resistance to sulfuric acid where little water is present. Various curing agents can be used to yield desirable characteristics, so suppliers should be informed about the intended use in order to recommend suitable products.

Asphalt

Various properties can be obtained by combining petrochemical asphalts and a natural asphalt known as gilsonite in differing proportions. Alphalt enamels are made of high softening-point asphalts. Microscopic particle-size asphalts are dispersed in a water phase with clay as an emulsifier and are known as asphalt emulsions. Cutback asphalt is a solution in a volatile solvent. Asphalt can be thickened with mineral fillers or fibers and can be applied by spreading with a trowel or by high-pressure spray guns.

Rubber Linings

Natural rubber, compounded with other materials, is used in linings at least 1/8-in. (3 mm) thick to protect tanks and other equipment in water treatment plants. Sheets are calendered, precut-edged, and attached to sand-blasted surfaces. Hard rubber linings are used where rigid pieces are to be protected. Soft rubber is used where abrasion is likely to be encountered. Such linings were often used in ion exchange reactors, but nowadays the protection is either epoxy or polyvinyl chloride (PVC).

Vinylidene Chloride

These materials are composed of vinylidene chloride and some other monomer such as acrylonitrile or vinyl chloride. Their properties are similar to vinyl. They may be used in 4- to 10-mil (0.1- to 0.25-mm) films laid down in solutions or as linings. Their function is to protect against water penetration and inorganic acids. Experience has shown that these materials vary considerably from time to time. Therefore, performance should *always* be tested under service conditions.

COATING FAMILIES FOR WATER TREATMENT PLANTS

Table III identifies 24 vehicle families and types of coatings. The chemicals likely to be used in a water treatment plant are listed in Table IV with the identification of the coating families that are useful for protection. The process temperatures, chemicals, concentration and length of exposure have a bearing on the selection of the best families. The material in Table IV was derived from various publications of the

Table III.   Index to Coatings and Membranes

Ident. No.	Name	Ident. No.	Name
1.	Rubber[a]	13.	Epoxy (Phenolic Cured and Baked)
2.	Butyl Rubber[b]	14.	Epoxy (Amine-Cured)
3.	Acrylonitrile-Butadiene[a]	15.	CTFE (Teflon®[c]) [b]
4.	Neoprene Lining[b]	16.	Vinylidine Fluoride (Kynar)[a]
5.	Vinyl	17.	Chlorinated Rubber
6.	Vinylidine Chloride	18.	Urethane (Air Dry)
7.	Phenolic (Baked)	19.	Urethane (Baked)
8.	Neoprene Coating[b]	20.	Chlorosulfonated Polyethylene (Hypalon)
9.	Coal Tar Epoxy	21.	Asphalt
10.	Coal Tar Urethane	22.	Vinyl Chloride (Plastisols)
11.	Furane Resins	23.	Chlorinated Polyether[b]
12.	Polyesters	24.	Metallized Coatings

[a] Membrane-linings only.
[b] Available as linings and coatings.
[c] Registered trademark of E. I. du Pont de Nemours and Company, Inc., Wilmington, Delaware.

National Association of Corrosion Engineers, Pennwalt Corporation, Metco, Inc., Carboline, Inc., and Mobay Corporation.*

Proprietary tables available from accredited members of the National Association of Corrosion Engineers enable the specifier to determine which families and which members of the families to use. Protection against corrosion in the harsh environments of a water treatment plant demands specialized knowledge. These tables are not meant to encourage "do-it-yourself" specifying but to foster an understanding of the problems and to enable specifiers to recognize the need for competent technologists in the corrosion prevention field.

---

*National Association of Corrosion Engineers. Monolithic Organic Corrosion-Resistant Floor Surfacings. (A standard). 2400 West Loop South, Houston, Texas 77027.

Pennwalt Corporation, 3 Parkway, Philadelphia, Pennsylvania 19102.

Metco, Inc., 1101 Prospect Avenue, Westbury, New York 11590.

Carboline, Inc., 350 Hanley Industrial Court, St. Louis, Missouri.

Mobay Corporation, Penn Lincoln Parkway West, Pittsburgh, Pennsylvania 15205.

Table IV. Coatings and Linings for Corrosive Water Treatment Chemicals

Chemical	Coating or Membrane (from Table III)
Alum-Fluoride Solution	1, 2, 3, 4, 11, 16, 19, 23
Chlorine	1, 2, 16, 17, 19, 23
Chlorine Dioxide	16, 24
Copper Sulfate	1, 2, 3, 4, 5, 6, 8, 11
Hydrofluoric Acid	16, 24
Hydrofluorosilicic Acid	16, 24
Ozone	16, 24
Sodium Bisulfite	1, 2, 3, 4, 6, 18, 19, 20, 21, 23
Sodium Chlorite	1, 2, 20, 23
Sodium Fluoride	1, 2, 3, 4, 20, 23
Sulfur Dioxide	16, 21
Recarbonation Stage	
Calcium Hydroxide	1, 2, 3, 4, 5, 6, 16, 17
Flocculation Stage	
Aluminum Sulfate	2, 3, 4, 9, 18, 19, 22, 23
Caustic Solution	16, 24
Ferric Chloride	1, 2, 3, 4, 7, 8, 9, 10, 13, 15
Ferric Sulfate	1, 2, 3, 4, 5, 16, 23
Ferrous Sulfate	1, 2, 3, 4, 5, 16, 20, 23
Magnesium Carbonate	16, 20, 23
Other Water Treatment Chemicals	
Brine	1, 2, 3, 4, 5, 9, 12, 13, 14, 16, 17, 24
Caustic Soda	16, 24
Lime	20, 23
Potassium Permanganate	1, 2, 3, 16, 20, 23
Soda Ash	1, 2, 3, 4, 6, 17, 18, 19, 23
Sodium Chloride	1, 2, 3, 4, 5, 6, 9, 10, 11, 13, 14, 16, 17, 24
Sulfuric Acid (50%)	1, 2, 3, 4, 6, 8, 11, 15, 20, 23, 24
Sulfuric Acid (98%)	15, 16, 24

SURFACE PREPARATION DETAILS

High-quality coatings used under the demanding conditions of a water treatment plant should be applied only to surfaces that have had suitable (and usually exacting) preparation. Coatings manufacturers serving these facilities are capable of recommending suitable techniques of surface treatment and preparation. In areas where corrosive chemicals are to be used, meticulous surface preparation is especially essential if intervals between future painting cycles are to be lengthened as much as possible. Almost invariably, the more durable coatings are likely to need time-consuming and costly surface preparation.

A simple alkyd or polyvinyl acetate paint can penetrate readily into surfaces that will not hold tough products such as epoxies, urethanes or vinyls. The latter have longer, heavily entwined molecules that need a surface that has been microscopically scarred by chemical etching or abrasive blasting. Mechanical cleaning or hand cleaning is usually inadequate for this kind of situation, so only abrasive blast cleaning and etching are discussed.

Abrasive Blast Cleaning

Particles of abrasives are driven against the surface under air pressure of 100 psi (690 kN/m^2) or more to remove any old paint or debris and to create what is known as an "anchor pattern" to serve as a perch for the high-quality coating to be applied. Blast cleaning can be specified for any one of six degrees.

1. Where optimum corrosion resistance is required and the finest paints are to be used, WHITE METAL BLAST should be specified. Steel is cleaned to a state that matches pictorial standards established by organizations such as the Steel Structures Painting Council.[3]

2. NEAR WHITE METAL BLAST can be carried out twice as fast as WHITE METAL BLAST and is adequate for all but the most corrosive environments. Everything is removed from the surface, but the whiteness is not as complete as the first degree.

3. COMMERCIAL BLAST is suitable only for mildly corrosive environments because some of the mill scale (a black iron oxide left on steel after the rolling process and which often leads to undercoat-rusting) is not removed. Commercial blast leaves the surface gray. It can be accomplished in half of the time required for near-white blast.

4. BRUSH-OFF or SWEEP BLAST only cleans the surface lightly and removes loose or dried paint, loose rust and loose mill scale. For a simple noncorrosive environment, this is all that is needed, particularly for concrete.

5. WET BLASTING may be used under circumstances where dust cannot be tolerated. The abrasive is wetted, so steel is subject to corrosion unless inhibitors are used in the water. Recent tests show that after wet blasting, special maintenance grades of acrylic latex are effective because they can be applied over the resulting wet surface.

6. WATER BLASTING can be used under pressure of 3000 psi (20,700 kN/m^2) or more with little or no abrasives. Everything but mill scale can be removed, and with a little abrasive this can also be removed.

In specifying any kind of abrasive cleaning, care must be taken that the anchor pattern is not too deep because this can lead to sharp peaks, which may result in inadequate film build over them. Paint is deposited in the deep valleys, where adequate film build occurs, but the sharp peaks are sometimes barely covered.

Shallow anchor patterns may be obtained with a blast of very fine steel grit, very fine Ottawa (or a similar clean, screened) sand, or a new effective material called Black Beauty. The maximum height of the profile should be 1.5 mils (0.038 mm) if a 5/16-in. Venturi nozzle at 80 psi (550 kN/m²) is held 18-24 in. (0.46-0.61 m) from mill scale-covered mild steel plate. Medium Ottawa sand and fine steel grit yield a 2-mil (0.05-mm) profile under the same conditions. Specifying more than 2 mils can result in too sharp a profile. Particular care must be observed in requiring the smoothing of weld seams to free them of slag and scale, because over-blasting can cause pitting, which results in thin paint coverage, a probable cause of rusting.

If rust and mill scale are to be removed from iron or steel, or if metallic oxides are to be removed from zinc, aluminum, copper or brass, acid pickling may be used. This is done preferably with sulfuric acid, but hydrochloric, nitric, or phosphoric may be used except for aluminum, which requires caustic soda. After oil and grease are removed from the surface by solvent cleaning, the acid is applied by spray or in a wash. It must be allowed to soak only until the rust and mill scale are removed; if it soaks longer, excessive metal is removed. Several clear water rinses are required to remove the acids and salts. The final rinse should have a weak alkali added to retard rusting until the application of the coating.

Etching

Extremely tight, fine-grained concrete often causes problems for paint adhesion. A 5% solution of hydrochloric acid, allowed to stand until bubbling stops, leaves an etch pattern to which paint will adhere. Alternatively, a solution of 3% zinc chloride and 2% phosphoric acid in water can be used.

Maintenance

Recoating an existing structure raises the question as to which coating should be applied over the existing system and how it should be done. Old coatings should be thoroughly cleansed by wiping with a solvent. This treatment should be followed by a brush-off blast to remove loose paint and provide an anchor pattern for the finish coat. There are special surface preparation solvents that should be used for some systems such as epoxy-coal tars. Always check with the manufacturer for the exact recommendations. It is always best to use a coating compatible with the existing coating. Table V should assist maintenance personnel to determine the compatibility of various coatings.

CORROSION UNDERGROUND

Very small electric currents flow between dissimilar metals linked to each other by water or by any other suitable conductor. This causes corrosion. Corrosion resulting from electrochemical action is minimized by protective coatings to insulate the metal to be safeguarded as well as possible or cathodic protection systems.[4]

Cathodic protection systems sacrifice a replaceable metal for the protected metal. They are particularly valuable on immersed metals or on metal that is buried in the earth where all sorts of currents may be flowing between buried objects or structures or even among the mineral elements in the soil.

Cathodic protection may be obtained by installing what is known as an "impressed-current" system, whereby an external source of direct current is imposed on a circuit set up between the anode (or positive terminal) of the power source and the protected metal, which serves as the cathode (or negative terminal). Current flows when the intervening earth is damp enough to become an electrical conductor. Without the impressed direct current, the dampness serves as a conductor between the metals in the adjacent soil and the endangered object, which leads to corrosion. But when the direct current is impressed, the positive terminal (anode) at the power source is sacrificed instead. When the anode is almost completely oxidized, it can easily be replaced. The imposed current not only provides electrical protection, but also deposits a calcareous or siliceous protective barrier, picked up from minerals in the damp earth or water, on the buried structure. A further benefit is that the barrier coating eventually reduces the amount of current needed for the system to operate, and thus reduces the drain on the power source.

Cathodic protection may also be obtained by a simpler "sacrificial anode" system. Suitably-sized pieces of metal that can establish a current (without an external power source) when connected to steel, or other metal is placed near the object to be protected and connected to it. The current thus produced has the same effect as the electricity imposed in the "impressed-current" system. However, this second method is more difficult to control accurately and since the current changes as the sacrificial metal is used up, frequent inspections are necessary to assure protection by replacing metal as needed. The more common sacrificial metals are magnesium, aluminum and zinc (or their alloys).

Table V. Paints and Coatings Compatibility Guide[a]

Topcoat	Acrylic	Alkyd	Bitumen	Chlorinated Rubber	Epoxy-Amine	Epoxy-Coal Tar	Epoxy-Ester	Epoxy-Polyamide	Latex	Oil-Base	Phenolic (Modified)	Polyester	Silicone	Silicone-Alkyd	Urethane (Modified)-Roughen Surface First	Vinyl	Zinc-Inorganic	Zinc-Organic
Acrylic	G[b]	M	M[c]	G	G	G[c]	M	G	G	G	G	G	G	M	G	G	G	G
Alkyd	M	G	G[c]	G	G	G[c]	G	G	M	G	G	G	G	G	G	G	NR	NR
Bitumen	G	G	G	G	G	G[c]	G	G	G	G	G	G	M	M	G	G	G	G
Chlorinated Rubber	M	G	M[c]	G	G	G[c]	G	G	G	G	G	G	G	G	G	G	G	G
Epoxy-Amine	NR	NR	NR	NR	G	G[c]	NR	G	NR	NR	G	G	NR	NR	G	NR	G	G
Epoxy-Coal Tar	NR	NR	M	NR	G	G[c]	M	G	NR	NR	G	G	NR	NR	G	NR	G	G
Epoxy Ester	G	M	M[c]	G	G	G[c]	G	G	G	G	G	G	M	M	G	G	M	M
Epoxy-Polyamide	NR	NR	NR	NR	M	M[c]	NR	M	G	M	G	G	NR	NR	M	NR	G	G
Latex	M	M	M[c]	M	M	G[c]	M	G	M	G	G	G	M	M	M	M	G	G
Oil-Base	M	M	G[c]	G	M	M[c]	G	G	NR	M	G	G	G	G	M	G	NR	NR
Phenolic (Modified)	NR	NR	NR	NR	M	M[c]	NR	M	NR	NR	M	G	NR	NR	M	NR	G	G
Polyester	NR	NR	NR	NR	M	M[c]	NR	M	NR	NR	G	G	NR	NR	M	NR	G	G
Silicone	G	M	M[c]	G	G	G[c]	G	G	M	M	G	G	G	M	G	G	G	G
Silicone-Alkyd	M	G	G[c]	G	G	G[c]	G	G	M	G	G	G	G	G	M	G	G	G
Urethane (Modified)	NR	M	M[c]	M	M	M[c]	M	M	M	M	G	G	M	M	M	M	G	G
Vinyl	M	M	M[c]	G	G	G[c]	M	G	M	M	G	G	M	M	G	G	G	G

[a]Reprinted with permission of the Carboline Company, 350 Handley Industrial Court, St. Louis, Missouri 63144.
[b]Key: G–Good, Recommended; M–Marginal; NR–Poor, Not Recommended.
[c]Indicates that "bleed through" may be a problem.

When these sacrificial metals are near steel in the ground, corrosion takes place in the metal that is higher in what is known as the "galvanic series" of metals, listed in Table VI. For example, lead is higher in the series than copper or brass, so if lead and copper are near each other in damp ground, lead is eroded whereas the copper is protected. Because silver, carbon, platinum and gold are lowest in the galvanic series, they are the most stable. Because zinc, aluminum and magnesium are higher in number than steel, they can all be used as sacrificial anodes to protect steel.

Table VI. Galvanic Series of Metals

Series	Metal	Series	Metal
10	Magnesium	5	Copper, Brass
9	Aluminum	4	Silver, Silver Solder
8	Zinc	3	Graphite, Carbon
7	Steel, Iron	2	Platinum
6	Lead, Lead-Tin Solder	1	Gold

Any buried structure of substantial size has a potential for setting up a galvanic current. The magnitude of the current depends on the resistivity of the soil connecting the metals. This resistivity can vary over a range of 1000 to 1, depending on the condition of the soil. The same soil can vary with season and rainfall.

Before building a water treatment plant, especially in wet ground, the soil resistivity should be checked. A test for resistivity consists of spacing four electrodes at varying distances, impressing static electric charges on them and then taking average resistivity readings with a megohm meter. Soils with a resistivity of about 1000 cm or less are considered severely corrosive. For comparison, the resistivity of sea water is 20 ohm-cm. Soils with resistivity in excess of 20,000 ohm-cm are rated as slightly corrosive. But even so, small, deep pits may develop in steel.

When steel is in contact with several layers of soil, the differences in resistivity of the various soils compound the problems, and such members corrode more rapidly than those in uniform soil. Hence, to determine the corrosion potential of buried steel, it is necessary to take borings and check the resistivity at different depths.

The moisture content in soils is important in resistivity. Well-drained surface soils are usually highly resistive because most of the soluble elements, which enhance conductivity, have been washed away. But damp soils that contain cinders cause deep pits on steel soon after placement.

Note that carbon, the main component of cinders, has a lower number than steel in the galvanic series of metals (Table VI).

CORROSION IN WATER

If treatment plants are built partly on water, corrosion occurs in steel piling unless coatings or cathodic protection are used. Unprotected piling usually shows scaling in what is called the "splash zone," just above the high tide or high water level. Corrosion is also severe just below the mean low water level. Corrosion in these zones results from "differential aeration," the difference in the oxygen in the water at the corrosion points compared to elsewhere. The name "differential aeration" describes the difference in available oxygen. The portion of the pile between mean low water and high water naturally receives more oxygen than those portions that are continually immersed. Therefore, corrosion currents are set up in the steel between areas of low oxygen (the immersed portion) and those above the immersed portion where oxygen is abundant. This creates a galvanic cell that corrodes the steel just below the water.

Buried and immersed metals are frequently protected by both coatings and cathodic protection. The National Association of Corrosion Engineers (NACE) reported[6] tests showing that extreme care is needed to avoid trouble when using cathodic protection and coatings. Selected coatings must be able to resist the alkali reaction that may cause blistering under some conditions that may take place at the protected metal surface. Vinyl coatings and chlorinated rubber are reported to have better properties for preventing this trouble, but even these can be damaged if excessive alkali accumulates. To avoid blistering, very modest voltages and currents must be used. NACE[6] says that cathodic protection voltages are usually less than 1 volt and current density is usually less than 10 milliamps/ft^2. In considering a cathodic protection system, it is recommended that a certified, accredited corrosion engineer be engaged.

SPECIFICATIONS

For specifications useful in designating ultra-high performance coatings material, necessary surface preparation, and product application in the shop or the field, the following companies have helpful publications.

Carboline, 350 Hanley Industrial Court, St. Louis, Missouri 63144
Gates Engineering, Inc., 100 S. West St., Wilmington, Delaware 19899
United Coatings, 1130 E. Sprague Avenue, Spokane, Washington 99202
Porter Paint Co., 400 South 13th St., Louisville, Kentucky 40201

REFERENCES

1. Banov, A. *Paints and Coatings Handbook,* 2nd ed. (Farmington, Michigan: Structures Publishing Co., 1978).
2. Levinson, S. B. "Painting," in *Facilities and Plant Engineering Handbook* (New York: McGraw-Hill Book Co., 1976).
3. "Color Photographic Standards for Surface Preparation" (SSPC-Vis 1), Steel Structures Painting Council, 4400 Fifth Avenue, Pittsburgh, Pennsylvania 15213.
4. Schmidt, H. J., Jr. "Preventing Underground and Underwater Corrosion," *Construction Specifier* (Washington, D.C.: October 1975).
5. Schmidt, H. J., Jr. "Protection of Ferrous Metals by Coating and/or Cathodic Protection," Het Ingenieursblad, Jan van Rijswijcklaar 58, 2000 Antwerp, Belgium.
6. Weaver, P. E. *Industrial Maintenance Painting,* 3rd ed. (Houston, Texas: National Association of Corrosion Engineers, 1967).

CHAPTER 28

DESIGN OF GRANULAR-MEDIA FILTER UNITS

A. Amirtharajah, Ph.D., C. Eng.
Associate Professor
Department of Civil Engineering
and Engineering Mechanics
Montana State University
Bozeman, Montana 59717

INTRODUCTION

Granular-media filtration is the final solid-liquid separation process in water treatment. Classical sedimentation theory of discrete particles predicts that imposing finite dimensions on sedimentation tanks limits the size of particles that may be removed in sedimentation. Consequently, particles with settling velocities smaller than the overflow rate are carried into the filter units. Thus, filters are almost always required to remove these particles and to ensure the final clarity of the water.

Even though depth filtration has been studied extensively over the last two decades,[1-6] it is still not possible in general to predict theoretically the performance of a filter in terms of: (a) the influent characteristics, (b) the characteristics of the media, and (c) the mode of operation of the filter. A significant part of current research[4,6] in filtration theory is directed towards understanding initial removal of particles on clean individual collectors. It is hoped that the future will provide a mathematical function describing particle removal during the progress of a run and that it could be formulated as an expression with the initial removal function incorporated in it. Semiconceptual models and their use for predicting the effluent quality and head loss profiles for purposes of engineering design have appeared in the recent literature[7,8] and seem

675

promising. However, these models need validation in a variety of conditions and require further refinement for general use.

In general it is not possible to design filters on theoretically valid models at the present time. However, the models do provide significant concepts and insights into the filtration process and also provide guidance for minimizing the range of pilot plant studies needed for design of filters.[8]

Baumann deals extensively with principles and current advances in application of filtration to production of potable water in Chapter 12. The present chapter summarizes some well-documented facts on filtration that are directly related to filter design and illustrates their use with typical process design calculations. Since hydraulic computations are an essential feature of filter hardware design, these calculations are also illustrated.

FILTRATION PRINCIPLES AFFECTING DESIGN

The following is a summary of significant facts in filtration that are reasonably well-documented, and have a major impact on filter design.

1. Pretreatment, Filtration Rate and Water Quality

In the early 1960s, Robeck and coworkers showed in several studies[9,10] that with adequate pretreatment, filtration rates from 2 to 6 gpm/ft^2 (1.36 to 4.08 mm/sec) produce the same effluent water quality. Recent work[5,11] has confirmed the concept that chemical destabilization is an essential prerequisite for effective filtration. Inadequate pretreatment may produce poor quality effluent even at filtration rates less than 2 gpm/ft^2 (1.36 mm/sec).

2. Filtration Rate Changes, Water Quality and Rate Control for Design

Any rate change during filtration causes significant deterioration of the effluent quality.[12,13] The degradation in quality can be qualitatively correlated directly with the magnitude of the rate change and inversely with the time for the rate change.

This fact forms the underlying basis for the superiority of variable declining rate filters,[5,14] over the traditional constant rate filters with automatic rate controllers actuated by mechanical devices. Omission of rate controllers not only results in capital cost savings but also produces a better quality effluent. In addition, increases in run length are achieved by head loss recovery both within the dirty filter and the underdrain system due to the declining rate. A minimum of four cells is recommended design practice for effective operation of a bank of declining rate filters.

3. Filter Media, Run Length and Backwashing

The filter media effectively controls the performance of a filter. The following competing and conflicting requirements control filter media design: (a) the finer the media, the better the quality but the head loss increases, which rapidly reduces the run length; (b) coarser media implies a balance between effluent quality degradation and head loss build-up, but requires higher backwash rates; (c) simultaneous air scour and sub-fluidization water wash is the most effective cleaning system,[15] but it destratifies single media and mixes dual and multimedia; (d) restratification of dual or multimedia requires that the filter media be fluidized.

The theoretical optimum in use of granular filters was defined by Mintz[16] as the condition where the time to reach the limit of quality degradation (t_1) and the time to reach the limit of available head loss (t_2) are reached simultaneously. Practical designs for filters constrain t_1 to be less than t_2 to avoid breakthrough during a run.

Theoretical and empirical models[17,18] and experimental studies[19-21] have indicated that optimum designs are approached with coarse uniform single media (> 1 mm) and extra deep beds (> 4 ft), with air scour for backwash to maintain destratification. For example, the design for the new direct filtration plant for Los Angeles[21] uses 2-mm unsized sand of 8 ft depth with air scour and media-retaining strainers. These designs approximate common European practice.[22] Present trends indicate that media design is often controlled by backwashing requirements. Instead of coarse single-media deep beds, dual- or multi-(mixed) media also provide the balance between run length and effluent quality. Baumann indicates in Chapter 12 that run lengths should be 15-30 hr. In general, dual-media filters produce better water quality for a longer run length than single-media filters.

4. Filtration and Polyelectrolytes

Polyelectrolytes can assist in improving effluent quality when added in small amounts as a filter aid (0.08 to 0.001 mg/l) just prior to filtration. However, they cannot substitute for effective pretreatment. Addition of polyelectrolytes makes backwashing more difficult due to the higher attachment forces,[23] and auxiliaries like surface wash or air scour are necessary. Recent studies[3,11,23] show considerable similarities between coagulation-flocculation and filtration when polymers are used. Optimum dosages, overdosing and stoichiometry, which are characteristics of destabilization in coagulation, also occur with polymer-aided filtration.

These concepts are of major importance in design of direct filtration plants, which are commonly the most economic treatment scheme for low-turbidity raw waters (< 50 TU), such as those from lakes needing low alum coagulant dosages (< 15 mg/l).

5. *Filter Backwashing*

The best criteria for backwash effectiveness are the quality of the filtered water and long-term absence of dirty filters and mudball formation. Amirtharajah[24,25] has indicated that few or no contacts occur between fluidized particles. Hence particulate fluidization with water alone is an intrinsically weak cleaning process. Air scour, which causes abrasions between particles throughout the depth of the bed, and surface wash, which causes collisions at the top of the bed, are effective auxiliaries for cleaning.[15,25]

For ordinary solids (low adhesive forces), water wash alone—which expands the bed by 30-40% to give expanded porosities around 0.70 at the top layers of the backwashed filter—provides optimum cleaning.[24] This is economically achieved with a total wash water usage of 75-100 gal/ft² (*i.e.*, 15 gpm/ft² for 6-7 min). In metric units these correspond to 3.06 to 4.07 m³ per m² (*i.e.*, 10.2 mm/sec for 6 to 7 minutes). An important consideration in estimating backwash requirements is to calculate the expansion of the media with the warmest temperature of the wash water. Simultaneous air scour (3-5 scfm/ft², in metric units 0.9-1.5 m³/m² min) and subfluidization water wash (6-8 gpm/ft², in metric units 4.08-5.44 mm/sec) provide the best cleaning characteristics for solids with higher adhesive forces (polyelectrolytes or wastewater solids).[15,26] Dual-media filters have the ever present danger of loss of media with simultaneous air and water wash and also need a fluidization wash at the end of the backwash cycle to restratify the media. If supporting gravel layers are used, their movement during air scour is possible. Loss of media can be avoided by: (a) lowering the water level to the media surface prior to wash, (b) using air scour alone and (c) providing greater depth between the surface of the media and the lip of the wash water troughs.

MODELS FOR DESIGN

In recent times several texts[5,26,27] detailing design procedures have appeared. This section summarizes some of the models appearing in recent literature that provide a framework for design of filters. Few of them are ideal, but most are the best available at the present time (1978) and give reasonably valid results for filter design. Their use is illustrated in the following design example.

Optimal Process Design of Filters

Adin and Rebhun[23] and Kavanaugh *et al.*[8] in two recent papers illustrate use of Mintz's[16,18] concept for optimal design of filters. Adin and Rebhun[23] showed the determination of optimal filter run in a comparative study using alum and polymer in direct filtration for various depths. Kavanaugh *et al.*[8] illustrated the concept for phosphorus removal by filtration for various media. Collection of data from a pilot plant or laboratory column experiments provides guidelines for rational decisions in optimal design.

Figure 1-A illustrates schematically in three variables the optimal design options. The t_1 and t_2 surfaces are the times needed to reach limits of effluent quality and head loss. The intersecting line of the two surfaces indicates the optimal condition. Figure 1-B shows the optimization diagram for three types of media obtained by Kavanaugh *et al.*[8] The results show that with a minimum run length constraint of 10 hr, filters 2 and 3 need to be operated at approximately 2.2 gpm/ft² (1.50 mm/sec) while filter 1 can be operated at approximately 3.8 gpm/ft² (2.58 mm/sec). The condition $t_1 > t_2$ can be satisfied by flow rates < 4 gpm/ft² (2.72 mm/sec) for filters 1 and 2. Such data can be developed with pilot plant or column studies by plotting head loss and effluent quality curves and would give rational guidelines for indicating the acceptable operating range of the filters and for selection of media size, media depth and filtration rate.

Media Design—Sizes

In the absence of pilot plant data for the selection of media size, depth and filtration rate as indicated above, the approach should be conservative and based on past experience. The size of the media must balance the conflicting requirements of effluent quality which improves with finer size, and solids storage and head loss requirements, which improve with coarser media. For dual media, the sizing of sand and coal can be based on rational theory.

The equation for minimum fluidization developed by Leva[28] from the Kozeny equation can be expressed as a superficial velocity as done by Amirtharajah[29] for graded sands to give,

$$V_{mf} = \frac{0.00381(d_{60\%})^{1.82} \ [\gamma(\gamma_m - \gamma)]^{0.94}}{\mu^{0.88}} \tag{1}$$

in which V_{mf} = minimum fluidization velocity in gpm/ft²

$d_{60\%}$ = diameter of particle in mm

γ, γ_m = fluid and media specific weights in lb/ft³

μ = viscosity in centipoises

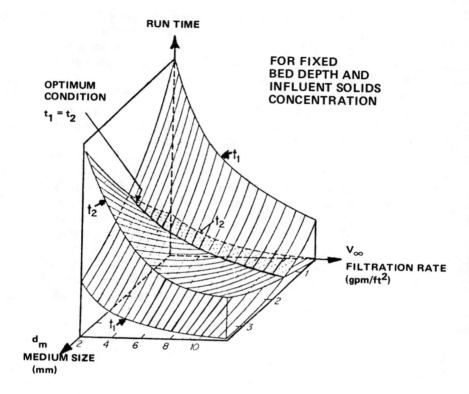

A. Schematic description of design options and optimal conditions.

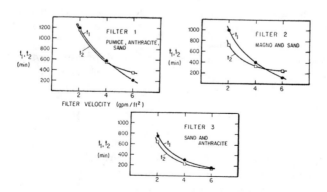

B. Optimization diagram for three filters.

Figure 1. Schematics of optimal process design of filters. (After Reference 8.)

If the minimum fluidization velocity of coal with characteristics ψ_c, d_c, γ_c is equal to that of sand with characteristics ψ_s, d_s, γ_s and assuming $d_{60\%} \cong \psi d$ where ψ = sphericity, then from Equation 1,

$$(\psi_c d_c)^{1.82}[(\gamma_c - \gamma)]^{0.94} = (\psi_s d_s)^{1.82}[(\gamma_s - \gamma)]^{0.94}$$

Simplifying,

$$\frac{d_c}{d_s} = \left(\frac{\psi_s}{\psi_c}\right)\left[\frac{Sg_s - 1}{Sg_c - 1}\right]^{0.52} \tag{2}$$

in which $\gamma_c = Sg_c \gamma_c$ and $\gamma_s = \gamma_s Sg_s$ where Sg = specific gravity. Using typical values in Equation 2,

$$\text{coal, } \psi_c = 0.70, \ Sg_c = 1.50$$
$$\text{sand, } \psi_s = 0.85, \ Sg_s = 2.65,$$

$$\frac{d_c}{d_s} = \left(\frac{0.85}{0.70}\right)\left[\frac{1.65}{0.50}\right]^{0.52}$$

$$= 2.3 \cong \frac{d_{c90\%}}{d_{s10\%}}$$

Thus, for simultaneous fluidization with a sharp interface and no intermixing, the maximum diameter of the bottom coal must be 2.3 times as large as the top sand. Cleasby and Sejkora[30] confirmed the validity of the size ratio at the interface as the most important factor in determining the extent of intermixing; their results are similar to those derived above.

The above theory underlies Baumann's recommendation in Chapter 12 of selecting the bottom of the coal layer ($d_{90\%}$ coal) to be 3 times the top of the sand layer ($d_{10\%}$ sand) for partial intermixing. In addition, the bottom of the coal ($d_{90\%}$ coal) and the bottom of the sand ($d_{90\%}$ sand) should be sized so that they fluidize simultaneously. This recommendation for intermixing evolves from recent work[30] on interface effects of dual media. Camp[1] and Ives[31] in publications of the 1960s discouraged intermixing; however, the design of a mixed interface necessitates use of substantially finer sand, and consequently produces a better quality filtrate.

For a sharp interface, ($d_{90\%}$ coal/$d_{10\%}$ sand) < 2.5

For substantial intermixing, ($d_{90\%}$ coal/$d_{10\%}$ sand) > 4.0

Recommended design for partial intermixing, ($d_{90\%}$ coal/$d_{10\%}$ sand) $\cong 3.0$

For a log probability distribution if the uniformity coefficient $(d_{60\%}/d_{10\%})$ is 1.5 then $(d_{90\%}/d_{10\%})$ equals 2.0 and these factors may be used for the selection of the appropriate media size.

The alternative approach to selection of media is the use of continental European designs.[22] These use single monosized sand media from 0.9 to 1.2 mm and deeper beds. The filters are used with media-retaining strainers with 0.35-mm or 0.70-mm slots. Simultaneous air and water wash keep the bed essentially unstratified.

Media Design—Depth

The determination of filter depth[8,19] can be related to current filtration theory.[3,4,6] The capture of particles involves transport and attachment. The forces of gravity, inertia and hydrodynamic effects transport particles to the collector with interception as the boundary condition for collection. The attachment is effected by double-layer interactions and London-van der Waals forces.

The particle capture by an individual collector[3,4,6] is given by

$$-\frac{\partial c}{\partial l} = \lambda c = \frac{\alpha\beta\ (1 - \epsilon)\eta}{d}\ c \qquad (3)$$

in which
- c = mass concentration of solids
- l = bed depth
- λ = $[\alpha\beta\ (1 - \epsilon)\eta]/d$ = filter coefficient
- α = collision efficiency factor
- β = form factor equal to 3/2 for spherical collectors
- η = single collector efficiency defined as ratio of particles striking collector to total particle flux to collector
- ϵ = porosity
- d = media size

The initial first-order expression was proposed by Iwasaki,[32] and the filter coefficient was determined in terms of media characteristics by Yao et al.[3] Spielman and Fitzpatrick[4] developed a theory for the single collector efficiency η for nondiffusive particles given by

$$\frac{\eta}{\eta_I}\ =\ f(N_{Ad},\ N_{Gr}) \qquad (4)$$

in which
- N_{Ad} and N_{Gr} = dimensionless adhesion and gravity groups
- η_I = $1.5\ BR_d^2$ = single collector efficiency for interception
- R_d = ratio of particle to media diameters
- B = dimensionless flow model parameter

Integrating Equation 3,

$$\ln \frac{c}{c_0} = -\alpha \beta (1 - \epsilon)\eta \frac{1}{d} \tag{5}$$

Equation 5 shows that removal efficiency is a function of $(1 - \epsilon)(1/d)$. It can be shown that $(1 - \epsilon)(1/d)$ expresses the total surface area of the grains. Consider a filter media of depth 1 and unit area. Let there be N media particles of size d and sphericity ψ per unit volume. Equating volumes of solids,

$$N\left(\frac{\pi(\psi d)^3}{6}\right) = 1(1 - \epsilon) \tag{6}$$

Total surface area of the media,

$$\Sigma A = N \pi (\psi d)^2 \tag{7}$$

From Equations 6 and 7,

$$\Sigma A = \frac{6(1 - \epsilon)}{\psi} \left(\frac{1}{d}\right) \tag{8}$$

These results indicate that provision of equal media surface area among different filters would provide approximately equivalent efficiencies.[8,19] Equation 8 may be used to design filter depths of different media as done by Kawamura.[19] For typical high-rate filters $\epsilon = 0.45$, $\psi = 0.8$ and $(1/d) = 680$, thus $\Sigma A \cong 2800$ m^2. Since the effective size is normally specified, $(1/d) = 680$ corresponds to $(1/d_{10\%}) = 950$ where $d_{10\%} =$ effective size. Figure 2 has been developed for $\Sigma A = 2800$ m^2 and can be used to estimate depths of various combinations of dual media. The lines in the figure determine the weighted average effective size d_e for duel media given by

$$\frac{(d_{10\%c} \, l_c + d_{10\%s} \, l_s)}{(l_c + l_s)} = d_e \tag{9}$$

in which the suffixes c and s denote coal and sand.

Alternatively, Equation 8 may be summed over the entire depth of the media by determining the parameters for layers. However, this degree of accuracy is unwarranted because other factors excluding surface area also influence filter performance.

Baumann in Chapter 12 recommends the following depths for dual media filters:

coal: 18-24 in. (0.46 to 0.61 m)
sand: 15-18 in. (0.38 to 0.46 m)

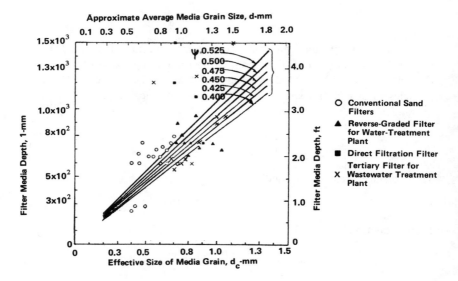

Figure 2.   Diagram for media depth design. (After Reference 19.)

Optimum Backwashing

The design of filter media is based on production of adequate filtrate quality and acceptable head loss as described.   It is also determined by backwashing constraints.

Amirtharajah[24,25] summarized the rationale indicating that significant collisional interactions do not occur in the fluidized state during backwash. Therefore, the principal mode of cleaning is by hydrodynamic shear, τ. Amirtharajah[24,25] showed theoretically (and supported with experimental evidence) that maximum hydrodynamic shear occurred at expanded porosities ϵ about 0.70.   Camp and Stein's equation,[33] applied to a backwashed filter, gives

$$\left(\frac{dV'}{dl} \right) = \left[\frac{gV'}{\nu} \left(\frac{dh}{dz} \right) \right]^{1/2} \tag{10}$$

in which (dV'/dl) =   velocity gradient within the pores

g =   acceleration due to gravity

ν =   kinematic viscosity

(dh/dz) =   head loss gradient

The Fair and Hatch[34] relation equating head loss across a fluidized bed to the buoyant weight of particles is, in differential form,

$$dh \; \rho_f \; g \; = \; dz(\rho_s - \rho_f) \; g \; (1 - \epsilon) \tag{11}$$

in which ρ_s, ρ_f = solid and fluid densities. The modified Richardson and Zaki equation[35] used by Amirtharajah and Cleasby[36] for predicting the expansion of a fluidized bed is,

$$V' = k\epsilon^{(n-1)} \tag{12}$$

in which k = constant and n = expansion coefficient.
Hydrodynamic shear is defined as

$$\tau \; = \; \mu \left(\frac{dV'}{dl} \right) \tag{13}$$

From Equations 10 and 13,

$$\tau \; = \left[\mu \; \frac{gV'}{\nu} \left(\frac{dh}{dz} \right) \right]^{1/2} \tag{14}$$

Substituting for V' and (dh/dz) from Equations 10 and 11 in Equation 14 and collecting terms,

$$\tau \; = \; K[\epsilon^{(n-1)} - \epsilon^n]^{1/2} \tag{15}$$

in which K = constant for a system. Differentiating and equating to zero indicates that the maximum hydrodynamic shear τ_{max} occurs at

$$\epsilon \; = \; \frac{(n-1)}{n} \tag{16}$$

For sand systems that have typical values of n = 3.3, the maximum shear and hence optimum cleaning occurs at expanded porosities of 0.70 to 0.72. For graded sands these expanded porosities occur at expansions of 30-40% in the top layers wherein most of the particles are deposited during filtration. The maximum shear occurs on a flat curve, indicating that washing at porosities different from the theoretical optimum does not cause a major decrease in cleaning efficiency. The significant result for practice is that lack of abrasion indicates that water backwash is an intrinsically weak cleaning process and auxiliaries such as surface wash or air scour, which create collisions in the media, are essential for effective cleaning.

Expansion During Backwash

A model to predict the expansion of graded sand and coal systems is required for filter media design. The procedures in standard textbooks, such as those by Fair *et al.*,[37] and Clark *et al.*,[38] for calculating expansions are based on the Carmen-Kozeny equation. This equation is founded on laminar flow conditions and has limited applicability under the transitional flow conditions that prevail during backwashing. These texts also use a Richardson and Zaki[35] type equation (Equation 12) with n = 4.5 to 5.0. Actual values of n for sand are approximately 3.0-3.5. Therefore, the procedures in the above texts[37,38] overestimate the expansions to a significant degree.

Amirtharajah and Cleasby[36] developed a method for predicting the expansion of graded sands and compared it with experimental results for several sands. Subsequent research[39] has indicated possible theoretical weaknesses in one of the equations of the model. However, the self-correcting nature of the calculations gives excellent expansion estimates. An advantage of the model is that use is made of the 60% finer size (effective size x uniformity coefficient) to represent the graded bed. This size is almost always available to designers of filters. The calculation procedure is summarized as follows.

Using Equation 1 given previously, calculate the minimum fluidization velocity V_{mf} for the $d_{60\%}$ sand size.

The dimensionless Reynolds number $Re_{mf} = (\rho_f V_{mf} d_{60\%})/\mu$, corresponding to a minimum fluidization velocity and the 60% finer sand size, is calculated. If Re_{mf} is greater than 10, a multiplying correction factor K_R must be applied to the above value of V_{mf} given by

$$K_R = 1.775 \; Re_{mf}^{-0.272} \tag{17}$$

The unhindered settling velocity V_s of the hypothetical average particle represented by $d_{60\%}$ is calculated from

$$V_s = 8.45 \; V_{mf} \tag{18}$$

The Reynolds number for this particle based on V_s is calculated as $Re_o = (\rho_f V_s d_{60\%})/\mu = 8.45 \; Re_{mf}$ for use in calculating the expansion coefficient n from the equation

$$n = 4.45 \; Re_o^{-0.1} \tag{19}$$

Using the values of V_{mf} and ϵ_{mf} (porosity of fixed bed) the constant k in the following equation is determined.

$$V = k \, \epsilon^n \qquad (20)$$

Equation 20 is identical to Equation 12 except that it is now expressed as the superficial velocity, V, which equals interstitial velocity multiplied by porosity, $V'\epsilon$.

The expanded bed depth l_e is calculated from the static bed l_{mf} using,

$$l_e \, (1 - \epsilon) = l_{mf} \, (1 - \epsilon_{mf}) \qquad (21)$$

The above sequence of calculations could be used to determine the expansion of a graded sand at any flow rate and temperature given the effective size and uniformity coefficient of the graded sand and its fixed bed porosity, which can be easily measured by water displacement.

There is currently (1978) no satisfactory mathematical model for graded coal. However, to assist designers, an approximate estimate of the expansion of graded coal may be obtained from Figure 3. The curves plotted are experimental results for three different commercially

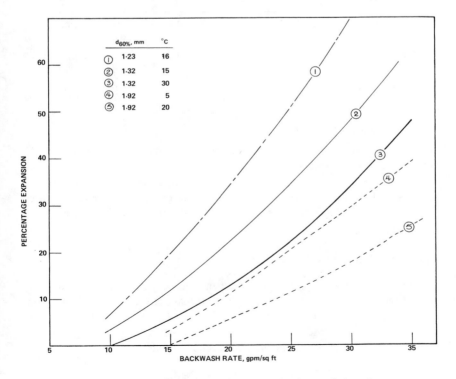

Figure 3. Expansion–flow rate characteristic for graded coal.

available anthracite coals used in filters. The expansion flow rate characteristics are shown for different temperatures to indicate the importance
of assessing expansions for the summer or warmest temperatures. If the
graded coal used in design is obtained from the particular supplier listed,
then the expansion obtained from Figure 3 is accurate if no significant
changes in production have occurred. If not, the values are approximate
and give a rough indication of the expected expansions for design
purposes.

Table I indicates the properties of the coals used for Figure 3 and
their sources of supply. Expansion of dual media filters is obtained by

Table I. Properties of Graded Coal

Curve No.	Name of Coal	Coal Properties			
		Effective Size $d_{10\%}$, mm	Uniformity Coefficient	$d_{60\%}$, mm	Supplier
1	Philterkol	0.90	1.37	1.23	Reading Anthracite Coal Co., Pottsville, PA
2,3	Philterkol No. 1	1.02	1.29	1.32	Reading Anthracite Coal Co., Pottsville, PA
4,5	Anthracite No. 2	1.24	1.55	1.92	Permutit Co., Inc., Paramus, NJ

adding the expansion of the separate media,[3,9] even under conditions
where intermixing occurs.

For selecting a backwash rate for 10% expansion of an unsized
media (sand, coal or garnet), Table V in Chapter 12 by Baumann is used.
Calculation procedures for selecting compatible media are shown by
Baumann.

Optimizing Filter Box Dimensions

The total cost of a filter box is dependent on (a) filter walls and
gallery and (b) the filter piping and appurtenances. Cleasby[5] shows how
a minimum cost design is obtained by differentiating the cost function in
terms of the length/width ratio (L/W) as a variable.

For a single row of filters on one side of a gallery,

$$\left(\frac{L}{W}\right)_{opt} = \frac{(N_f \, C_g + 2N_f \, C_w)}{(N_f + 1) \, C_w} \tag{22}$$

in which $(L/W)_{opt}$ =   optimum length to width ratio

N_f =   number of filters

C_g =   cost per unit length of gallery, which includes piping, floor slab and roof slab (excludes valves and appurtenances that are not a function of gallery length)

C_w =   cost per unit length of filter wall

For a two-row filter system with a central gallery,

$$\left(\frac{L}{W}\right) = \frac{(N_f\ C_g + 4N_f\ C_w)}{2(N_f + 2)C_w} \tag{23}$$

Common designs of filters are a square box. But typical cost data indicate that least-cost designs are approximated by length-to-width ratios of 3-6.

FILTER HYDRAULICS

An inseparable part of filter design is its hydraulics. This section collates and summarizes the data needed and the procedures required for design. An attempt has been made to be reasonably complete; however, individual designers would obviously have personal preferences and the literature, especially the commercial literature, would provide more specific information on particular items of equipment.

Porous Media Hydraulics

The classical head loss expression for laminar flow through porous media was formulated by Darcy as

$$V = K_p\left(\frac{h}{l}\right) \tag{24}$$

in which K_p = coefficient of permeability. The coefficient of permeability for laminar flow has been expressed by Kozeny in terms of the characteristics of the media. The expression was modified for graded media by Fair and Hatch[34] and is presented in Fair, Geyer and Okun[37] as the Kozeny equation:

$$\frac{h}{l} = k'\frac{\nu}{g}V\frac{(1-\epsilon)^2}{\epsilon^3}\left(\frac{6}{\psi}\right)^2\sum_{i=2}^{N}\frac{x_i}{d_i^2} \tag{25}$$

in which x_i = fraction of particles of diameter d_i and k' = 5.0. Equation 25 is used to estimate clean media head losses during filtration.

Equation 25 is adequate for head loss computations during the low velocities that prevail during filtration. However, the expression becomes increasingly in error at flow rates used for backwashing where calculations show that the Reynold's number of the particles lies in the transitional range. The best expression to calculate these head losses is the two-term Ergun equation,[40] which holds for all regimes of flow. With minor modifications to include graded particles and sphericity, the equation is,

$$\frac{h}{l} = 150 \, \frac{\nu}{g} \, \frac{(1-\epsilon)^2}{\epsilon^3} \, V \left(\frac{1}{\psi}\right)^2 \sum_{i=1}^{N} \frac{x_i}{d_i^2} + 1.75 \, \frac{(1-\epsilon)}{\epsilon^3} \, \frac{V^2}{g} \, \frac{1}{\psi} \sum_{i=1}^{N} \frac{1}{d_i} \quad (26)$$

Head loss computations for fluidized media during backwash are made using the integrated form of Equation 11,

$$h = l(Sg - 1)(1 - \epsilon) \quad (27)$$

in which Sg = specific gravity of solid.

Declining Rate Hydraulics

The design of hydraulic control systems for constant rate and declining rate filters are discussed by Arboleda.[41] Design methodology and layouts for minimizing use of equipment, effective use of common wall construction and even systems for eliminating use of backwash pumps and piping by washing a filter with the effluent from the others are presented.[41] A rigorous quantitative study of declining rate filtration has still to be completed, even though several plants are being designed and operated in the declining rate mode.

Reported operating data[26,41] suggest that declining rate filters should be hydraulically designed to operate from 50 to 150% of average flow. Figure 10 in Chapter 12 by Baumann shows the typical operating saw tooth type curve for a four-cell system. The literature does not sufficiently stress the fact that the available head loss, which is the height above the control weir in the clear water tank, is approximately 6-8 ft (1.83 to 2.44 m), while the actual water level fluctuations during operation are only 1-2 ft (0.30 to 0.61 m). In fact, these fluctuations are a function of: (a) the number of filters and (b) the volume of the influent conduit and its backflow characteristics. A large number of filters and/or a slow backflow response due to a large influent channel and settling tank effluent channels causes negligible water level fluctuations, approximately 4 to 6 in. (100-150 mm).

An important factor in the design of the hydraulics of declining rate filters is to provide sufficient head loss in an orifice meter so that at the high water level the flow through the clean filter is limited to the maximum design rate of flow.

Consider a declining rate filter with an orifice restricting flow. Let h = the available head (difference between control weir level and water level in filters), h_f = head loss due to clogging, h_c = head loss through clean filter, supporting gravel and underdrains, and h_o = head loss through orifice.

$$(h_f + h_c + h_o) = \text{function of flow rate} = f(Q)$$

at design flow rate Q_D (discharge per unit area)

$$h_D = (h_f + h_c + h_o)_D \qquad (28)$$

If $h_D > (h_f + h_c + h_o)_D$, then $Q > Q_D$, and when $h_D < (h_f + h_c + h_o)$, then $Q < Q_D$, which illustrates the declining filtration system. In general, h_D will lie in the range $h_{low} < h_D < h_{high}$ where h_{low} and h_{high} represent the fluctuations in water level, during flow rates that change from Q_{high} to Q_{low}. The designer can essentially choose the fluctuations in water level and then design the orifice to control the range of Q's chosen.

The above computations are easily performed by plotting head loss versus flow rate lines on logarithmic coordinates as shown in Figure 4. Since, in general, head loss is a power function of Q, these curves plot as straight lines on logarithmic coordinates. The line for h_c versus Q is first calculated. The designer chooses the h_{low} value and the corresponding maximum filtration rate Q_{high}, fixing point A and the orifice loss h_o at Q_{high}. The orifice dimensions are thus fixed. The line ($h_c + h_o$) vs Q can be developed from the orifice characteristics. The end of the filter run is fixed by the designer's choice of two of the three variables (Q_{low}, h_{high} or h_f max) which fixes point B.

The dotted curve from A to B represents the range of water level changes during the course of a run, 2-3 ft (0.61-0.91 m) and the ordinate between the ($h_c + h_o$) line and the dotted curve A-B represents the head loss through the dirty filter due to clogging [approximately 4-8 ft (1.22-2.44 m)]. The dotted line AB represents smoothly declining flow rate in a single declining rate filter from start to end of a run, and the boundary conditions represented by points A and B are actual limiting conditions even in a bank of filters. An important difference can be noted in variable declining rate filters as compared to influent flow splitting or constant rate filters: for declining rate filters the highest flow rate occurs at the lowest water level with the filter clean and vice versa.

The actual head and flow characteristics during declining rate are given in Figure 10 of Chapter 12. The actual head loss vs flow rate characteristic

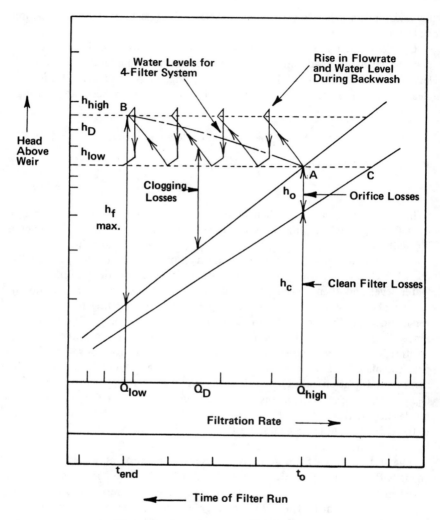

Figure 4. Declining rate hydraulics.

is a sawtooth-like curve with the number of teeth being equal to the number of filters. This is also shown in Figure 4 for a bank of four filters. Figure 4 clearly illustrates the options available to the designer. The point C corresponds to the very high initial filtration rate if no orifice is provided, which leads to degradation of effluent quality. However, by designing an effluent channel with weir control h_o units above the previous design, the h_c line is increased by an ordinate distance h_o; hence no orifice need be provided.

Figure 5 shows two typical declining rate filter arrangements, one with orifice control and an influent pipe and another with an influent channel and without orifice control. The second filter is designed for backwash from the other filters, which eliminates the backwash pump and piping systems.

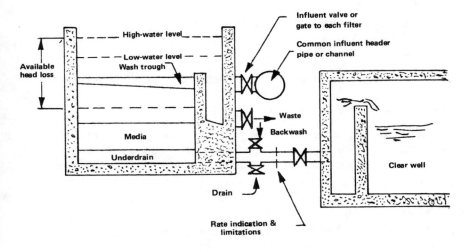

A. Orifice controlled.

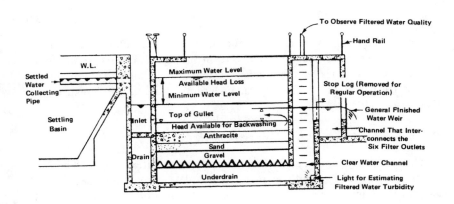

B. Backwashed with filtrate from other filters.

Figure 5.   Declining rate filters.

Wash Water Trough Hydraulics

These designs are illustrated in several texts.[5,37] For a rectangular horizontal level trough (*i.e.*, no slope) with a critical discharge at the downstream end, the maximum depth of flow D at the upstream end is given by,

$$D = 1.73 \left(\frac{Q^2}{g\,W^2} \right)^{1/3} \tag{29}$$

in which   W = width of rectangular channel in ft
         Q = cfs
         g = ft/sec^2

For channels of other cross sections, similar expressions are available.[5,37]

Underdrain Hydraulics

Details of underdrain design are given by Cleasby.[5] Most old filters are designed with a manifold and lateral system and a gravel layer to support the filter media. Two important points need to be noted in designing these systems during backwash: (a) the losses through the orifices are kept comparatively high—approximately 15 ft of water (4.57 m) to maintain uniform discharge during backwashing through all the orifices and (b) head loss recovery occurs along both the manifold and laterals due to the decreasing flow rate. Thus the pressure is higher at the downstream ends of the conduits (*i.e.*, highest pressures are at the furthermost orifice from the inlet end during backwash). The discharge equation used for the orifices is

$$Q = C_0\,A\sqrt{2\,gh} \tag{30}$$

Whenever gravel layers are used for supporting the filter media, the symmetrical (reverse-graded) system shown in Figure 6 is recommended. Experience has indicated that such a system is superior to the common asymmetrical system. Some typical underdrain systems are shown in Figures 6 and 7.

Hydraulics for Conduits and Appurtenances

This section summarizes an important facet of filter design. The material is treated in depth in many standard texts on fluid mechanics[42,43] and water engineering.[44] The purpose of this section is to provide sufficient information and computational methodology for the particular calculations required in designing filters.

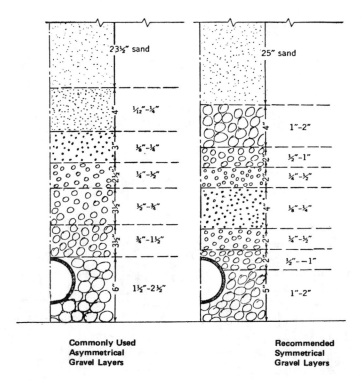

Figure 6. Supporting gravel layers for filter media.

Open Channel Designs

The most efficient trapezoidal or rectangular channel section has the hydraulic radius R (area of flow/wetter perimeter) equal to D/2 where D is the depth of channel. For a rectangular channel of width W, the result simplifies to

$$W = 2D \tag{31}$$

For a V-shaped channel, the section giving the highest discharge is with the apex angle equal to 90°. Since the wetted perimeter is a minimum for the most efficient section, it also leads to the least-cost design. Practical limitations may obviate the use of these sections.

The discharge for uniform flow with normal depth can be obtained from Manning's equation and the area of cross section, A as,

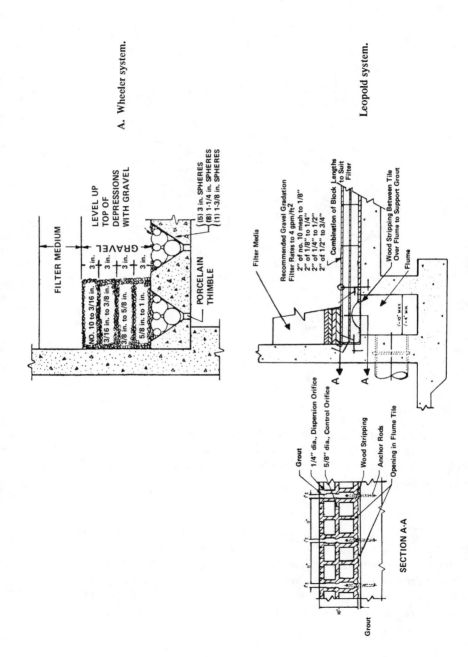

A. Wheeler system.

Leopold system.

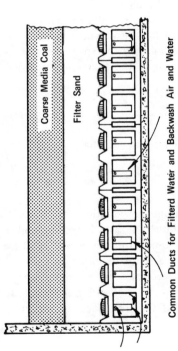

C. Walker system—
combination
air-water wash.

Coarse Media Coal

Filter Sand

Common Ducts for Filterd Water and Backwash Air and Water

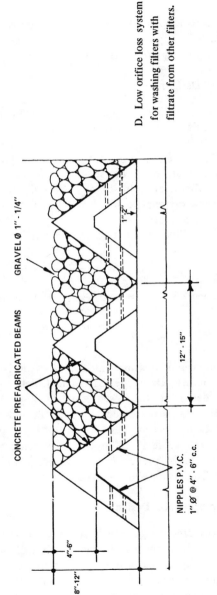

D. Low orifice loss system
for washing filters with
filtrate from other filters.

CONCRETE PREFABRICATED BEAMS

GRAVEL Ø 1″ - 1/4″

NIPPLES P.V.C.
1″ Ø @ 4″ - 6″ c.c.

12″ - 15″

1″-2″

4″-6″

8″-12″

Figure 7. Filter underdrain systems.

$$Q = A \ V = A \ \frac{1.486}{n_c} \ R^{2/3} \ S^{1/2} \qquad (32)$$

in which n_c = coefficient of roughness = 0.013 for concrete channels and S = slope. In metric units, the expression holds with the same values of n_c if 1.486 in the numerator is replaced by 1.0.

Pipe Designs

Most designs for pipe flow problems are based on the Hazen-Williams formula. Jain[45] *et al.* have recently reviewed the weaknesses of the Hazen-Williams equation and pointed out the possibility of errors up to ± 50% in estimation of frictional resistance. The chief weakness of the Hazen-Williams formulation is that the coefficient C is independent of pipe diameter, velocity of flow and viscosity, whereas in fact it must depend on relative roughness and Reynolds number to be truly representative as a friction coefficient. Jain *et al.*[45] propose the following modified Hazen-Williams equation for pipe friction.

$$V = \frac{3.83 \ C_R \ d^{0.6575} \ (g \ S)^{0.5525}}{\nu^{0.105}} \qquad (33)$$

in which C_R =   coefficient   =   1 for smooth pipes and <1 for non-smooth pipes

   d   =   diameter of pipe

   S   =   slope

The explicit relationship has been derived using the equations of continuity, Darcy-Weisbach and Colebrook-White[46,47] and is recommended for design at the present time (1978). The nomograph developed by these authors[45] is presented in Figure 8, and the tables for determining C_R values for use in filter design with cast iron, steel or concrete pipes are given in Table II. The effects of temperature are small and are neglected. The table is prepared for different values of k_s, the roughness height in the Colebrook-White equation.[46,47]

Appurtenance Design

In filter design, the head losses in the appurtenances such as bends, tees, valves and venturi meters control the hydraulics of the conduits rather than friction losses along the length of the conduit. In order to facilitate these computations, Table III gives head loss factors for typical appurtenances used in filters. The data have been collated from several sources. Head losses are given in terms of the velocity head ($V^2/2g$). Use of the data with typical calculation sheets is illustrated in the design example.

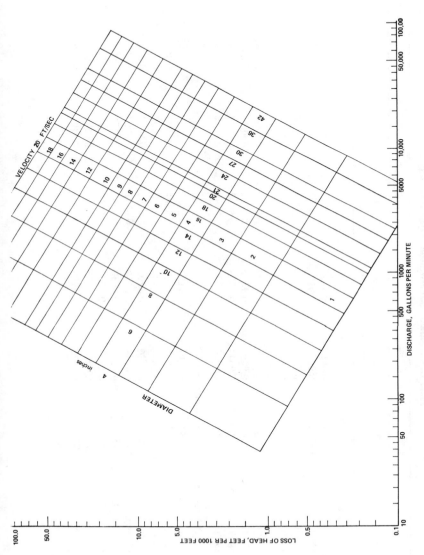

Figure 8. Modified Hazen-Williams nomograph for water (20°C) for smooth pipes ($C_R = 1$). [For $C_R < 1$: multiply given Q or V by ($1/C_R$) to find S; or multiply found Q or V by ($C_R/1$) for given S.]

Table II. Modified Hazen-Williams Coefficient, C_R

Nominal Diameter, in in. (mm)	Velocity, in ft/sec (m/sec)						
	1.0 (0.3)	3.0 (0.9)	3.9 (1.2)	5.9 (1.8)	7.9 (2.4)	9.8 (3.0)	19.7 (6.0)
1. Cast-Iron (new coated) Pipes, k_s = 0.006 in. (0.00015 m)							
4 (100)	0.9524	0.9089	0.8922	0.8659	0.8459	0.8299	0.7786
8 (200)	0.9760	0.9291	0.9117	0.8810	0.8641	0.8477	0.7953
12 (300)	0.9865	0.9379	0.9202	0.8927	0.8720	0.8554	0.8025
16 (400)	0.9927	0.9428	0.9249	0.8972	0.8763	0.8596	0.8064
20 (500)	0.9967	0.9459	0.9279	0.9000	0.8790	0.8623	0.8089
24 (600)	0.9950	0.9480	0.9299	0.9019	0.8809	0.8641	0.8106
28 (700)	1.0000	0.9495	0.9313	0.9032	0.8821	0.8653	0.8117
32 (800)	C_R=1.0	0.9514	0.9323	0.9042	0.8837	0.8662	0.8125
36 (900)	C_R=1.0	0.9524	0.9330	0.9049	0.8842	0.8668	0.8131
2. Cast-Iron (old) Pipes, k_s = 0.095 in. (0.0024 m)							
4 (100)	0.6552	0.6886	0.7170	0.5485	0.5325	0.5203	0.4841
8 (200)	0.7025	0.6311	0.6130	0.5881	0.5709	0.5578	0.5190
12 (300)	0.7261	0.6532	0.6335	0.6078	0.5900	0.5765	0.5364
16 (400)	0.6657	0.6466	0.6203	0.6203	0.6022	0.5884	0.5475
20 (500)	0.7518	0.6753	0.6559	0.6293	0.6093	0.5969	0.5554
24 (600)	0.7599	0.6826	0.6630	0.6361	0.6175	0.6034	0.5614
29 (700)	0.7664	0.6884	0.6687	0.6415	0.6227	0.6085	0.5661
32 (800)	0.7718	0.6932	0.6733	0.6459	0.6270	0.6127	0.5701
36 (900)	0.7762	0.6972	0.6772	0.6497	0.6307	0.6163	0.5734

3. Concrete (new) Pipes, k_S = 0.0014 in. (0.000035 m)

4 (100)	1.0	1.0	0.9960	0.9821	0.9694	0.9581	0.9156
8 (200)	1.0	1.0	1.0	0.9909	0.9775	0.9657	0.9223
12 (300)	1.0	1.0	1.0	0.9935	0.9798	0.9677	0.9239
16 (400)	1.0	1.0	1.0	0.9943	0.9803	0.9682	0.9241
20 (500)	1.0	1.0	1.0	0.9944	**0.9802**	0.9681	0.9237
24 (600)	1.0	1.0	1.0	0.9936	**0.9798**	0.9675	0.9231
28 (700)	1.0	1.0	1.0	0.9930	**0.9795**	0.9668	0.9223
32 (800)	1.0	1.0	1.0	0.9923	0.9785	0.9661	0.9215
36 (900)	1.0	1.0	1.0	0.9916	0.9778	0.9653	0.9207

4. Steel (new) Pipes, k_S = 0.0024 in. (0.00006 m)

4 (100)	0.9942	0.9766	0.9658	0.9466	0.9304	0.9166	0.8687
8 (200)	1.0	0.9905	0.9789	0.9587	0.9420	0.9279	0.8790
12 (300)	1.0	0.9973	0.9838	0.9631	0.9461	0.9318	0.8828
16 (400)	1.0	0.9985	**0.9861**	0.9651	0.9480	0.9336	0.8842
20 (500)	1.0	1.0	**0.9873**	0.9661	0.9488	**0.9343**	**0.8848**
24 (600)	1.0	1.0	**0.9879**	0.9665	0.9491	**0.9346**	**0.8850**
28 (700)	1.0	1.0	**0.9881**	0.9666	0.9492	**0.9346**	**0.8849**
32 (800)	1.0	1.0	**0.9881**	0.9662	0.9490	0.9344	**0.8847**
36 (900)	1.0	1.0	0.9880	0.9662	0.9487	0.9341	0.8844

Table III. Hydraulic Head Losses for Appurtenances

Appurtenance —Alphabetically	Head Loss as Multiple of $(V^2/2g)$
1. Butterfly Valves	
Fully open	0.3
Angle closed, $\theta = 10°$	0.46
$\theta = 20°$	1.38
$\theta = 30°$	3.6
$\theta = 40°$	10
$\theta = 50°$	31
$\theta = 60°$	94
2. Check (Reflux) Valves	
Ball Type (fully open)	2.5-3.5
Horizontal Lift Type	8-12
Swing Check	0.6-2.3
Swing Check (fully open)	2.5
3. Contraction—Sudden	
4:1 (in terms of velocities of small end)	0.42
2:1	0.33
4:3	0.19
also see Reducers	
4. Diaphragm Valve	
Fully open	2.3
¾ open	2.6
½ open	4.3
¼ open	21.0

Appurtenance —Alphabetically	Head Loss as Multiple of $(V^2/2g)$
5. Elbow—90°	
Flanged—Regular	0.21-0.30
Flanged—Long Radius	0.18-0.20
Intersection of two cylinders (welded pipe—not rounded)	1.25-1.8
Screwed—Short Radius	0.9
Screwed—Medium Radius	0.75
Screwed—Long Radius	0.60
6. Elbow—45°	
Flanged—Regular	0.20-0.30
Flanged—Long Radius	0.18-0.20
Screwed—Regular	0.30-0.42
7. Enlargement—Sudden	
1:4 (in terms of velocities of small end)	0.92
1:2	0.56
3:4	0.19
also see Increasers	
8. Entrance Losses	
Bell mouthed	0.04
Pipe flush with tank	0.5
Pipe projecting into tank (Borda Entrance)	0.83-1.0
Slightly rounded	0.23
Strainer and foot valve	2.50

9. *Gate Valves*

Open	0.19
¼ closed	1.15
½ closed	5.6
¾ closed	24.0
also see Sluice Gates	

10. *Increasers*

$0.25\ (V_1^2/2g - V_2^2/2g)$
where V_1=velocity at small end

11. *Miter Bends*

Deflection angle, θ

5°	0.016-0.024
10°	0.034-0.044
15°	0.042-0.062
22.5°	0.066-0.154
30°	0.130-0.165
45°	0.236-0.320
60°	0.471-0.684
90	1.129-1.265

12. *Obstructions in Pipes* (in terms of pipe velocities) Pipe to Obstruction Area Ratio

1.1	0.21
1.4	1.15
1.6	2.40
2.0	5.55
3.0	15.0
4.0	27.3
5.0	42.0
6.0	57.0
7.0	72.5
10.0	121.0

13. *Orifice Meters* (in terms of velocities of pipe) Orifice to Pipe Diameter Ratio

0.25 (1:4)	4.8
0.33 (1:3)	2.5
0.50 (1:2)	1.0
0.67 (2:3)	0.4
0.75 (3:4)	0.24

14. *Outlet Losses*

Bell mouthed outlet	$0.1\left(\dfrac{V_1^2}{2g} - \dfrac{V_2^2}{2g}\right)$
Sharp cornered outlet	$\left(\dfrac{V_1^2}{2g} - \dfrac{V_2^2}{2g}\right)$
Pipe into still water or air (free discharge)	1.0

15. *Plug Globe or Stop Valve*

Fully open	4.0
¾ open	4.6
½ open	6.4
¼ open	780.0

16. *Reducers*

Ordinary (in terms of velocities of small end)	0.25
Bell mouthed	0.10
Standard	0.04
Bushing or coupling	0.05-2.0

17. *Return Bend (2 nos. 90°)*

Flanged–Regular	0.38
Flanged–Long Radius	0.25
Screwed	2.2

Table III. continued,

Appurtenance –Alphabetically	Head Loss as Multiple of $(V^2/2g)$	Appurtenance –Alphabetically	Head Loss as Multiple of $(V^2/2g)$
18. *Sluice Gates*		Short Tube Type–Throat-to-inlet diameter ratio	
Contraction in conduit	0.5	0.33 (1:3)	2.43
Same as conduit width without top submergence	0.2	0.50 (1:2)	0.72
Submerged port in 12 inch wall	0.8	0.67 (2:3)	0.32
		0.75 (3:4)	0.24
19. *Tees*			
Standard–bifurcating	1.5-1.8		
Standard–90° turn	1.80		
Standard–run of tee	0.60		
Reducing–run of tee			
2:1 (based on velocities	0.90		
4:1 of smaller end)	0.75		
20. *Venturi Meters*			

The head loss occurs mostly in and downstream of throat, but losses shown are given *in terms of velocities at inlet ends to assist in design.*

Long Tube Type–Throat-to-inlet diameter ratio			
0.33 (1:3)	1.0 -1.2		
0.50 (1:2)	0.44-0.52		
0.67 (2:3)	0.25-0.30		
0.75 (3:4)	0.20-0.23		

DATA FOR DESIGN

The previous sections of this chapter indicate the models that are available for the design of filters at the present time (1978). It is evident that for optimum design of filters in terms of run length and quality and for selection of media and filtration rate, pilot plant tests with the particular water to be treated is the best approach for large (> 2 mgd) filtration plants. The collection of detailed data from predesign studies and pilot plant design for filtration are described by Trussell in Chapter 3. With adequate data from pilot plant operation, design parameters can be selected with reasonable confidence, and the constraints imposed by standards and typical designs need not be exactly followed in design. With decreasing data the conservativeness of the design should increase and adequate safety factors should be incorporated in the design. A typical design example is presented where pilot plant data is unavailable.

ILLUSTRATIVE DESIGN EXAMPLE

Design a set of filters for a flow of 3.0 mgd (131.4 l/sec). Let the filters be of the variable declining rate type and assume an average filtration rate of 4 gpm/ft^2 (2.72 mm/sec). Use dual media consisting of anthracite coal/silica sand, but consider the alternative of single media. The layout of the plant suggests a single row of filters with a gallery along one side. Typical temperatures are: summer = 25°C and winter = 8°C. Design:

1. The filter media.
2. Optimum overall dimensions of the filter.
3. The variable declining rate system and indicate high and low water levels in the filter.
4. Dimensions of all pipes or channels in the filter gallery. Assume velocities with typical ranges suggested in standard texts[37,38] for filter influent, filter effluent, backwash influent and effluent, and waste conduits. Show an elevation and section of the filters.
5. Head and flow requirements for backwash pumps based on designed underdrain. Specify backwash procedures.
6. Dimensions of washwater troughs.

Summarize the design.

Design of Filters (in English Units)

Media Design—Dual Media

Two alternative approaches are possible, selecting the coal size and determining the compatible sand as illustrated in Chapter 12, or selecting the sand and determining the compatible coal.

Select a silica sand (Sg = 2.65) with an effective size ($d_{10\%}$) of 0.6 mm and a uniformity coefficient of 1.4.

For partial intermixing:

$$\text{coarse coal size } (d_{90\%}) \;=\; 3 \times \text{fine sand size } (d_{10\%})$$
$$=\; 3 \times 0.6 = 1.8 \text{ mm}$$

If ($d_{60\%}/d_{10\%}$) = 1.5, then ($d_{90\%}/d_{10\%}$) = 2.0.
Therefore,

$$\text{effective coal size } (d_{10\%}) \;=\; \frac{1.8 \text{ mm}}{2.0} = 0.9 \text{ mm}$$
$$\text{and uniformity coefficient} \;=\; 1.5$$

For a uniformity coefficient of 1.5, the $d_{90\%}$ sand is 1.2 mm. Since the uniformity coefficient is 1.4, the $d_{90\%}$ sand is approximately 1.1 mm. Using Table V, Chapter 12, flow rates for 10% expansion at 25°C are:

$$d_{90\%} \text{ coal, } 1.8 \text{ mm} = 21.5 \text{ gpm/ft}^2$$
$$d_{90\%} \text{ sand, } 1.1 \text{ mm} = 18.8 \text{ gpm/ft}^2$$

Thus the selected media are compatible for backwashing, since the backwashing rates are reasonably close. Also the coarse sand may fluidize slightly before the coarse coal. Refer to Chapter 12, page 272. Try a sand depth of 15 in. and a coal depth of 18 in. Using Equation 9,

$$\frac{(0.6 \times 15 + 0.9 \times 18)}{(15 + 18)} \;=\; 0.76$$

From Figure 2, for d_e = 0.76 and ϵ = 0.475,

$$\text{Total depth of media} = 2.3 \text{ ft} = 27.6 \text{ in.}$$

The assumed depth of 33 in. is higher than the calculated depth of 27.6 in. Try sand = 12 in. and coal = 18 in.

$$\frac{(0.6 \times 12 + 0.9 \times 18)}{(12 + 18)} \;=\; 0.78$$

From Figure 2, for d_e = 0.78 and ϵ = 0.475,

Total depth of media = 2.4 ft = 28.8 in.

These depths of media are the minimum required and are adequate. Using Baumann's recommendations as a guideline too, the following media are specified:

Design Media: Sand = 0.6 mm effective size, 1.4 uniformity coefficient and 15 in. depth.

Coal = 0.9 mm effective size, 1.5 uniformity coefficient and 18 in. depth.

The designed media are plotted on logarithmic probability coordinates in Figure 9.

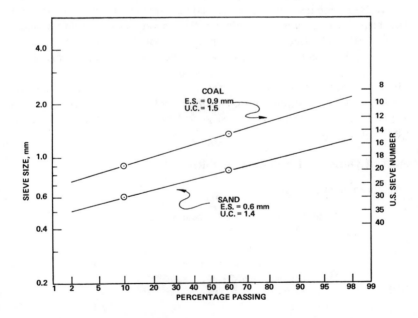

Figure 9. The designed media.

Media Design—Single Media

An alternative design is to specify a coarse monosized sand, 1.00-1.20 mm, that is 16-20 mesh U.S. sieve range. From Table V, Chapter 12, the backwash rate for this sand for fluidization and 10% expansion is,

$$\text{at } 25^\circ C = 21 \text{ gpm/ft}^2$$

This media would be used with 0.7-mm slot media retaining strainers and simultaneous air and water backwash.

From Figure 2, for d_e = 1.1 mm and ϵ = 0.425,

$$\text{depth of media required} = 3 \text{ ft}$$

With a 1-ft allowance for safety, the media depth required is 4 ft. Typical European designs[22] with 1.0 mm sand have media depths of 5 ft.

Design Media: Sand = 1.00-2.00 mm (16-20 mesh) unsized media and 48 in. depth

The decision to use dual or single media would be based on (a) competitive prices for media and underdrain equipment, (b) filter influent water characteristics; if coagulation-sedimentation is very effective and sudden increased turbidity fluctuations are not anticipated, monosized media are adequate, (c) if mudballs and backwash have been difficult in previous experience, use air scour and monosized media.

In the present design example, the dual-media filters typical of U.S. designs are used. But the above unsized sand with the deeper bed is an acceptable alternative too, and least cost designs in the future will tend to be of this type.[21]

Design of Optimum Dimensions of Filter

Required depth of filter:

Underdrain depth (*i.e.,* Leopold Type)*	10 in.
Depth of gravel in 4.2-in. layers	8 in.
Filter media – sand	15 in.
– anthracite coal	18 in.
Maximum operating depth of water above media	75 in.
Freeboard	12 in.
Total depth of filter	138 in. = 11 ft, 6 in.

A tentative estimate of the maximum operating depth, 75 in. is made at this stage. Subsequent calculations on delining rate hydraulics would finalize

---

*The writer uses a Leopold underdrain for purposes of illustration. Any recognized underdrain system with proven effectiveness may be used.

this depth. Some typical cost data (1977) for concrete construction of filters are: (a) walls of filter box–$14/ft², (b) gallery roof, gallery slab, or outside wall–$9/ft², (c) total piping costs–$170/ft. These unit cost figures need to be checked for the particular town where the plant is to be constructed.

Since a minimum of four filter units are required for effective declining rate operation, assume six filter units for the flow of 3.0 mgd; the filters are 11.5 ft deep. Assuming a gallery width of 10 ft,

$$
\begin{aligned}
C_g &= \text{cost/ft of gallery} = \text{(cost of gallery roof + cost of gallery slab}\\
&\qquad\qquad\qquad\qquad\qquad\; \text{+ cost of outside wall + cost of piping)}\\
&\qquad\qquad\qquad\qquad\qquad\; \text{per ft of gallery}\\
&= \$9 \times 10 + \$9 \times 10 + \$9 \times 11.5 + \$170 = \$453.50/\text{ft of gallery}\\
C_w &= \text{cost per ft of wall} = \$14 \times 11.5 = \$161/\text{ft}.
\end{aligned}
$$

Using Equation 22,

$$
\begin{aligned}
\left(\frac{L}{W}\right)_{opt} &= \frac{(N_f\, C_g + 2N_f\, C_w)}{(N_f + 1)C_w}\\[2mm]
&= \frac{(6 \times 453.50 + 2 \times 6 \times 161)}{(6 + 1)161} = 4.13
\end{aligned}
$$

The filter units are designed so that the design flow is met with one filter out of service (*i.e.,* 5 filters) to allow for backwash time and possible emergency repairs.

$$
\begin{aligned}
\text{Design flow} = 3 \text{ mgd} &= 2083.3 \text{ gpm for 24 hr operation}\\
\text{Total filter surface}&= \frac{2083.3 \text{ gpm}}{4 \text{ gpm/ft}^2} = 520.8 \text{ ft}^2\\
\text{area required}&\\
\text{Surface area of each filter} &= \frac{520.8}{5} = 104.2 \text{ ft}^2\\
\text{Therefore } (L_{opt})(W_{opt}) &= 104.2\\
(4.13 W_{opt})(W_{opt}) &= 104.2\\
W_{opt} &= \left(\frac{104.2}{4.13}\right)^{1/2} = 5.02 \text{ ft}
\end{aligned}
$$

Use design width = 5 ft and design length = 20 ft.

The area provided (100 ft² per filter) is slightly lower than that required, but the design is satisfactory since there is an extra filter.

Design Dimensions of Each of Six Filters

Length = 20 ft, width = 5 ft, depth = 11 ft 6 in.

With a width of 2 ft 3 in. for the gullet into which the wash water troughs discharge, the internal dimensions of the filter box are 22 ft 3 in. long by 5 ft wide.

Variable Declining Rate Hydraulics

Head Losses During Filtration

Head losses through the media, gravel and underdrain system need to be calculated to define the variable declining rate hydraulics.

The calculations for head losses through sand, coal and gravel are shown in Table IV based on the media shown in Figure 9. The gravel

Table IV. Calculations for Head Losses Through Clean Media

Between U.S. Std. Sieves				Geometric Mean Size ($\sqrt{d_1 d_2}$) (d_i mm)	Fraction Between Sieves (x_i)	$\dfrac{x_i}{d_i^2}$ (mm^{-2})
Passing		Retained				
No.	(mm)	No.	(mm)			
1. Sand						
30	(0.595)	40	(0.420)	0.50	0.10	0.400
25	(0.707)	30	(0.595)	0.65	0.20	0.473
20	(0.841)	25	(0.707)	0.77	0.30	0.506
18	(1.00)	20	(0.841)	0.92	0.26	0.307
14	(1.41)	18	(1.00)	1.19	0.14	0.099
					1.00	1.785
2. Coal						
20	(0.841)	25	(0.707)	0.77	0.07	0.118
18	(1.00)	20	(0.841)	0.92	0.12	0.142
16	(1.19)	18	(1.00)	1.09	0.23	0.194
14	(1.41)	16	(1.19)	1.30	0.25	0.458
12	(1.68)	14	(1.41)	1.54	0.19	0.080
10	(2.00)	12	(1.68)	1.83	0.10	0.030
8	(2.38)	10	(2.00)	2.18	0.04	0.008
					1.00	1.030
3. Gravel						
1/8 in.	(3.18)	10	(2.00)	2.52	1.00	0.157
1/4 in.	(6.35)	1/8 in.	(3.18)	4.49	1.00	0.50
1/2 in.	(12.7)	1/4 in.	(6.35)	8.98	1.00	0.012
3/4 in.	(19.1)	1/2 in.	(12.7)	15.57	1.00	0.004
						0.223

layers are those recommended by the manufacturer of the underdrain system. The average filtration rates are:

When six filters are operating $= \dfrac{2083.3}{6 \times 100} = 3.47$ gpm/ft^2

When five filters are operating $= \dfrac{2083.3}{5 \times 100} = 4.17$ gpm/ft^2

Let us assume declining rate operation ranges from 2 to 6 gpm/ft$_2$ (i.e., 1.36 to 4.07 mm/sec). Using Equation 25,

$$\frac{h}{l} = k' \frac{\nu}{g} V \frac{(1-\epsilon)^2}{\epsilon^2} \left(\frac{6}{\psi}\right)^2 \Sigma \frac{x_i}{d_i^2}$$

For sand, $\epsilon = 0.40$, $\psi = 0.8$, at 25°C $\nu = 0.90$ centistokes or mm^2/sec, at 2 gpm/ft^2 (1.36 mm/sec), V = 1.36 mm/sec, from Table IV,

$$\Sigma \frac{x_i}{d_i^2} = 1.785$$

Therefore

$$\frac{h}{l} = 5.0 \times \frac{0.90}{9810} \times 1.36 \times \frac{(1 - 0.40)^2}{0.40^3} \times \left(\frac{6}{0.80}\right)^2 \times 1.785 = 0.352$$

at 6 gpm/ft^2, (h/l) = 1.056. Similarly, at 8°C, $\nu = 1.39$ centistokes, and (h/l) = 0.544 at 2 gpm/ft^2 and 1.631 at 6 gpm/ft^2. These calculations show that the highest head losses occur at the lowest temperatures (winter conditions).

For design, head losses h through 15 in. of clean sand are:

$$h = 0.544 \times \frac{15}{12} \text{ ft} = 0.68 \text{ ft at 2 gpm/ft}^2 \text{ and}$$
$$= 2.04 \text{ ft at 6 gpm/ft}^2$$

For coal, similar calculations with $\epsilon = 0.50$ and $\psi = 0.70$ give head losses for 18 in. of clean coal at 8°C of:

$$h = 0.146 \times \frac{18}{12} \text{ ft} = 0.22 \text{ ft at 2 gpm/ft}^2 \text{ and}$$
$$= 0.66 \text{ ft at 6 gpm/ft}^2$$

For gravel, similar calculations with $\epsilon = 0.40$, $\psi = 0.8$ give head losses for the four 2-in. layers at 8°C of:

$$h = 0.068 \times \frac{2}{12} \text{ ft} = 0.01 \text{ ft at 2 gpm/ft}^2 \text{ and}$$

$$= 0.03 \text{ ft at 6 gpm/ft}^2$$

For underdrains, assume Leopold-type underdrain bottom with clay blocks of overall dimensions 2 ft x 1 ft. Each square foot of the block has 18 1/4-in. diam orifices in the surface and 2 5/8-in. diam control orifices in middle web.

At a filtration rate of 6 gpm/ft^2,

$$Q_N = \frac{6}{60 \times 7.48} = 0.0134 \text{ ft/sec or cfs/ft}^2$$

Flow rate through each 1/4-in. diam orifice,

$$Q = \frac{0.0134}{18} = 7.44 \times 10^{-4} \text{ cfs}$$

Area of orifice,

$$A = \frac{\pi}{4} (1/4)^2 \times \frac{1}{144} = 3.41 \times 10^{-4} \text{ ft}^2$$

Using Equation 30,

$$Q = C_o A \sqrt{2gh}$$
$$7.44 \times 10^{-4} = 0.6 \times 3.41 \times 10^{-4} \sqrt{64.4h}$$

Therefore, $h = 0.21 \text{ ft}$

Similarly for each 5/8-in. orifice

$$Q = 6.70 \times 10^{-3} \text{ cfs,}$$
$$A = 2.13 \times 10^{-3} \text{ ft}^2$$

Therefore, $h = 0.43 \text{ ft}$

At a filtration rate of 2 gpm/ft^2, the corresponding head losses are:

for 1/4-in. diam orifice $h = 0.02 \text{ ft}$
for 5/8-in. diam orifice $h = 0.05 \text{ ft}$

During filtration the conduit losses will be negligible since these are designed to carry the backwash flow which is several times the filtration flow.

Total losses through clean media and underdrains are:

at 6 gpm/ft^2 $h_c = 2.04 + 0.66 + 0.03 + 0.21 + 0.43 = 3.37 \text{ ft}$
at 2 gpm/ft^2 $h_c = 0.68 + 0.22 + 0.01 + 0.02 + 0.05 = 0.98 \text{ ft}$

The computations show the head recovery in the system that occurs during declining rate operation that extends the filter run. The head recovery from the clogged filter at declining rates is much more than the above values.

The calculations for each flow rate of 6 gpm/ft² and 2 gpm/ft² need not be made as above. The computation for head losses through the clean filter are easily made at any flow rate once the calculations are made for one particular rate, if the proportionality between head loss h and flow rate Q is known. For example, for losses through media, $h \propto Q$ and for losses through orifices, $h \propto Q^2$. Using the values of h for 6 gpm/ft² the head losses for 4 gpm/ft² are easily computed as,   at 4 gpm/ft²

$$h_c = (2.04 + 0.66 + 0.03) \times \frac{4}{6}$$
$$+ (0.21 + 0.43) \times \left(\frac{4}{6}\right)^2 = 2.10 \text{ ft.}$$

The h_c vs Q curve is plotted in Figure 10 on logarithmic coordinates. The ordinate h represents the height of the water level above the effluent control weir in the clear water tank. The designer can essentially set the operating water levels, and design could either include or exclude an orifice on the effluent line to control the starting flow rate. The present design includes an orifice to illustrate the calculations.

Assume that the h_{low} value is 4.0 ft and the h_{high} value is 6.0 ft, hence the water level fluctuations during operation from 6 gpm/ft² to 2 gpm/ft² is 2.0 ft. For illustration of design, the points are shown as A and B and connected by a dotted line. For a filter run there are six oscillations between these operating levels since there are six filters.

From Figure 10 it is seen that the filter effluent pipe and orifice on each filter should be designed to dissipate a head loss of (4.0 - 3.37) = 0.63 ft at a flow rate of 6 gpm/ft₂. The design of the orifice is presented with the design of the effluent pipe from the filters in Table VIII, pp. 720-721.

Declining Rate Operation

Assume that a typical filter run is 24 hr, that all filters have identical filtration characteristics and that the filtration rate uniformly declines from 6 gpm/ft² to 2.0 gpm/ft². Hence every filter is washed at a fixed time of day. Two limiting conditions arise for a particular filter: (a) just after a filter is washed and (b) just before a filter is washed. In designing and operating the declining rate filters it is best to have the cleanest and dirtiest filters adjacent to each other to minimize spatial variation of head

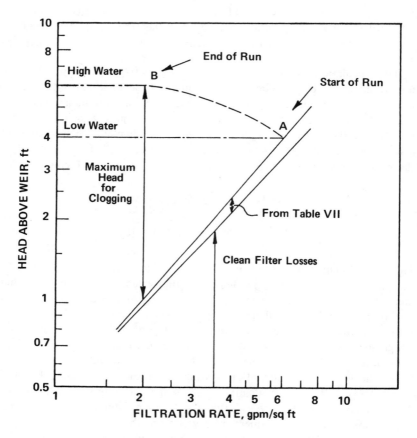

Figure 10. Design of declining rate system.

along the influent conduit. Table V, a two-way table, illustrates the flow rates in the filters at particular times.

Similar flow rate tables can be prepared for different filter runs (for example, if the run length is 36 hr instead of 24 hr) as long as the run time is divided by the number of filters. In Table V it is assumed that a filter to be washed is taken out of service about 20 min before the hour. During the washing period of Filter Number 1, the filtration rate in the other filters increases to accommodate the flow rate of 2 gpm/ft² lost from Filter Number 1, with a minor increase in operating water level. Assuming increases in flow rate proportional to existing flow rates, the flow rates are as given in the last row of Table V.

Table V. Declining Filtration Rates

Time (hr)	Filter Number						Mean Overall
	1	2	3	4	5	6	
	Filtrate Rate (gpm/ft^2)						
0 (after washing 1)	6.0	2.5	3.2	3.9	4.6	5.3	4.25
4	5.3	2.0-6.0	2.5	3.2	3.9	4.6	
8	4.6	5.3	2.0-6.0	2.5	3.2	3.9	
12	3.9	4.6	5.3	2.0-6.0	2.5	3.2	
16	3.2	3.9	4.6	5.3	2.0-6.0	2.5	
20	2.5	3.2	3.9	4.6	5.3	2.0-6.0	
23.7 (before washing 1)	2.0	2.7	3.4	4.1	4.8	5.5	3.75
23.7-24	–	3.0	3.7	4.5	5.2	6.0	4.48

Design of Filter Conduits

Commonly the conduits in a filter gallery are pipes. For declining rate operation it is necessary that the operating levels in all the filters are essentially the same. Hence head losses in the influent conduit must be kept quite low. In order to provide for this low head loss and also illustrate a channel design calculation, the influent is carried in a channel and all other conduits are pipes.

For preliminary estimates of sizes to be used in the conduits and as a check for typical designs, Table VI is used.

Table VI. Typical Velocities in Filter Conduits[37,38]

Conduit	Velocity (ft/sec)	(m/sec)
1. Influent with raw water	3-6	(0.91-1.83)
2. Influent with flocculated water	2-6	(0.61-1.83)
3. Effluent with filtered water	3-6	(0.91-1.83)
4. Waste with spent wash water	4-8	(1.22-2.44)
5. Wash water with clean wash water	8-12	(2.44-3.66)
6. Drain from filters	12-15	(3.66-4.57)

Influent Header Channel

An influent channel would probably be cheaper than a pipe but with a sluice valve it is not possible to shut off the valve with an actuator. Therefore, a combined channel-pipe arrangement, which combines the best features of both, is used as shown in Figure 11. Designing the channel is a problem in spatially varied flow. However, a constant channel cross section is used.

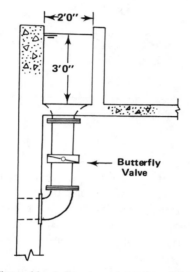

Figure 11. Influent conduit arrangement.

$$
\begin{aligned}
\text{Design flow} &= 3 \text{ mgd} \times 1.547 = 4.64 \text{ cfs} \\
\text{Area A} &= \text{width W} \times \text{depth D}
\end{aligned}
$$

For the best hydraulic section, $R = D/2$ and $W = 2D$ (see Equation 31). Using Manning's equation 32,

$$
Q = A \frac{1.486}{n_c} R^{2/3} S^{1/2}
$$

Therefore,

$$
Q = 2D^2 \frac{1.486}{n_c} \left(\frac{D}{2}\right)^{2/3} S^{1/2}
$$

$$
4.64 = 2D^2 \frac{1.486}{0.013} \left(\frac{D}{2}\right)^{2/3} S^{1/2}
$$

$$D = 0.2758 \ \frac{1}{S^{0.1875}}$$

Possible designs are

S	= 0.001	0.002	0.003	0.005	0.007
D (ft) =	1.01	0.88	0.82	0.75	0.70

Choose a depth of 0.75 ft and a slope of 0.005. In actual construction the floor of the channel may be constructed level since the slope required is so minimal. For the most efficient hydraulic section the width of the channel required is 1.5 ft. Calculations show that the depth of flow varies spatially from 0.8 ft to 0.15 ft. To reduce this variation still further, increase the width of the channel to 2.0 ft. Using Equation 32,

$$Q = 2D \ \frac{1.486}{0.013} \left(\frac{D \times 2}{2 + 2D} \right)^{2/3} (0.005)^{1/2}$$

$$= 16.17 \ \frac{D^{1.667}}{(1 + D)^{0.667}}$$

Select values of D and calculate values of Q as tabulated below or alternatively use a Manning's nomograph.

D (ft)	= 0.8,	0.6,	0.4,	0.2,	0.1
Q (cfs)	= 7.53,	5.04,	2.8,	0.98,	0.33
V (ft/sec)	= 4.7,	4.2,	3.5,	2.5,	1.7

Consider the spatially varied flow at time, 16 hr in Table V. Assuming the flow enters close to Filter Number 1, the cumulative flows along the influent channel are:

Filter Number	1	2	3	4	5	6
Filtration Rate (gpm/ft²)	= 3.2	3.9	4.6	5.3	6.0	2.5
Flow rate through each filter (cfs)	= 0.71	0.87	1.02	1.18	1.34	0.56
Cumulative flow along channel (cfs)	= 5.68	4.67	4.10	3.08	1.90	0.56
Depth of flow (ft)	= 0.64	0.59	0.52	0.43	0.31	0.14

The above calculations indicate that the total depth of flow variation along the channel is 0.5 ft for the above flow rates. The maximum difference of depth between two adjacent filters is 0.17 ft. The above is the worst operating condition. At any other time indicated in Table V, the variations of depth of flow are less than the above. In practice the cumulative flow is 4.64 cfs (3 mgd) and the depth variations are still smaller. The design is satisfactory.

Influent Pipe from Header to Each Filter

Figure 11 and the line diagram in Table VII show the appurtenances in the influent pipe. For all hydraulic computations in a filter, a tabular type of analysis shown in Table VII is recommended. In a design office a standard tabular form with columns similar to Table VII are prepared and used. Calculation procedures for design are by trial and error. A pipe size is selected and the head losses are computed. If the head losses are too large, the next larger pipe size is selected and the losses are recalculated.

As seen in Table VII, a first trial choice of 6 in. diam pipe gave a head loss of 1.60 ft, which was deemed too high for the influent conduit; the pipe was redesigned for 10 in. diam. The length of the pipe is determined from layout plans and a line diagram is included in the computation sheet. The total flow of 600 gpm is at the highest filtration rate of 6 gpm/ft^2. Using a discharge of 600 gpm and a diameter of 6 in., the velocity is determined from Figure 8 as 6.8 ft/sec. Using Table II for old cast iron and a velocity of 6.8 ft/sec gives a C_R value of 0.56, by interpolation:

$$Q/C_R = \frac{600}{0.56} = 1071 \text{ gpm}$$

From Figure 8 for discharge of 1071 gpm and 6 in. diam pipe, the head loss is read as 66 ft/1000 ft. A standard Hazen-Williams nomograph with C = 100 gives a head loss of 45 ft/1000 ft for Q = 600 gpm and d = 6 in. This illustrates the considerable inaccuracies that may result from use of the standard Hazen-Williams formulation.

The head loss factors are determined from Table III. Hence, the head loss in all the appurtenances are calculated by multiplying the head loss factor by V^2/2g from the velocity already shown in column 3. Thus, the design requires a 10 in. diam header line from the influent channel to each filter.

Filter Effluent Pipe

In order to have a similar response from all filters for declining flows, the effluent pipes are identical in layout and discharge into a common weir box spilling into the clear well. The hydraulic computations are similar to those made previously for the influent pipes, and calculations are shown in Table VIII. The designed pipe is shown in Figure 14, p. 733. In the table the orifice meter is selected so that the head loss available in Figure 11 (0.63 ft) is dissipated. The orifice meter designed has an orifice-to-pipe diameter ratio of 1:3.

Table VII. Hydraulic Analysis Computation Sheet

Location: Influent Headers to Filters 1 to 6

Line Diagram

4 ft 6 in.

1 ft 10 in.

Pipeline		Flow, gpm Velocity (ft/sec)	C_R Value Head Loss, H_f (ft)	Appurtenances			Total Head Loss, $H_f + H_v$ (ft)
Diameter (in.)	Length (ft-in.)			Type or Formula	Head Loss Factor	Head Loss, H_v (ft)	
6	4-6 1-10	600 gpm 6.8 ft/sec	0.56 66 ft/1000 ft	Entrance-bell mouth Butterfly valve 90° elbow Outlet	0.04 0.3 0.30 1.0	Factor x $V^2/2g$	
Totals 6-4			0.42		1.64	1.18	1.60
Redesign 10		2.4 ft/sec	0.66 4.2 ft/1000 ft				
Totals 6-4			0.03		1.64	0.15	0.18

Table VIII. Hydraulic Analysis of Effluent Pipes

Location: Effluent Pipes from Underdrain to Common Weir Box

Line Diagram

| Pipeline | | Flow, gpm | C_R Value | Appurtenances | | | Total Head |
Diameter (in.)	Length (ft-in.)	Velocity (ft/sec)	Head Loss, H_f (ft)	Type or Formula	Head Loss Factor	Head Loss, H_v (ft)	Loss, $H_f + H_v$ (ft)
10	14-6	600 gpm	0.66	Entrance-pipe flush with tank	0.50		
		2.4 ft/sec	4.2 ft/1000 ft	Run of tees–2 nos.	1.20		
				Orifice meter (1:3)	2.50		
				Butterfly valve	0.30		
				Outlet into still water	1.00		
Totals	14-6		0.06		5.50	0.49	0.55
10	14-6	400 gpm	**0.69**				
		1.7 ft/sec	1.9 ft/1000 ft				
Totals	14-6		0.03		5.50	0.25	0.28
10	14-6	200 gpm	0.72				
		0.8 ft/sec	0.5 ft/1000 ft				
Totals	14-6		0.01		5.50	0.05	0.06

The calculations for the flow rates of 400 gpm and 200 gpm are also included in Table VIII to complete the declining rate analysis of Figure 10. The head losses through the two-step lateral channels in the Leopold block and the common header channel connecting all the lateral channels are negligible in relation to losses in the orifices and media. Under declining rate hydraulics, the required orifice loss to limit the maximum rate to 6 gpm/ft² is 0.63 ft. The designed orifice and effluent piping provide a loss of 0.55 ft and are sufficiently close to be adequate. The orifice design is critical for declining rate hydraulics and manufacturer's data on head loss must be checked with the design calculations. The orifice can also be used to meter the flow.

Design of Backwashing Systems

The media designed have the following sizes. The sphericity and static bed porosities are assumed as follows:

Sand: $d_{60\%}$ = 0.6 x 1.4 = 0.84 mm, depth = 15 in.

        ψ = 0.8, ϵ_{mf} = 0.40, Sg = 2.65

Coal: $d_{60\%}$ = 0.9 x 1.5 = 1.35 mm, depth = 18 in.

        ψ = 0.7, ϵ_{mf} = 0.50, Sg = 1.5

A backwash rate of 21.5 gpm/ft² at 25°C is selected for the design based on 10% expansion of the $d_{90\%}$ coal. The expanded height of the graded sand at this backwash rate is estimated as outlined in Equations 17 to 21. Using Equation 1 with μ = 0.895 centipoise = 1.87 x 10^{-5} lb sec/ft² at 25°C,

$$V_{mf} = \frac{0.00381(d_{60\%})^{1.82}\ [\gamma(\gamma_m - \gamma)]^{0.94}}{\mu^{0.88}}$$

$$= \frac{0.00381(0.84)^{1.82}\ [62.4(2.65 \times 62.4 - 62.4)]^{0.94}}{(0.895)^{0.88}}$$

$$= 11.61\ \text{gpm/ft}^2 = 0.0259\ \text{ft/sec.}$$

$$Re_{mf} = \frac{\rho_f\ V_{mf}\ d_{60\%}}{\mu}$$

$$= \frac{1.94\ \text{lb sec}^2\ \text{ft}^{-4} \times 0.0259\ \text{ft/sec}\ (0.84 \times 3.28 \times 10^{-3})\ \text{ft}}{0.895 \times 2.088 \times 10^{-5}\ \text{lb sec ft}^{-2}}$$

$$= 7.41$$

Since $Re_{mf} < 10$, no correction is required.

$$Re_o = 8.45 \ Re_{mf}$$
$$= 8.45 \times 7.41 = 62.61$$
$$n = 4.45 \ Re_o^{-0.1} = 4.55 \times 62.61^{-0.1}$$
$$= 3.01$$

Using Equation 20 at minimum fluidization,

$$V_{mf} = k\epsilon_{mf}^{n}$$
$$11.61 = k(0.40)^{3.01}; \text{ therefore } k = 183.08.$$

For a backwash rate of 21.5 gpm/ft^2,

$$21.5 = 183.08(\epsilon)^{3.01}$$

Therefore

$$\epsilon = 0.491.$$

Using Equation 21,

$$l_e (1 - \epsilon) = l_{mf}(1 - \epsilon_{mf})$$
$$l_e(1 - 0.491) = 15(1 - 0.40)$$
$$l_e = 18.0 \text{ in.}$$

The expanded height of graded sand during backwash is 18 in. or an expansion of 20%.

The coal has a $d_{60\%}$ of 1.35 mm. Assume that it has characteristics similar to the coal illustrated by curves 2 and 3 in Figure 3, which has a $d_{60\%}$ size of 1.32 mm. Interpolating for a water temperature of 25°C, the estimated expansion at 21.5 gpm/ft^2 is 19%.

Expanded height of coal layer	$= 18 \times 1.19 = 21.4$ in.
Total expanded height of dual media at 25°C	$= 18 + 21.4 = 39.4$ in.
Overall expansion of dual media	$= \dfrac{(39.4 - 33)}{33} \times 100\% = 19.4\%$

This is satisfactory and indicates the reasonably accurate design procedures used to select compatible sand and coal for backwashing, as well as the methodology used for estimating expansion.

Since colder waters expand media further and may lead to loss of media over the troughs, the expansion is also estimated at the lowest possible temperature of 8°C:

$$\mu \text{ at } 8°C = 1.39 \text{ centipoise} = 2.91 \times 10^{-5} \text{ lb sec/ft}^2$$

Similar calculations as above give:

$$\text{Expanded height of graded sand layer} = 19.9 \text{ in.}$$

From Figure 3 by extrapolating for temperature,

Expanded height of graded coal layer $\quad = 18 \times 1.31 = 23.6$ in.
Total expanded height of dual media at $8°C \quad = 43.5$ in.

Overall expansion $\qquad\qquad\qquad\qquad\qquad = \dfrac{10.5}{33} \times 100 = 31.8\%$

The design is satisfactory.

The calculations above dramatically illustrate the effects of temperature. Minimum expansion is given by warmest temperature, which controls the effectiveness of the backwash. The maximum expansion is given at coldest temperatures and controls the height of the gullet to prevent loss of media.

Head Losses During Backwash

During backwash the coal and sand media are fluidized and the head losses are estimated from Equation 27:

$$h = l(S_g - 1)(1 - \epsilon)$$
for coal, $\quad h = 18(1.5 - 1)(1 - 0.50) = 4.5$ in.
for sand, $\quad h = 15(2.65 - 1)(1 - 0.40) = 14.9$ in.
Total head loss $= 4.5 + 14.9$ in. $= 1.62$ ft

The head loss through the gravel, which is a static bed, is estimated using Equation 26. Table IV gives $\Sigma \, (x_i/d_i^2)$ for gravel $= 0.223 \text{ mm}^{-2}$.

$$\Sigma \frac{x_i}{d_i} = \frac{1}{2.52} + \frac{1}{4.49} + \frac{1}{8.98} + \frac{1}{15.57}$$
$$= 0.397 + 0.223 + 0.111 + 0.064 = 0.795 \text{ mm}^{-1}$$

For gravel, $\epsilon = 0.40$, $\psi = 0.8$, ν at $8°C = 1.39$ centistokes, at 21.5 gpm/ft² $V = 14.6$ mm/sec. Using Equation 26,

$$\frac{h}{l} = 150 \frac{\nu}{g} \frac{(1 - \epsilon)^2}{\epsilon^3} V \left(\frac{1}{\psi}\right)^2 \sum_{i=1}^{N} \frac{x_i}{d_i^2} + 1.75 \frac{(1 - \epsilon)}{\epsilon^3} \frac{V^2}{g} \frac{1}{\psi} \sum_{i=1}^{N} \frac{x_i}{d_i}$$

$$\frac{h}{l} = 150 \times \frac{1.39}{9810} \times \frac{(1 - 0.40)^2}{0.4^3} \times 14.6 \times \left(\frac{1}{0.8}\right)^2 \times 0.223$$

$$+ 1.75 \times \frac{(1 - 0.4)}{0.4^3} \times \frac{14.6^2}{9810} \times \frac{1}{0.8} \times 0.795 = 0.608 + 0.354 = 0.962$$

It is seen that the second term due to kinetic effects is a significant fraction of the first term due to viscous effects. However, in terms of losses through the orifices these head losses are negligible.

Total head loss due to four 2-in. layers at $8°C$,

$$= 0.962 \times \frac{2}{12} = 0.16 \text{ ft}$$

The head losses through the underdrain orifices during backwash may be calculated by proportion ($h \propto Q^2$) from those computed previously for filtration.

Head losses at 21.5 gpm/ft^2,

$$\text{for 1/4-in. diam orifice} = \frac{0.21}{6^2} \times 21.5^2 = 2.70 \text{ ft}$$

$$\text{for 5/8-in. diam orifice} = \frac{0.43}{6^2} \times 21.5^2 = 5.52 \text{ ft}$$

The underdrain block is laid on 12 in. centers. It has two primary feeder channels per ft width with a total area of 30.5 in.^{2} which feeds the secondary channels via 2 5/8-in. diam orifices. For each ft width of filter 20 ft in length,

$$\text{Total backwash flow} = 1 \times 20 \times 21.5 \text{ gpm} = 0.958 \text{ cfs}$$

$$\text{Velocity at entrance to feeder channel} = \frac{0.958}{30.5} \times 144 \text{ ft/sec}$$

$$= 4.52 \text{ ft/sec}$$

The primary feeder channels are 4 in. by 3.8 in.

$$\text{Hydraulic radius} = \frac{4 \times 3.8}{2(4 + 3.8)} = 0.97 \text{ in.} = 0.08 \text{ ft}$$

Using Equation 32,

$$V = \frac{1.486}{n_c} R^{2/3} S^{1/2}$$

$$4.52 = \frac{1.486}{0.013} (0.08)^{2/3} S^{1/2}$$

$$S = 0.045$$

Head loss along 20 ft of channel where the flow is continuously decreasing

$$= \frac{1}{3} \times \text{full flow head loss} = \frac{1}{3} \times 0.045 \times 20 = 0.30 \text{ ft}$$

Velocity head at entrance to channel $= \dfrac{V^2}{2g} = \dfrac{4.52^2}{64.4} = 0.32$ ft

Velocity head in channel near last orifice $= 0$

If p_1 = pressure head at entrance and p_2 = pressure head at end of channel,

$$p_1 + 0.32 - 0.30 = p_2 + 0$$
$$p_1 + 0.02 = p_2$$

Therefore, the pressure head is essentially constant along the length of the channel since the velocity head recovery almost exactly balances the head loss due to friction.

Similar conditions prevail along the top secondary feeder, causing reverse flow to balance pressures if the discharges at the primary orifices are unequal. Thus, the pressure in the secondary feeders are essentially constant, which results in equal discharge through the orifices.

The dual-media expansion during backwash at 8°C is 10.5 in. Allow 6 in. of free depth between the top of the fluidized coal and the bottom of the wash water trough. For preliminary estimates assume a wash water trough depth of 15 in.

Total static head required from bottom of underdrain

= wash trough depth + clearance + media expansion + coal and sand depths
+ gravel depth + underdrain depth.

= 15 in. + 6 in. + 10.5 in. + 18 in. + 15 in. + 8 in. + 10 in.

= 82.5 in. = 6.9 ft.

Head required at top of the flume feeding the primary channels of the underdrain = 6.9 + 5.52 + 2.70 + 0.16 + 1.62 = 16.9 ft.

The primary channel of the underdrain tiles are supplied from a flume. The width of the flume is less than the maximum length of the underdrain block, which is 2 ft.

Assume width of flume   = 1 ft 6 in.
Total backwash rate   = 21.5 gpm/ft^2 x 100 ft^2 = 2150 gpm
= 4.79 cfs

Assume flow from backwash pipe supplied at center of filter. Therefore, flow carried by flume on either side =(4.79/2)= 2.40 cfs.

If velocity is limited to 2 ft/sec,

minimum depth of flume required $= \dfrac{2.40}{1.5 \times 2} = 0.8$ ft

At these velocities, the head losses in the flume are negligible. The backwash pipe supplies the flume via the filter effluent pipe designed in Table VIII. Size of pipe from flume is 10 in. diam. Use a flume of internal dimensions 1 ft 6 in. x 1 ft.

Design of Backwash Piping System

A tabulation of the hydraulic analysis for the backwash piping system is shown in Table IX. Manufacturers' catalogs need to be consulted to determine the sizes of suction and discharge piping diameters and for dimensions and types of pumps. In this design a horizontal split case type pump or a vertical in-line booster type both having the motor on top of the pump and horizontal suction and discharge ends is suitable. Manufacturers' catalogs indicate that for pumps in the range of discharge Q = 2150 gpm and head, H = 30 ft, the suction and discharge diameters are 10-8 in.

The pump-piping layout provides for 100% standby capacity in the wash water pumps and arrangements for removing, repairing or replacing the pumps. In the diagram the highest head loss occurs when backwash pump No. 1 is used to wash Filter No. 6. Using 10-in. diam pipes the hydraulic head required for the piping system is 16.09 ft. Make an allowance of 1.0 ft for the static head in the flume.

$$\begin{aligned}
\text{Total head for backwash pumps} \quad &= \text{losses in media + depth fof flume} \\
&\quad + \text{losses in piping} \\
&= 16.9 + 1.0 + 16.09 \\
&= 33.99 \text{ ft, say 34 ft}
\end{aligned}$$

Pumps required for backwash are specified as:

Discharge = 2150 gpm, at a head = 34 ft.

At first glance it may seem that the 10-in. piping system with velocities of 9 ft/sec is too small and a designer may prefer to use a 12-in. diam pipes to reduce the velocity to 6.3 ft/sec and the appurtenance losses to 7.2 ft. This leads to a total head of approximately 25-26 ft. Sometimes pumps of high discharge and low heads are difficult to obtain. The extra capital cost of the pump and power due to the use of the smaller pipe (10-in. diam) is often less than the cost of all the fittings that are needed for the entire filter block at the larger pipe diameter (12 in.). A final decision may be made after assessing the costs of the two systems. In the example design, the 10-in. diam system is used.

Table IX. Hydraulic Analysis of Backwash Pipes

Location: Backwash Piping with Backwash Pumps

Line Diagram

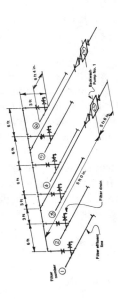

| Pipeline | | Flow, gpm | C_R Value | Appurtenances | | Head | Total Head |
Diameter (in.)	Length (ft-in.)	Velocity (ft/sec)	Head Loss, H_f (ft)	Type or Formula	Head Loss Factor	Loss, H_v (ft)	Loss, $H_f + H_v$ (ft)
10	5 6	2150	0.574	Entrance-pipe flush with tank	0.50		
	5 0	8.8 ft/sec	54 ft/1000 ft	Gate valve, 2 nos.	0.38		
				Check valve	2.50		
	3 0			Tee-90° turn	1.80	$V^2/2g = 1.20$	
	18 0			Tee-run, 4 nos. @ 0.60	2.40		
				Elbow-90°	0.30		
	3 0			Butterfly valve	0.30		
				Tee-90 turn	1.80		
	6 4			Tee-run	0.60		
				Outlet into still water	1.00		
Totals 40	10		2.19		11.58	13.90	16.09

Design of Backwash Water Waste Drain

The backwash water waste drain discharges freely into a channel running along the side of the filters. The hydraulic computations are shown in Table X. Since the available head for free discharge is nearly 7 ft, a 10-in. waste line is adequate.

The waste channel is also able to carry the wastewater from the filter effluent line when there is a need to drain the filter.

The waste channel has to discharge 2150 gpm = 4.79 cfs. The discharge is approximately equal to that of the influent to the filters of 4.64 cfs (see page 716). Therefore, for the most efficient section, a channel 0.75 ft deep and 1.5 ft wide with a slope of 0.005 is adequate. Use design dimensions of 2 ft deep and 1 ft 6 in. width for channel.

Design of Wash Water Troughs

A preliminary estimate of the wash water trough level has been made previously, namely 10.5 in. for media expansion, 6 in. for free depth and 15 in. for trough depth, giving a total height of 31.5 in. above the media.

Tesarik[48] has shown that the spacing between collecting launders or troughs should be less than twice the height above the fluidized bed. Therefore, **maximum** distance between troughs is 2 x 31.5 in. or 5 ft 3 in. Since the filters are only 5 ft wide, a single trough is adequate.

Assume a trough width of 16 in. (= 1.33 ft) to carry the backwash flow of 2150 gpm (4.79 cfs). Using Equation 29, for a rectangular trough,

$$D = 1.73 \left(\frac{Q^2}{gW^2} \right)^{1/3}$$

$$= 1.73 \left(\frac{4.79^2}{32.2 \times 1.33^2} \right)^{1/3} = 1.28 \text{ ft}$$

$$= 15.4 \text{ in.}$$

Area of flow required = 16 x 15.4 = 246 in.² Use a trough with a semicircular bottoms as shown on Figure 12. Allow a freeboard of 1.0 in. in the trough.

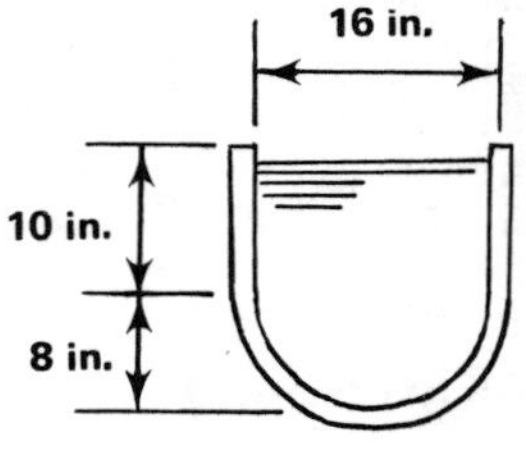

Figure 12.   Section of wash water trough.

Table X. Hydraulic Analysis of Backwash Water Waste Drain

Location: Wash water Gullet to Channel

Line Diagram

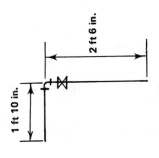

2 ft 6 in.

1 ft 10 in.

Pipeline				Flow, gpm Velocity (ft/sec)	C_R Value Head Loss, H_f (ft)	Appurtenances			Head Loss, H_v (ft)	Total Head Loss, $H_f + H_v$ (ft)
Diameter (in.)	Length (ft-in.)					Type or Formula	Head Loss Factor			
10	1-10			2150	0.574	Entrance-pipe flush with wall	0.50			
	2-6			8.8 ft/sec	54 ft/1000 ft	Elbow-90°, flanged	0.30		$V^2/2g=1.20$	
						Butterfly valve	0.30			
						Outlet-free discharge	1.0			
Totals	4-4				0.23		2.10		2.52	2.75

$$\text{Area of flow} = 16 \times 9 + \frac{1}{2} \times \frac{\pi}{4} \times 16^2$$

$$= 245 \text{ in.}^2$$

Therefore, the design is adequate. A small slope on the trough improves the discharge characteristics further. The free clearance between the expanded level of the media and the bottom of the trough is now 3 in., but is adequate since the depth of the trough is now 18 in.

Backwash Procedures

The filters are designed for average runs of 24 hr. Thus, one of the filters is washed every 4 hr. The recommended total usage of backwash water is 75 to 100 g/ft². [26,39] At a washing rate of 21.5 gpm/ft², this results in a backwash time of 5 min. Suggested backwash procedure is quick opening of backwash valve, fluidization wash at 21.5 gpm/ft² for 5 min and slow closure of valve over 1 min, decreasing flow gradually from 21.5 gpm/ft² to zero.

$$\text{Total water consumption per filter} = 100 \text{ ft}^2 \times 21.5 \text{ gpm/ft}^2 \times 5.5 \text{ min}$$
$$= 11,825 \text{ gal}$$
$$\text{Total backwash water used per day} = 6 \times 11,825 = 70,950 \text{ gal}$$

If the filter units are to produce a finished water volume of 3 mgd, then the flow into the filters should be 3.08 mgd.

$$\text{The percentage of backwash water usage} = \frac{70,950 \times 100}{3 \times 10^6} \%$$

$$= 2.4\%$$

This falls within the typical range of 2-5%. It is recommended that the backwash water be recycled through the chemical treatment system.

The filtration rates including the backwash volume of 80,000 gpd are:

$$\text{With six filters operating} = \frac{3.08 \times 10^6}{24 \times 60 \times 600} = 3.57 \text{ gpm/ft}^2$$

$$\text{With five filters operating} = \frac{3.08 \times 10^6}{24 \times 60 \times 500} = 4.28 \text{ gpm/ft}^2$$

Hydraulic Overdesign Capacity

The calculations shown above do not incorporate an overdesign capacity. It is, however, strongly recommended that at least a 50% overdesign capacity be provided in the hydraulic carrying capacity of the system. This provision enables the production of more water after operational experience. In addition, if future development in treatment permit higher filtration rates, the existence of higher throughput capacity allows increased production of water with minimal capital costs.

Design Summary

The following summarizes the design. A plan and section of the filter block designed are shown in Figures 13 and 14.

1. Design flow = 3 mgd

   Filter influent flow = 3.08 mgd.

2. Filtration rate: declines from 6 gpm/ft^2 to 2 gpm/ft^2

   The filter units are designed to give 3.0 mgd with one filter out of service for emergency repair or maintenance.

   Average filtration rates   = 3.57 gpm/ft^2 for 6 filters
                                      = 4.28 gpm/ft^2 for 5 filters

3. Filter media:

Coal	=	18 in. deep, $d_{10\%}$ = 0.9 mm, u.c. = 1.5, Sg = 1.5
Sand	=	15 in. deep, $d_{10\%}$ = 0.6 mm, u.c. = 1.4, Sg = 2.65
Gravel	=	8 in. deep, 4 nos./2 in. layers, U.S. Sieve No. 10 to 3/4 in.

4. Underdrain system:

Leopold system	=	10 in. deep
Surface orifices	=	18 nos./(1/4-in. diam)/ft^2
Control orifices	=	2 nos./(5/8-in. diam)/ft^2
Backwash flume	=	1 ft 6 in. wide x 1 ft deep

5. Filter dimensions:

   6 filters each with the following dimensions
   Length   = 20 ft, width = 5 ft, depth = 11.5 ft
   Gullet   = 1 ft 6 in. wide with 9 in. wall.

6. Filter operational data:

   Effluent weir in clear water tank = 3 in. above media
   High water level = 6 ft above effluent weir
   Low water level = 4 ft above effluent weir

7. Backwash system:

   Fluidization wash at a rate = 21.5 gpm/ft^2
   Expansion of media at warmest temperature of 25°C
      = 19.4% or 6.4 in.

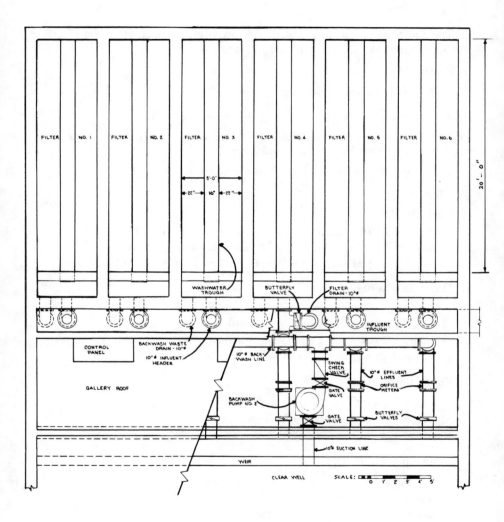

Figure 13. Plan of filter units.

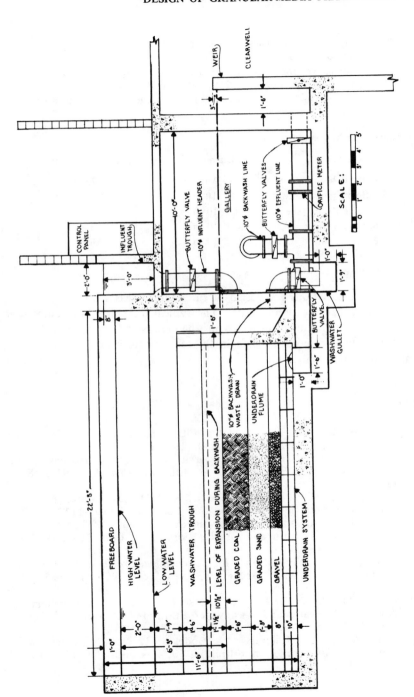

Figure 14. Section of filter units.

Expansion of media at coldest temperature of $8°C$
= 31.8% or 10.5 in.

Wash water drawn from clear well by a pump (with 100% standby) having the specifications Q = 2150 gpm and H = 34 ft

Wash water trough—one is provided with 16-in. wide section 18 in. deep. The bottom is semicircular.

8. Gallery conduit system:

	Conduit	Diameter or Size	Velocity, ft/sec
(a)	Influent header channel	2 ft wide	4.7-1.7
(b)	Influent pipe to filter	10 in.	2.4
(c)	Filter effluent pipe	10 in.	2.4-0.8
(d)	Backwash piping	10 in.	8.8
(e)	Backwash waste drain	10 in.	8.8

Design of Filters in Metric Units

Since the calculations for design of filters are so extensive, they are not illustrated here in metric units. The typical units and procedures used for design in metric units illustrated in Chapters 8 and 11 need to be followed. Wherever possible, data in the metric equivalents are provided to assist design in metric units.

REFERENCES

1. Camp, T. R. "Theory of Water Filtration," *J. Environ. Eng. Div., Proc. Am. Soc. Civil Eng.* 90:1-30 (1964).
2. Ives, K. J., and I. Sholji. "Research on Variables Affecting Filtration," *J. Environ. Eng. Div., Proc. Am. Soc. Civil Eng.* 91:1-18 (1965).
3. Yao, K. M., M. T. Habibian and C. R. O'Melia. "Water and Wastewater Filtration," *Environ. Sci. Technol.* 5:1105-1112 (1971).
4. Spielman, L. A., and J. A. Fitzpatrick. "Theory of Particle Collection under London and Gravity Forces," *J. Coll. Interface Sci.* 42:607-623 (1973).
5. Cleasby, J. L. "Filtration," *Physicochemical Processes for Water Quality Control,* W. J. Weber, Ed. (New York: Wiley-Interscience, 1972).
6. Ghosh, M. M., T. A. Jordan and R. L. Porter. "Physicochemical Approach to Water and Wastewater Filtration," *J. Environ. Eng. Div., Proc. Am. Soc. Civil Eng.* 101:71-86 (1975).
7. Adin, A., and M. Rebhun. "A Model to Predict Concentration and Head-Loss Profiles in Filtration," *J. Am. Water Works Assoc.* 69:444-452 (1977).

8. Kavanaugh, M. *et al.* "Contact Filtration for Phosphorus Removal," *J. Water Poll. Control Fed.* 49:2157-2176 (1977).
9. Robeck, G. G., K. A. Dostal and R. L. Woodward. "Studies of Modifications in Water Filtration," *J. Am. Water Works Assoc.* 56:198-213 (1964).
10. Dostal, K. A., and G. G. Robeck. "Studies of Modifications in Treatment of Lake Erie Water," *J. Am. Water Works Assoc.* 58:1489-1504 (1966).
11. Habibian, M. T., and C. R. O'Melia. "Particles, Polymers and Performance in Filtration," *J. Environ. Eng. Div., Proc. Am. Soc. Civil Eng.,* 101:567-584 (1975).
12. Cleasby, J. L., M. M. Williamson and E. R. Baumann. "Effect of Filtration Rate Changes on Quality," *J. Am. Water Works Assoc.* 55:869-880 (1963).
13. Tuepker, J. L., and C. A. Buescher, Jr. "Operation and Maintenance of Rapid Sand and Mixed Media Filters in a Lime Softening Plant," *J. Am. Water Works Assoc.* **60:1377-1388** (1968).
14. Hudson, H. E., Jr. "Declining Rate Filtration," *J. Am. Water Works Assoc.* 51:1455-1463 (1959).
15. Cleasby, J. L. *et al.* "Backwashing of Granular Filters, Committee Report," *J. Am. Water Works Assoc.* 69:115-126 (1977).
16. Mintz, D. M. "Modern Theory of Filtration," Special Subject No. 10, International Water Supply Association Congress, Barcelona (1966).
17. Huang, J. Y. C., and E. R. Baumann. "Least Cost Sand Filter Design for Iron Removal," *J. Environ. Eng. Div., Proc. Am. Soc. Civil Eng.* 97:171-190 (1971).
18. Ives, K. J. "Theory of Filtration," Special Subject No. 7, International Water Supply Congress and Exhibition, Vienna (1969).
19. Kawamura, S. "Design and Operation of High-Rate Filters—Part I, *J. Am. Water Works Assoc.* 67:535-544 (1975).
20. Jung, H., and E. S. Savage. "Deep-Bed Filtration," *J. Am. Water Works Assoc.* 66:73-78 (1974).
21. McBride, D. M. "Facilitating Plant Operation and Control—Los Angeles Water Treatment Plant," presented at the April 25-29, 1977, Filtration Society Meeting, held at Valley Forge, Pennsylvania.
22. Degremont. *Water Treatment Handbook,* English edition. (Distributed in U.S. by Taylor and Carlisle, New York, 1973).
23. Adin, A., and M. Rebhun. "High-Rate Contact Flocculation-Filtration with Cationic Polyelectrolytes," *J. Am. Water Works Assoc.* 66:109-116 (1974).
24. Amirtharajah, A. "Optimum Expansion of Sand Filters During Backwash," unpublished Ph.D. Thesis, Iowa State University, Ames, Iowa (1971).
25. Amirtharajah, A. "Optimum Backwashing of Sand Filters," *J. Environ. Eng. Div., Proc. Am. Soc. Civil Eng.,* in press (October 1978).
26. Cleasby, J. L., and E. R. Baumann. "Wastewater Filtration-Design Considerations," U.S. EPA, Tech. Transfer Seminar Publ., Washington, D.C. (1974).

27. Hazen and Sawyer, Engineers. "Process Design Manual for Suspended Solids Removal," U.S. EPA, Tech. Transfer, Washington, D.C. (1975).
28. Leva, M. *Fluidization* (New York: McGraw-Hill Book Co., Inc., 1959).
29. Amirtharajah, A. "Expansion of Graded Sand Filters During Backwashing," unpublished M.S. Thesis, Iowa State University, Ames, Iowa (1970).
30. Cleasby, J. L., and G. D. Sejkora. "Effect of Media Intermixing on Dual Media Filtration," *J. Environ. Eng. Div., Proc. Am. Soc. Civil Eng.* 101:503-516 (1975).
31. Ives, K. J. "Progress in Filtration," *J. Am. Water Works Assoc.* 56:1225-1232 (1964).
32. Iwasaki, T. "Some Notes on Sand Filtration," *J. Am. Water Works Assoc.* 29:1591-1597 (1937).
33. Camp, T. R., and P. C. Stein. "Velocity Gradients and Internal Work in Fluid Motion," *J. Boston Soc. Civil Eng.* 30:219-237 (1943).
34. Fair, G. M., and L. P. Hatch. "Fundamental Factors Governing the Streamline Flow of Water Through Sand," *J. Am. Water Works Assoc.* 25:1551-1565 (1933).
35. Richardson, J. F., and W. N. Zaki. "Sedimentation and Fluidization," *Trans. Inst. Chem. Eng.* 32:35-53 (1954).
36. Amirtharajah, A., and J. L. Cleasby. "Predicting Expansion of Filters During Backwash," *J. Am. Water Works Assoc.* 64:52-59 (1972).
37. Fair, G. M., J. C. Geyer and D. A. Okun. *Water and Wastewater Engineering*, Vol. 2 (New York: John Wiley and Sons, Inc., 1968).
38. Clark, J. W., W. Viessman, Jr. and M. J. Hammer. *Water Supply and Pollution Control*, 3rd ed. (New York: IEP—A Dun-Donnelley Publisher, 1977).
39. Cleasby, J. L., and E. R. Baumann. "Backwash of Granular Filters Used in Wastewater Filtration," Environmental Protection Technology Series, EPA-600/2-77-016, U.S. EPA, Cincinnati, Ohio (1977).
40. Ergun, S. "Fluid Flow Through Packed Columns," *Chem. Eng. Prog.* 48:89-94 (1952).
41. Arboleda, J. "Hydraulic Control Systems of Constant and Declining Flow Rate in Filtration," *J. Am. Water Works Assoc.* 66:87-93 (1974).
42. Chow, V. T. *Open Channel Hydraulics* (New York: McGraw-Hill Book Co., 1959).
43. Peterson, A. C. *Applied Mechanics: Fluids* (Boston: Allyn and Bacon, Inc., 1971).
44. Institution of Water Engineers. *Manual of British Water Engineering Practice, Volume II: Engineering Practice*, 4th ed. (Cambridge, England: W. Heffer & Sons, Ltd., 1969).
45. Jain, A. K., D. M. Mohan and P. Khanna. "Modified Hazen-Williams Formula," *J. Environ. Eng. Div., Proc. Am. Soc. Civil Eng.* 104:137-146 (1978).

46.   Barlow, J. F., and E. Markland. "Converting Hazen-Williams' Equation to Colebrook Function," *International Water Power and Dam Construction* 27:331-336 (1975).
47.   Ackers, P. "Tables for the Hydraulic Design of Storm-drains, Sewers and Pipe-lines," Hydraulic Research Paper No. 4, 2nd ed., Dept. of the Environment, London, England (1969).
48.   Tesarik, I. "Flow in Sludge Blanket Clarifiers," *J. Environ. Eng. Div., Proc. Am. Soc. Civil Eng.* 93:105-120 (1967).

CHAPTER 29

LABORATORY DESIGN

Albert H. Ullrich, P.E.
Freese and Nichols, Inc.
Consulting Engineers
314 Highland Mall Blvd.
Austin, TX 78752

OBJECTIVE

The purpose of this chapter is to discuss some of the important and desirable features that should be incorporated into the design of laboratory facilities for small- to medium-sized water treatment plants. Emphasis is directed toward the testing facilities required for efficient treatment control and for monitoring the more common water quality parameters. It is not the purpose of this chapter to discuss designs for laboratory facilities needed for treatment control and water quality monitoring for very large water treatment plants (*e.g.*, 20 mgd or larger) that employ highly trained technical personnel who are competent to advise the design engineer regarding the facilities needed, including layouts, space requirements, and water quality parameters to be determined.

Facilities required for determination of heavy metals, pesticides, herbicides, chlorination products ("trihalomethanes") or other suspected toxic substances are not discussed for two reasons. In some states, the State Department of Health Laboratory performs the required number of analyses for these substances for both large and small water utilities. Where the state does not provide this service, the smaller plants (and some of the larger plants also) are likely to find it more economical to contract with a certified or state-approved private or commercial laboratory to provide the needed services rather than try to make these determinations in-house.

Emphasis is placed on the testing facilities for efficient treatment control and for monitoring the more common water quality parameters rather than on performing routine complete chemical analyses. Although water utilities should have fairly current complete chemical or mineral analyses available upon request, the smaller systems are likely to find it more economical to contract with a certified commercial laboratory to make complete analyses on a periodic basis rather than make such analyses in-house. If this type information for a given source of supply is of wide public interest it may be possible to enter into a cooperative agreement with the U.S. Geological Survey (U.S.G.S.) to perform this service.

CLASSIFICATION OF WATER TREATMENT PLANT LABORATORIES

Water treatment plant process control laboratories may be classified with respect to the analytical determinations for which they are equipped and these, in turn, depend on the type or method of water treatment that is to be practiced. In some states laboratories may also be classified by size, which may be dictated by the number of tests to be made during a given time. And in some states (such as Texas, where the State Health Department Laboratory performs the required bacteriological analytical work for all domestic water systems serving populations of less than 25,000 persons) the laboratory may be further classified by whether bacteriological testing facilities are included.

In order to provide the necessary testing facilities the design engineer must consider the treatment to be used and the in-house treatment control and monitoring tests to be made. From this information he must then determine the facilities and space required to perform the tests. Tables I and II list the more common methods of water treatment currently in use and the minimum laboratory tests that need to be made for control purposes.

References to Laboratory Models A, B and C are included in Tables I and II. Plans and facilities for these laboratories are detailed elsewhere in this chapter. Laboratory Model A is a relatively complete but compact facility suitable for plants treating about 20 mgd (0.9 m³/sec) or less, and it includes equipment for bacteriological testing and algae counting and identification. Depending upon the circumstances, Laboratory Model A might be suitable for larger water treatment plants. Laboratory Models B and C are *minimal* types and suitable only for limited work by a single technician. The laboratory models presented herein are intended for guides and suggestions, and the *design engineer is advised to modify plans, equipment, and supplies to meet the specific needs* of his particular water treatment plant.

Table I. Tests for Ground Water Supplies

| | | Laboratory Model | |
| | | Population Served | |
Treatment	Control Tests[a]	100-25,000	> 25,000
Chlorination	Chlorine Residual	C	B
pH adjustment	pH determination	C	B
Chlorination	Chlorine residual		
pH adjustment	pH determination	C	B
Sodium hexametaphosphate (SHP) treatment for iron and/or manganese sequestering	Note: chemical supplier will usually monitor SHP residual		
Chlorination	Chlorine residual		
Fluoridation	Fluoride determination		
Iron Removal			
Aeration	pH determination	C	B
pH adjustment	Chlorine residual		
Chlorination	P and M alkalinity		
Settling	Iron		
Filtration	Turbidity		
Fluoridation	Fluoride determination		
	Chloride determination		
Lime Softening	pH measurement	B	A
Mixing	Carbonate alkalinity		
Flocculation	Bicarbonate alkalinity		
Settling	Total hardness		
Recarbonation or sodium hexametaphosphate stabilization	Chlorine residual		
	Turbidity		
Filtration	Fluoride determination		
Chlorination	Chloride determination		
Fluoridation			

[a]For all large systems (pop. > 25,000) add bacteriological testing and specify **Laboratory A** unless all bacteriological testing is done by state or private laboratories.

Table II. Tests for Surface Water Supplies

| | | Laboratory Model | |
| | | Population Served | |
Treatment	Control Tests[a]	100-25,000	> 25,000
Prechlorination with or without preammoniation Coagulation Settling Filtration Fluoridation	Chlorine residual pH P and M alkalinity Turbidity Jar test Fluoride determination	B	A
Prechlorination with or without preammoniation Coagulation with pH adjustment Settling Filtration Fluoridation Sodium hexametaphosphate corrosion control	Chlorine residual pH P and M alkalinity Turbidity Jar test SHP residual	B	A
Prechlorination with or without preammoniation Taste and odor control with powdered activated carbon Color removal Coagulation with pH adjustment Settling Filtration Fluoridation	Chlorine residual pH Odor threshold P and M alkalinity Turbidity Jar test Calcium carbonate stability Fluoride Color Chloride	B	A

Process	Test	
Lime Softening	pH measurement	
Prechlorination with or without preammoniation	Carbonate alkalinity	
Taste and odor control with powdered activated carbon or permanganate	Bicarbonate alkalinity	A
Color removal	Total hardness	
Mixing	Chlorine residual	
Flocculation	Turbidity	
Settling	Fluoride determination	
Recarbonation or sodium hexametaphosphate	Odor threshold	A
Filtration	Jar test	
Post chlorination	Calcium carbonate stability	
Fluoridation	Color	A
	Chloride	

---

^aFor all large plants (pop. > 25,000)
 a. Add bacteriological testing and specify Laboratory A unless all bacteriological testing is done by state or private laboratories.
 b. Add algae counting and identification.

LABORATORY APPARATUS, EQUIPMENT, CHEMICALS AND SUPPLIES

After determining the tests to be made, the design engineer should determine and list the apparatus, equipment, chemicals and supplies needed to perform the tests. The list should describe each item in detail and show the quantity required. Since there are a number of suppliers that specialize in furnishing laboratory apparatus, equipment and supplies, it is also necessary for the design engineer to specify standards of quality. Perhaps the best way to do this is to obtain current catalogues from two or more reputable laboratory supply firms and specify each item by description, the supplier's catalogue numbers and the quantity needed. Thus, if we assign the letter "X" to represent one acceptable supplier and the letter "Y" to represent another acceptable supplier the specifications may be written in accordance with Table III.

Table III. Specifications for Laboratory Equipment

Item No.	Description	"X" Cat. No.	"Y" Cat. No.	Quantity
1	Fluorescent Titration Illuminator	10-550-2	9-665-3	1 ea.
2	Balance, Harvard Trip, Single Beam	3-125	5-750	1 ea.

The following tables give lists of typical items of glassware, chemicals, reagents, apparatus, equipment and supplies that the design engineer may find useful for selecting the items needed for his project. In order to equip the most complete laboratory considered in this chapter, identified as Laboratory A and shown in Figure 1, the engineer needs to specify (as a minimum) all of the items listed in Tables IV-IX. For smaller laboratories he may select items given and described in Figure 2 and Tables X-XII and to supplement these with additional items required for special tests. The final list to be used for specifying and bidding purposes should be checked against the apparatus, equipment, reagents and supplies listed in the latest edition of *Standard Methods for Examination of Water and Wastewater* for each of the tests that is to be performed.

LABORATORY FURNITURE, SPACE REQUIREMENTS AND LAYOUT

After determining the necessary tests and the apparatus, equipment, chemicals, reagents and supplies that are needed to perform the required tests, the design engineer must determine the units of laboratory furniture and the space requirements for these tests. A number of reputable laboratory furniture manufacturers offer unitized furniture that is readily adaptable for multiple component installation. It is suggested that the design engineer base his space requirements and layout on the use of this type of furniture. Figures 1, 2 and 3 and accompanying Tables IV, X and XIII show how unitized furniture may be used to advantage in designing a laboratory layout.

Laboratory furniture may be obtained in either metal or wood (generally plywood) construction. Only the highest quality of material and workmanship should be accepted. For wood construction, the design engineer should specify minimum acceptable thickness as well as quality. Likewise for metal construction, he should specify minimum metal thickness. For either, the specifications should be very specific with respect to finishes, drawer suspensions, hinges, table-top material and finish. It is suggested that current catalogues from several reputable laboratory manufacturers be obtained and that the detailed specifications be based on their first-line offerings.

HEATING, AIR CONDITIONING, VENTILATION AND LIGHTING

Special attention should be given to the design of the heating, air conditioning and lighting systems, particularly for medium- to large-size laboratories. Maintenance of a uniform temperature is desirable and drafts or air currents should be avoided, particularly in areas where balances and gas burners are located. In Figure 1 attention is called to the window located near the analytical balance. In this particular layout this window is for outside light and not for ventilation. It is desirable to provide an exhaust system above units 8, 9 and 10 (in Figure 1) to exhaust hot, humid air during hot weather.

Utilities

Where natural gas as well as electric power is available it is desirable to provide both gas cocks and electric outlets at strategic locations along and above the table tops. These facilities can be furnished by all reputable

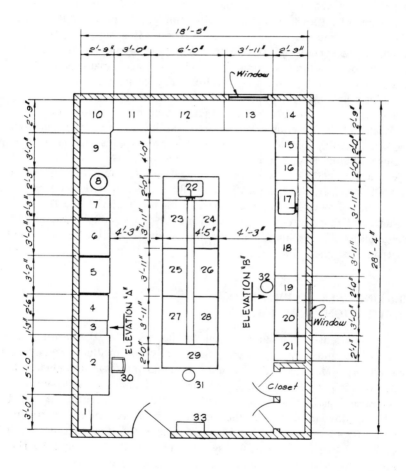

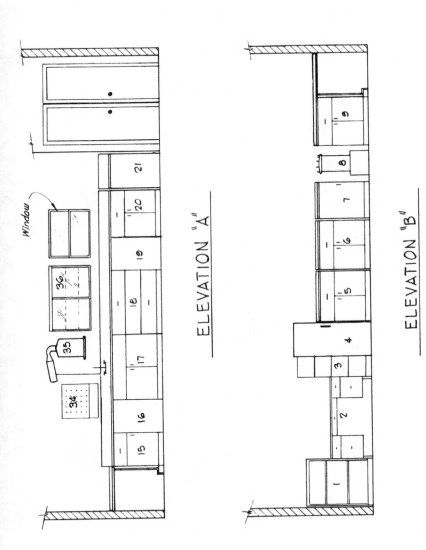

ELEVATION "A"

ELEVATION "B"

Figure 1. Laboratory A.

Table IV. Description of Units Shown on Figure 1 for Laboratory A

1. Book case
2. Desk
3. Filing cabinet
4. Refrigerator, 17 ft^3
5. Drawer, shelf base unit for centrifuge
6. Drawer, shelf base unit for open top space
7. Incubator
8. Autoclave, floor mounted
9. Drawer, shelf base unit for hot air sterilizer corner unit
10. Corner unit
11. Drawer, shelf base unit for muffle furnace
12. Fume hood
13. Drawer, shelf base unit for 6-paddle sample stirrer
14. Corner unit
15. Drawer, shelf base unit for water bath
16. Knee space filler unit
17. Sink base unit
18. Drawer base unit
19. Knee filler unit for microscope
20. Drawer, shelf base unit
21. Knee filler unit for chemical balance
22. Sink with drain boards
23. Drawer, shelf base unit
24. Drawer, shelf base unit
25. Drawer, shelf base unit
26. Drawer, shelf base unit
27. Drawer, shelf base unit
28. Drawer, shelf base unit
29. Throw table
30. Office chair
31. Side chair
32. Laboratory stool
33. Hat and coat rack and shelf
34. Pegboard
35. Still
36. Wallcase
37. Cup sink for vacuum aspirator unless vacuum lines supplied

Table V. Appartus and Equipment for Laboratory A

	Description	Quantity
1.	Balance, with chain attachment, notched beam, horizontal vernier and magnetic damper	1 ea.
2.	Balance, Harvard trip, single beam	1 ea.
3.	Balance weights, 50 g	1 st.
4.	Balance weights, std.−A−weigh, 1000 g	1 st.
5.	Burner, natural gas model	1 ea.
6.	Burner wingtop	1 ea.
7.	Distilling apparatus, electrically heated, with low water cut-off and automatic shut-off, 1 gph	1 ea.
8.	Fluorescent titration illuminator	1 ea.
9.	Stirrer, magnetic	1 ea.
10.	Comparator, Hellige with color discs, reagents and accessories−Set	1 ea.
	a. Chlorine residual (0-tolidine disc)	
	b. pH disc (specify range)	
	c. pH disc (specify range)	
11.	Autoclave, electric, sz. 1	1 ea.
12.	Sterilizing box, Petri dish	2 ea.
13.	Sterilizing box, Petri	2 ea.
14.	Quebec colony counter	1 ea.
15.	Incubator, Mdl. 200A	1 ea.
16.	Sterilizer, hot air	1 ea.
17.	Microscope, with binocular head	1 ea.
18.	Illuminator	1 ea.
19.	pH meter, Accumet, Mdl. 220	1 ea.
20.	Centrifuge, International clinical mdl.	1 ea.
21.	Centrifuge head, 4 place, 50 ml	1 ea.
22.	Metal shield, 50 ml	4 ea.
23.	Trunnion ring, 50 ml tube	4 ea.
24.	Turbidimeter, Jackson	1 ea.
25.	Turbidimeter tube, 75 cm, white	1 ea.
26.	Colorimeter, B&L, Spectronic "20"	1 ea.
27.	Infrared sensitive phototube and filter	1 ea.
28.	Distilling apparatus, 1000 ml	1 ea.
29.	Desiccator, 250 mm, 1 lb	1 ea.
30.	Oven, drying	1 ea.
31.	Water bath	1 ea.
32.	Interval timer, spring-wound	1 ea.
33.	Stop watch	1 ea.
34.	Refrigerator, 17 ft^3	1 ea.
35.	Sample stirrer, 6-paddle	1 ea.
36.	Muffle furnace, electric	1 ea.
37.	Hot plate	1 ea.
38.	Dissolved oxygen meter	1 ea.
39.	Torch (burner), propane	1 ea.

Table VI. Supplies for Laboratory A

Description	Quantity
1. Tongs, beaker	1 ea.
2. Bottles, 16 oz capacity, polyethylene	1 dz.
3. Bottle, washing, polyethylene, 500 ml (6/pk)	
4. Clamp utility	3 ea.
5. Forceps	1 ea.
6. Airejector	1 ea.
7. Splashgon	1 ea.
8. Pipet filler, safety	1 ea.
9. Ring, support, cast iron, with clamp, 2 in.	1 ea.
10. Ring, support, cast iron, with clamp, 4 in.	1 ea.
11. Scoopula, stainless steel (6/pk)	1 pk.
12. Spatula, stainless steel	6 ea.
13. Support stand	1 ea.
14. Buret support	2 ea.
15. Support, rack	1 ea.
16. Tongs, crucible (3/bg)	1 bg.
17. Tongs, pick-up (3/bg)	1 bg.
18. File, round, 6 in. (3/bg)	1 bg.
19. File, triangular, 4 in. (6/bg)	1 bg.
20. Tripod, iron, with concentric rings, 10 in.	1 ea.
21. Brush, bottle and flask (12/pk)	1 pk.
22. Brush, test tube	1 dz.
23. Filter paper, Whatman #1 unwashed, medium, 12 1/2 cm (100 sh/bx)	1 bx. 1 bx.
24. Filter paper, Whatman # 3 unwashed, thick, 12 1/2 cm (100 sh/bx)	1 bx.
25. Tubing, rubber, 3/16 in. (12 ft/ln)	1 ln.
26. Tubing, rubber, 5/16 in. (12 ft/ln)	1 ln.
27. Tubing, rubber, vacuum 1/4 in. (10 ft/pk)	1 pk.
28. Wire gauze squares, 4 in. x 4 in. (12/pk)	1 pk.
29. Wire gauze squares, 6 in. x 6 in. (12/pk)	1 pk.
30. Stoppers, rubber, sz. 0-8 assorted	3 lb.
31. Cotton, aseptic, nonabsorbent, 1 lb	1 ea.
32. Needle holder	1 ea.
33. Innoculating loop, 3 mm	1 ea.
34. Pipet, water analysis, 11 ml (12/pk)	2 pk.
35. Support, aluminum test tube rack	12 ea.
36. Buret, Squibb, type B, 50 ml	1 ea.
37. Thermometer, general lab engraved stem $-10°$ to $+260°C$, $1°$ subdivision	2 ea.
38. Thermometer, precision$-1°$ to $+101°C$, $1/10°$ subdivision	1 ea.
39. Color tube support for APHA std. tubes 50 ml	1 ea.
40. Candles, paraffin (24/pk) for Jackson turbidimeter	1 pk.
41. Casserole, Coors porcelain, 500 ml	1 ea.
42. Buret micro, with stopcock and reservoir	3 ea.
43. Drierite, indicating 8 mesh, 1 lb	1 ea.
44. Crucible, Gooch, F, 30 ml (12/cs)	1 cs.
45. Crucible holders, Gooch (B)	3 ea.
46. Rubber adapter (12/pk)	1 pk.
47. Dish, evaporating, 115 mm	6 ea.
48. Tongs, crucible	1 ea.
49. Asbestos, long fiber for Gooch crucibles, 1 lb	1 ea.
50. Crucible, filtering, Coors, 40 ml	2 ea.
51. Mortar, porcelain, 100 mm	1 ea.

Table VII. General Glassware for Laboratory A

	Description	Quantity
1.	Beaker, Griffin, Pyrex with spout, 50 ml (12/pk)	1 pk.
2.	Beaker, Griffin, Pyrex with spout, 250 ml (12/pk)	1 pk.
3.	Beaker, Griffin, Pyrex with spout, 600 ml (6/pk)	1 pk.
4.	Beaker, Griffin, Pyrex with spout, 1000 ml (6/pk)	1 pk.
5.	Beaker covers, 75 mm Pyrex (12/pk)	1 pk.
6.	Bottles, wide mouth without caps, 32 oz	1 dz.
7.	Bottles, 4 oz capacity glass with narrow mouth and glass stopper	6 ea.
8.	Bottles, 1000 ml capacity, Pyrex with narrow mouth and glass stopper	6 ea.
9.	Buret, blue line Exax., 25 ml grad., in 1/10 ml	1 ea.
10.	Buret, blue line Exax., 50 ml grad., in 1/10 ml	1 ea.
11.	Cylinder, graduated, Pyrex, 50 ml	2 ea.
12.	Cylinder, graduated, Pyrex, 100 ml	2 ea.
13.	Cylinder, graduated, Pyrex, 500 ml	2 ea.
14.	Cylinder, graduated, Pyrex, 1000 ml	2 ea.
15.	Flask, Erlenmeyer, Pyrex, 250 ml	12 ea.
16.	Flask, Erlenmeyer, Pyrex, 500 ml	6 ea.
17.	Flask, Filtering, with side-type, Pyrex brand glass, 1000 ml	4 ea.
18.	Flask, volumetric, Pyrex with stopper, 100 ml	2 ea.
19.	Flask, volumetric, Pyrex with stopper, 500 ml	4 ea.
20.	Flask, volumetric, Pyrex with stopper, 1000 ml (6/cs)	1 cs.
21.	Funnel, chemical filtering, Kimax, 65 mm	2 ea.
22.	Funnel, chemical filtering, Kimax, 90 mm	2 ea.
23.	Glass rod, Kimble standard Flint glass, 4 mm, 4 ft lengths	1 lb.
24.	Glass tubing, Kimble standard Flint glass, 8 mm, 4 ft lengths	1 lb.
25.	Glass stirring rods (72/pk), 150 mm	1 pk.
26.	Pipet, volumetric, Pyrex, 25 ml	2 ea.
27.	Pipet, volumetric, Pyrex, 50 ml	2 ea.
28.	Pipet, volumetric, Pyrex, 100 ml	2 ea.
29.	Pipet, Mohr. 1/10 ml, grad., 1 ml (12/pk)	1 pk.
30.	Pipet, Mohr. 1/10 ml, grad., 2 ml (12/pk)	1 pk.
31.	Pipet, Mohr. 1/10 ml, grad., 5 ml (12/pk)	1 pk.
32.	Pipet, Mohr. 1/10 ml, grad., 10 ml (12/pk)	1 pk.
33.	Pipet, Mohr. 1/10 ml, grad., 25 ml (12/pk)	1 pk.

Table VIII. Special Glassware for Laboratory A

	Description	Quantity
1.	Bottles, dropping, 30 ml (6/pk)	1 pk.
2.	Glass beads, 1/2 lb	1 ea.
3.	Bottles, square jar, 2 oz	2 dz.
4.	Bottles, Pyrex, 160 ml (12/pk), coliform analysis	1 pk.
5.	Bottles, dilution (48/cs), coliform analysis	1 cs.
6.	Dish, Petri, 100 x 15 mm	48 ea.
7.	Cover glass, micro, 18 mm	1 oz.
8.	Slide, microscope, 75 x 25 mm	1 gr.
9.	Pipet, water analysis, 11 ml (12/pk)	2 pk.
10.	Test tube, 75 x 10 mm (72/pk)	1 pk.
11.	Test tube, 150 x 25 mm	1 pk.
12.	Nessler tubes, APHA std., 50 ml (2 st/cs)	1 cs.
13.	Test tubes, selected, 1/2 in. diam	2 dz.
14.	Separatory funnels, Squibb, 500 ml (4/cs)	3 cs.
15.	Flask, filtering	3 ea.
16.	Bottles, narrow mouth, amber glass coin stopper, 32 oz	4 ea.

Table IX. Chemicals and Reagents for Laboratory A

	Description	Quantity
1.	Acetic acid, glacial, 99.7% (6 x 5 pt/cs)	1 cs.
2.	Sulfuric acid conc., 6 x 5 pt	1 cs.
3.	Iodine, resublimed, 1 lb	1 ea.
4.	Potassium bi-iodate, purified, 1/4 lb	1 ea.
5.	Potassium iodide, crystals, 5 lb	1 ea.
6.	Potassium dichromate, N/10 soln., 1 qt	1 ea.
7.	Sodium thiosulfate, 0.1 N, 5 gal	1 ea.
8.	Starch indicator solution, 1 qt	1 ea.
9.	O-tolidine, 0.1 % APHA, 1 qt	6 ea.
10.	Mercaptosuccinic acid, practical, 500 g	1 ea.
11.	Potassium dichromate, 1 lb	1 ea.
12.	Potassium chromate, 1 lb	1 ea.
13.	Potassium phosphate, monobasic, 5 lb	1 ea.
14.	Sodium arsenite, meta, 1 lb	1 ea.
15.	Sodium phosphate, dibasic, 1 lb	1 ea.
16.	Trypton glucose agar extract, 1 lb	1 bx.
17.	Potassium hydroxide, soln., 1 N, 1 qt	1 ea.
18.	Brilliant green lactose bile broth 2%, 1 lb	1 ea.
19.	Eosin methylene blue agar, 1 lb	1 ea.
20.	Lactose broth, 1 lb	1 ea.
21.	Nutrient agar, 1 lb	1 ea.
22.	Sodium hydroxide, soln., 1 N, 1 gal	1 ea.
23.	Ethyl alcohol, denatured, 1 gal	1 ea.
24.	Ammonium oxalate, 1 lb	1 ea.
25.	Crystal violet, 100 g	1 ea.

Table IX (Continued)

Description	Quantity
26. Safranin soln., aqueous 1%, gram counter stain, 2 1/2 gal	1 ea.
27. Potassium hydrogen phthalate, 1 lb	1 ea.
28. Sodium borate, tetra, 1 lb	1 ea.
29. Buffer soln., 4.0 at 25°C, 1 pt	1 ea.
30. Buffer soln., 9.0 at 25°C, 1 pt	1 ea.
31. Bromceresol green, sodium salt, 5 g	1 ea.
32. Methyl red, sodium salt, 10 g	1 ea.
33. Sodium carbonate, anhydrous, 1 lb	1 ea.
34. Sulfuric acid soln., 0.02 N, 1 gal	1 ea.
35. Methyl orange soln., 0.5% APHA, 1 pt	1 ea.
36. Phenolphthalein indicator soln., APHA, 1 qt	1 ea.
37. Sodium hydroxide soln., 0.02 N, 1 gal	1 ea.
38. Sodium bicarbonate, powder, 1 lb	1 ea.
39. Buffer soln. (phosphate), pH 7.2 APHA, 1 qt	1 ea.
40. Platinum cobalt color std. soln., No. 500 APHA, 1 qt	1 ea.
41. Kaolin, 5 lb	1 ea.
42. Mercuric chloride, 1/4 lb	1 ea.
43. Chloroform, ACS, 1 qt	1 ea.
44. Methylene blue, water soluble, 1/4 lb	1 ea.
45. Sodium phosphate, monobasic, 5 lb	1 ea.
46. Sulfuric acid, soln., 1 N, 1 gal	1 ea.
47. Alkyl benzene sulfonate std. soln., 1 qt	2 ea.
48. Hydrochloric acid, conc. (6 x 5 pts/cs)	1 cs.
49. Silver sulfate, crystals, 1/4 lb	1 ea.
50. Sodium fluoride, 1 lb	1 ea.
51. Zirconyl chloride, 1/4 lb	1 ea.
52. 4.5 dihydroxy-3-(p-sulfophenylazo)2.7 acid trisodium salt naphthalene-ditilfonic, 25 g	1 ea.
53. Ammonium chloride, 5 lb	1 ea.
54. Ammonium hydroxide (6 x 5 pt/cs)	1 cs.
55. Calcium carbonate, anhydrous, 1 lb	1 ea.
56. Hydroxylamine hydrochloride, 500 g	1 ea.
57. Magnesium disodium ethylenediamino tetraacetate, 1 lb	1 ea.
58. Eriochrome black T soln	1 ea.
59. Methyl red soln., 0.02%, 1 pt	1 ea.
60. Sodium (di) ethylenediamine tetraacetate std. soln., 1 gal	1 ea.
61. Murexide (ammonium purpurate acid indicator tablets (100/bt)	1 bt.
62. 1 amino-2-naphol-4-sulfonic acid, 25 G	1 ea.
63. Nitric acid, 5 pt	1 ea.
64. Ammonium molybdate, 5 lb	1 ea.
65. Sodium meta-bisulfate, 5 lb	1 ea.
66. Sodium sulfite, 1 lb	1 ea.
67. Silver nitrate solution, 0.0141 N, APHA	1 qt.
68. Barium chloride, crystal 20-30 mesh, 1 lb	1 ea.
69. Glycerin, 1 gal	1 ea.
70. Sodium chloride, crystal, 5 lb	1 ea.

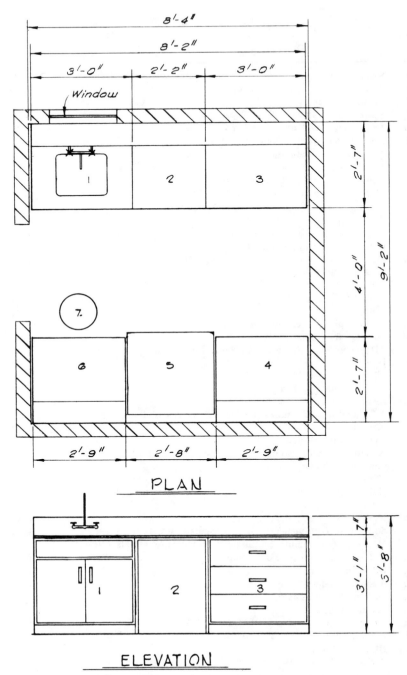

Figure 2. Laboratory B.

Table X. Description of Units shown in Figure 2 for Laboratory B.

1. Sink base unit
2. Drawer base unit for general "wet" tests
3. Drawer base unit for 6-paddle stirrer
4. Shelf base unit for general work space
5. Refrigerator
6. Office desk with file space
7. Chair

Equipment and Supplies for Laboratory B

Description	Quantity
1. Balance, with chain attachment, notched 1-916	1 ea
2. Balance, Harvard trip, single beam	1 ea
3. Balance weights 50 g	1 st
4. Balance weights, std.−A−weigh, 1000 g	1 st
5. Tongs, beaker	1 ea
6. Bottles, 16 oz capacity, polyethylene	1 dz
7. Bottles, dropping, 30 ml (6/pk)	1 pk
8. Bottle, washing, polyethylene, 500 ml (6/pk)	1 pk
9. Burner, natural gas model	1 ea
10. Burner, wingtop	1 ea
11. Clamp, utility	3 ea
12. Forceps	1 ea
13. Distilling apparatus, electrically heated, 1 gph	1 ea
14. Airejector	1 ea
15. Fluorescent titration illuminator−Fisher	1 ea
16. Pipet filler, safety	1 ea
17. Ring, support, cast iron, with clamp $2''$	1 ea
18. Ring, support, cast iron, with clamp $4''$	1 ea
19. Scoopula, stainless steel (6/pk)	1 pk
20. Spatula, stainless steel	6 ea
21. Stirrer, magnetic	1 ea
22. Support stand	1 ea
23. Buret support	2 ea
24. Support, ready rack	1 ea
25. Tongs, crucible (3/bg)	1 bg
26. Tongs, pick-up (3/bg)	1 bg
27. File, round, $6''$ (3/bg)	1 bg
28. File, triangular, $4''$ (6/bg)	1 bg
29. Tripod, iron, with concentric rings, $10''$	1 ea
30. Brush, bottle and flask (12/pk)	1 pk
31. Brush, test tube	1 dz
32. Filter paper, Whatman #1 unwashed medium, 12-½ cm (100 sh/bx)	1 bx
33. Filter paper, Whatman #3 unwashed thick, 12-½ cm (100 sh/bx)	1 bx
34. Tubing, rubber, 3/16″ (12 ft/ln)	1 ln
35. Tubing, rubber, 5/16″ (12 ft/ln)	1 ln

Table XI. Continued

Description	Quantity
36. Tubing, rubber, vacuum 1/4″ (10 ft/pk)	1 pk
37. Wire gauze squares 4″ x 4″ (12/pk)	1 pk
38. Wire gauze squares 6″ x 6″ (12/pk)	1 pk
39. Stoppers, rubber, sz 0-8 assorted	3 lb
40. Acetic acid, glacial, 99.7% (6 x 5 pt/cs)	1 cs
41. Sulfuric acid conc. 6 x 5 pt	1 cs
42. Iodine, resublimed 1 lb	1 ea
43. Potassium bi-iodate, purified 1/4 lb	1 ea
44. Potassium iodide, crystals 5 lb	1 ea
45. Potassium dichromate, N/10 soln., 1 qt	1 ea
46. Sodium thiosulfate $0.1N$ 5 gal	1 ea
47. Starch indicator solution 1 qt	1 ea
48. Comparator, chlorine (D.O.−1 oppm)	1 ea
49. O-tolidine, 0.1% APHA 1 qt.	6 ea
50. Mercaptosuccinic acid, practical 500 g	1 ea
51. Bottles, square jar, 2 oz	2 dz
52. Potassium dichromate 1 lb	1 ea
53. Potassium chromate 1 lb	1 ea
54. Potassium phosphate, monobasic 5 lb	1 ea
55. Sodium arsenite, meta 1 lb	1 ea
56. Sodium phosphate, dibasic 1 lb	1 ea
57. pH meter, Accumet, mdl. 220	1 ea
58. Potassium hydrogen phthalate 1 lb	1 ea
59. Sodium borate, tetra 1 lb	1 ea
60. Sodium phosphate, dibasic 1 lb	1 ea
61. Buffer soln., 4.0 at 25°C 1 pt	1 ea
62. Buffer soln., 9.0 at 25°C 1 pt	1 ea
63. Buret, Squibb, Type B, 50 ml	3 ea
64. Bromcresol green, sodium salt 5 g	1 ea
65. Methyl red, sodium salt 10 g	1 ea
66. Sodium carbonate, anhydrous 1 lb	1 ea
67. Sulfuric acid soln., $0.02N$ 1 gal	1 ea
68. Methyl orange soln., 0.5% APHA 1 pt	1 ea
69. Phenolphthalein indicator soln., APHA 1 qt	1 ea
70. Sodium hydroxide soln., $0.02N$ 1 gal.	1 ea
71. Sodium bicarbonate, powder 1 lb	1 ea
72. Sodium-carbonate, anhydrous 1 lb	1 ea
73. Phenolphthalein indicator soln. 1 qt	1 ea
74. Buffer soln., (Phosphate) pH 7.2 APHA 1 qt	1 ea
75. Sodium hydroxide soln., $1N$ 1 gal	1 ea
76. Thermometer, general lab engraved stem−10° to +260°C 1° subdivision	2 ea
77. Thermometer, precision−1° to +101°C, 1/10° subdivision	1 ea
78. Centrifuge, International Clinical model	1 ea
79. Centrifuge head, 4 place, 50 ml	1 ea
80. Metal shield, 50 ml	4 ea
81. Trunnion ring, 50 ml tube	4 ea
82. Centrifuge tube, 50 ml	12 ea

Table XI. Continued

Description	Quantity
83. Nessler tubes, APHA std. 50 ml	1 cs
84. Color tube support for APHA std. tubes 50 ml	1 ea
85. Platinum cobalt color std. soln., No. 500 APHA 1 qt	1 ea
86. Candles, paraffin (24/pk)	1 pk
87. Turbidimeter, Jackson	1 ea
88. Turbidimeter tube, 75 cm, white line exax.	1 ea
89. Kaolin 5 lb	1 ea
90. Mercuric chloride 1/4 lb	1 ea

Table XII. General Glassware for Laboratory B

Description	Quantity
1. Beaker, Griffin, Pyrex with spout 50 ml (12/pk)	1 pk
2. Beaker, Griffin, Pyrex with spout 250 ml (12/pk)	1 pk
3. Beaker, Griffin, Pyrex with spout 600 ml (6/pk)	1 pk
4. Beaker, Griffin, Pyrex with spout 1000 ml (6/pk)	1 pk
5. Beaker covers 75 mm Pyrex (12/pk)	1 pk
6. Bottles, wide mouth without caps 32 oz	1 dz
7. Bottles, 4 oz capacity glass with narrow mouth and glass stopper	6 ea
8. Bottles, 1000 ml capacity, Pyrex with narrow mouth and glass stopper	6 ea
9. Buret, blue line exax., 25 ml grad in 1/10 ml	1 ea
10. Buret, blue line exax., 50 ml grad in 1/10 ml	1 ea
11. Cylinder, graduated, Pyrex, 50 ml	2 ea
12. Cylinder, graduated, Pyrex, 100 ml	2 ea
13. Cylinder, graduated, Pyrex, 500 ml	2 ea
14. Cylinder, graduated, Pyrex, 1000 ml	2 ea
15. Flask, Erlenmeyer, Pyrex, 250 ml	12 ea
16. Flask, Erlenmeyer, Pyrex, 500 ml	6 ea
17. Flask, filtering, with side-type, Pyrex Brand Glass 1000 ml	4 ea
18. Flask volumetric, Pyrex with stopper, 100 ml	2 ea
19. Flask volumetric, Pyrex with stopper, 500 ml	4 ea
20. Flask volumetric, Pyrex with stopper, 1000 ml (6/cs)	1 cs
21. Funnel, chemical filtering, Kimax 65 mm	2 ea
22. Funnel, chemical filtering, Kimax 90 mm	2 ea
23. Glass rod, Kimble standard flint glass 4 mm 4' lengths	1 lb
24. Glass tubing, Kimble standard flint glass 8 mm 4' lengths	1 lb
25. Glass stirring rods (72/pk) 150 mm	1 pk
26. Pipet, volumetric, Pyrex 25 ml	2 ea
27. Pipet, volumetric, Pyrex 50 ml	2 ea
28. Pipet, volumetric, Pyrex 100 ml	2 ea
29. Pipet, Mohr. 1/10 ml grad. 1 ml (12/pk)	1 pk
30. Pipet, Mohr. 1/10 ml grad. 2 ml (12/pk)	1 pk
31. Pipet, Mohr. 1/10 ml grad. 5 ml (12/pk)	1 pk
32. Pipet, Mohr. 1/10 ml grad. 10 ml (12/pk)	1 pk
33. Pipet, Mohr. 1/10 ml grad. 25 ml (12/pk)	1 pk

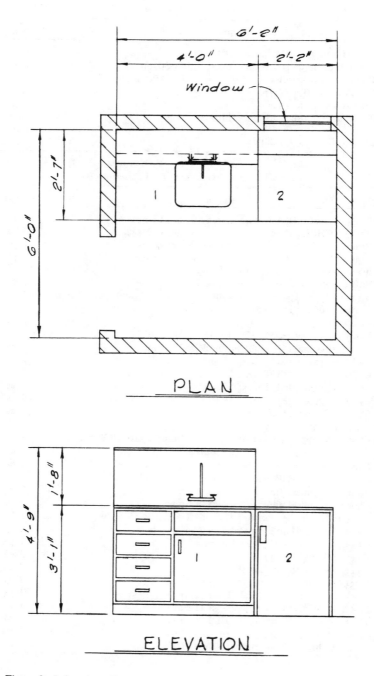

Figure 3. Laboratory C.

Table XIII. Description of Units Shown on Figure 3 for Laboratory C

1.	Sink-drawer base unit
2.	Refrigerator

Table XIV. Equipment and Supplies for Laboratory C

	Description	Quantity
1.	Balance, harvard trip, single beam	1 ea
2.	Balance weights 50 g	1 st
3.	Balance weights, std.–A–weigh, 1000 g	1 st
4.	Tongs, beaker	1 ea
5.	Bottles, 16 oz capacity, polyethylene	1 dz
6.	Bottles, dropping, 30 ml (6/pk)	1 pk
7.	Bottle, washing, polyethylene, 500 ml (6/pk)	1 pk
8.	Burner, natural gas model	1 ea
9.	Burner, wingtop	1 ea
10.	Clamp, utility	3 ea
11.	Forceps	1 ea
12.	Fluorescent titration illuminator	1 ea
13.	Ring, support, cast iron with clamp 2$''$	1 ea
14.	Ring, support, cast iron with clamp 4$''$	1 ea
15.	Scoopula, stainless steel (6/pk)	1 ea
16.	Spatula, stainless steel	6 ea
17.	Stirrer, magnetic	1 ea
18.	Support stand	1 ea
19.	Buret support	2 ea
20.	File, round, 6$''$ (3/bg)	1 bg
21.	File, triangular, 4$''$ (6/bg)	1 bg
22.	Tripod, iron, with concentric rings, 10$''$	1 ea
23.	Brush, bottle and flask (12/pk)	1 pk
24.	Brush, test tube	1 dz
25.	Filter paper, Whatman #1 unwashed medium, 12-½ cm (100 sh/bx)	1 bx
26.	Filter paper, Whatman #3 unwashed thick, 12-½ cm (100 sh/bx)	1 bx
27.	Tubing, rubber, 3/16$''$ (12 ft/ln)	1 ln
28.	Tubing, rubber 5/16$''$ (12 ft/ln)	1 ln
29.	Tubing, rubber, vacuum 1/4$''$ (10 ft/ln)	1 pk
30.	Wire gauze squares 4$''$ x 4$''$ (12/pk)	1 pk
31.	Wire gauze squares 6$''$ x 6$''$ (12/pk)	1 pk
32.	Stoppers, rubber, sz. 0-8 assorted	3 lb
33.	Acetic acid, glacial, 99.7% (6 x 5 pt/cs)	1 cs
34.	Sulfuric acid conc. 6 x 5 pt	1 cs
35.	Iodine, resublimed, 1 lb	1 ea
36.	Potassium bi–iodate, purified	1 ea
37.	Potassium iodide, crystals 6 lb	1 ea
38.	Potassium dichromate, N/10 soln. 1 qt	1 ea
39.	Sodium thiosulfate 0.1N 5 gal	1 ea
40.	Starch indicator soln., 1 qt	1 ea
41.	Comparator, chlorine (D.O.–1 oppm)	1 ea

Table XIV. Continued

Description	Quantity
42. O-tolidine, 0.1% APHA 1 qt	6 ea
43. Mercaptosuccinic acid, practical, 500 g	1 ea
44. Bottles, square jar, 2 oz	2 dz
45. Potassium dichromate 1 lb	1 ea
46. Potassium chromate 1 lb	1 ea
47. Potassium phosphate, monobasic, 5 lb	1 ea
48. Sodium arsenite, meta 1 lb	1 ea
49. Sodium phosphate, dibasic 1 lb	1 ea

Table XV. General Glassware for Laboratory C

Description	Quantity
1. Beaker, Griffin, Pyrex with spout 50 ml (12/pk)	1 pk
2. Beaker, Griffin, Pyrex with spout 250 ml (12/pk)	1 pk
3. Beaker, Griffin, Pyrex with spout 600 ml (6/pk)	1 pk
4. Beaker, Grififn, Pyrex with spout 1000 ml (6/pk)	1 pk
5. Beaker covers 75 mm Pyrex (12/pk)	1 pk
6. Bottles, wide mouth without caps 32 oz	1 dz
7. Bottles, 4 oz capacity glass with narrow mouth and glass stopper	6 ea
8. Bottles, 1000 ml capacity, Pyrex with narrow mouth and glass stopper	6 ea
9. Buret, blue line exax., 25 ml grad. in 1/10 ml	1 ea
10. Buret, blue line exax., 50 ml grad. in 1/10 ml	1 ea
11. Cylinder, graduated, Pyrex, 50 ml	2 ea
12. Cylinder, graduated, Pyrex, 100 ml	2 ea
13. Cylinder, graduated, Pyrex, 500 ml	2 ea
14. Cylinder, graduated, Pyrex, 1000 ml	2 ea
15. Flask, Erlenmeyer, Pyrex, 250 ml	12 ea
16. Flask, Erlenmeyer, Pyrex, 500 ml	6 ea
17. Flask, filtering, with side-type, Pyrex brand glass 1000 ml	4 ea
18. Flask, volumetric, Pyrex with stopper 100 ml	2 ea
19. Flask, volumetric, Pyrex with stopper 500 ml	4 ea
20. Flask, volumetric, Pyrex with stopper 1000 ml (6/cs)	1 cs
21. Funnel, chemical filtering, Kimax 65 mm	2 ea
22. Funnel, chemical filtering, Kimax 90 mm	2 ea
23. Glass rod, Kimble standard flint glass 4 mm 4' lengths	1 lb
24. Glass tubing, Kimble standard flint glass 8 mm 4' lengths	1 lb
25. Glass stirring rods (72/pk) 150 mm	1 pk
26. Pipet, volumetric, Pyrex 25 ml	2 ea
27. Pipet, volumetric, Pyrex 50 ml	2 ea
28. Pipet, volumetric, Pyrex 100 ml	2 ea
29. Pipet, Mohr. 1/10 ml grad. 1 ml (12/pk)	1 pk
30. Pipet, Mohr. 1/10 ml grad. 2 ml (12/pk)	1 pk
31. Pipet, Mohr. 1/10 ml grad. 5 ml (12/pk)	1 pk
32. Pipet, Mohr. 1/10 ml grad. 10 ml (12/pk)	1 pk
33. Pipet, Mohr. 1/10 ml grad. 25 ml (12/pk)	1 pk

laboratory furniture manufacturers. In some laboratories it is also desirable to provide a vacuum system with vacuum outlets. Care must be taken to provide electric outlets with the proper voltage and amperage.

MISCELLANEOUS CONSIDERATIONS

There are some miscellaneous items not covered in the text which the engineer may or should incorporate in the design of a water treatment plant laboratory. Examples of some such items are as follows:

1. The building housing the laboratory should be as sturdy and free from vibration as possible. It is particularly important that analytical balances be located and mounted so as to be subjected to a very minimum of vibration.

2. Small laboratories (such as B or C, Figures 2 and 3), which may use only small quantities of distilled water, might be more economically supplied with purchased distilled water instead of an in-house still. Hence, the optional still is not shown in Figures 2 and 3. However, commercial distilled water must be of the quality required. Usually, a simple chloride and/or hardness test is adequate to determine whether the quality is satisfactory.

SELECTED REFERENCES

1. *Standard Methods for the Examination of Water and Wastewater*, 14th ed. (New York: American Public Health Assoc., 1975).
2. *Simplified Procedures for Water Examination*. AWWA Manual M-12 (Denver, CO: Amer. Water Works Association, 1964).
3. Shugar, G. J., R. A. Shugar and L. Baumann. *Chemical Technicians Ready Reference Handbook* (New York: McGraw-Hill Book Co., 1973).
4. The Texas Water Utilities Assoc. *Manual of Water Utility Operations*, 6th ed. (Austin, Texas: State Department of Health, 1975).

CHAPTER 30

ESTIMATING WATER SYSTEM COSTS

R. Dewey Dickson, P.E.
Senior Vice President—Engineering
James M. Montgomery
Consulting Engineers, Inc.
Pasadena, California   91101

INTRODUCTION

Prior to making a final selection of the components of a water system, whether they be treatment plants, reservoirs, pumping stations, pipelines or a combination of these, it is necessary to have an idea of the cost of the components. Certainly a proposed water system improvement program is virtually worthless in the absence of a reliable cost estimate for the recommended improvements. Experience has shown that in the conceptual or predesign stage of a project, time does not usually permit preparation of a detailed cost estimate, but rather a quick rule-of-thumb method must be used.

HISTORY OF COST INDICES

Within the water works industry there are two systems of indices that are normally used: *Engineering News-Record* and Handy-Whitman.

Engineering News-Record Construction Cost Index

This system was created in 1931 to diagnose the erratic construction costs that occurred immediately following World War I. The base year is 1913, where the index is designated as 100. The ENR index is based upon the cost at any particular time of constant quantities of structural steel, Portland cement, lumber and common labor. The indices have been developed for 20 different cities in the United States and 2 cities in Canada, where each represents a local geographical area. The average of the indices for the 20 U.S. cities is commonly considered as the overall national average.

Handy-Whitman Index of Water Utility Construction Costs

The Handy-Whitman system was designed by Whitman, Requardt and Associates and established in 1957. The base year for Index 100 is 1949, when construction costs were in a relatively stable condition. The index numbers have been developed for six geographical divisions of the United States. Cost indices are given for five main classes of construction: source of supply plant, pumping plant, water treatment plant, transmission and distribution plant, and miscellaneous. The index numbers are published[*] semiannually, on January 1 and July 1 of each year.

The most common index and the one universally used within the construction industry is the *Engineering News-Record* Construction Cost Index (ENR). This index is published monthly in the "Scoreboard" section of the *Engineering News-Record* magazine.[1] A curve of the index between 1913 and publication date appears every year in the March issue.

Table I tabulates historical indices for ENR-LA and ENR-20 cities, together with Handy-Whitman Pacific area indices for large and small water treatment plants and pipelines. Both the ENR and H-W cost indices differ widely across the nation. For the past ten years the ENR index has shown that construction cost is least in Atlanta and greatest in the San Francisco area with Los Angeles not far behind. Figure 1 is a plot of the ENR Construction Cost Index for the years 1966 through 1976 for Atlanta, Denver, Los Angeles, Philadelphia and Seattle. Figure 2 shows the 40-year trend of ENR-LA and San Francisco and the Handy-Whitman Pacific Division indices.

DEVELOPMENT OF COST CURVES

The cost curves included in this chapter are based on actual construction costs of projects, most of which were designed and supervised by James M. Montgomery, Consulting Engineers, Inc. The majority of the projects awere constructed in the southwestern United States, primarily in the Los Angeles area. For this reason, the *Engineering News-Record*

---

*Compiled and published by Whitman, Requardt & Associates, Engineers-Consultants, 1304 Saint Paul Street, Baltimore, Maryland 21202.

Table I. Construction Cost Indices

Year	ENR-LA (September)	ENR-20 (September)	H-W Pacific L. W. T. P.[b]	H-W Pacific S. W. T. P.[b]	H-W Pacific All Type Pipes[b]
1935	107[a]		47	47	45
1936	209		52	52	50
1937	241		54	53	52
1938	233		53	52	51
1939	235		53	52	50
1940	244		57	56	54
1941	263		62	60	58
1942	282		63	61	59
1943	294		64	62	60
1944	301		66	64	61
1945	309		76	74	71
1946	360		90	88	85
1947	426		99	99	97
1948	478		100	100	100
1949	480		106	105	102
1950	552	531	113	112	108
1951	558	544	114	113	111
1952	610	587	117	116	115
1953	627	611	123	121	123
1954	665	640	128	125	128
1955	705	673	135	133	135
1956	736	705	139	138	142
1957	778	738	144	141	149
1958	822	774	151	149	155
1959	877	812	156	153	162
1960	908	831	156	152	164
1961	940	854	159	155	166
1962	969	881	163	158	168
1963	997	914	169	162	171
1964	1019	947	171	165	173
1965	1071	982	179	173	174
1966	1139	1038	187	179	176

LEGEND:

ENR = *Engineering News-Record* Construction Cost Index, based on 1913
U.S. average = 100
LA = Los Angeles
H-W Pacific = Handy-Whitman Index, Pacific Division
L. W. T. P. = Large Water Treatment Plant
S. W. T. P. = Small Water Treatment Plant

[a]Prior to 1950, U.S. national index data available only (based on 20 cities average).
[b]January following year shown.

Table I. Continued

Year	ENR-LA (September)	ENR-20 (September)	H-W Pacific L. W. T. P.[b]	H-W Pacific S. W. T. P.[b]	H-W Pacific All Type Pipes[b]
1967	1198	1097	197	188	179
1968	1272	1184	209	199	183
1969	1292	1285	216	204	189
1970	1482	1421	233	220	208
1971	1694	1654	257	241	217
1972	1907	1786	268	250	219
1973	2093	1929	291	268	225
1974	2290	2089	357	351	322
1975	2568	2272	391	392	347
1976	2807	2468			

Construction Cost Index—Los Angeles (ENR-LA) has been used as the reference for all cost curves. For ease of comparison, an ENR-LA of 2500 was used in developing all curves, but at the time of this writing the ENR-LA was almost 3000.

Cost curves are shown for reservoirs, pumping stations, water treatment plants and pipelines. Data for these curves were obtained from actual construction costs, wherein all nonapplicable cost components were subtracted, leaving only the construction cost of the facility for which the curves were drawn. Preceding each set of curves is a tabulation of pertinent data for each facility. Each numbered item in the tabulation corresponds to the number plotted on the graphs. Best fit curves were then drawn through the appropriate points.

Construction costs are affected by numerous factors other than merely the construction process; all of these must be considered in preparing a preliminary cost estimate. Some of these factors are location, geographic limitations, availability of labor and materials, transportation, seasonal considerations, level of competition and delivery time for critical items. Obviously, the degree of accuracy of these cost-estimating curves does not warrant detailed investigations of factors that might affect the final construction cost. However, any known condition that would substantially affect the cost should be considered and appropriate adjustments made to the cost data.

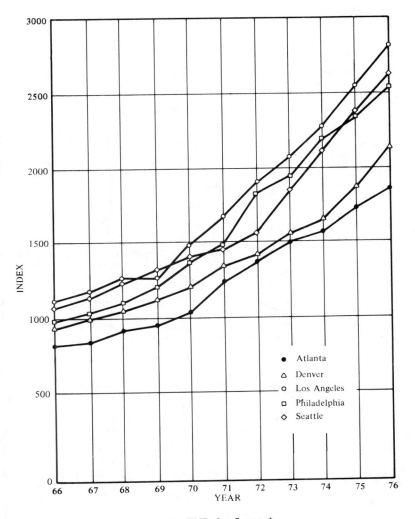

Figure 1. ENR for September.

USE OF COST CURVES

In using the curves for rule-of-thumb cost estimating, the estimator should exercise judgment in all matters that might be unique to his project or items that might affect the actual construction cost. In addition, wherever possible the estimator should refer to the table

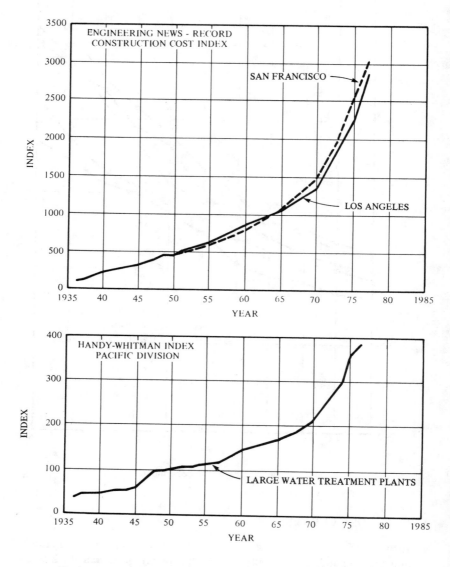

Figure 2. Historical plot of selected *Engineering News-Record* and Handy-Whitman construction cost indices.

preceding the particular curve to determine construction conditions, location, details of construction materials and equipment that might affect the estimate. Once the appropriate unit cost has been determined, the estimator must then adjust this cost to reflect the ENR index for the area in which the facility is to be constructed and extrapolate the cost to the year when construction is anticipated to occur.

OPERATION AND MAINTENANCE COSTS

In order that a cost estimate be truly complete, annual operations and maintenance (O & M) costs should be calculated and added to the annual costs of the capital investment. O & M costs can vary from very low to very high depending upon a multitude of factors, such as the location and attitude of the agency in charge, the amount of money devoted to these activities and the sophistication of the facilities. Further, the operating costs of a treatment plant depend to a great degree upon energy and chemical costs, both of which are extremely sensitive to the total fuel problems presently being experienced throughout the United States.

Because of the highly variable items and the difficulty in projecting rational O & M costs, no attempt has been made to include operations and maintenance cost curves in this chapter. If the estimator wishes to include O & M costs, he should conduct his own investigation concerning those conditions that would affect these costs and extrapolate them over the anticipated life of the facility.

ESTIMATING COSTS OF RESERVOIRS

Reservoir construction cost data in Tables II and III and cost curves in Figures 3, 4 and 5 have been developed from actual construction costs of over 100 reservoirs, ranging in capacity from 0.2 to 990 million gallons (757 m^3 to 3.75 x 10^6 m^3). Costs of unrelated items included in the actual bids were omitted. Curves were plotted for the following types of reservoirs: buried reinforced concrete (Figure 3), partially buried concrete (Figure 4), and steel reservoirs (Figure 5). Three curves are shown for each type of reservoir: high, low and median. The tabulated and plotted costs include those items directly appurtenant to the reservoirs, such as earthwork, underdrainage system and piping. Except for unusually sophisticated controls or treatment of the walls or roof, the costs shown would generally include all items normally included in a reservoir project.

Determination of the appropriate curve should be based upon anticipated degree of difficulty in constructing the project being estimated. Principal items affecting cost are location, site limitations, earthwork

Table II. Concrete Reservoir Data

No.	Reservoir Location	Capacity (million gallons)	Type of Construction			Construction Cost		Date Bid	ENR Construction Index	Cost at ENR-LA Index of 2500 (¢/gal)
			Walls	Floor	Roof	Total	¢/gal			
1	Pasadena-Jones	50	B-C	Reinf. conc.	U	$ 950,000	1.90	Nov. 1949	480.0	9.6
2	Pasadena-Coronet	1.4	C	Reinf. conc.	U	100,000	7.1	June 1951	543.4	32.9
3	Torrance	10	B-C	Reinf. conc.	U	422,000	4.2	Sept. 1952	586.7	18.1
4	Inglewood-Morningside	15	B-C	Asph. conc.	V	388,000	2.6	Jan. 1953	587.9	11.5
5	LVVWD-Charleston Heights	30	A-C	Asph. conc.	S	617,000	2.1	Oct. 1954	641.5	7.9
6	Beverly Hills-Sunset	6	C	Reinf. conc.	U	422,000	7.0	Jan. 1956	676.4	26.0
7	Ontario-Boles	2	C	Reinf. conc.	U	103,000	5.2	Mar. 1956	680.2	19.0
8	Anaheim	4	A-C	Asph. conc.	T	186,000	4.7	May 1956	688.4	16.9
9	Santa Ana-Walnut St.	7	C	Reinf. conc.	U	362,000	5.2	Sept. 1956	704.9	18.4
10	Monrovia-Mountain Ave.	5	C	Reinf. conc.	U	265,000	5.3	Oct. 1956	703.9	18.8
11	Monrovia-Cloverleaf	6.5	C	Reinf. conc.	U	353,000	5.4	Mar. 1957	708.6	19.3
12	Cucamonga	4	C	Asph. plank	X	75,000	1.9	May 1957	715.7	6.5
13	Newport Beach-Big Canyon	196	a	a	X	968,000	0.48	Dec. 1957	723.9	1.71
14	Ontario-Totman	3.75	C	Reinf. conc.	U	184,000	4.9	June 1958	757.3	16.2
15	Ontario-Fern Ave.	20	C	Reinf. conc.	U	693,000	3.5	Nov. 1958	774.1	11.2
16	Fullerton-Laguna	2	E	Reinf. conc.	U	131,000	6.6	Jan. 1959	778.3	21.0
17	Long Beach	14	C	Unreinf. conc.	X	344,000	2.5	Apr. 1959	784.4	7.8
18	Santa Monica-Riviera	25	C	Reinf. conc.	U	998,000	4.0	Jan. 1960	811.8	12.4
19	Arcadia-Santa Anita	4.1	C	Reinf. conc.	U	319,000	7.8	Jan. 1960	811.8	24.0
20	Bellflower	2	C	Reinf. conc.	U	177,000	8.9	Mar. 1960	813.3	27.2
21	Beverly Hills, Lower La Cienega	0.2	C	Reinf. conc.	Y	50,000	25.0	Mar. 1962	861.5	72.5
22	CCCWD-Bailey	2	C	Reinf. conc.	U	152,000	7.6	Apr. 1962	863.2	22.1
23	Santa Barbara	5.0	C	Reinf. conc.	U	373,000	7.5	Oct. 1962	880.2	21.2
24	CCCWD-San Miguel	2	C	Reinf. conc.	U	178,000	8.9	Nov. 1962	879.9	25.3

#	Name								
25	Irvine RWD-San Joaquin	990	a	a	X	4,800,000	0.49	Dec. 1962	879.9	1.4
26	Pomona	1	F	Reinf. conc.	V	61,000	6.1	Mar. 1963	884.2	17.2
27	CCCWD-Newhall	3	C	Reinf. conc.	U	259,020	8.6	Mar. 1964	922.4	23.4
28	Buena Park	20.0	C	Reinf. conc.	U	810,276	4.05	June 1964	935.4	10.9
29	Santa Ana-South	2-6	F	Steel & conc.	V	840,000	7.0	Mar. 1965	957.7	18.2
30	EBMUD-Oakland-Dunsmuir	65	C	Reinf. conc.	U	3,512,000	5.4	Sept. 1966	1037.0	13.1
31	T. P. Clearwell	10	C	Reinf. conc.\	U	745,000	7.4	Sept. 1966	1139.0	16.3
32	Vernon-Civic Center	10	C	Reinf. conc.	U	975,000	9.75	Feb. 1967	1143.0	21.3
33	Pomona-5B	10	C	Reinf. conc.	U	499,000	4.99	May 1967	1194.0	10.4
34	LVVWD-Flamingo (Campbell)	40	C	Reinf. conc.	U	1,995,000	5.0	Aug. 1967	1198.0	16.4
35	Encinitas-San Dieguito	13.3	C	Reinf. conc.	U	920,000	6.9	Sept. 1967	1198.0	14.4
36	Beverly Hills-Greystone	19.3	C	Reinf. conc.	U	2,093,000	10.7	Nov. 1967	1203.0	22.5
37	CCCWD-North Lime Ridge	3.0	C	Reinf. conc.	U	372,600	12.42	Mar. 1968	1191.0	26.0
38	Monrovia-Ivy Avenue	3.2	C	Reinf. conc.	Y	379,300	11.85	July 1968	1249.0	23.7
39	Pomona Res. #9	5.3	C	Reinf. conc.	U	429,499	8.25	Aug. 1968	1272.0	15.9
40	LVVWD-Leonard Fayle	40.0	C	A. C. w/butyl	S	2,069,158	5.17	Jan. 1969	1275.0	10.1
41	Redlands-Country Club (Round)	2.0	C	Reinf. conc.	Y	202,618	10.13	Apr. 1969	1288.0	19.7
42	San Diego	15.0	C	Reinf. conc.	S	826,588	5.51	May 1969	1331.0	10.3
43	San Francisco-San Andreas	6.5-	C	Reinf. conc.	U	598,662	9.21	Aug. 1969	1292.0	17.8
44	CCCWD-Bailey #2 (3-wall)	4.5	C	Reinf. conc.	U	419,000	9.31	Jan. 1970	1309.0	17.8
45	Sacramento-Florin	15.0	C	Reinf. conc.	U	1,250,000	8.33	Jan. 1971	1495.0	14.0
46	LVVWD-Pico	20.0	C	Reinf. conc.	U	1,527,660	7.63	Dec. 1970	1479.0	12.9
47	Redlands-Dearborn	10.6	C	Reinf. conc.	Z	710,730	6.71	June 1971	1596.0	10.4
48	Ontario-Totman	3.0	C	Reinf. conc.	U	362,200	12.1	Sept. 1971	1694.0	17.8
49	Duarte Water Co.-Scott Res.	1.5	E	Reinf. conc.	Dome	149,000	9.92	Apr. 1960	818.0	30.3
50	Santa Ana-Cambridge	1.3	C	Reinf. conc.	Y	141,800	10.9	Dec. 1966	1125.0	24.1
51	CCCWD-Pine Hollow #2 (3-wall)	2.0	C	Reinf. conc.	U	288,560	14.4	Oct. 1972	2078.0	17.4
52	Valley W. Co.-Canada #2	1.4	C	Reinf. conc.	Builtup	223,750	16.0	May 1973	1979.0	20.2
53	Redlands-5th Ave.	5.0	C	Reinf. conc.	Y	542,003	10.8	July 1973	1984.0	13.6
54	Fairfield-Waterman	10.0	C	Reinf. conc.	U	1,172,300	11.7	Sept. 1973	2224.0	13.1

Table II. Continued

No.	Reservoir Location	Capacity (million gallons)	Type of Construction Walls	Type of Construction Floor	Roof	Construction Cost Total	Construction Cost ¢/gal	Date Bid	ENR Construction Index	Cost at ENR-LA Index of 2500 (¢/gal)
55	CCCWD-Lime Ridge #2 (3-wall)	7.2	C	Reinf. conc.	U	$ 743,850	10.3	Oct. 1973	2224.0	11.7
56	LVVWD-West Central	20.0	C	Reinf. conc.	U	2,228,610	11.1	Jan. 1974	2100.0	13.2
57	La Habra-Coyote Hills	5.0	C	Reinf. conc.	Y	639,380	12.8	Feb. 1974	2110.0	15.23
58	San Dieguito I. D.-Balour	2.5	C	Reinf. conc.	U	483,000	19.3	June 1974	2121.0	22.7
59	NMCWD-Pacheco	5.0	C	Reinf. conc.	Y	1,034,000	20.7	July 1974	2148.0	24.1
60	LVVWD-Harvey E. Luce	20.0	C	Reinf. conc.	U	2,370,000	11.9	Jan. 1975	2285.0	12.96
61	Oceanside-El Camino	3.0	C	Reinf. conc.	Y	576,600		July 1975	2438.0	
62	Redlands-Highland Ave.	10.0	C	Reinf. conc.	U	1,615,000	16.2	Aug. 1975	2438.0	16.56

LEGEND:

Type of wall

A = Sloping—asphaltic concrete
B = Sloping—concrete
C = Vertical—concrete
D = Steel—exposed
E = Prestressed concrete
F = Arch panels
G = Steel—partially buried

Type of roof

S = Steel frame with aluminum sheeting
T = Steel frame with galvanized steel sheeting
U = Monolithic concrete—buried
V = Precast concrete—not buried
W = Steel—not buried
X = Open
Y = Monolithic concrete—not buried
Z = Post tensioned concrete roof—not buried

a Asphaltic concrete lined open reservoir with compacted earth embankment. Cost includes control building and chlorine facilities.

Table III. Steel Reservoir Data

No.	Reservoir Location	Capacity (million gallons)	Type of Construction			Construction Cost		Date Bid	ENR Construction Index	Cost at ENR-LA Index of 2500 (¢/gal)
			Walls	Floor	Roof	Total	¢/gal			
1	Arcadia-Orange Grove	2-3.5	D	Steel	W	$ 262,000	3.7	Mar. 1953	588.00	15.9
2	FMWD-La Canada	1	G	Steel	W	60,000	6.0	Apr. 1954	615.4	24.4
3	FMWD-Paschall	1	G	Steel	W	61,000	6.1	Apr. 1954	615.4	24.9
4	Monterey Park	3	D	Steel	W	76,000	2.5	May 1954	618.7	10.3
5	Fullerton-Acacia	5	D	Steel	W	180,000	3.6	Oct. 1955	673.9	13.4
6	Pomona-Nos. 2B & 7B	2-3	G	Steel	W	212,500	3.5	Feb. 1956	680.2	12.9
7	Monrovia-Norumbega	1	D	Steel	W	69,000	6.9	Apr. 1956	683.3	25.3
8	Monrovia-Emerson Flts.	0.5	D	Steel	W	40,000	8.0	Apr. 1956	683.3	29.3
9	Beverly Hills-5	1	D	Steel	W	64,000	6.4	June 1956	692.1	23.1
10	Beverly Hills-4B	1	D	Steel	W	64,000	6.4	June 1956	692.1	23.1
11	Pomona-No. 8A	3	D	Steel	W	144,000	4.8	Oct. 1956	703.9	17.1
12	Beverly Hills-7	1.2	D	Steel	W	69,000	5.8	Oct. 1958	775.3	18.5
13	Beverly Hills-6	1	D	Steel	W	59,000	5.9	Oct. 1958	775.3	19.0
14	Sierra Madre-Auburn	1.3	D	Steel	W	84,000	6.5	June 1959	794.4	20.3
15	Yorba Linda	2-2	D	Steel	W	192,000	4.8	Mar. 1960	813.3	14.7
16	Ventura-Sexton	2.6	G	Steel	W	177,000	6.8	May 1960	822.6	20.7
17	Ventura-Willis	1	D	Steel	W	96,000	9.6	May 1960	822.6	29.1
18	Ventura-Foothill	0.75	D	Steel	W	45,000	6.0	May 1960	822.6	18.2
19	Ventura-Golf Course	0.2	G	Steel	W	35,000	16.6	May 1960	822.6	53.2
20	Crescenta VCWD-Oakland	1.5	D	Steel	W	110,000	7.3	July 1960	829.3	22.1
21	Crescenta VCWD-Goss	0.5	D	Steel	W	45,000	9.0	July 1960	829.3	27.1
22	Crescenta VCWD-Smith	0.5	D	Steel	W	54,000	10.8	July 1960	829.3	32.5
23	Sierra Madre-Mira Monte	1	D	Steel	W	80,000	8.0	Feb. 1961	834.3	24.0
24	Azusa	0.6	D	Steel	W	70,000	11.7	Nov. 1961	855.4	34.1
25	Covina	3	D	Steel	W	130,000	4.3	Jan. 1962	855.4	12.6
26	Fullerton	5	D	Steel	W	205,000	4.1	Aug. 1962	881.4	11.6

Table III. Continued

| No. | Reservoir Location | Capacity (million gallons) | Type of Construction | | | Construction Cost | | Date Bid | ENR Construction Index | Cost at ENR-LA Index of 2500 (¢/gal) |
			Walls	Floor	Roof	Total	¢/gal			
27	Santa Ana-West	6	D	Steel	W	$ 245,000	4.1	Nov. 1962	879.9	11.6
28	LVVWD-Southwest	3.2	D	Steel	W	152,000	4.8	Aug. 1963	914.8	12.9
29	LVVWD-West-Central	2.1	D	Steel	W	136,000	6.5	Aug. 1963	914.8	17.5
30	Upland-15th Street	4	D	Steel	W	222,000	5.6	Sept. 1963	913.8	15.1
31	Upland-19th Street	5.4	D	Steel	W	210,000	3.9	Sept. 1963	913.8	10.6
32	Pomona-No. 2C	3	D	Steel	W	132,000	4.4	Nov. 1963	913.5	12.1
33	Pomona-No. 8B	3	D	Steel	W	133,000	4.5	Nov. 1963	913.5	12.1
34	Fountain Valley	5	D	Steel	W	266,000	5.3	Sept. 1964	947.4	14.0
35	Chino Basin-Feeder	3	D	Steel	W	171,000	5.7	Jan. 1965	947.6	15.0
36	Chino Basin-Fontana	3	D	Steel	W	183,000	6.1	Jan. 1965	947.6	16.0
37	LVVWD-Charleston Heights	2-10	D	Steel	W	1,005,400	5.9	May 1965	969.0	12.9
38	Fullerton-Acacia #2	5	D	Steel	W	172,910	3.5	May 1967	1170.0	7.4
39	Fullerton-2D	5-6.4	D	Steel	W	1,492,800	4.7	May 1967	1170.0	10.0
40	Covina	1.1	D	Steel	W	103,551	9.4	Oct. 1967	1180.0	19.9
41	Santa Ana	6.0	D	Steel	W	233,390	3.8	June 1968	1249.0	7.8
42	Cucamonga CWD	1.0	D	Steel	W	82,920	8.2	Nov. 1969	1305.0	15.9
43	Umark-West Covina	7.0	D	Steel	W	332,652	4.75	July 1970	1475.0	8.1

LEGEND: Type of wall
A = Sloping–asphaltic concrete
B = Sloping–concrete
C = Vertical–concrete
D = Steel–exposed
E = Prestressed concrete
F = Arch panels
G = Steel–partially buried

Type of roof
S = Steel frame with aluminum sheeting
T = Steel frame with galvanized steel sheeting
U = Monolithic concrete–buried
V = Precast concrete–not buried
W = Steel–not buried
X = Open
Y = Monolithic concrete–not buried
Z = Post tensioned concrete roof–not buried

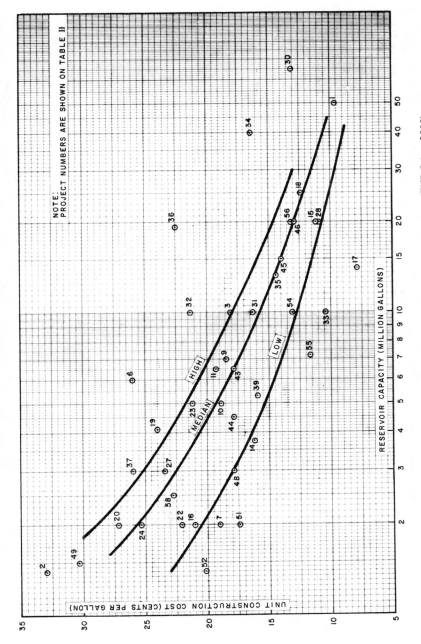

Figure 3. Construction cost of buried reinforced concrete reservoirs. (ENR-LA = 2500)

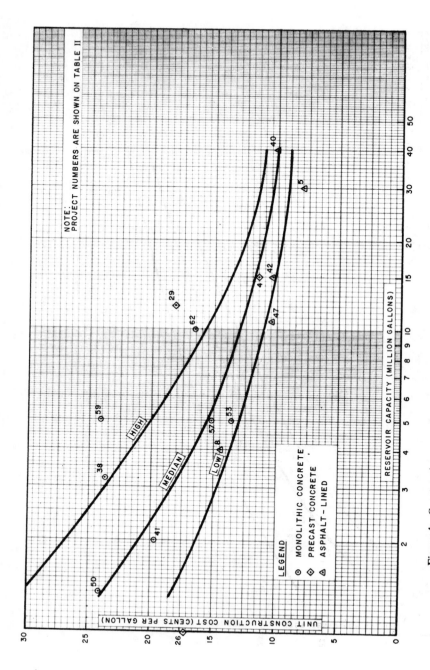

Figure 4. Construction cost of partially buried reinforced concrete reservoirs. (ENR-LA = 2500)

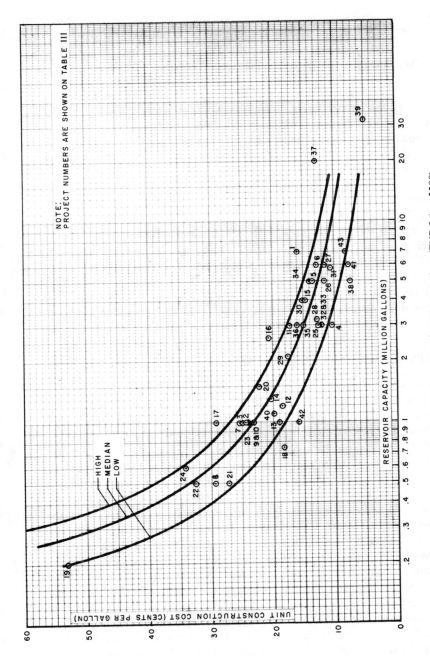

Figure 5. Construction cost of steel reservoirs. (ENR-LA = 2500)

quantities, foundation and seismic conditions, utility relocation, architectural treatment and roof use and loading.

ESTIMATING COSTS OF PUMPING STATIONS

Table IV presents a list of pumping stations recently constructed in the southwestern United States. For some of those listed, it was not possible to develop cost data or such data were not readily available. Costs for pumping stations located in the San Francisco Bay area were developed using the ENR-SF index, while installations in southern California were based on the ENR-LA index.

Figure 6 shows a family of cost curves for pumping stations of varying degrees of complexity, including additions to existing pumping stations as well as gas or diesel engine-driven stations. These curves relate the unit cost in dollars per horsepower to the total installed horsepower, which itself is a direct measure of the pumping capacity and total dynamic head. Capacities of the pumping stations range from 230 to 250,000 gpm (14.5 l/sec to 15.77 x 10^3 l/sec) and 20 to 90,000 (15 to 67,000 kW) installed horsepower. The curves were developed for a complete pumping station, including all earthwork, structures, metering and controls. The numbers shown in Figure 6 are identified in Table IV.

ESTIMATING COSTS OF PIPELINES

The costs presented in Table V are based on data furnished by pipe manufacturers augmented by historical construction information. The data of Table VI are actual historical construction costs of transmission pipelines commonly used in water projects. These data are limited to mortar-lined and mortar-coated steel or mortar-lined and coal tar-coated steel pipe (ML and CSP), steel cylinder concrete pipe (SCCP) and pre-stressed concrete cylinder pipe (PCCP). The actual costs are updated to an ENR-LA index of 2500, and these updated costs are plotted in Figure 7 as a family of curves for pipelines at different depths and for diameters of 10-60 in. (25-152 cm). The cost curves include all trenching and backfilling, piping materials, line (isolation) valves, blow-offs, air release valves, and other appurtenances, but they exclude costs of major control features such as pressure-reducing or rate-of-flow control stations. In general, pipe materials represent approximately one-half of the total construction cost.

Costs for other pipeline materials, such as ductile iron and asbestos cement, should be quite similar to the costs shown in Figure 7, with some possible variations due to material costs. Pipe commonly used in

Table IV. Pumping Station Data

No.	Owner	Type	Pump Data No. (Fut.)	Pump Data Design Capacity (gpm)	Horsepower Installed	Horsepower Future (Est.)	Horsepower Total	Total Construction Cost ($)	Bid Date	ENR[a] @ Bid Date	Construction Cost—$/hp @ ENR·LA[b] 2500 Installed	Future (Est.)	Total	Remarks
1	Ventura-Seawater Intake	V.T.	1	350	20	–	20	5,000	9/49	480	1,300	–	1,300	Salt water
2	Ventura-Modella	V.T.	3	1,500	75	–	75		9/49	480				
3	Ventura-Buena Vista	Horiz.	4	1,500	140	–	140		9/49					
4	Pasadena								1949					
5	Torrance	V.T.	3	2,100	150	–	150		10/52					
6	Pomona								1952					
7	El Centro	Horiz.		6,200			115		6/53					
8	Inglewood-Morningside			10,000			635		1953					
9	Foothill MWD - La Canada	Sub	3(1)	5,600	600	200	800	128,260	4/54	667	808	432	714	
10	Foothill MWD - Main	Sub	6(4)	20,000	1,325	800	2,125	309,530	4/54	667	883	261	649	
11	LVVWD - Whitney	Sub	4(2)	35,000	1,400	700	2,100	277,800	12/54	667	750	300	600	
12	LVVWD - BMI No. 2	Sub	5(1)	30,000	2,000	400	2,400	297,500	3/55	708	525	261	481	
13	Crescenta Valley CWD - Paschall	Sub	2(1)	4,000			500		4/55					
14	LVVWD - BMI Intake	Sub	5	30,000	4,000	–	4,000		4/55					
15	LVVWD - Bonanza	V.T.	3(1)	8,000			525		5/55					
16	LVVWD - Charleston Blvd.	V.T.	4(7)	17,200			1,250		5/55					
17	Vernon - No. 2	V.T.	3	4,600	320		320		6/55					
18	Vernon - No. 3		3	4,600	320		320		6/55					
19	LVVWD - Manganese	V.T.	3	4,050	300		300		2/56					
20	Cucamonga CWB - #1 - #5	Sub	10	14,100	1,475		1,475		3/56					
21	Santa Ana - Bristol St.	V.T.		15,100	300	550	850		1/56					
22	Monrovia - San Gabriel	V.T.	2(2)	12,000	600	600	1,200	254,390	7/56	731	1,450	333	892	
23	Beverly Hills - No. 4-A	V.T.	1	2,000	20	–	20	17,800	8/56	731	3,042	–	3,042	
24	Beverly Hills - No. 4-B	Sub	2	3,000	250	–	250	50,600	8/56	731	692	–	692	
25	Beverly Hills - No. 8	Horiz.	4	2,850	280	–	280	62,800	8/56	731	767	–	767	
26	Santa Ana - 1st Street		5	12,800			850		9/56					
27	Newport Beach - San Gabriel								1956					
28	Monrovia-Ridgeside	Sub	4(1)	2,400	200	50	250	84,610	5/57	778	1,358	796	1,245	
29	Monrovia - May Ave.	Sub	4(2)	6,000	250	250	500	99,300	5/57	778	1,275	378	827	
30	Laguna Beach - No. 1-11	Vert.	Various	4,500			180		12/57					
31	Beverly Hills - No. 5	Sub	2	1,900	150	–	150	65,600	9/58	824	1,325	–	1,325	
32	Beverly Hills - No. 6	Sub	2	2,000	150	–	150	52,820	9/58	824	1,067	–	1,067	

Table IV. Continued

No.	Owner	Pump Data			Horsepower			Total Construction Cost ($)	Bid Date	ENR[a] @ Bid Date	Construction Cost—$/hp @ ENR - LA[b] 2500			Remarks
		Type	No. (Fut.)	Design Capacity (gpm)	Installed	Future (Est.)	Total				Installed	Future (Est.)	Total	
33	Beverly Hills - No. 3	Sub	2(1)	5,700	250	125	375	118,235	12/58	824	1,433	557	1,141	
34	Beverly Hills - No. 2	Sub	3(2)	11,000	525	350	875	160,270	12/58	824	925	318	682	
35	Newport Beach - Sierra Drive	V.I.	2	400			30		5/59					
36	Fullerton-Laguna	Sub	1(2)	2,400	40	80	120	62,500	1959	868	4,500	636	1,924	
37	Yorba Linda - No. 2	Sub	3(1)	7,600			500		3/60					
38	Duarte Water Co. - New Scott	V.I.	2(2)	1,700			200		4/60					
39	Bellflower - Flora Vista	V.I.	3	5,100	310	–	310	102,625	5/60	901	917	–	917	
40	Yorba Linda - No. 3	V.I.	4	4,000	235	–	235		8/60					
41	Ventura - Golf Course	V.I.	3(1)	5,700			725		8/60					
42	LVVWD - Charleston Heights	V.I.	2(8)	16,000			1,250		10/60					
43	Ontario - Sewage	V.I.	2(2)	3,500	60	80	140	33,225	4/61	940	1,473	637	995	
44	Yorba Linda - No. 1	Horiz.	3(1)	8,000			500		4/61					
45	Sacramento - Riverside Hi-lift	Sub	5	25,000	1,050		1,050		5/61					
46	Sacramento - Riverside Lo-lift	V.I.	3	19,200	125		125		5/61					
47	Newport Beach - Big Canyon	V.I.	4	2,750	115	–	115	156,375	6/61	940	3,617	–	3,617	
48	Azusa - Res. No. 1	Horiz.	2(1)	2,700	200	100	300	59,020	8/61	940	785	525	698	
49.	Sacramento - American River Hi-lift	V.I.	6	105,000	4,200	–	4,200		11/61					
50	Sacramento - American River Lo-lift	V.I.	5	73,000	1,050		1,050		11/61					
51	LVVWD - Pittman	Horiz.	–						1961					
52	Beverly Hills - Lower La Cienega	Sub	3(1)	6,000			240		3/62					
53	Sacramento - No. 1	Horiz.	3	12,600	150		150		1/62					
54	Sacramento - No. 2	Horiz.	3	12,600	150		150		6/62					
55	Sacramento - No. 3	Horiz.	3	12,600	150		150		6/62					
56	Fullerton - 3 BC - 4	Horiz.	2	1,200	50		50	26,900	8/62	881	1,528	–	1,528	
57	Avalon - Sewage		2	3,600	40		40	51,500	10/62	881	3,656	–	3,656	
58	Contra Costa CWD - San Miguel	Sub	5	8,120			450		11/62	966				

Table IV. Continued

No.	Name	Type											
59	Santa Ana - West Res.	V.T.	3(3)	7,800	450	450	900	98,976	1962	966	569	235	402
60	Avalon - Seawater	V.T.	2	3,400	200	–	200	63,490	1/63	991	800	–	800
61	Contra Costa CWD - Govt. Ranch	V.T.	3	7,500	600	–	600	107,140	8/63	1019[a]	450	–	450
62	Upland - 15th Street	V.T.	2(2)	6,000	250	250	500	67,760	9/63	991	683	383	533
63	Upland - 19th Street	V.T.	2(2)	5,000	250	250	500	88,490	9/63	991	892	383	638
64	Upland - 22nd Street	V.T.	2(2)	5,000	250	250	500	73,610	9/63	991	742	383	563
65	Fountain Valley - Euclid	V.T.	2(1)	6,000	150	75	225	58,460	9/64	1017	958	750	888
66	Goleta San. Dist. Sewage	V.T.	2	6,000	120	–	120		10/64	1017			
67	Santa Ana - So. Res.	V.T.	3(1)	16,000	705	470	1,175	507,060	3/65	1068	1,683	250	1,110
68	Inglewood - North	V.T.	2(1)	14,400	300	100	400	290,640	12/65	1068	2,267	617	1,855
69	Valley Water Co. - Flintridge	V.T.		900	80	40	120	14,158	1964	1017			
70	Valley Water Co. - Main		6	6,450	750	–	750	137,306	1964	1017	450	–	450
71	San Diego - No. 1 Sewage	Horiz.		170,000	1,800	1,800	3,600	1,011,978	1962	966	1,455	500	978
72	San Diego - No. 2 Sewage	Horiz.		250,000	9,000	9,000	18,000	5,059,892	1962	966	1,455	500	978
73	Contra Costa CWD - Nobhill	V.T.	2	4,000			200		1965				
74	Contra Costa CWD - Pine Hollow	Horiz.	3	6,000	225	150	375	170,000	1/66	1133[a]	1,667	517	1,206
75	Contra Costa CWD - Seminary	V.T.	2	4,400			300		1965				
76	Contra Costa CWD - Bailey	Horiz.	3	7,500	375	250	625	75,000	1/66	1133[a]	441	331	397
77	Santa Monica - San Vicente	Sub	6	9,100	545	–	545		1958				
78	Santa Monica - Arcadia	Sub	5	12,100	675	–	675		1958				
79	Santa Monica - Charnock	V.T.	6(2)	13,760			770						
80	Beverly Hills - La Cienega	V.T.	4(1)	12,000	800	200	1,000	231,640	3/66	1088	665	433	532
81	Govt. of Jamaica - Beacon	V.T.	2	340			45		1962				
82	Govt. of Jamaica - Palmeto Pen	V.T.	2	940			50						
83	Govt. of Jamaica - Ewarton	Horiz.	2	840					1963				
84	Govt. of Jamaica - Miranda Hill	V.T.	1	1,900			60		1962				
85	Govt. of Jamaica - Salt Springs	V.T.	2	320			65						
86	Govt. of Jamaica - Reading Springs	V.T.	1	230			75						

Table IV. Continued

No.	Owner	Pump Data			Horsepower			Total Construction Cost ($)	Bid Date	ENR[a] @ Bid Date	Construction Cost—$/hp @ ENR - LA[b] 2500			Remarks
		Type	No. (Fut.)	Design Capacity (gpm)	Installed	Future (Est.)	Total				Installed	Future (Est.)	Total	
87	Govt. of Jamaica - Spanish Town	V.T.	3	2,500			75							
88	Govt. of Jamaica - Appleton Hill	V.T.	2	1,100			100							
89	Govt. of Jamaica - Hounslow	V.T.	2	840			110							
90	Govt. of Jamaica - Nutshell	Horiz.	3	1,100			120							
91	Govt. of Jamaica - Barnstaple	V.T.	6	1,460			120							
92	Govt. of Jamaica - Tollgate	Horiz.	3	2,300			180							
93	Govt. of Jamaica - Jericho	Horiz.	3	1,920			240							
94	Govt. of Jamaica - New Forest	V.T.	4	1,000			260							
95	Govt. of Jamaica - Falmouth	V.T.	5(4)	7,100			275							
96	Govt. of Jamaica - Hopewell - Sandy Bay	V.T.	7(4)	11,600			480							
97 / 100	Not used													
101	Santa Ana - Bristol St. P.S. Add.	V.T.	3	8,000	550	–	550	84,200	5/62	863	444	–	444	
102	LVVWD - Charleston Heights Add.	V.T.	4	6,400	500	–	500	36,972	12/62	880	210	–	210	
103	Ontario - Magnolia Ave. Sewage P.S.	V.T.	2	1,200	30	–	30	35,400	1/66	1135	2,596	–	2,596	
104	Coachella Valley CWD - Hayes St. P.S. & Res.	Sub	3	2,250	300	–	300	144,300	7/66	1135	1,058	–	1,058	
105	Contra Costa CWD - Clayton Valley P.S.	V.T.	3(2)	8,400	300	200	500	131,345	9/66	1210[a]	905	433	716	
106	Contra Costa CWD - San Miguel P.S. Add.	Sub	2	8,700	350	–	350	49,845	9/66	1210[a]	294	–	294	
107	Santa Ana - Cambridge P.S.	V.T.	3(1)	6,600	225	75	300	65,300	12/66	1135	638	570	621	

Table IV. Continued

No.	Name	Type	Units						Date					Notes
108	Vernon - Civic Center P.S.	V.T.	5(1)	13,700	960	300	1,260	285,000	2/67	1182	628	350	561	Pumps furnished by District
109	Santa Fe Irrigation Dist. San Dieguito P.S. Add.	V.T.	3	10,500	1,250	-	1,250	109,000	3/67	1182	184	-	184	Pumps furnished by District
110	Pomona - Recl. Water P.S. Hi-Lift	V.T.	2(1)	2,700	125	75	200	37,000	6/67	1182	626	636	629	
111	Pomona - Recl. Water P.S. Lo-Lift	V.T.	2(2)	6,000	40	40	80	35,000	6/67	1182	1,850	820	1,335	
112	LVVWD - Flamingo (Campbell) P.S.	V.T.	11(2)	85,000	6,350	1,200	7,550	1,656,000	10/67	1182	551	125	483	
113	Chino Basin MWD - P.S. No. 1	Vert.	2	3,000	200	-	200	291,000	5/68	1249	2,910	-	2,910	Gas engines
114	LVVWD - West Central P.S. Zone A	V.T.	3(5)	13,500	210	350	560	101,000	7/68	1275	943	300	541	
115	Santa Ana - East P.S.	V.T.	2(2)	9,200	250	250	500	221,600	10/68	1275	1,737	486	1,112	
116	LVVWD - Fayle P.S.	V.T.	5(7)	49,000	1,500	2,100	3,600	540,000	4/69	1309	688	167	384	
117	Contra Costa CWD - Mallard P.S. Eqmt.	V.T.	2	11,200		-	500		7/69	1525a	2,190	-	2,190	
118	Reedy Creek I.D. - P.S. A	Cent.	5	12,000	750	-	750	853,000	2/70	1298				
119	Santa Rosa Ranches W.D. - Grand Ave.	V.T.	3	1,950	600	-	600	80,107	2/70	1298	257	-	257	Pumps by District
120	San Dieguito Irrigation Dist.	V.T.	2	8,400	1,000	-	1,000	47,862	7/71	1595	75	-	75	Pumps by District
121	City of Covina - Charter Oak P.S.	V.T.	2(2)	5,200	150	150	300	40,697	1/71	1474	460	-		
122	City of Upland - Sec. Ph. Imp. at E. 16th	V.T.	2(1)	4,950	200	100	300	119,160	4/71	1524	977	602	852	
123	City of Los Angeles - Vicksburg P.G.	Vert.	3	4,700	120	-	120	264,000	3/71	1524	3,608	-	3,608	
124	Contra Costa CWD - Lime Ridge P.S.	V.T.	3(1)	16,000	600	200	800	139,000	7/71	1709a	338	338	338	
125	Rancho California W.D. - Del Portola P.S.	V.T.	1	1,380	150	-	150		1/72	1705				
126	Rancho California W.C. - Anza P.S.	V.T.	2	3,000	300	-	300	163,000	1/72	1705				
127	Santa Rosa Ranches W.D. - E. Bluff P.S.	V.T.	1	740	125	-	125		1/72	1705	582	830	604	
			1	685	125	-	125							
			1(1)	1,225	160	40	200							
128	Yorba Linda CWD - P.S.	V.T.	2	4,600	225	-	225	157,500	3/72	1723	1,015	-	1,015	
129	City of Del Mar - Sewage P.S.	Vert.	3	2,880	210	-	210	155,000	4/73	1983	930	-	930	
129A	Clark CWD - #1 East Las Vegas L.S.							223,739	1/74	2100				
130	Santa Rosa Ranches W.D. - Del Oro P.S.	V.T.	1	1,000	75	-	75	132,500	3/74	2076	2,120	-	2,120	

Table IV. Continued

No.	Owner	Type	No. (Fut.)	Pump Data Design Capacity (gpm)	Horsepower Installed	Horsepower Future (Est.)	Horsepower Total	Total Construction Cost ($)	Bid Date	ENR[a] @ Bid Date	Construction Cost—$/hp @ ENR - LA[b] 2500 Installed	Future (Est.)	Total	Remarks
131	Rancho California Road P.S.	V.T.	3	4,050	375	–	375		1974					Pumps & meters only
132	City of Anaheim - Westridge P.S.	Cent.	1	1,500	150	–	150							
133	Otay MWD - La Presa P.S. Modifications	Horiz.	2(2)	700 (2,000)	60	200	260	237,732	1/74	2131	1,325	432	889	
134	Fairfield - Fairfield-Suisun W. W. P.S.	V.T.	1	760	125	–	125		2/74	2219[a]				
135	Cordelia W.W. P.S.	Horiz.	3(1)	22,800	900	300	1,200	1,802,000	9/74	2449[a]	2,042	305	1,608	Diesel engines
136	Sacramento - Florin P.S.	Horiz.	2(1)	2,000	200	840	1,680	918,000	9/74	2449[a]	4,682			Diesel engines
		Horiz.	3(3)	23,100	840	50	100	1,709,000	12/70	1598	2,996	160	1,578	Underground structure with gas engines
	P.S.		2	2,400	50									
137	Systems # 3 Inc. - Sewage P.S.	Vert.	2(1)	7,000	400	200	600							
138	Cucamonga CWD - W.T. Pump Sta. No. 1A	V.T.	1	1,000	150	–	150	153,000	2/75	2299	1,112	–	1,112	

[a] ENR - SF used for projects in San Francisco area.
[b] ENR - LA used for projects in Los Angeles area.

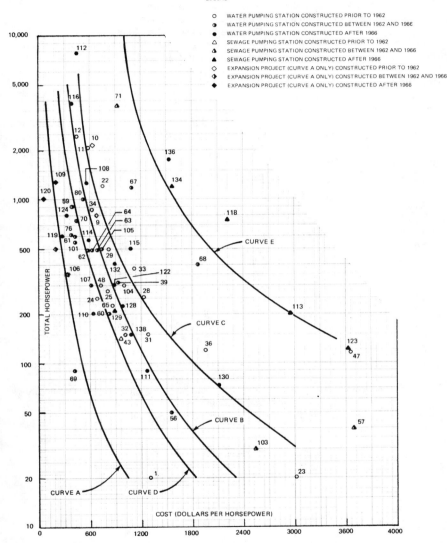

LEGEND

O	WATER PUMPING STATION CONSTRUCTED PRIOR TO 1962
◑	WATER PUMPING STATION CONSTRUCTED BETWEEN 1962 AND 1966
●	WATER PUMPING STATION CONSTRUCTED AFTER 1966
△	SEWAGE PUMPING STATION CONSTRUCTED PRIOR TO 1962
◮	SEWAGE PUMPING STATION CONSTRUCTED BETWEEN 1962 AND 1966
▲	SEWAGE PUMPING STATION CONSTRUCTED AFTER 1966
◇	EXPANSION PROJECT (CURVE A ONLY) CONSTRUCTED PRIOR TO 1962
◈	EXPANSION PROJECT (CURVE A ONLY) CONSTRUCTED BETWEEN 1962 AND 1966
◆	EXPANSION PROJECT (CURVE A ONLY) CONSTRUCTED AFTER 1966

NOTE:
CURVE A: Estimating curve for additions to existing pumping stations designed for expansion, multistaged construction
CURVE B: Estimating curve for new pumping stations (average complexity) for ultimate station hp, multistaged construction
CURVE C: Estimating curve for new pumping stations for ultimate station hp, multistaged construction (complex)
CURVE D: Estimating curve for new pumping stations (average complexity) single-stage construction
CURVE E: Estimating curve for new complex pumping stations including gas or diesel engines for ultimate station hp, multistaged construction

Figure 6. Construction cost of water pumping stations (sewage lift stations included for reference only). (ENR-LA = 2500)

Table V. Average Construction Costs (Material and Installation)
For Class 150 Mortar-Lined and Coated Steel (ML & CP)
or Steel Cylinder Concrete Pipe (SCC) Pipelines[a]

Pipe Diam. (in.)	Cost/Linear Foot Trench Depth (feet)				
	4	6	8	10	12
6	$ 7.10	$ 7.80	$16.30	$18.90	$23.00
8	8.80	9.30	17.90	20.50	26.00
10	11.40	12.20	20.80	23.50	29.30
12	14.00	15.00	24.50	27.00	32.30
14	19.80	20.70	29.30	32.20	35.10
16	22.60	23.50	32.30	35.20	38.00
18	25.30	26.40	35.20	38.00	41.00
21	–	32.00	40.90	45.50	47.00
24	–	36.20	45.10`	48.30	51.40
27	–	41.80	50.80	54.10	57.30
30	–	47.80	56.90	60.20	63.40
33	–	54.10	63.30	66.70	70.10
6	–	60.00	69.20	72.60	76.00
39	–	69.00	83.50	88.30	93.30
42	–	–	91.50	96.30	101.30
48	–	–	110.00	115.00	120.00
54	–	–	131.00	136.00	141.00
60	–	–	156.40	161.70	167.30
66	–	–	184.00	190.00	195.40
72	–	–	217.40	223.20	229.00

[a]Note: Prices include 20% for contractor's profit and overhead and 10% for miscellaneous fittings (exclusive of main line valves and other major appurtenances). Cost of shoring in trenches deeper than 5 feet in compliance with OSHA requirements is also included. Add $1.10/ft^2 for paving.

Table VI. Pipeline Data

No.	Client	Pipe Diam. (in.)	Length (ft)	Type or Class	Construction Cost ($)	Bid Date	ENR-LA Index	Cost/lin ft ($)	Cost/lin ft @ENR-LA of 2500 ($)
1	Oceanside	24	16,000	ML & CSP	421,410	8/75	2568	26.33	26.33
2	Costa Mesa	18	5,170	ML & CSP	58,473	1/65	1014	11.31	27.82
3	Huntington Beach	42	9,695	SCCP	363,803	8/63	914	37.52	102.43
		42	5,002	SCCP	170,802	8/63	914	34.14	93.20
		42	1,000	SCCP	35,000	8/63	914	35.00	95.55
		42	10,508	SCCP	339,883	8/63	914	32.34	88.29
		36	3,318	SCCP	109,822	8/63	914	33.09	90.34
		30	3,488	SCCP	87,699	8/63	914	25.14	68.63
		24	7,446	SCCP	94,000	8/63	914	12.62	34.45
		16	5,391	SCCP	59,064	8/63	914	10.95	29.89
		24	5,637	SCCP	105,000	8/63	914	18.62	50.83
4	Covina	18	2,200	SCCP	37,000	8/69	1292	16.81	32.44
		16	1,750	SCCP	43,000	8/69	1292	24.57	47.42
		16	2,393	SCCP	53,800	8/69	1292	22.48	43.39
5	CMCWD	42	4,389	SCCP	640,000	9/74	2290	145.81	158.93
6	CMCWD	30	4,285	SCCP	231,750	8/74	2147	54.08	62.73
7	Covina	12	2,115	ML & CSP	21,659	9/64	947	10.24	26.93
8	Escondido	30	9,070	ML & CSP	413,380	6/70	1473	45.57	77.01
		42	3,885	ML & CSP	175,235	6/70	1473	45.10	76.22
9	Fairfield	36	15,800	ML & CSP	1,447,400	9/74	2290	91.60	99.84
		18	16,747	ML & CSP	1,221,000	9/74	2290	72.90	79.46
10	Fullerton	42	2,810	ML & CSP	102,960	10/65	1068	36.64	85.74
		36	3,287	ML & CSP	120,202	10/65	1068	36.56	85.55
		30	7,543	ML & CSP	188,782	10/65	1068	25.02	58.55
		24	6,510	ML & CSP	137,453	10/65	1068	21.11	49.40
		18	1,236	ML & CSP	25,848	10/65	1068	20.91	48.93

Table VI. Continued

No.	Client	Pipe Diam. (in.)	Length (ft)	Type or Class	Construction Cost ($)	Bid Date	ENR-LA Index	Cost/lin ft ($)	Cost/lin ft @ENR-LA of 2500 ($)
11	Goleta	33	11,202	ML & CSP	298,536	8/60	830	26.65	79.95
		33	9,458	ML & CSP	261,901	8/60	830	27.69	83.07
		33	4,329	ML & CSP	124,406	8/60	830	28.73	86.19
		33	2,765	ML & CSP	77,876	8/60	830	28.16	84.48
		33	5,435	ML & CSP	164,539	8/60	830	30.27	90.81
		33	2,450	ML & CSP	76,287	8/60	830	31.13	93.39
12	Inglewood	30	2,092	ML & CSP	75,675	2/68	1181	36.17	76.32
13	LVVWD	24	39,040	ML & CSP	570,500	1/64	918	14.61	39.74
14	LVVWD	24	2,543	ML & CSP	32,000	11/64	948	12.58	33.09
15	LVVWD	24	16,919	SCCP	284,560	5/63	894	16.81	46.90
16	LVVWD	24	18,480	ML & CSP	273,611	11/63	914	14.80	40.40
17	LVVWD	60	5,536	ML & CSP	350,000	1/68	1181	63.22	133.39
		60	8,301	ML & CSP	550,000	1/68	1181	66.25	139.79
		60	5,500	ML & CSP	360,000	1/68	1181	65.45	138.10
		60	7,940	ML & CSP	560,000	1/68	1181	70.52	148.80
		42	5,488	ML & CSP	266,000	1/68	1181	48.46	102.25
		30	3,259	ML & CSP	117,000	1/68	1181	35.90	75.75
18	LVVWD	60	7,258	ML & CSP	446,002	5/67	1192	61.44	128.41
		48	2,498	ML & CSP	136,067	5/67	1192	54.47	113.84
		48	2,312	ML & CSP	124,829	5/67	1192	53.99	112.84
		42	5,520	ML & CSP	237,934	5/67	1192	43.10	90.08
		36	1,745	ML & CSP	58,782	5/67	1192	33.68	70.39
		36	7,346	ML & CSP	249,585	5/67	1192	33.97	71.00
		36	1,700	ML & CSP	61,509	5/67	1192	36.18	75.62

No.	Agency								
19	LVVWD	60	5,603	ML & CSP	460,118	7/68	1249	82.11	170.79
		60	1,800	ML & CSP	170,892	7/68	1249	94.94	197.48
		54	2,325	ML & CSP	189,045	7/68	1249	81.30	169.10
		54	5,101	ML & CSP	336,666	7/68	1249	66.00	137.28
		48	2,126	ML & CSP	111,317	7/68	1249	52.35	108.89
		48	2,969	ML & CSP	166,264	7/68	1249	56.00	116.48
20	LVVWD	24	8,594	ML & CSP	199,973	1/67	1135	23.26	51.17
21	Monterey Park	18	11,135	ML & CSP	244,970	3/71	1524	22.00	36.08
22	Oceanside	24	16,200	ML & CSP	415,300	8/75	2568	25.63	25.63
23	Pomona	36	3,522	ML & CSP	94,400	3/65	1018	26.80	65.66
		30	6,910	ML & CSP	132,700	3/65	1018	19.20	47.04
24	Pomona	18	7,785	ML & CSP	96,300	4/62	863	12.36	35.72
25	Pomona	24	7,825	ML & CSP	222,666	11/70	1479	28.45	50.36
26	Pomona	16	1,326	ML & CSP	47,736	9/73	2092	36.00	42.84
27	RCWD	24	5,851	ML & CSP	245,742	9/73	2092	42.00	49.98
		24	8,937	ML & CSP	118,572	4/66	1088	13.26	30.37
		16	10,193	ML & CSP	98,507	4/66	1088	9.66	22.12
		12	10,774	ML & CSP	67,892	4/66	1088	6.30	14.43
		10	2,655	ML & CSP	15,878	4/66	1088	5.98	13.69
28	Santa Ana	24	2,490	ML & CSP	63,030	6/63	899	25.31	70.36
29	SanDieguito I.D.	54	1,595	ML & CSP	73,370	9/67	1198	46.00	95.68
		54	2,400	ML & CSP	112,800	9/67	1198	47.00	97.76
		54	4,365	ML & CSP	231,345	9/67	1198	53.00	110.24
30	SGVMWD	30	17,825	SCCP (200)[a]	1,403,125	1/73	1962	78.71	99.96
		30	7,850	SCCP (225)	400,350	1/73	1962	51.00	64.77
		30	3,100	SCCP (275)	206,700	1/73	1962	66.67	84.67
		30	2,200	SCCP (250)	122,845	1/73	1962	55.83	70.90
		30	5,300	SCCP (225)	287,207	1/73	1962	54.19	68.82
		30	2,950	SCCP (200)	141,055	1/73	1962	47.81	60.72
		30	3,315	SCCP (175)	151,584	1/73	1962	45.72	58.06
		30	3,629	SCCP (150)	161,157	1/73	1962	44.40	56.39

Table VI. Continued

No.	Client	Pipe Diam. (in.)	Length (ft)	Type or Class	Construction Cost ($)	Bid Date	ENR-LA Index	Cost/lin ft ($)	Cost/lin ft @ENR-LA of 2500 ($)
30	SGVMWD	30	2,066	SCCP (275)	127,800	1/73	1962	61.85	78.55
		30	2,350	SCCP (250)	138,300	1/73	1962	58.85	74.74
		30	2,250	SCCP (225)	119,500	1/73	1962	53.11	67.45
		41	4,700	SCCP (200)	381,700	1/73	1962	81.21	103.14
		41	1,700	SCCP (225)	160,800	1/73	1962	94.58	120.12
		41	2,800	SCCP (175)	205,000	1/73	1962	73.21	92.98
		41	12,670	SCCP (150)	845,000	1/73	1962	66.69	84.70
		41	8,961	SCCP (125)	561,000	1/73	1962	62.61	79.51
		41	7,143	SCCP (100)	423,500	1/73	1962	59.28	75.29
		41	3,669	SCCP (150)	215,003	1/73	1962	58.59	74.41
		41	3,000	SCCP (175)	175,800	1/73	1962	58.60	74.42
		41	14,244	SCCP (200)	897,372	1/73	1962	63.00	80.01
		41	8,101	SCCP (225)	534,666	1/73	1962	66.00	83.82
		41	5,189	SCCP (200)	304,853	1/73	1962	58.74	74.60
		41	2,765	SCCP (300)	232,200	1/73	1962	83.97	106.64
		41	677	SCCP (300)	39,500	1/73	1962	58.34	74.09
		41	3,200	SCCP (275)	176,800	1/73	1962	55.25	70.17
		41	2,450	SCCP (250)	129,800	1/73	1962	52.97	67.27
		41	2,540	SCCP (225)	131,200	1/73	1962	51.65	65.60
		41	2,300	SCCP (200)	111,000	1/73	1962	48.26	61.29
		41	2,150	SCCP (175)	102,000	1/73	1962	47.44	60.25
		41	3,617	SCCP (150)	158,800	1/73	1962	43.90	55.75
		41	8,500	SCCP (125)	353,100	1/73	1962	41.54	52.76
		41	8,000	SCCP (100)	315,900	1/73	1962	39.48	50.14
		54	6,960	PCCP (150)[a]	554,900	1/73	1962	79.72	101.24

54	1,088	PCCP (175)	86,700	1/73	1962	79.68	101.19
54	1,300	PCCP (325)	109,900	1/73	1962	84.53	107.35
54	1,200	PCCP (425)	110,000	1/73	1962	91.66	116.41
54	1,400	PCCP (450)	132,300	1/73	1962	94.50	120.02
54	2,200	PCCP (450)	204,400	1/73	1962	92.90	117.98
54	6,448	PCCP (325)	542,500	1/73	1962	84.13	106.85
54	1,132	PCCP (225)	100,100	1/73	1962	88.42	112.29
54	1,600	PCCP (125)	121,900	1/73	1962	76.18	96.75
54	1,970	PCCP (150)	152,300	1/73	1962	77.30	98.17

[a]Pipe class in psi.

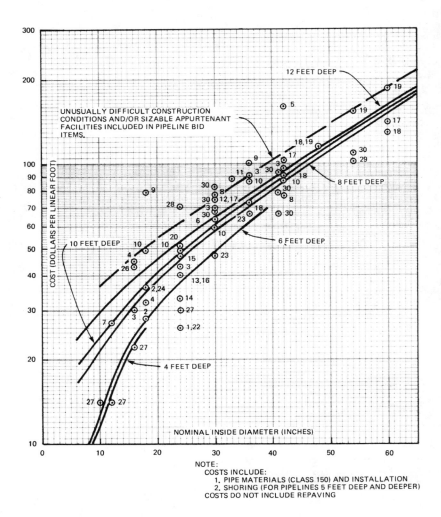

Figure 7. Construction cost of steel cylinder concrete pipeline or mortar-lined and coated steel pipeline. (ENR-LA = 2500)

water works include several classes in compliance with American Water Works Association (AWWA) standards, American Society for Testing Materials (ASTM) standards or other standards.

Pipeline installation costs are affected by complex factors such as soil conditions, traffic control, above and below ground interferences, trenching and backfilling requirements and repaving requirements. All of these factors must be considered to the greatest extent possible when preparing a preliminary cost estimate for the pipeline project.

ESTIMATING COSTS OF WATER TREATMENT PLANTS

This section contains construction cost information on water treatment plants designed by Montgomery Engineers. In addition, three plants designed by the Metropolitan Water District of Southern California (MWD) are also listed. The water treatment plant construction costs were adjusted to exclude costs of the following items: land, rights-of-way, raw-water reservoirs, treated-water reservoirs, intake pumping stations, and treated-water pumping stations. These items were excluded from the total actual construction costs to focus on the costs for the portion of the plant directly associated with the water treatment operations, which usually includes operations and administration buildings.

A summary of the adjusted water treatment plant costs is contained in Table VII. Construction cost curves are shown in Figure 8. Although the curves do not show separate costs for major treatment plant components, the construction costs for average conditions can be apportioned as follows:

· Earthwork, general site work and yard piping	15-20%
· Sedimentation and flocculation basins	20-30%
· Filters and appurtenant system, including wash water tank, pumps and wash water reclamation facilities	20-35%
· Operations and administration building	10-25%
· Electrical and telemetry	10-20%
· Miscellaneous chemical tanks, small structures	10-15%

In determining the above figures, two items normally included in a treatment plant complex were not included because of their highly variable costs: (a) raw water pumping station and high service pumping station, and (b) clear well.

It is to be noted that the cost of pumping stations and clear wells are not included in the cost data tabulated in Table VII and shown in Figure 8.

Conventional Water Treatment Plants

Conventional water treatment plants are defined as full treatment facilities involving flocculation, sedimentation and filtration unit processes. The construction cost curve for conventional plants includes the following plants (refer to index numbers shown in Table VII): 4, 6, 7, 8, 10, 11, 12, 14, 15, 16, 20, 21, 22, 23, 24, 25, 26, 28, 29 and 31. The cost of

Table VII. Water Treatment Plant Costs

Index No.	Owner	Plant	Maximum Hydraulic Capacity Used as Basis for Unit Cost (mgd)	Construction Cost per mgd of Maximum Hydraulic Capacity for ENR-LA=2500 ($/mgd)	Adjusted Const. Cost for ENR-LA=2500 ($)	Remarks
1	Alameda County Water District	Manual J. Bernardo Water Soft, Plant	18.4	114,000	2,100,000	
2	Azusa Valley Water Co.	Water Filtration Plant	7.5	94,000	704,000	
3	Chino Basin Municipal Water District	Carl B. Masingale Reg. Tertiary Treatment Plant No. 1	20	127,000	2,540,000	
4	Contra Costa County Water District	Ralph D. Bollman Water Treatment Plant	80	116,000	9,280,000	
5	Covina Irrigating Co.	Water Treatment Plant	15	86,000	1,280,000	
6	City of Escondido-Vista Irrigation District	Joint Water Treatment Plant	37.5	224,000	8,400,000	
7	City of Fairfield	Waterman Treatment Plant	15	293,000	4,400,000	
8	Helix Water District	R. M. Levy Filtration Plant (initial)	50	129,000	6,430,000	
9	Helix Water District	R. M. Levy Filtration Plant (expansion)	67	43,000	2,870,000	
10	Helix Water District	R. M. Levy Filtration Plant (composite)	67	139,000	9,300,000	
11	MWD of Southern California	Weymouth Softening & Filtration Plant (composite)	600	138,000	83,000,000	$56,000,000 without softening

12	MWD of Southern California	Diemer Filtration Plant (initial)	400	87,000	34,850,000	
13	MWD of Southern California	Diemer Filtration Plant (expansion)	400	63,000	25,000,000	
14	MWD of Southern California	Diemer Filtration Plant (composite)	800	75,000	60,000,000	
15	MWD of Southern California	Jensen Filtration Plant	550	95,000	52,000,000	
16	MWD of Southern California	Skinner Filtration Plant	150	125,000	18,700,000	
17	City of Ontario	Water Purification Plant	7	86,000	603,000	
18	City of Pasadena	John L. Beyner Water Treatment Plant	7.5	186,000	1,390,000	
19	City of Pomona	Canyon Filter Plant	5	109,000	544,000	
20	City of Sacramento	Riverside Water Treatment Plant	20	208,000	4,150,000	
21	City of Sacramento	American River Water Treatment Plant	72	201,000	14,460,000	
22	City of San Diego	Alvarado Water Treatment Plant	100	204,000	20,400,000	$17,340,000 without softening
23	City of San Diego	Miramar Water Filtration Plant	75	121,000	9,070,000	
24	San Dieguito Irrigation District-Santa Fe Irrigation District	Joint Filtration Plant	27	189,000	5,100,000	
25	San Francisco, City & County of	San Andreas Water Filtration Plant	80	155,000	12,370,000	
26	San Francisco, City & County of	Sunol Valley Water Filt. Plant (initial)	80	118,000	9,470,000	

Table VII. Continued

Index No.	Owner	Plant	Maximum Hydraulic Capacity Used as Basis for Unit Cost (mgd)	Construction Cost per mgd of Maximum Hydraulic Capacity for ENR-LA=2500 ($/mgd)	Adjusted Const. Cost for ENR-LA=2500 ($)	Remarks
27	San Francisco, City & County of	Sunol Valley Water Filt. Plant (expansion)	80	71,000	5,660,000	
28	San Francisco, City & County of	Sunol Valley Water Filt. Plant (composite)	160	95,000	15,130,000	
29	City of Santa Barbara	William B. Cater Filtration Plant	15	209,000	3,140,000	
30	City of Santa Monica	Arcadia Water Softening Plant	7.3	123,000	900,000	
31	Stockton-East Water District	Stockton-East Water Treatment Plant	30	239,000	7,170,000	

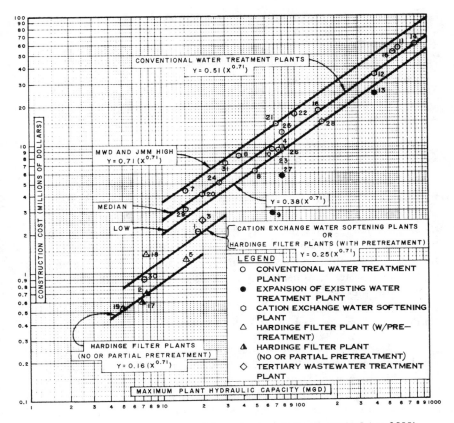

Figure 8. Construction costs of water treatment plants. (ENR-LA = 2500)

the initial Helix Water District plant (8) is included under conventional plants even though the initial construction did not include cost of the sedimentation basins. The costs for cation exchange softening were subtracted from the total costs for the MWD Weymouth plant (11) and the San Diego Alvarado plant (22). The data for conventional water treatment plants were curve-fitted with the resulting best-fit equation:

$$Y = 0.51 \ (X^{0.71})$$ (1)

where

Y = construction cost (million dollars)
X = maximum plant hydraulic capacity (mgd)

Equation 1 includes several composite costs (10, 11, 14, 28). A composite cost is defined as the present-worth cost of initial plus expansion costs. Therefore, a composite cost is the total present worth of the facility less adjustments listed above. Equation 1 does not include the individual expansion costs for the MWD Diemer (13), Helix Water District (9) and San Francisco Sunol (27) plants. As can be seen from the scattering of points in Figure 8, the costs for the expanded portion of the above three plants are significantly below the best-fit curve.

Cation-Exchange Water Softening Plants

The curve for cation-exchange water softening plants is based solely on two plants: Alameda County Water District (1) and City of Santa Monica (30). The two plants are based on completely different methods of construction. Hence this curve should be used with discretion. Since both plants soften ground waters, the curve does not include the cost of filtration. The curve was drawn utilizing the same slope as that used for conventional water treatment plants. Costs of ion exchange units are also given in Chapter 25.

Hardinge (Automatic Backwash) Filter Plants

Cost data have been plotted for five water treatment plants, each of which includes a proprietary automatic backwashing filter manufactured by the Environmental Systems Division, a subsidiary of Koppers Company, Inc.* The filter was originally marketed by the Hardinge Co. and is still commonly referred to as the "Hardinge Filter."

The Hardinge plants vary from no pretreatment to full pretreatment. The costs for Hardinge plants with no or partial pretreatment (2, 17, 19) naturally are less than comparable Hardinge plants employing full pretreatment (5, 18). Therefore, separate cost curves are shown. The cost curves are also based, in part, on the same slope as determined for conventional water treatment plants. A single cost curve is shown for Hardinge plants (with full pretreatment) and for cation exchange water softening plants merely because, based on the limited data available, the curves were coincident.

Comparison of Cost Curves

Figure 8 also shows a water treatment plant construction cost curve, which was originally prepared[2] in June 1969. At that time, the

---

*P.O. Box 298, Baltimore, Maryland 21203

ENR-LA index was 1300. However, the MWD curve was indexed for a construction cost of ENR-LA 1400.

The MWD curve is applicable to treatment plants of the size and nature of the Weymouth and Diemer plants (excluding water softening facilities). The MWD curve has been updated to reflect an ENR-LA index of 2500, shown in Table VIII.

Table VIII. Metropolitan Water District Treatment Plant Costs

Rated Plant Capacity (mgd)	Construction Cost at ENR-LA 1400 (million dollars)	Construction Cost at ENR-LA 2500 (million dollars)
10	2.0	3.57
100	10.0	17.86

The equation for the MWD curve at ENR-LA 2500 is thus $Y = 0.72 (X^{0.70})$. Since the curve is nearly identical to the JMM high curve, a single line is used to represent both the MWD and JMM high curves. The equation of the line representing both curves is $Y = 0.71 (X^{0.71})$, as shown in Figure 8.

REFERENCES

1. *Engineering News-Record* (New York: McGraw-Hill).
2. Brown and Caldwell/Robert A. Skinner. "Water Pricing Policy Study," Report to Metropolitan Water District of Southern California (June 1969).

RECOMMENDED REFERENCES

1. "Building Construction Cost Data." Robert Snow Means Co., Inc., 100 Construction Plaza, Duxbury, Massachusetts 02332
2. *Dodge Guide* (New York: McGraw-Hill).
3. "National Construction Estimator." Craftsman Book Co. of America, 542 Stevens Ave., Solana Beach, California 92075.

CHAPTER 31

DESIGN BLUNDERS

James R. Wright, P.E.

Project Engineer
Black and Veatch Consulting Engineers
1500 Meadow Lake Parkway
Kansas City, Missouri 64114

Everyone has seen the cartoon that depicts two construction crews building a bridge, each crew cantilevering the structure from opposite banks of the river. The crews meet at midspan with the respective halves of the bridge off-set by the width of the structure. Most design blunders are not that spectacular, but they all have the common denominator that everybody involved in committing the blunder knew better; the blunder was obvious from the beginning, but no one caught it in time.

There was a flume that was to carry a raw water supply across a coulee. Soil conditions were good and the designer sized the footings under the piers accordingly, which meant that they were fairly small. The piers were high, as it was a deep coulee. The structure would have been quite stable once the flume was in place and tied to the piers. Unfortunately, the small footings and tall piers were a very unstable structure by themselves, and several fell over before the flume was constructed.

The only sure protection against design blunders is never to design anything. That statement is axiomatic in terms of Murphy's first law which says, "In any field of scientific endeavor, anything that can go wrong will go wrong."

Some years ago when drawings were prepared initially in pencil and then traced onto linen, a tracer was working on a pumping station piping system that included several pumps with a suction and discharge header. Each of the pumps had a gate valve on the suction side. Gate valves were denoted by an "X" symbol and check valves denoted by a single diagonal. The tracer went down the line tracing one diagonal of the "X" with the intention of letting the ink dry and then going back down the line completing the "X" with the other diagonal. Something distracted him and he never got back to complete the job. Consequently, the piping system was installed with check valves on the suction side of the pumps. According to Murphy's fourth law, which states that "Nature always sides with the hidden flaw," neither the designer, checker, project engineer or project manager detected the missing diagonal. Of course, Murphy's second law, which states, "Left to themselves, things always go from bad to worse," dictated that the check valves were installed to prevent flow to the pump.

In another instance, a pump discharge piping system included a harnessed Dresser coupling arrangement. The Dresser coupling was for piping flexibility and the harnessed connection was to take the hydraulic thrust. The harness bolts were designed and specified to be high tensile bolts. In the oversight process that is typical of blunders, mild steel harness bolts were installed in lieu of the high tensile harness bolts. When the pumps were tested, the harness bolts failed, resulting in the pump base bending and the anchor bolts failing with consequent movement of the entire pumping unit away from the header. If the header had been of the type that "floats," the failure of the harness bolts would have resulted only in movement of the pipe header with little consequent damage. This illustrates Murphy's third law: "If there is a possibility of several things going wrong, the one which will go wrong is the one that will do the most damage."

In another example, vertical diffusion vane pumps, designed for a pumping station, were to be installed above a suction header. The suction header had tees with the branch vertical. A butterfly shutoff valve was installed on the tee branch, and a short flexible spool piece placed between the valve and bottom pump flange. The lower bearing of the pump extended below the pump flange, which escaped notice of the detailer of the piping shop drawings, the shop drawing checker, the mechanical contractor and the resident engineer. When the time came to put the station into operation, the butterfly valve could not be opened because the valve disc hit the tail bearing of the pump. This illustrates Murphy's sixth law: "If everything seems to be going well, you have obviously overlooked something."

A steel chemical tank was designed to provide pump suction and storage. The chemical feed pumps connected directly to the tank. The designer neglected to vent the tank, and as the chemical feed pumps withdrew the liquid, atmospheric pressure collapsed the tank. This occurrence proves the most frightening of Murphy's laws, his fifth, which declares, "Mother nature is a bitch."

In a sanitary engineering office, concrete settling basins with cantilevered effluent launders are so common that their repetition can cause details to be overlooked. For one such basin, the detailer neglected to show the cantilevered bars hooked into the basin wall. Of course, the checker, steel fabrication detailer, shop drawing checker and resident engineer, all of whom knew it was wrong, also overlooked it. When the basin was put into service, the launder dropped off the side of the basin.

Another blunder involving a settling basin launder occurred on a design that had a steel launder supported on corbels inside the basin. The designer correctly designed the corbels to support the weight of the launder full of water, a loading condition that could result if the slide gates on the basin effluent were closed and the basin drained. He neglected, however, to anchor the launder to the corbel against uplift. When the basin was initially filled, a buoyant force was exerted on the launder as the water level came up, causing it to float off its supports.

A package sewage pumping station was designed to be installed in a flood plain. An effort was made to protect the station from flooding by sealing the entrance tube with a submarine-type hatch, sealing electrical connections to the station, and extending vents to above-flood stage. The pumps were controlled through a pneumatic system employing a bubbler tube to sense wetwell levels. The bubbler tube was not extended above flood stage, and, yes indeed, the station was flooded through the bubbler tube.

A basin complex consisting of several integral, independent basins had an underdrain system beneath the basin floors. Each basin also had a drain valve. Both the underdrain system and basin drain system were connected to a common header. A basin was drained for maintenance and before it was refilled, the drain valve on another basin was opened. There was sufficient head loss in the drain header to cause enough hydrostatic pressure in the empty basin's underdrain system to damage the basin floor structurally. The designer was no doubt aware of the potential danger, but failed to transmit the critical operational information to the owner.

The myriads of details required in plans and specifications almost defy the statistical probability of the absence of error. There are many horror stories about design blunders in providing details on drawings. For example, a steel joist roof system might be designed as 24 joists on 16-in. centers and be shown on a drawing as 16 joists on 24-in. centers. The only defense against this type of blunder is constant vigilance and awareness. An attitude of responsibility must be instilled in engineers (especially young engineers as they begin work in a firm) because once an error is committed, it can go undetected through the process. The Murphy law that "Nature always sides with the hidden flaw" was cited herein to provide a bit of levity, but the fundamental facts of human nature make it much more than a humorous declaration.

A designer, and more particularly a design team, can suffer the myopic phenomenon of not being able to see the forest because of the trees. It pays to pause, step back and look at the big picture occasionally. This sounds very simple, but in the intense concentration on details individuals can forget to do it. For example, a piece of equipment might be designed to go into a structure that has no doors or openings large enough to permit its passage. Since such equipment is often installed initially before the structure is completed, the original installation may be completed with no problem. But what about the poor, unsuspecting owner who discovers some years later that the piece of equipment must be removed for repairs?

How often have bolted pipe fittings been cast into walls without enough clearance to install the bolts? How many times have valves been shown in pipe lines without sufficient clearances for handwheels, operators or operating nuts?

A safe driving slogan says "The most dangerous part of the automobile is the nut on the steering wheel." That slogan could be paraphrased to apply to the engineering profession by stating that the individual most responsible for design blunders is the person holding the pencil. If you need any more incentive to help remember that, meditate on O'Toole's law which states that "Murphy's laws are overly optimistic."

CHAPTER 32

O & M MANUALS AND OPERATOR TRAINING

Terry M. Regan, P.E.

President
T. M. Regan, Inc.
377 Waller Avenue
Lexington, Kentucky 40504

INTRODUCTION

In recent years we have heard increasing complaints concerning EPA's required paperwork. While we do not always agree with our friends at EPA on the value of some of their paperwork, our experience makes us support the concept of a properly prepared O & M manual. The key words are: "properly prepared."

The first O & M manual we prepared was an act of desperation. On entering municipal service in 1961, our first assignment was to prepare to place in operation a new 12-mgd activated sludge plant with Kraus modification. In itself this is not an overly complicated task. However, this facility was replacing a 6-mgd standard-rate trickling-filter facility, and all operating personnel in the old facility were assigned to the new facility. In order to train these men and familiarize them with the new plant, we prepared an O & M manual and conducted a 200-hr training course, with the O & M manual as the basic text. The results were excellent. In fact, later on, the plant was able to meet all state and local water quality standards under a 25% overload condition. This could not have been done without good line operators.

THE O & M MANUAL AS A DESIGN TOOL

Our experience has shown that O & M manual development concurrent with design has several advantages. In the first place, it emphasizes the importance of O & M to the client as well as to design personnel. As O & M considerations are studied early in the project, the designer can formulate realistic O & M cost estimates. Unfortunately, most of us have calculated O & M costs based on total electrical requirements, estimated chemical needs and various utility costs, and then neglected to estimate the personnel requirements for that specific plant. How many of us have sat down during the design of the project and defined the specific operational requirements?

The single most important item standing between design performance success or failure is operation or, ultimately, the operator. Yet we, as engineers, have in the past paid less attention to operational requirements than we have to calculating the horsepower requirements of a raw sewage pump.

An unfortunate example of this was a small (0.36 mgd) privately owned wastewater treatment plant. The owner was told, by both the engineer and the supplier, that the treatment plant was "automatic" and the owner's handyman could operate it in his spare time. Some years later, after the plant had never met its designed effluent requirements, the owner was ordered to expand and upgrade his facility. After he entered into a $150,000 contract to expand and upgrade the facility, he employed a trained and certified operator. Within two months of his hiring and eight months prior to the completion of the expansion, the operator had the original plant producing an effluent that met stringent local effluent limitations (10 mg/l BOD_5 and 15 mg/l SS).

Development of the O & M manual concurrently with design also makes the client aware early that he is facing a significant cost in operating the facility and he had better fund for it. An increase in rates is much easier for the public to accept during the publicity of a building program than two years later, after it is discovered that the system is underfunded. The examples of this situation are too numerous to mention.

Second, and more importantly, the preparation of an O & M manual brings into focus the importance of operational and maintenance criteria in the design phase of a project. To be operated at maximum efficiency, equipment and valves must be conveniently located. We may argue that if a certain series of mechanical functions must be performed to assure proper operation the operators should perform these functions even if they are inconvenient. That is excellent philosophy, but it is not realistic.

The operator of your facility is a normal red-blooded human being, with all of the inherent strengths and weaknesses of the species. If the operational functions are difficult, adjustment schedules will vary from seldom to never, and you, as designer, will catch all the blame from the client.

Also, the preparation of an O & M manual during the design stage focuses the designer's attention on process *control* as well as process design. By process control we are not speaking about the physical location of these controls, but the actual existence of them. How many of us have discovered that a critical valve, pipe interconnection or motor control had been omitted on the plans?

More and more emphasis is now being placed on emergency operating procedures. We can no longer arrange a series of unit processes together and provide minimum flexibility in operations. In preparing both the manual and the design we must attempt to provide for emergencies.

The designer and manual writer must realize that a critical pump or control *will* malfunction and that the delivery time for the replacement part is at least 16 weeks. How can the plant perform its designed function without this pump or control? The engineer not only needs to provide the flexibility, he must show the operator that the flexibility is there.

WHEN DO YOU START PREPARATION OF THE O & M MANUAL?

Our experience has shown that preparation of the O & M manual should begin with the design phase. This enables a cooperative effort between the two tasks. The manual should be completed and delivered to the client's operating personnel at least two months prior to plant completion.

Once the document is delivered, the operator should become familiar with its use and contents. This will enable the operating personnel to understand how the plant is *supposed* to operate based on conditions existing and anticipated at the time the plant was designed. It is extremely doubtful that this manual is totally valid after a year of actual operation. We would suggest that the manual be reviewed and updated after one year to fit actual conditions.

BLUNDERS LARGE AND SMALL (THE DIRTY DOZEN)

Previously mentioned were the various advantages of using the O & M manual as a design tool. In this section we will describe some of the blunders we have encountered in about a quarter of a century of inspecting and operating treatment facilities. These blunders are not included to embarrass or harass anyone, but they are included to illustrate the need for considering basic operational functions at the initial stage of design. Had the designers considered O & M functions throughout the design, these blunders would never have occurred.

1. After construction was almost complete on a relatively large activated sludge plant, we were called in to write the O & M manual. The basic design was excellent except that the return sludge pumps did not have variable flow controls and the effective rate of return was either 0 *or* 100% of the pump capacity. The designer stated that controlling the rate of return could be accomplished by simply turning the pump on and off as required. Unfortunately, the pump controls were located one level below the operation control floor. After some discussion, a change order was issued to install variable speed controls on the pumps and controls on the operation control floor.

2. Again after construction, we were preparing a manual for an 8-mgd plant with anaerobic digesters. The condensate traps on all sludge gas lines were neatly located 5.5 ft above floor level (face high) with no provision for drainage to the floor, an extremely minor point. However, how often would anyone drain the condensate traps, if he got a face full of that mess every time he turned the handle? The solution was obvious and inexpensive and was accomplished prior to plant start-up. Now, the first thing the operator invites visitors to the plant to see are the condensate traps.

3. On reviewing a set of plans for a proposed water treatment plant, we noticed that the chemical storage floor was located two stories above the ground-level loading dock. This is not unusual and is an excellent method to store chemicals since the feeder hoppers could be filled from the storage floor. The absence of an elevator or material hoist was the unusual aspect of this design.

4. A large pump required major maintenance. When the service crew attempted to remove the unit from the concrete structure, they discovered it could not pass through the door. Further investigation revealed that the pump was placed in the structure before the walls were poured.

5. Two pumps were located in a pit 20 ft deep with access by a ladder. In addition, the location of the pumps, pipe configuration, valve locations and other equipment in the pit provided for only about 12 in. of clearance around the pumps. To add to this almost "perfect blunder," no provision was made to hoist the pumps out of the pit.

This situation is such a classic blunder we should examine it in more detail. In one small structure the designer managed to violate most, if not all, of the basic principles of operation, maintenance and safety. First, the pumps were standard raw sewage pumps of the "nonclog" variety that clogged at least twice a day and required manual cleaning. The operator had requested recessed impeller pumps to eliminate this *known* clogging problem. He was informed by the designer that the recessed impeller pumps were inefficient and could not be used. This brilliant decision was based on the fact that the recessed impeller pumps required 2.5 additional horsepower each. No thought was given to labor costs or down-time involved in the constant unclogging of the "nonclog" units.

Second, the ladder violated safety principles in at least four ways:

- It was too high for safe usage.
- It is virtually impossible to carry tools up or down a ladder of any height.
- It is unsafe for anyone to work under a ladder while someone is carrying tools up or down it.
- After a man has suffered a concussion from a falling wrench or collapsed from lack of oxygen or toxic gas in a deep pit, it is incredibly difficult to carry him up a ladder.

Third, the manner in which the pumps were positioned violated O & M and safety principles in the following ways:

- When one has to contort his frame to move between two moving shafts, or shafts with the potential of moving, located 12-14 in. apart, he stands a good chance of leaving part of his frame wrapped around the shaft or shafts.
- Most standard heavy-duty wrenches have a span of 18-24 in. Trying to maneuver an 18-in. wrench in a 12-in. opening is a hazard to the knuckles.
- If the equipment is difficult to maneuver around, routine maintenance will be performed seldom, if ever. In most tight areas, maintenance will be performed only after total failure of the equipment.

Fourth, the lack of an equipment hoist merely underlines and emphasizes the lack of thought toward operations in this particular design. Aside from the capability such a hoist would have had to remove the pumps, it could have been used to lower and raise tools and equipment and, if necessary, injured workmen.

6. A standard problem seems to be the control panel door that is placed 6 in. from (and facing) a concrete wall.

7. Another standard problem is furnishing flow measuring devices and chemical feed devices sized for the ultimate size of the facility. In

one facility, we found a flow-controlled chlorinator (designed for a 24-mgd plant) attempting to maintain an acceptable chlorine residual at 6-mgd flow.

8. In another facility, we discovered the vent in the chlorine room near the ceiling. This we thought was the ultimate until we saw another facility with the vent fan switch and gas masks inside the chlorine room.

9. In a 9-mgd facility, we discovered several control valves placed approximately 12 ft above the floor with no provisions for chain operation. In order to adjust these control valves, the operator had to use a step ladder.

10. We witnessed the effect of furnishing nonexplosion-proof electrical controls for a gas compressor room. The effect was quite similar to that seen in World War II movies of the London Blitz.

11. Again, while reviewing a set of plans for a 5-mgd facility, we noted that the flow through the clarifiers was controlled by a series of slide gates. Access to these gates was by wading the effluent trough.

12. Blunder Number 5 (above) occurred in a wastewater treatment plant raw sewage pumping station. Based on normal design criteria, it would be impossible to duplicate this situation in a water treatment plant. Blunder Number 12 is a description of the impossible!

The only method of entering and exiting from a 14-ft deep pit containing filters, some filter controls, wash water lines, and some chlorine lines was by a ladder. As in Blunder Number 5, no hoist was provided. Access between the filters was provided. Unfortunately, the space provided required the operator, or the visiting engineer, to turn sideways and take a deep breath to squeeze in. It is most difficult to move, let alone try to use tools in this position.

While this blunder was 6 ft less in height than Number 5, the designer compensated for the lack of height. By careful planning it was possible to locate a header pipe directly over the 14-ft ladder. Further planning lowered the pipe to an elevation slightly lower than a man's head. Now when the operator climbs out of the pit he must wear a hard hat and carefully duck under the header. In the event he should forget either of these items, the following series of events could occur in chronological order: concussion, broken foot, shattered femur, broken hip, permanent back injury, and death by starvation.

However, if the man fell at the correct angle he could eliminate the pain associated with all these items. He could manage to break a chlorine line on the way down and avoid a long painful wait for rescuers.

To round out this "dirty dozen" we will add one more for a baker's dozen. The most trouble we encountered in a water plant start-up was in a plant that was well designed and quite functional from an operator's point of view. The plant is located adjacent to a dam, and raw water is automatically fed from the impoundment into the raw water storage basin when the water level in the basin reaches a certain level.

Each morning the system was started and chemical feed units were adjusted and excellent flocculation and settling was obtained. But 2 hr later solids literally flowed over the weirs out of the clarifier. Jar tests were again run and entirely different chemical feed rates were indicated. After these adjustments were made, the plant again produced an excellent water.

Investigation revealed that three springs in the vicinity of the dam had been piped to the raw water basin. When the plant was shut down no surface water was fed into the basin and it gradually filled with ground water. When the plant was started each day, initial chemical calculations for raw water were based on the ground water, but after the plant ran for a couple of hours the ground water in the storage basin was replaced with surface water with totally different characteristics.

This was not a classic blunder, but it was a situation that need not have occurred. Any plant design should be based upon the actual water to be treated, and we should remember that all waters, like people, have different characteristics.

WHO SUPPLIES INPUT FOR THE O & M MANUAL?

The designer *is, or should be, responsible* for the O & M manual. Unfortunately, this does not mean he is always the best one to write the manual. We engineers have a tendency to write technical papers under the assumption that everyone reading them is as miseducated as we are.

Another problem encountered with those O & M manuals requiring state or federal approval is the review process. All of us, as consultants, have learned how to write reports to pass state and federal agency review. To avoid a hassle and delay, some engineers write reports to indicate their knowledge of the subject and the English language. This practice tends to please some reviewers who may hold a Master's degree but have little, if any, experience. This can produce an excellent paper or college text on the theory of treatment plant operation, but also produces a useless O & M manual. An example of this was shown at a recent EPA meeting where the following excerpt from an actual wastewater treatment plant O & M manual was presented:

> "Anaerobic digestion of domestic sewage sludge is the process whereby anaerobic and facultative bacteria liquefy and gasify the organic portion of the sludge in order to obtain energy and certain elements, such as carbon, for the synthesis of their protoplasmic production. Anaerobic

digestion is not a complete stabilization process since many of the organic constituents resist biological decomposition; for example, carbohydrates and short-chained fatty acids are catabolized more rapidly in the process than fats and oils.

During liquefaction, saprophytic bacteria secrete extracellular enzymes which in turn liquefy and hydrolyze the complex molecules of the solids into simpler compounds. These simpler compounds are used in turn by microbial cells as a source of energy and for cell production. The often-referred-to volatile acids, produced in the liquefaction-hydrolysis stage, are converted to organic salts and then are acted upon by organisms which are capable of catabolizing them into methane gas and carbon dioxide."

How many operators of small treatment plants could read, much less understand, this technically correct description of the anaerobic digestion process?

The basic objective of an O & M manual is not to provide an outlet for the literary talent of the writer. Neither is it written to satisfy the editorial urge of the reviewer. Each O & M manual is a *specific* text for a *specific* plant to assist the *personnel* of that plant to operate it efficiently. The designer, if he elects to write the O & M manual, must remember the basic objective of the O & M manual and the audience that he wants to reach. To include extra verbiage simply to increase the size of the manual defeats its purpose. The time a manual spends on a shelf is directly proportional to its thickness.

We have found that consulting with the facility's operating personnel greatly increases the acceptance of both the design and the O & M manual. Some of our fellow engineers raise their eyebrows at this point. What could they possibly learn from a high school drop-out who has to scratch out a living in a treatment plant? The good treatment plant operator has a background of practical knowledge that could save the designer from some theoretical pitfalls. He also has several built-in prejudices. He has certain pieces of equipment he likes and certain pieces he dislikes. He also feels he knows more about day-to-day operations than any "out-of-town expert." In most cases, he is right about his particular facility.

To ignore this practical knowledge is folly. Also, it is poor public relations. Remember, the operator will eventually have the last word about the ability of the plant to meet design standards and he is the "local expert." If he does not like the operational aspects of the design, the plant may never reach its full efficiency and, usually, it will be the engineer's fault, the contractor's fault, the manufacturer's fault, the regulatory agency's fault, or all of the above. We have all seen examples of this at one time or another. So the designer should become familiar with practical operations and should actively seek, not just paternally accept, the input of operating personnel.

OPERATOR TRAINING

After you have spent many hours designing the facility and have completed the O & M manual, you have *almost* finished the job. A final and important phase of the project remains, that is to train the operator to operate the facility as it was designed to operate. Unfortunately, some engineers believe at this point that the job is complete when they leave a copy of the O & M manual with the city clerk. This does not work. A copy of the manual at city hall does not help the operator.

If you have properly prepared the O & M manual, that document is the most authoritative text in existence on the operation of that specific facility. Unless you are ashamed of it, do not drop it and run. Use it to train the operator. We would recommend one to two months training prior to start-up of the facility, followed by another one to two months training immediately after start-up. The actual time spent on pre- and post-start-up training would vary according to plant size and complexity. This should be actual on-the-job training under the supervision of an experienced certified operator. The basic text for this training program should be the O & M manual.

After these training sessions, the instructor should visit the facility on a regularly scheduled basis (one day per month) for at least 12 months. This permits revision of the O & M manual and allows the operator the opportunity for technical assistance during the critical first year of operation.

SUMMARY

In summary, we feel that the O & M manual is an effective design tool if it is started during the early stages of the design process. It focuses attention on the physical operational problems of the facility. It assists the designer in preparing more valid manpower and staffing requirements. It enables the engineer and owner to estimate realistically O & M costs for the facility and makes planning for adequate future funding possible.

At the least, it can involve the facility operator early in the project for his input. If a man feels that he has had input into a project, he will work much harder to make sure it works.

Write the O & M manual for an operator at a specific facility. Do not write the O & M manual to please a reviewer. If the designer does not feel he has the practical operating knowledge to prepare a manual for operating personnel to understand, then he should find someone else to prepare the O & M manual.

Finally, when you deliver the manual to the owner, don't just drop it at city hall and run. Provide an instructor, train the operator, and use the O & M manual as your training text.

If properly prepared and understood by the operating personnel, the O & M manual can be the greatest single factor affecting the efficient operation of the facility. If it is poorly prepared or written for other engineers to read, it will join countless other plans, studies, evaluations and reports on a forgotten shelf and be a total waste of time, money and paper.

CHAPTER 33

HOW TO AVOID DESIGN BLUNDERS

Robert L. Sanks, Ph.D., P.E.

Professor
Department of Civil Engineering
and Engineering Mechanics
Montana State University
Bozeman, Montana 59717

Senior Engineer
Christian, Spring, Sielbach & Associates
2020 Grand Avenue
Billings, Montana 59102

INTRODUCTION

The perfect set of plans has not yet been drawn, and some blunders inevitably appear in all work, as Wright states in Chapter 31. But blunders can be greatly reduced (and the serious ones can be virtually eliminated) by the use of several carefully programmed reviews of the plans and specifications. Probably all large, established consulting firms have well-developed systems for plan reviews, which no doubt work well. But many small firms may not, and it is principally to them that this chapter is addressed.

Most engineers would agree in principle to the procedure recommended herein. There may, perhaps, be some disagreement about specific details. Of course, each firm should develop a procedure that suits its purpose best, although even so the procedure might vary somewhat depending on the job. Those who reviewed this chapter agreed with the recommendations about space, flexibility and special considerations. One reviewer expressed reservations concerning provisions for future expansion, because he believes designers are not adept at predicting future needs or future technology.

When the plans are reviewed, each review should be made with reference to a limited number of considerations to make the task manageable, because otherwise some important aspects may be overlooked and some blunder escape detection. This chapter presents both a checklist of review considerations and a rationale for them.

SCHEMATICS

One of the first sheets in the set of plans, especially for large plants, should be a schematic diagram of the entire plant. The schematic ought to include all: (a) basins and their volumes, (b) piping, (c) valves, (d) mechanical equipment related to the process (such as feeders), (e) the hydraulic profile, and (f) the critical elevations. It is helpful to tabulate all critical design factors such as: (a) minimum, average and peak flow rates, (b) overflow and weir loading rates, (c) minimum and maximum chemical feed rates, (d) operating pressures, if applicable, (e) power ratings for pumps, feeders and mixers and (f) any other parameters of importance.

In addition to presenting an overall view (which is most useful for the operator), it provides the designer with a ready reference for checking details on other sheets. The schematic can be made even more useful by numbering every valve or control device for identification and using these numbers in the Operation and Maintenance (O & M) manual for describing operational procedures. Finally, the schematics form a good basis at an early stage in design for major decisions on design and operation.

REVIEWING PLANS

The key for the elimination of blunders is several reviews or checks of the plans, and they ought to be made at several stages of the design process. A preliminary review or "brainstorming" session should be made of the alternatives to ensure that the process selected is the best and that it will function as the designer intends. By this time the predesign studies given in Chapter 3 (or, at the very least, chemical analyses and jar tests for coagulant selection and feed rates) should have been completed. A second review should be made about midway in the design while it is still easy to make changes in space and flexibility requirements. A third and overall check when the plans are 90% complete should give special consideration to machinery, equipment and consonance between plans and specifications.

Fresh viewpoints are important in reviewing plans, because otherwise something overlooked once may be overlooked again. Without a fresh viewpoint, major decisions are unlikely to be questioned, so reviews should be made by those not intimately involved nor completely absorbed in the design. The principals and experts in the firm should make the early decisions on alternatives and should be involved in the preliminary review. If possible, a team that has not been involved in the design should make the subsequent reviews, and many large consulting firms utilize this approach. But small firms may not have the back-up in competent manpower to make this possible. Such firms could arrange for a competent consultant to check the plans, but to obtain the most satisfaction, consultant and designer should agree explicitly on the extent of the checking. Although the consulting fee creates an added engineering expense, it saves money (for the client), embarrassment (for the designer), and change orders. Gnaedinger[1] states that all public projects should involve peer review concepts.

In Chapter 32, Regan emphasizes the need to write the O & M manual during the course of designing, and he prefers that it be written by someone other than the designer (although the designer should supervise the work and must ensure that the plant will be operated as designed). But a designer in a small firm may have no choice, and this might be an asset because the point of view required to write the manual is so different from that of design that it brings the fresh approach needed for reviewing.

One large engineering firm has halved the rate of claims over the last five years and has one of the lowest liability insurance rates in the nation, which results from the following program.[2]

- The project manager prepares a work plan detailing all responsibilities, time schedules, budgets and reviews or checks.
- The experts confer with the project team to select and discard alternatives, thereby eliminating many blind alleys. They confer again before the conceptual plan is finalized. Once approved, the design concept is not changed.
- A quality control team checks the first plans for mechanical and structural function before it is too late to make changes.
- The same quality control team makes the final check at the 90% completion stage.

SPACE REQUIREMENTS

It seems axiomatic to consider that expansion of any plant might be required, so it is inexcusable to place a plant on a building site in such a way that further expansion is blocked. And yet there are many examples of just such construction.

Consideration for expansion of facilities ought to be considered at the outset. Future extension of buildings, extra tanks or basins, and other appurtenances can be shown on the plot plan in dashed lines and labelled "future." Buildings can be lengthened or widened easily if reinforcing bar dowels are left protruding. After the forms are removed, the dowels can be bent up and covered with lean concrete to protect them from corrosion. When it is necessary to lengthen the building, the lean concrete can be chipped away, and the dowels can be straightened to lock foundation walls solidly to the existing building. Even water stops can be installed in this fashion.

Hydraulically, future expansion can be facilitated at little cost by utilizing tees and blind flanges instead of elbows (or crosses and blind flanges instead of tees). When expansion is required, the flanges can be removed and the new pipes attached with no difficulty.

Whether provision for future expansion is actually to be incorporated into the plans depends upon cost, the probability of a large increased need for water, the possibility for improved technology that might alter future plans, and, perhaps, the designer's faith in his ability to plan for the future. At least, the possibility for expansion (in some manner) should not be blocked.

Adequate space ought to be provided everywhere inside the plant for access by personnel. Headroom (at least 6 ft 6 in. or 2 m), freedom from a clutter of pipes, and space between units should not hamper operators. In one plant, several prefabricated vertical filters were spaced so closely that one must squeeze sideways to get between them. The plans should allow adequate space to get a wrench on every nut as well as adequate room for a small hoist for lifting heavy pieces. Ceiling hooks cast in the concrete at appropriate places for chain hoists cost little but pay big dividends when needed. One manager stated that it takes about twice the originally estimated equipment area to house auxiliary equipment such as plumbing, instruments, motor control panels and miscellaneous fixtures.

If "Murphy's Law" is as applicable as Wright so humorously states in Chapter 31, every piece of machinery in a plant will fail at some time, so there must be access large enough to remove it from the plant site and to substitute another. Still, there are plenty of examples of plants in which units had to be installed before the concrete could be placed! Without fail, there should be enough room around machinery for adequate servicing. An especially common blunder is failure to provide access to the back of control panels, which makes it necessary to unbolt the panel and swing it out to replace a fuse. In humid climates, control panels must be protected against corrosion by a dehumidifier or by inserting dessicants inside the air-tight panels. At a plant in Washington, each face plate must be removed to replace the dessicant. A simple, screw-type canister in the back of the panel would have made the job quick and simple. If packs are not easily replaceable, they won't be replaced.

Since the quality of the water source might deteriorate (or quality requirements might change), it may be necessary in the future to provide extra treatment such as coagulant aids, potassium permanganate, carbon, or (for water conditioning), lime or polyphosphate. So there should be space on the floor to add extra feeders if they become essential. Coagulant aids require extra rapid-mix tanks (or devices) as explained by O'Melia (Chapter 4).

Chlorine feeders are universally isolated, but at many plants dry feeders and dry storage are not. It is impossible to keep such plants clean. Isolated space should always be provided for any operation involving dust. In some cases vacuum dust covers may be adequate for isolation. Carbon feeders, especially, should be isolated. Some states require a separate room for fluoridation equipment. Furthermore, dry feeder equipment should be designed for easy filling. It is not easy to lift a 100-lb bag shoulder high, open it, and fill a hopper without causing dust or an injured back. Instead of preaching to the operator in the O & M manual about housekeeping, design the plant to ensure that maintenance and good housekeeping are made easy.

Isolation of noisy and heat-producing machinery (such as air compressors or large motors) by sound-resistant walls makes the plant more comfortable for operators and, according to some pundits, contributes to better workmanship.

As compared with stairs, ladders reduce space requirements, but they are an abomination as Regan points out in Chapter 32. If they must be used, there should be cages around the ladders and handrails should extend 3 ft (1 m) above the landings. It is well to install a nearby ceiling hook for a hoist to lift tools, heavy machinery, or even a stretcher for an injured worker.

FLEXIBILITY

Perhaps the most important feature in assuring flexibility is designing interconnecting pipes, channels, or other hydraulic structures for at least 150% of maximum hydraulic flow. Larger pipes and channels have an insignificant effect upon overall cost, although larger valves are, of course, more expensive. Equipment such as flocculation basins, sedimentation tanks and sand filters can usually be pushed far beyond their normal design capacity, so in most plants the capacity is limited by the hydraulics of the piping or other interconnecting structures. Oversizing the piping is also compatible with future plant expansion.

The extra feeders (and extra mixing basins and devices), discussed in "Space Requirements" might well be shown by dashed lines labelled "Future Feeders." Most plants are designed for applying chlorine at several different points, and this practice should be universal. It costs little to add such flexibility in the planning stages, but it may be expensive on a change order and even more expensive if added after the plant is constructed.

It may be necessary to bypass flocculation basins, sedimentation basins, or parts of the sand filter for painting, desludging, intermittent maintenance, or replacement of worn parts. It should not require a complete shutdown to do so, and the plans should provide for it. One way is to provide two entirely separate trains so that during periods of low demand, one train can be taken out of service entirely. If there is the slightest chance that, under some conditions, direct filtration may be preferred over flocculation-sedimentation-filtration, there should be enough flexibility to make such operation possible if it is allowed by law or regulation.

With the increasing emphasis on the removal of chlorinated hydrocarbons, it may become necessary to add activated carbon filters at many plants. Such requirements might be met with a layer of activated carbon granules on top of the sand filters. If so, it would be well to add an extra foot of freeboard to the filters at the time of construction in anticipation of the need.

SPECIAL CONSIDERATIONS

Freezing

If the plant is located in a severe climate, there must be sufficient cover over pipes to prevent freezing. This is particularly true for pipes that carry only intermittent flow, such as sludge lines or drain lines. Such lines might also be protected by making them self-draining.

The effect of freezing on outdoor basins could be disastrous. A sedimentation basin in a small Rocky Mountain town freezes 2 ft deep, which is almost enough to stop the chain scraper mechanism. That designer was just lucky. Had not the basin been grossly oversized, its performance would have been affected as well.

Draining and Flushing

Basins with desludging equipment need to be drained occasionally for maintenance, and without desludging equipment, they need to be drained regularly. Drain lines ought to be large enough to dewater a basin, clean it, fill it, and return it to service within a reasonable time without undue drawdown of the distribution reservoirs, which may receive no inflow during shutdown. Heavy sludge at low head moves slowly and the drain line should also be large enough to prevent ponding when the sludge is sluiced out of the basin.

Sludge lines should either drain by gravity or, better yet, be equipped for a freshwater (or air) flush. Some sludges (see Chapter 10) settle into a hard mass that is difficult to move, so backup with a positive displacement pump ought to be considered.

The velocities in flocculation basins should be high enough to prevent sludge accumulation, but if they are not, a means for desludging should be provided. At a plant in Wyoming, a lime feeder was added to improve flocculation and to increase the weight of the floc. But as the flocculation basin was not equipped with a desludging mechanism, the basin had to be dewatered and flushed several times per year.

Maintenance

All plants ought to be designed for ease of maintenance, so writing the O & M manual simultaneously with design can be of great help. Some items that contribute to easy maintenance (*e.g.,* isolating dusty operations and room to work) have been mentioned. One feature that often receives little or no attention is the location of the floor drains and the floor slope toward these drains. At an otherwise well-designed plant in Washington, much of the floor is always covered by a deep puddle of water stemming from condensation of the moist air on the sides of cold metal basins. It is good design practice to slope all floors so that, after the entire area is flushed with water, it can drain and dry quickly. Incidentally, hatches for clear wells should be raised above floor level to prevent contamination due to floor washing. It is hard to believe that this was not done at a small plant in the northern U.S.

Complexity

In general, simple systems are preferred over complex ones and this is particularly true of plants in small or isolated communities. Furthermore, complexity increases exponentially with the number of processes involved. One example is a medium-sized western town partially served by well waters of high hardness and high dissolved solids. A desalination plant was built to reduce the high solids, but the water had to be softened first, so a lime-softening plant was constructed. Seven years after the construction of this expensive plant, no desalinized water has yet been produced. Either process alone would tax a small community.

Both together constitute an almost insurmountable obstacle. The lesson is: complex processes (even those readily operated in industry) are quite difficult for a small municipality. The designer might better have advised acceptance of the high solids water with only softening and blending with low solids water for limited quality improvement. More attention to operational ease, flexibility and backup equipment in the softening plant would have been a better trade-off.

There are exceptions. Package plants with their "off-the-shelf" controls are entirely suitable for small towns. These plants are reliable, operate with a minimum of attendance (perhaps even as little as one or two hours per day), and are so obviously complicated that the treatment plant operators do not (and, of course, should not) tamper with them. Hence, the plants are usually maintained by factory representatives and receive the best of care.

Engineering

At the outset, it was stated that plan reviews could virtually eliminate serious blunders. Perhaps not. There is simply no substitute for thoughtful judgment and competent engineering. For example, a plant in the West has two high-service 60-hp pumps and one 75-hp pump. The operator complains that neither the two 60s nor the 75 are adequate and when he runs the 75 with either one or two 60s, the 60-hp motors overheat. When the pump station curve was plotted, the pumps were found to be incompatible. Either or both 60-hp pumps operate near shutoff head when the 75-hp pump is running. This blunder was made by one of the oldest and most respected consulting firms in the U.S., which shows that anyone can be a victim. Perhaps the designer never intended all the pumps to operate simultaneously, but nobody knows because there is no O & M manual—a blunder in itself. The conclusion is that even good engineering might give rise to blunders unless the operator knows (by the O & M manual) how the designer intended the plant to be operated.

Predesign studies, especially pilot plant tests as described by Trussell in Chapter 3 are necessary wherever the conditions are out of the ordinary. Good, or at least adequate, engineering and sound judgment are essentials. Talks with operators of treatment plants are often valuable. Finally, several careful checks of the plans (particularly by those not intimately involved in the design) can reduce the blunders to a minimum.

CHECKLIST

The sundry preachments, homilies and prejudices discussed are presented below as a checklist for plan reviews. Each review should be concerned with only a few items on the checklist lest some be missed. The checklist is intended only as a guide. It is too complete for the first check and not complete enough for the last. A firm can, however, use this presentation as a basis for developing the specialized checklists that best suit the purpose.

SCHEMATICS

____ Flow diagrams

____ Process instrumentation diagrams

____ Alternate operation for seasonal use, maintenance or equipment breakdown

____ Basin volumes, detention times

____ Valves and controls, numbered for reference in the O & M manual

____ Mechanical: sizes, maximum and minimum rates, hp

____ Hydraulic profiles, critical elevations

____ Design criteria

SPACE

____ Site: allow for future expansion, other facilities

____ Structures: architecture, steel reinforcement, pipe to allow future expansion

____ Sufficient space allowed for pipe connections, valves, vents, drains, conduits and servicing clearances: 30 in. clear and 6 ft 6 in. headroom

____ Operating clearance for hand wheels and valve operators

____ Working room and access (especially doors and hatches) for removal and replacement or repair of largest component of all machinery; ceiling hooks for hoists

____ Access to backs of electrical panels; screw-in containers for dessicants in moist climates (or use dehumidifiers)

____ Space for future feeders and extra mixing basins

____ Clearance conflicts between disciplines (e.g., electrical vs hydraulic)

____ Isolation: chlorine, dry chemicals and feeders

____ Sound and heat isolation: noisy and heat-generating machinery and processes

____ Stairs: straight (best), spiral (poor), ladder (only for access, not operations); nearby hoist

____ Laboratory: adequate work benches, sinks and other facilities; consider effect of changes in regulations on monitoring and instruments required

____ Workshop, tool room and garage

____ Lavatory and safety shower

FLEXIBILITY

_____ Excess hydraulic capacity: oversize piping

_____ Extra feeders and mixing basins for provision for future; alternate application points if appropriate; allowance for seasonal variations

_____ Water conditioning for corrosion control; feeders at high service main

_____ Bypasses: alternate operations, *e.g.,* direct filtration, use of one train in two- (or more) train system, dewatered basins

_____ Process instrumentation

_____ O & M manual: all modes of operation including valve and control settings fully explained

SPECIAL CONSIDERATIONS

_____ Freezing: sufficient bury, increased bury for nonflow conditions, self-draining, covered basins

_____ Draining and flushing; adequate drain lines, monitors or fire hoses and hydrants for flushing

_____ Desludging: ease of equipment inspection and repair, provision for future addition of desludging equipment, adequate hoppers, proper drain flow, timer-controlled drain valves, proper velocities to prevent deposition in flocculation basins

_____ Ventilation, especially around covered basins

_____ Safety considerations, *e.g.,* railings, rescue, chlorine and chemical handling

_____ Proper access to all working areas by stairs, walkways, etc.

_____ All (potential) cross-connections eliminated

MAINTENANCE AND OPERATION

_____ Slope floors to drains

_____ Good housekeeping easy; complete the O & M manual and imagine yourself to be the operator

_____ Valves and controls: access, location and position for easy use

_____ Recorders: easy serviceability, means for checking accuracy

_____ O & M manual: all operations (normal, alternative and emergency) described simply

CONCLUSIONS

The obvious nature of these considerations would make this chapter ridiculous were it not for the fact that so many plants, particularly the smaller ones, fail to meet so many requisites.

The discussion and the checklist are intended to be a guide. Designers should not expect either to be entirely complete.

ACKNOWLEDGMENTS

I am indebted to Messrs. T. M. Regan, R. Dewey Dickson, Carl W. Reh, and Drs. A. Amirtharajah and H. S. Peavy for critical reviews and helpful suggestions. The manuscript was typed by Mrs. Jean Julian.

REFERENCES

1. Gnaedinger, J. P. "Peer Review: Old Concept in New Situations," *Civil Eng.* 48(2):45-47 (1978).
2. "Quality Control." *Professional Services Management J.* 5(1):4 (1978).

APPENDIX A

ABBREVIATIONS

A	ampere	ea	each
Å	angstrom, 10^{-10} meters	ed.	edition
ac	alternating current	Ed.	editor
ac-ft	acre feet	ENR-LA	Engineering News-Record
ANPRM	EPA Advanced Notice of		cost index—Los Angeles
	Proposed Rule Making	ENR-20	Engineering News-Record
aq	aqueous		cost index—for 20 cities
atm	atmosphere pressure	EPA	U.S. Environmental Pro-
AVGF	automatic valveless		tection Agency
	gravity filter	eq	equivalent
avg	average	ES	effective size
bg	bag	°F	degree Fahrenheit
BOD_5, BOD-5	five-day biochemical	fob	freight on board
	oxygen demand	FRP	fiberglas reinforced plastic
BV	bed volume	ft	feet
BV/hr	bed volumes per hour	ft^2	square feet
bx	box	ft^3	cubic feet
°C	degree celsius	g	gram
cfs	cubic feet per second	GAC	granular activated carbon
CH	carbonate hardness	gal	gallon
cm	centimeter	$gal/min/ft^3$	gallons per minute per
cm^2	square centimeters		cubic foot (of resin)
C-L	Caldwell-Lawrence	$gal/min/ft^2$	gallons per minute per
cs	case		square foot
CU	standard APHA color unit	gpg	grains per gallon
cu ft	cubic feet	gph	gallons per hour
dc	direct current	gpm	gallons per minute
DDT	dimethyldichlorodiphenyl-	gr	grain
	trichloroethane, a per-	gr $CaCO_3$	grains as calcium carbonate
	sistent, insoluble	gr/gal	grains as calcium carbonate
	pesticide		(except as noted) per
diam	diameter		gallon
DPD	a method for testing	hp	horsepower
	chlorine residual	hr	hour
dz	dozen	H-W	Handy-Whitman cost index

827

Hz	hertz	psi	pounds per square inch
i.d.	inside diameter	psia	pounds per square absolute
ISA	Instrument Society of America	psig	pounds per square gage
kg	kilogram	Rd	design filtration rate
kgr	kilograin	Re	net effective filtration rate (Also Reynold's number)
kW	kilowatt		
l	liter		
lb	pound	RO	reverse osmosis
ln	length	rpm	revolutions per minute
m	meter	scfm	standard cubic feet per minute
m³	cubic meter		
ma	milliampere	SDWA	Safe Drinking Water Act
max	maximum	sec	second
MCL	maximum contaminant level	Sg	specific gravity
		sh	sheet
MG	million gallon	SS	suspended solids
mg	milligram	st	set
mil	1/1000 inch	SWD	side water depth
meq	milliequivalent	T	short ton (2000 lb)
ml	milliliter	t	metric ton
mm	millimeter	Tan	total anions
mol	mole	Tcat	total cations
NORS	EPA National Organics Reconnaissance Survey	TDS	total dissolved solids
		TH	total hardness
NPSH	net positive suction head	THM	trihalomethane
o.d.	outside diameter	TOC	total organic carbon
O&M	operation and maintenance	TON	threshold odor number
		Tr	trace
PAC	powdered activated carbon	TU	turbidity units
		UBWV	unit backwash volume
PCB	polychlorinated biphenyl	UC	uniformity coefficient
pc/l	picocuries per liter	UFRV	unit filter run volume
pfu	plaque-forming unit	WC	water column
PIR	Interim Primary Regulations	wt	weight
		WHO	World Health Organization
pk	pack	μl	microliter
ppm	parts per million $\cong$ mg/l	μm	micrometer, micron

APPENDIX B

CONVERSION FACTORS—ENGLISH TO METRIC

| English Units | | | Metric Units |
Description	Symbol	Multiplier	Symbol
acre	ac	0.04047	ha (hectare)
cubic foot	cu ft	0.02832	m^3
cubic foot	cu ft	28.32	l
cubic feet per minute	cfm	0.4719	l/sec
cubic feet per second	cfs	0.02832	m^3/sec
cubic inch	cu in.	0.01639	l
cubic yard	cu yd	0.7646	m^3
foot	ft	0.3048	m
feet per hour	ft/hr	0.08467	mm/sec
feet per minute	fpm	0.00508	m/sec
gallon	gal	3.785	l
gallons per acre	gal/ac	0.00935	m^3/ha
gallons per day per linear foot	gpd/lin ft	0.01242	m^3/m day
gallons per day per square foot	gpd/sq ft	0.04074	m^3/m^2 day
gallons per minute	gpm	0.06308	l/sec
gallons per minute per square foot	gpm/sq ft	0.04074	m^3/m^2 min
grain	gr	0.06480	g
grains per gallon	gr/gal	17.12	mg/l
horsepower	hp	0.7457	kW
inch	in.	25.40	mm
mile	mi	1.609	km
million gallons	MG	3785.0	m^3
million gallons per acre per day	mgad	0.935	m^3/m^2 day
million gallons per day	mgd	43.81	l/sec
million gallons per day	mgd	0.04381	m^3/sec
pound (force)	lbf	4.448	N
pound (mass)	lb	0.4536	kg
pounds per cubic foot	lb/cu ft	16.02	kg/m^3
pounds per square foot	lb/sq ft	4.882	kgf/m^2
pounds per square inch	psi	703.1	kgf/m^2

| English Units | | Metric Units | |
Description	Symbol	Multiplier	Symbol
pounds per square inch	psi	6.895	kN/m^2
square foot	sq ft	0.09290	m^2
square inch	sq in.	645.2	mm^2
square yard	sq yd	0.8361	m^2
ton, short	ton	0.9072	t